Nutraceuticals Production from Plant Cell Factory

Tarun Belwal • Milen I. Georgiev •
Jameel M Al-Khayri
Editors

Nutraceuticals Production from Plant Cell Factory

Springer

Editors
Tarun Belwal
Biosystems Engineering and Food Science
Zhejiang University
Hangzhou, China

Milen I. Georgiev
Institute of Microbiology
Bulgarian Academy of Sciences
Plovdiv, Bulgaria

Jameel M Al-Khayri
Agricultural Biotechnology
King Faisal University
Al-Hassa, Saudi Arabia

ISBN 978-981-16-8857-7 ISBN 978-981-16-8858-4 (eBook)
https://doi.org/10.1007/978-981-16-8858-4

This Springer imprint is published by the registered company Springer Nature Singapore Pte Ltd.
The registered company address is: 152 Beach Road, #21-01/04 Gateway East, Singapore 189721, Singapore

Preface

Plant cells have been effectively utilized over the past few decades to produce valuable natural bioactive compounds under artificial conditions. Nutraceutical compounds, which encompass nutrients and pharmaceuticals, are gaining higher market demands due to their health-promoting properties, added value to food products, and mitigation potential of various diseases. Considering their high demand and the limitations of natural resources, biotechnological tools based on cell culture techniques provide effective means of scaling up the production of these natural products. However, considering the complexity of cell types, genetic factors, and targeted nutraceutical compounds, the optimum cell culture conditions vary and hence require empirical determination.

The purpose of this book is to highlight the in vitro techniques, current status, and challenges of producing nutraceutical compounds. In addition, it provides an overview of different biosynthesis pathways and their modulation through cell culture technique for the production of nutraceutical compounds in high quantity and quality. The book also emphasizes assessment of the factors influencing production and advances in cell culture techniques, including scale-up approach using bioreactors. Overall, this book will provide the current status, methods, research, advances, and challenges of in vitro production of nutraceutical compounds along with recommendations for future research.

The book comprises different parts, namely theory and technology, in vitro production of nutraceutical compounds, and strategic advances and challenges. The theory and technology part covers the general description of nutraceutical compounds, plant cell culture technology, bioreactors, and factors affecting in vitro production of nutraceuticals. The in vitro production of nutraceutical compounds part mainly deals with the in vitro production of important nutraceutical compounds, namely polyphenols, alkaloids, coumarins, terpenoids, anthocyanins, carotenoids, saponins, steroids, tocopherols, phytosterols, and quinones. The last part of the book covers the strategic advances and challenges, comprises chapters dealing with optimization strategies for in vitro nutraceutical production, genetic engineering, and microbial cell factory for nutraceutical production, and highlights the challenges of in vitro nutraceutical production.

The book is an excellent reference source for researchers working in the area of in vitro biosynthesis of nutraceutical compounds, food science, plant biotechnology, nutraceutical research, and pharmacological activities. Also, it will be useful for industries working on plant biotechnology, especially the in vitro biosynthesis of nutraceutical compounds.

The editors appreciate chapter authors for their contributions towards the success and quality of this book, which represents the efforts of 56 scientists from five countries. We are also grateful to Springer for giving us an opportunity to compile this book.

Hangzhou, China — Tarun Belwal
Plovdiv, Bulgaria — Milen I. Georgiev
Al-Ahsa, Saudi Arabia — Jameel M. Al-Khayri

Contents

Part I

Theory and Technology

1 Nutraceutical Compounds, Classification, Biosynthesis, and Function

Hari Prasad Devkota

Abstract

In recent years, there is an increasing interest on the plant-based nutraceuticals, functional foods, and food supplements as potential agents for the maintenance of good health and the prevention and treatment of diseases. Phytochemicals, especially the polyphenols including flavonoids, phenolic acids, curcuminoids, and stilbinoids, are widely studied for their health beneficial properties. Among many other issues, one important issue is the continuous supply of active components in nutraceuticals to meet the market demand. As many phytochemicals present in nutraceuticals are specific to certain plant species, the conservation, cultivation, and sustainable utilization are equally important. Newer biotechnological tools such as tissue and cell culture have potential to provide the necessary amount of the specific nutraceutical compounds in future. For wider application, understanding their chemical classification, biosynthetic routes, potential health beneficial activities, and market trends must be well understood. This chapter focuses mainly on the classification of these compounds, their biosynthesis in plants and role in human health.

Keywords

Nutraceuticals · Functional foods · Phytochemicals · Polyphenols · Classification · Bioactivity

H. P. Devkota (✉)
Graduate School of Pharmaceutical Sciences, Kumamoto University, Kumamoto, Japan
e-mail: devkotah@kumamoto-u.ac.jp

T. Belwal et al. (eds.), *Nutraceuticals Production from Plant Cell Factory*,
https://doi.org/10.1007/978-981-16-8858-4_1

1.1 Introduction

Medicinal plants have been used in primary healthcare by more than 80% of the world's population (Fitzgerald et al. 2019; Devkota and Watanabe 2020), and plant-derived natural products have also played an important role in the discovery and development of modern drugs (Atanasov et al. 2015; Newman and Cragg 2020). Besides that, medicinal plants are also widely used as foods, spices, food supplements, food preservatives, food colorants, and sweeteners and also in cosmetics, aroma, and perfumes (Pawar et al. 2013; Sarkic and Stappen 2018; Negi 2012; Voon et al. 2012). In recent years, there is an ever-increasing interest on plant-based nutraceuticals, functional foods, and food supplements as potential agents in the maintenance of good health and the prevention and treatment of diseases (Gul et al. 2016; Martin et al. 2011; Tsao 2010). Phytochemicals, especially the polyphenols including flavonoids, phenolic acids, curcuminoids, and stilbinoids, are widely studied for their health beneficial properties (Tsao 2010; Lin et al. 2018; Shahidi and Ambigaipalan 2015; Yeung et al. 2019).

Various epidemiological studies have shown the important role of diet in the maintenance of healthy state of body and mind (Shahidi 2004). Foods rich in plant-based products including fruits and vegetables are well known for their health benefits. Evidences show that their high intake reduces the risk of metabolic diseases, cardiovascular problems, and cancers, which has contributed to the increased scientific and public interest in diet and their impact on health and diseases (Shahidi 2004; Day et al. 2009). Foods, nowadays, are consumed not only to satisfy hunger and to get nutrients but also to maintain and protect physical and mental health. Growing public awareness toward healthy lifestyle, expensive healthcare cost, increasing life expectancy, and the desire for improved health and better lifestyle by older people, all these factors have increased the market demand for health beneficial foods such as nutraceuticals and functional foods (Bigliardi and Galati 2013). Considering the already known evidences about the relationship between food and human health, researchers including food and nutrition scientists, health professionals, and food manufacturing companies are now focusing on the development of nutraceuticals that can fulfill the market demand (Bigliardi and Galati 2013).

In recent years, there are many nutraceuticals and related products available in the market, which are intended to use for the improvement of the health; however, the long-term goal of these products is the prevention of chronic diseases and hence increasing the healthy life span. Nutraceuticals are the food or food-derived products with health beneficial properties and are often formulated as capsules, tablet, powder, and solution. The term "nutraceutical" was defined by Stephen DeFelice, in 1989, and consists of two words "nutrition" and "pharmaceutical." Hence, nutraceutical can be defined as the products obtained from food or food itself which can be used for the treatment as well as for the prevention of diseases (Shahidi 2004; Andrew and Izzo 2017; Tapas et al. 2008). Nutraceuticals cover a wide range of natural products such as fortified products, functional foods, food supplements, etc. (Andrew and Izzo 2017). On the other hand, functional foods are defined as foods with additional health promoting functions and provide health beneficial effects beside nutrition and calories only (Day et al. 2009). The term "functional food" was first used in Japan in the 1980s to define the food products fortified with special constituents that contain advantageous physiological effects (Siró et al. 2008). There are now hundreds of functional foods in market that can be categorized under different groups, such as food to promote gut health, food to promote tooth and gum health, food to facilitate mineral absorption, food to promote bone strength, food for those who are concerned about hypertension, food for those who are concerned about blood sugar level, food for those who are concerned about blood cholesterol level, food for those who are concerned about body fat accumulation, etc. Most of the functional foods available in the market are probiotics, prebiotics, functional drinks, functional cereals, oat bran fiber, bakery products, spreads, functional meat, soy protein, fish oil fatty acids, and functional eggs (Day et al. 2009; Bigliardi and Galati 2013; Siró et al. 2008). Some examples of emerging nutraceuticals and functional foods from plants source are tea, flaxseed, tomato, soybean, citrus fruits, berries, garlic, grapes and wines, cruciferous vegetables, and their active constituents. Phytochemicals and other functional components from plant source are the potential candidates in the field of nutraceuticals. They provide high opportunity of minimizing the borderline between the food and medicine.

There is continuous rise in the development and launching of nutraceuticals in the market; however, there are many challenges to overcome. One of the most important issues is the scientific evidences for the claimed health benefits of these products, where the actual pharmacokinetic and pharmacodynamics properties of the active compounds in nutraceuticals are not well defined in many cases. Other important issues are the consumers' acceptance, the changing market trend, and the increasing trend of aging population. The long-term effects of these nutraceuticals are not always well studied. As nutraceuticals incorporate various bioactive ingredients, the processing and stability of the formulations is very important (Day et al. 2009). Among many other issues for raw materials, one important issue is the continuous supply of active components in nutraceuticals to meet the market demand. As many phytochemicals present in nutraceuticals are specific to certain plant species, the conservation, cultivation, and sustainable utilization are equally important. Newer biotechnological tools such as tissue and cell culture have the potential to provide the necessary amount of specific nutraceutical compounds in the future. For wider application, understanding their chemical classification, biosynthetic routes, potential health beneficial activities, and market trends is necessary. This chapter focuses mainly on the classification of these compounds, their biosynthesis in plants, and their role in human health.

1.2 Current Trend of Nutraceutical Research and Nutraceutical Market

Nutritional supplement provides adequate amounts of essential nutrients vital for the proper functioning of human bodies. They include vitamins, proteins, herbs, meal supplements, sports nutrition, and other related products. They are consumed to maintain health and provide support to the immune system and to reduce the risk of illness. The global nutritional supplement market is expected to expand at a CAGR of 10.01%, leading to global revenue of USD 245.43 Bn by 2023. Asia-Pacific is the fastest growing region for the nutritional supplement market due to the growing demand in China, Japan, and India (Research and Markets 2018). Another research showed that the global nutraceuticals market was valued at 252,535.4 million USD in 2018 and is projected to grow to 465,709.8 million USD by 2027 having CAGR of 7.1% (Nutraceuticals Market 2019). North America was reported to be the largest market which was mainly driven due to increased prevalence of lifestyle-related diseases (Nutraceuticals Market 2019). Similarly, the global market of raw or processed foods rich in bioactive phytochemicals such as dried fruit was valued at USD 8.94 billion in 2019 and is expected to witness significant growth in the forthcoming years. There is growing market for nutritious packaged food and wide application of dried fruit in the dairy, bakery, snack, and confectionery industries (Market Research 2020). With increasing life expectancy, aging population, and change in lifestyles, the market of nutraceuticals will grow more in the future.

1.3 Classification of Plant-derived Nutraceuticals

Most of the nutraceutical components in the market are derived from plants and their active constituents such as polyphenols, amino acids, peptides and derivatives, carotenoids, alkaloids, phytosteroids, triterpenoids and related compounds, non-starchy carbohydrates, etc. (Abuajah et al. 2015). In many cases, these compounds are present in various forms such as glycosylated, esterified, thiolated, and hydroxylated derivatives. They can be classified into various classes based on its chemical structure and functional activities. Some of the major classes of nutraceutical compounds are explained in following sections.

1.3.1 Polyphenols

Polyphenolic compounds are among the largest groups of plant secondary metabolites that are made of one or more aromatic rings with one or more hydroxyl groups. They are also among the most widely studied phytochemicals for their various pharmacological activities. Polyphenolic compounds can be classified into different groups such as phenolic acids, flavonoids, tannins, coumarins, stilbenes, curcuminoids, xanthones, and lignans, among others, based on their structures. The basic classification of polyphenolic compounds is represented in Fig. 1.1.

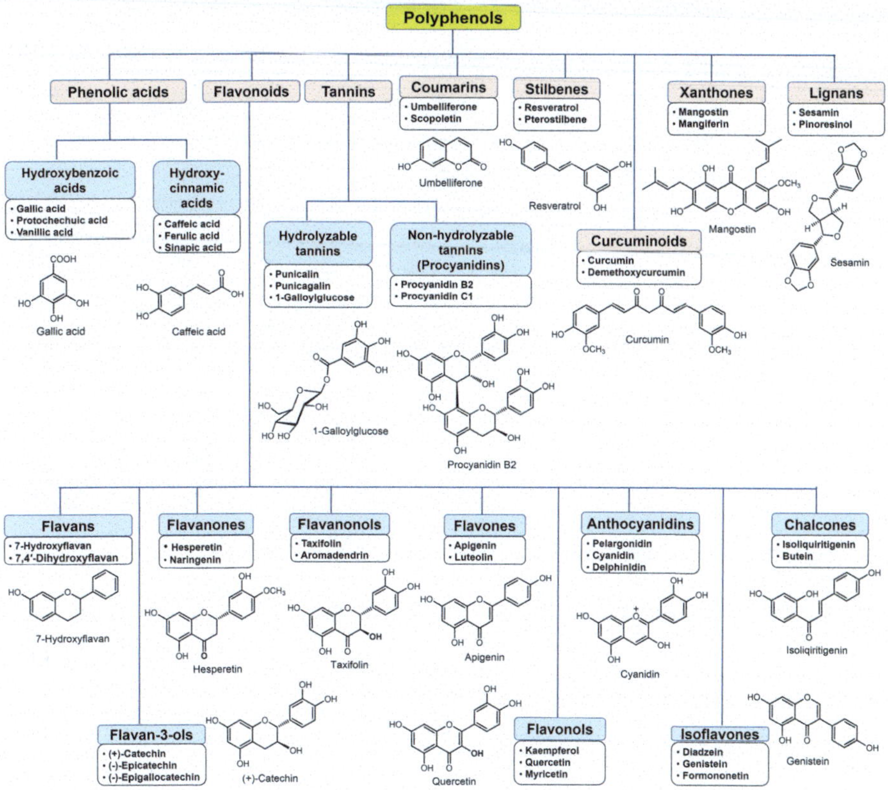

Fig. 1.1 Basic classification of polyphenols based on their structures

1.3.1.1 Phenolic Acids

Phenolic acids belong to the class of polyphenolic compounds and are commonly found in various plant species and abundant in vegetables, berries, fruits, and beverages (Zadernowski et al. 2005; Barros et al. 2009; Mattila et al. 2006; Nile and Park 2014). Phenolic acids are synthesized in plants during normal growth and development, as well as in response to stress conditions and against adverse factors such as drought, infections or physical damage, UV radiation, wounding, etc. (Dietrich 2004; Carl 2000). Based on their chemical structures, phenolic acids can be broadly divided into two groups, i.e., benzoic acid derivatives and hydroxycinnamic acid derivatives. The most common benzoic acid derivatives are gallic acid, protocatechuic acid, vanillic acid, and syringic acid (Fig. 1.2). These compounds mainly occur in plants either as free or as conjugated (bound) forms such as hydrolyzable tannins and lignins. Some of these hydroxybenzoic acid derivatives are also bound to flavonoid aglycones or flavonoid glycosides such as in case of epigallocatechin gallate (EGCG) or galloyl derivatives of flavonoid glycosides. The most common hydroxybenzoic acid derivatives are *p*-coumaric acid, caffeic acid, ferulic acid, sinapic acid, etc. (Fig. 1.2) (Barros et al. 2009; Mattila et al. 2006).

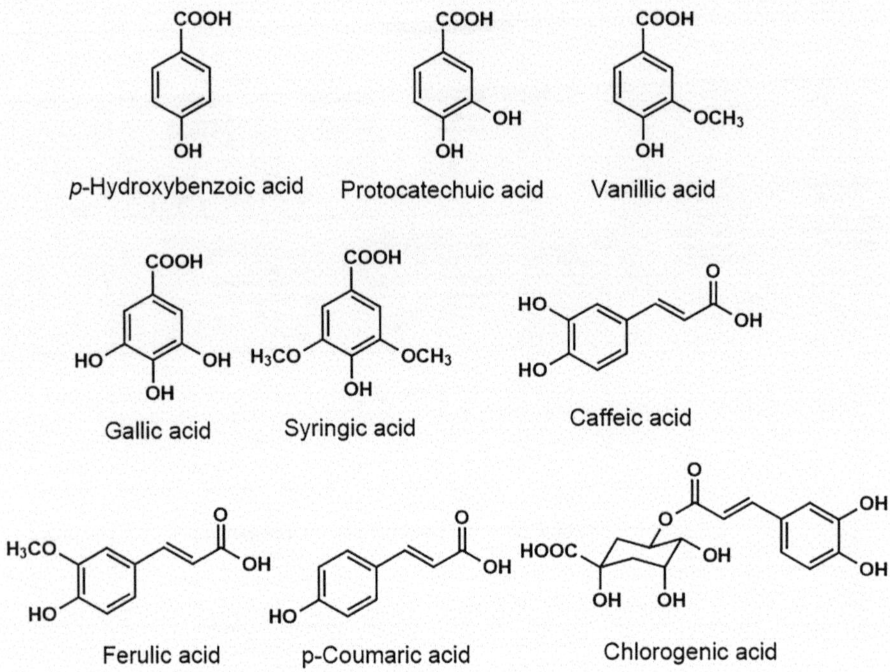

Fig. 1.2 Structures of common phenolic acids

Hydroxycinnamic acids are usually found in bound form with quinic acid, tartaric acid, or sugars. The most common examples are mono-caffeoylquinic acid, dicaffeoylquinic acids, and tricaffeoylquinic acid. Due to their common natural abundance, phenolic acids are widely studied for their antioxidant, anticancer, anti-inflammatory, and other pharmacological activities.

1.3.1.2 Flavonoids

Flavonoids are another class of polyphenolic compounds widely studied for their abundance and biological activities. Flavonoids contain two aromatic rings connected by a three-carbon bridge (C_6-C_3-C_6), and they are further divided into different classes based on the presence of double bond between 2 and 3 position carbons and the presence of hydroxyl group in position 3 and ketone in position 4. These classes are like flavans, flavanols, flavanones, flavanonols, flavones, flavonols, anthocyanidins, isoflavones, and chalcones (Fig. 1.1). Flavonoids are widely found in fruits, vegetables, legumes, wine, tea, and other beverages. These flavonoids usually present either in free form or their methylated, acylated, or glycosylated forms. The number of hydroxylation of the flavonoids is what makes them a diverse group of polyphenolic compounds.

Flavans (Fig. 1.3) do not have double bond in between carbon position 2 and 3 and also do not possess the hydroxyl or carbonyl group in position 3 and 4. They exist either in free form or glycosylated forms.

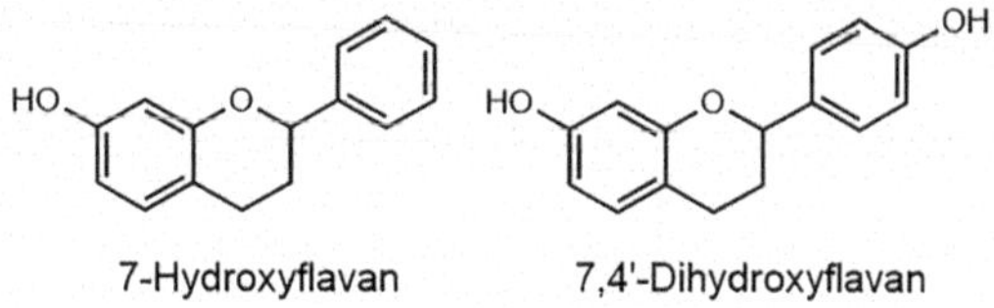

Fig. 1.3 Structures of common flavans

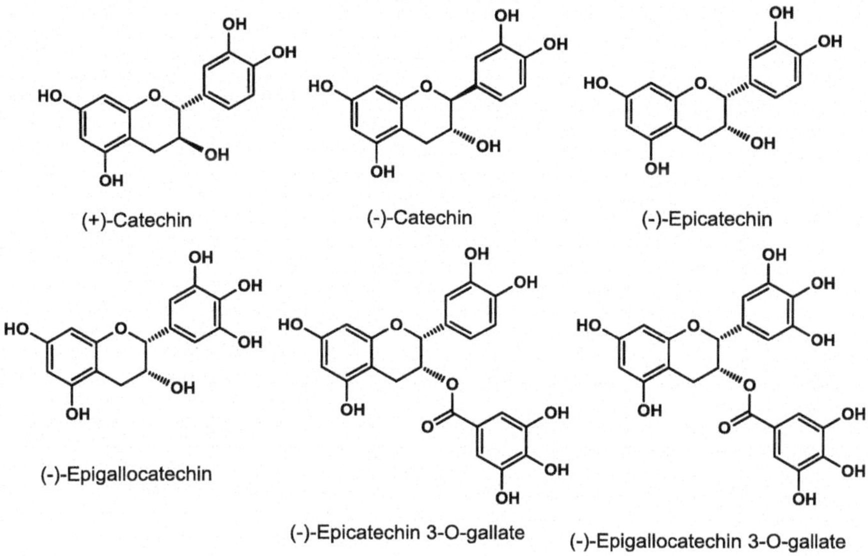

Fig. 1.4 Structures of common flavan-3-ols

Flavan-3-ols, also known as catechins, are one of the main constituents in tea and other common functional food materials (Carloni et al. 2013). They belong to one of the most widely studied plant natural products for chemical and pharmacological aspects such as antioxidant, cancer chemopreventive, anti-inflammatory, immunomodulatory activities (Khan et al. 2019; Wai et al. 2018). Structures of most common flavan-3-ols are given in Fig. 1.4.

Flavanones, such as hesperetin, naringenin, and their glycosylated derivatives such as hesperidin, naringin, and neohesperidin (Fig. 1.5), are commonly found in *Citrus* plants and commonly known as potent antioxidant and anti-inflammatory compounds (Adhikari-Devkota et al. 2019). Most of the flavanones are hydroxylated in B-ring; however, some B-ring non-hydroxylated flavanones such as pinocembrin and pinostrobin are also found in different plant families (Rasul et al. 2013).

Flavanonols, also known as dihydroflavones, are found in many plants. Structures of common members of this group are represented in Fig. 1.6. Silibinin, a flavanolignan derivative isolated from milk thistle (*Silybum marianum* (L.) Gaertn.), is another common derivative of this group.

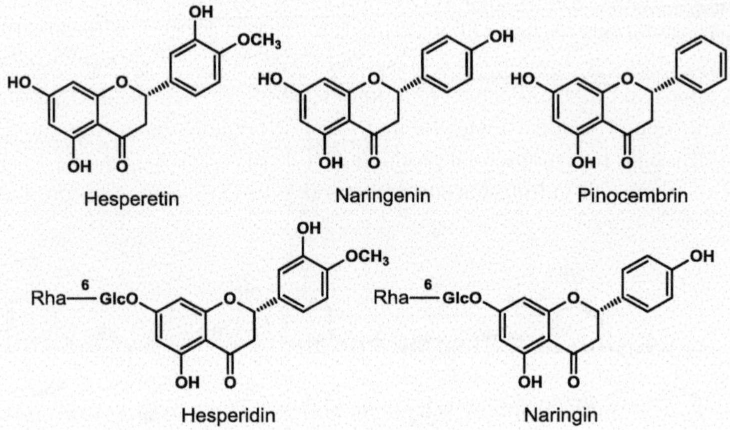

Fig. 1.5 Structures of common flavanones and their glycosides

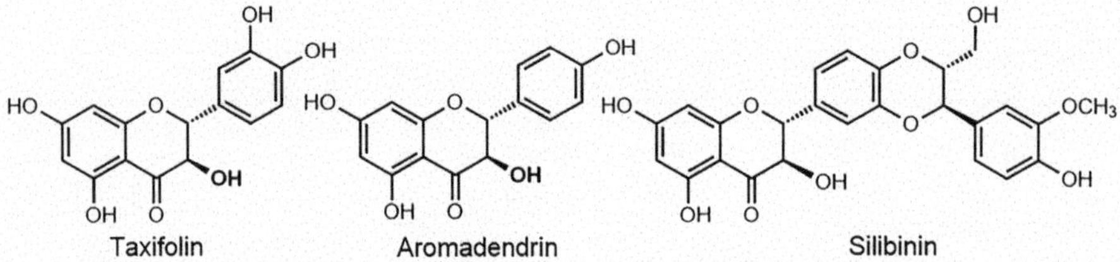

Fig. 1.6 Structures of common flavanonols

Flavones are another class of common flavonoids such as apigenin, luteolin, and their derivatives. Other derivatives, widely found in *Citrus* plants, are polymethoxyflavones such as nobiletin, tangeretin (Fig. 1.7), and their derivatives which are potent anti-inflammatory agents.

Flavanols such as quercetin, kaempferol, myricetin, and their derivatives such as rutin, nicotiflorin, quercitrin, myricitrin (Fig. 1.8) are common in many leafy vegetables and flowers having potent bioactivities (Li et al. 2016; Wang et al. 2016; Boots et al. 2008).

Anthocyanidins (e.g., cyaniding, pelargonidin, delphinidin) (Fig. 1.9) are aglycones of anthocyanins, which are responsible for the beautiful colors of many colorful flowers, fruits, and vegetables (Iwashina 2015).

Isoflavones differ from other common flavonoids having the B-ring linked in C-3 position. Genistein, daidzein, puerarin, and their glycosylated derivatives (Fig. 1.10) are commonly found in soybean and many other plants such as fava bean, lupine, kudzu and psoralea, etc. (Odegaard et al. 2011).

Chalcones, are another class of flavonoids having open C-ring (Fig. 1.11). Various chalcones are reported from nature having diverse chemical characteristics and

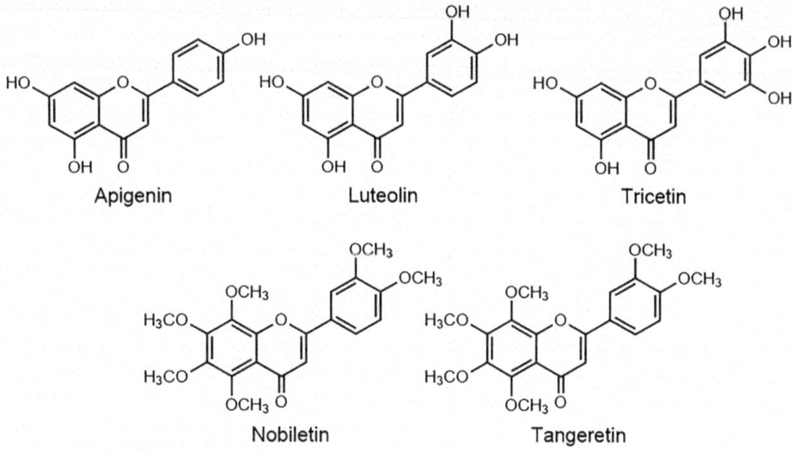

Fig. 1.7 Structures of common flavones

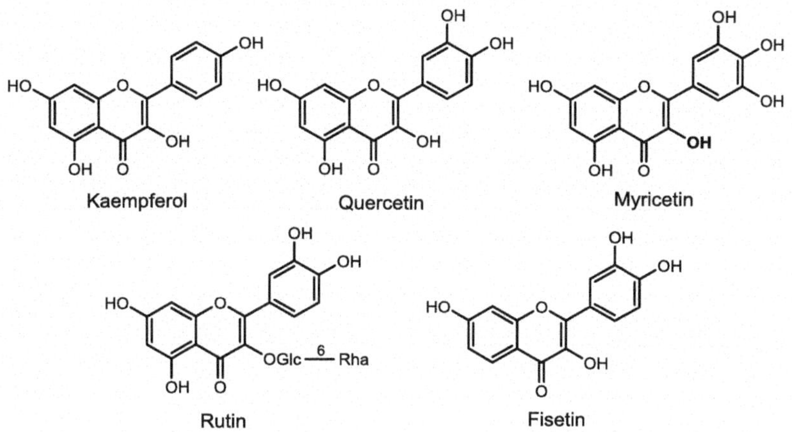

Fig. 1.8 Structures of common flavonols

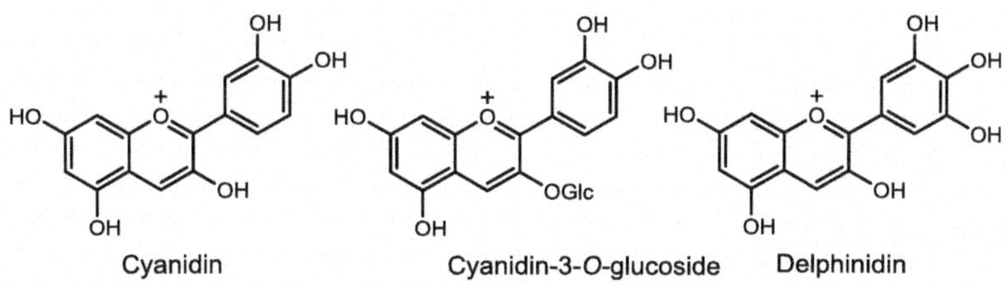

Fig. 1.9 Structures of common anthocyanidins

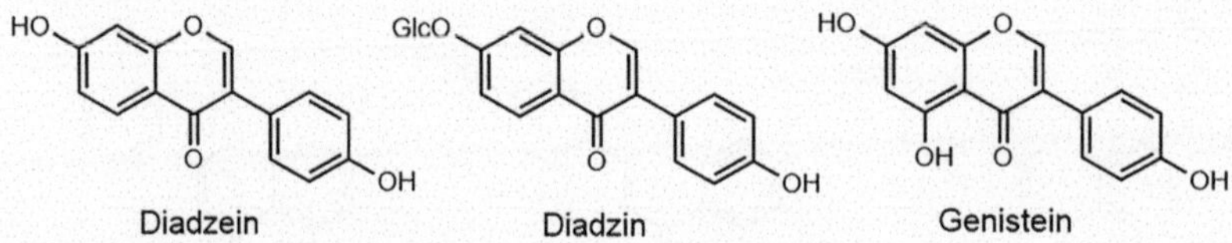

Fig. 1.10 Structures of common isoflavones

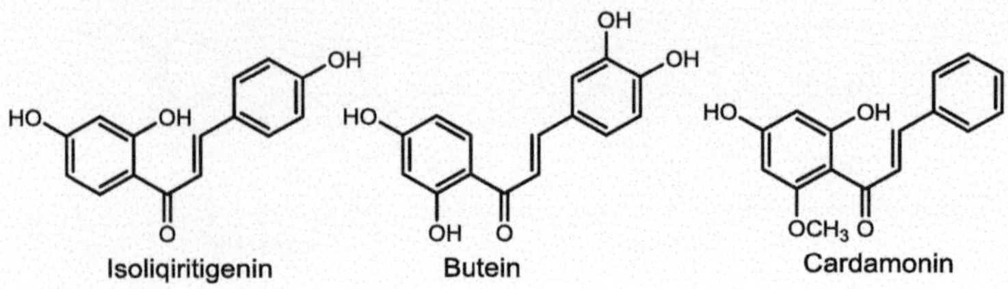

Fig. 1.11 Structures of common chalcones

pharmacological activities. They usually exist as glycosides or prenylated derivatives.

1.3.1.3 Tannins

Tannins are another class of polyphenolic compounds. They are mainly divided into two groups, i.e., hydrolyzable tannins and non-hydrolyzable tannins or condensed tannins (procyanidins). Hydrolyzable tannins are usually the glucosyl ester of gallic acids, and condensed tannins are polymers of catechins (Fig. 1.12). They are abundant in many plants and fruits.

1.3.1.4 Coumarins

Coumarins are another class of polyphenolic compounds having benzopyran-2-one or chromen-2-one rings (Fig. 1.13). As in the case of flavonoids, they also exist as their dimers or methylated, acylated, or glucosylated derivatives.

1.3.1.5 Stilbenes

Stilbenes or stilbinoids are another class of widely studied polyphenols mostly present in blueberries, grapes, raspberries, mulberries, pistachios, and peanuts. Resveratrol, oxyresveratrol, and pterostilbene (Fig. 1.14) are common monomers; however, they also exist as dimers, trimers, tetramers, or their respective glycosylated derivatives.

1.3.1.6 Curcuminoids

Curcuminoids are a group of phenolic compounds present in the rhizomes of turmeric (*Curcuma longa* L.) and other plants of Zingiberaceae family. The most common member of this group, curcumin, also known as diferuloylmethane, is a member of the diarylheptanoid class of chemical compounds having two aryl

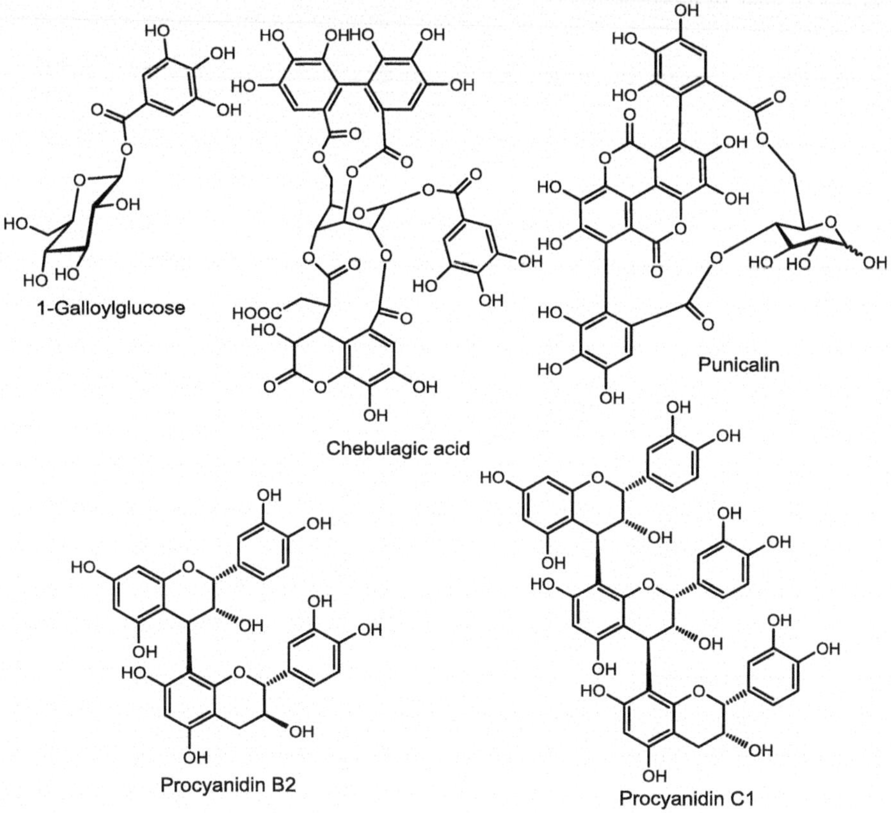

Fig. 1.12 Structures of common tannins

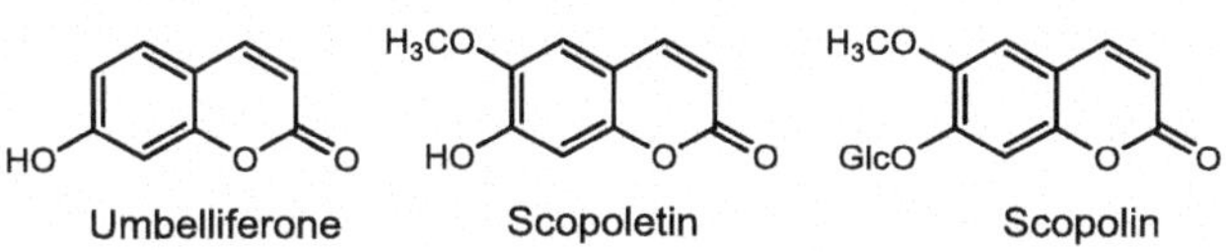

Fig. 1.13 Structures of common coumarins

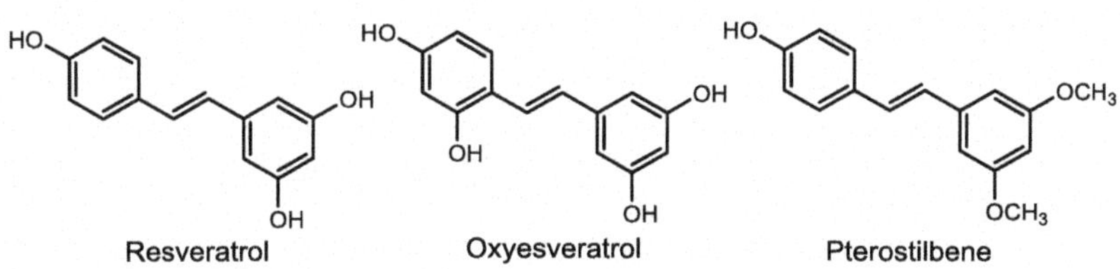

Fig. 1.14 Structures of common stilbenes

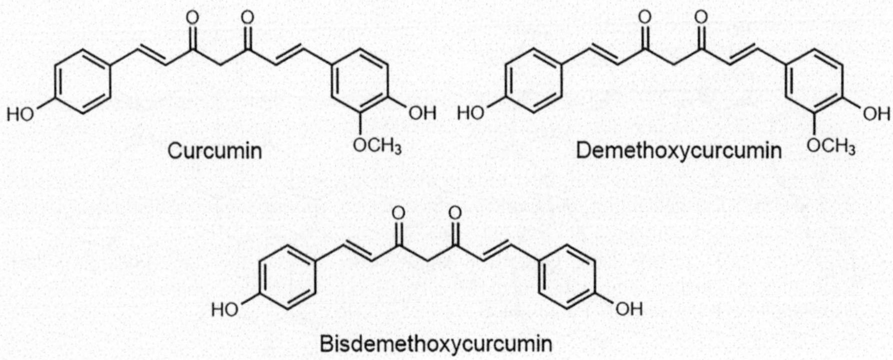

Fig. 1.15 Structures of common curcuminoids

Fig. 1.16 Structures of mangostin

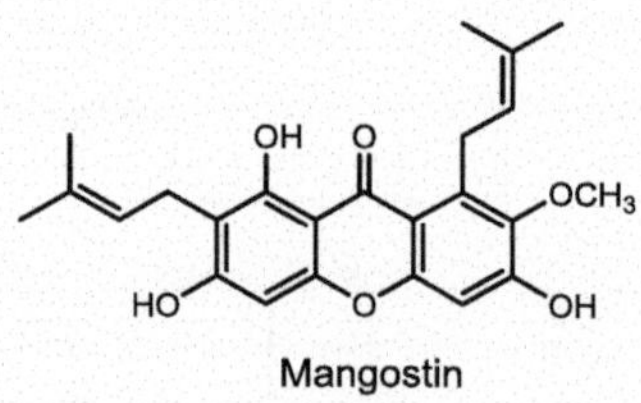

(phenyl) rings attached to heptane derivative. Demethoxycurcumin (DMC) and bisdemethoxycurcumin (BDMC) (Fig. 1.15) are other two common natural curcuminoids. The content of curcuminoids in dried turmeric rhizomes is reported to be around 1–5% and is also responsible for the yellow color of the rhizome (Li et al. 2019; Prasad et al. 2014; The Ministry of Health Labor and Welfare of Japan 2016).

1.3.1.7 Xanthones

Various xanthones are also reported as nutraceutical compounds. The most common example is mangostin (Fig. 1.16) obtained from mangosteen (*Garcinia mangostana* L.). Xanthones are widely reported as antioxidant and anticancer compounds.

1.3.1.8 Lignans

Lignans are another class of polyphenolic compounds formed by the conjugation of two phenylpropanoid units. Sesamin and sesaminol (Fig. 1.17) found in sesame oil are widely reported as potent antioxidant, anti-inflammatory, and anticancer lignans. Based on the chemical structures, lignans are further classified into neolignans, lignan glucosides, and other derivatives.

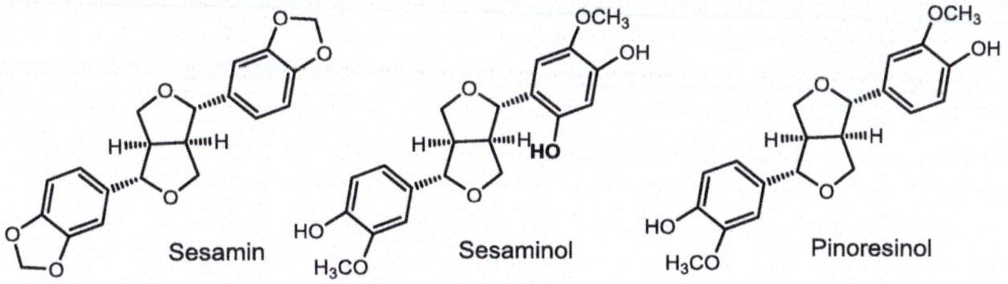

Fig. 1.17 Structures of common lignans

1.3.2 Carotenoids

Carotenoids are another class of common nutraceuticals which are responsible for various health beneficial effects in humans as potent antioxidants, anticancer compounds, and precursors of vitamin A (Khoo et al. 2011; Sereno et al. 2018). They are derivatives of isoprenoids usually having 40 carbons with polyene chains (Tanaka et al. 2008). They play an essential role in plant physiology such as photoprotection during photosynthesis, providing substrates for biosynthesis of plant hormones, etc. They are also responsible for providing colors to many fruits and flowers (Tanaka et al. 2008). Various acylated and glycosylated derivatives are also abundant in nature. Carotenoids such as α-carotene, β-carotene, and lycopene (Fig. 1.18) are widely used as nutraceuticals along with their plant sources such as tomatoes and carrots. Many carotenoids are also used as food colorants. Crocin, a glycoside of carotenoid, crocetin, found in saffron and gardenia is widely used as nutraceutical compound and food colorant.

1.3.3 Amino Acids and Related Compounds

Various branch chain amino acids such as isoleucine, leucine, and valine are also used as nutraceuticals. Many short chain peptides obtained from the hydrolysis of food proteins obtained from soybeans, wheat, vegetables, etc. are also marketed as nutraceuticals (Andrew and Izzo 2017).

1.3.4 Alkaloids

Alkaloids are another group of phytochemicals that contain nitrogen atom in ring structure and are found in various plants where they play an important role in plant protection from various insects and other organisms. Alkaloids are divided into several groups: pyrrolidine, pyrrolizidine, pyridine, quinoline, isoquinoline, imidazole, etc. Various plant-derived alkaloids such as vinblastine, vincristine, taxol, etc. are further developed as drugs in anticancer therapy. Many plant-derived alkaloids are also used as nutraceutical agents. Structures of some of the common alkaloids used in nutraceutical market are represented in Fig. 1.19. Berberine, a yellow-

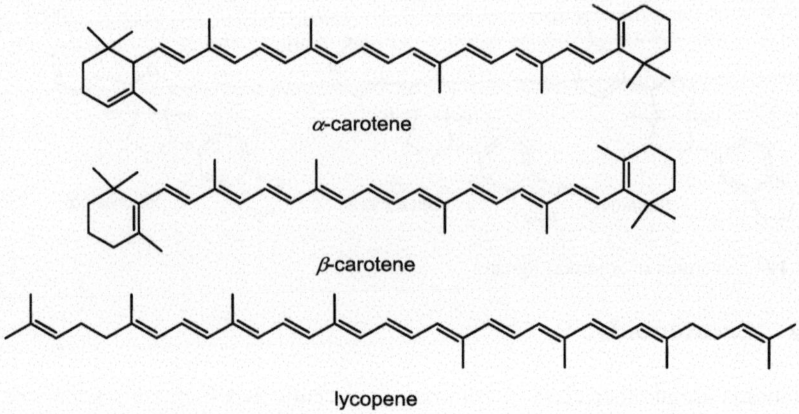

Fig. 1.18 Structures of common carotenoids

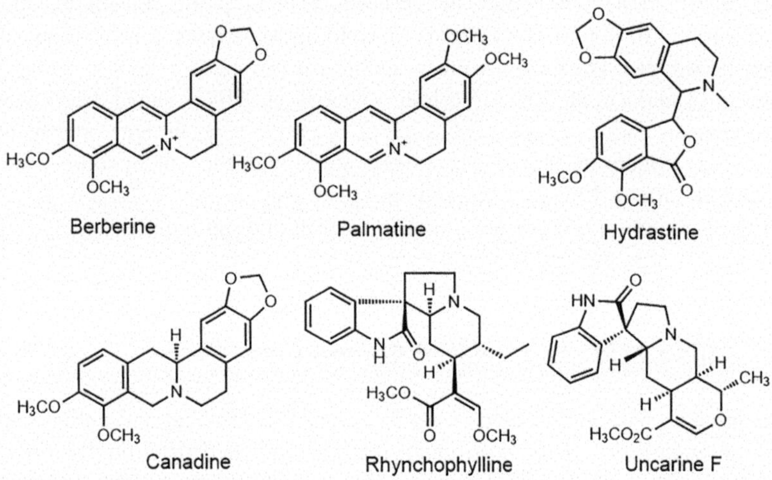

Fig. 1.19 Structures of common alkaloids

colored isoquinolone alkaloid present in many medicinal plants such as *Berberis* spp., *Coptidis* spp., *Hydrastis spp., Mahonia* spp., *Phellodendron amurense, Tinospora cordifolia*, and many others (Belwal et al. 2020a; Yeung et al. 2020a; Mandal et al. 2020), is used widely as a nutraceutical compound. Berberine and palmatine-rich extracts from *Berberis* plants (Belwal et al. 2020a), berberine and hydrastine-rich goldenseal (*Hydrastis canadensis* L.) (Mandal et al. 2020), and rhynchophylline and uncarine F-rich extracts of cat's claw (*Uncaria tomentosa* (Willd. ex Schult.) DC.) (Batiha et al. 2020) are widely marketed as potent nutraceuticals.

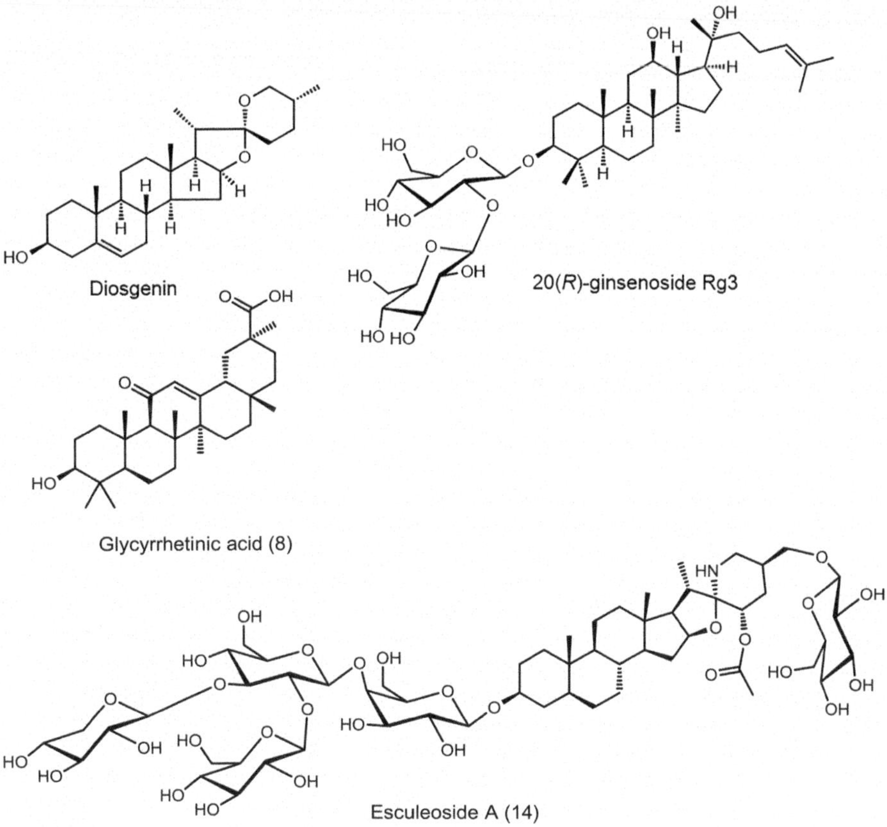

Fig. 1.20 Structures of common steroid and triterpenoid derivatives

1.3.5 Steroids, Triterpenoids, and Saponins

Phytosterols and triterpenoids and their glycosides (saponins) are among the most abundant chemical compounds in plants (Mozos et al. 2018; Piironen et al. 2003). Various phytosteroids, triterpenoids, and their glycosides (saponins) are marketed as cancer chemopreventive and cholesterol-lowering agents. Some examples of these compounds are presented in Fig. 1.20.

1.4 Biosynthesis of Nutraceuticals

Biosynthetic pathways of several phytochemicals have received attention in recent years as biotechnological targets for the production of leads for drug discovery and nutraceuticals. In this section, the biosynthesis of polyphenols is discussed in brief.

In plants, polyphenols such as phenolic acids and flavonoids play an important role in plant physiology as antioxidants, defense regulators, and signaling molecules

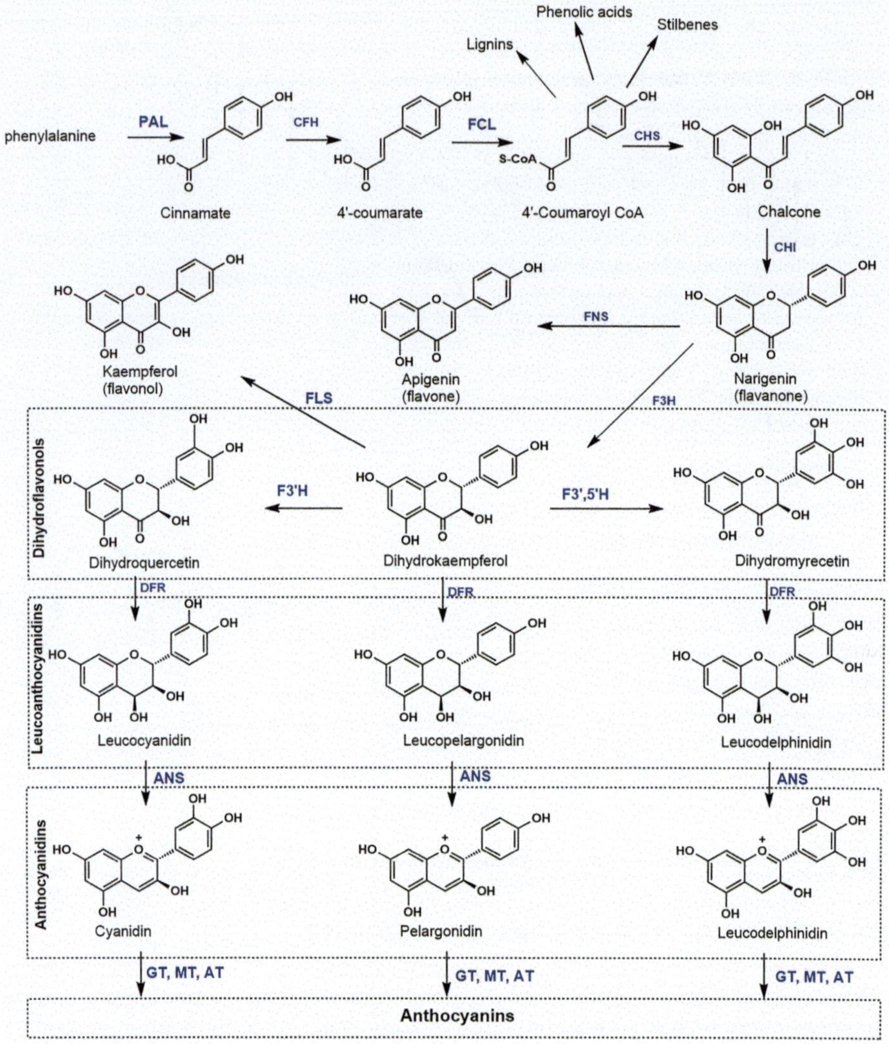

Fig. 1.21 Biosynthesis of flavonoids including anthocyanidins and anthocyanins. PAL: phenylalanine ammonia lyase, CFH: cinnamate-4-hydroxylase, FCL: 4-coumarate CoA ligase, CHS: chalcone synthase, CHI: chalcone isomerase, FNS: flavone synthase, F3′H: flavonoid 3′-hydroxylase, F3′5′H: flavonoid 3′,5′-hydroxylase, DFR: dihydroflavonol-4-reductase, ANS: anthocyanidin synthase, GT: glucosyltransferase, MT: malonyltransferase, AT: acyltransferase (Tanaka et al. 2008)

(Agati et al. 2012; Brown et al. 2001; Harborne and Williams 2000; Khalid et al. 2019; Liu and Murray 2016; Nakabayashi and Saito 2015; Weston and Mathesius 2013); thus, many plants synthesize these compounds.

The schematic representation of synthesis of flavonoids and other polyphenolic compounds is provided in Fig. 1.21 (Tanaka et al. 2008; Forkmann 1991). It is one of the most widely studied phytochemical synthesis pathway. Flavonoids are synthesized in cytosol where various enzyme interactions occur. Formation of 4-′-coumaroyl coenzyme A (CoA) from phenylalanine through multistep enzymatic catalysis is the initial step in the synthesis of various phenolic compounds such as phenolic acids, lignins, stilbenes, and flavonoids. Chalcone synthase converts 4-′-coumaroyl CoA to tetrahydroxychalcone (THC) using three molecules of malonyl CoA (Tanaka et al. 2008). Thus, formed THC is then isomerized to naringenin which further is converted to dihydrokaempferol, a flavanonol, or dihydroflavonol. Different specific hydroxylation enzymes convert dihydrokaempferol to dihydroquercetin and dihydromyricetin. These dihydroflavonols are converted to respective leucocyanidins by dihydroflavonol-4-reductase. These leucocyanidins are then converted to anthocyanidins by anthocyanidin synthase. Various UDP-glucosyltransferases then conjugate glucose and other sugars to anthocyanidins resulting in the synthesis of anthocyanins.

1.5 Biological Functions and Pharmaceutical Activities of Nutraceuticals

Many epidemiological studies in vivo and clinical studies have demonstrated that these nutraceutical phytochemicals can be effective for the prevention of various chronic diseases. Most of the research related to the health-promoting activities of nutraceuticals were conducted after 2000, and there is a vast growth in these studies after 2010. Nutraceuticals have been widely studied for their antioxidant activities or anticancer and antiaging activities in general. A simple Scopus search (www.Scopus.com, December 25, 2020) with the keyword “nutraceutical OR nutraceuticals AND immunomodulatory” and similar activity words (as given in Fig. 1.22) showed that antioxidant and cancer-activities are two of the most widely studied activity areas in nutraceutical bioactivity. As shown in Fig. 1.22, antioxidant and cancer-related activities are most studied ones.

Tea catechins are among the widely studied nutraceuticals well studied for their health beneficial activities such as antioxidant, antihyperlipidemic, antiagenins, antidiabetic, antiobesity, and cancer chemopreventive activities (Zaveri 2006). Tea is not only used as a beverage but also in various forms as functional food (Kurauchi et al. 2019; Namal Senanayake 2013; Hara 2011; Sanna et al. 2015). Many review articles are also published in these aspects of tea catechins (Khan and Mukhtar 2010; Yang and Wang 2016; Boehm et al. 2009). Various review articles have been already published which extensively cover these activities such as antioxidant activity (Gramza and Korczak 2005; Higdon and Frei 2003); potential use for prevention and treatment of obesity, diabetes, and other metabolic diseases (Zaveri 2006; Gramza and Korczak 2005; Higdon and Frei 2003; Park et al. 2009; Masterjohn and Bruno 2012; Kao et al. 2006; Legeay et al. 2015); cardiovascular diseases (Hodgson and Croft 2010); cognitive functions (Weinreb et al. 2004; Da Silva 2013; Pervin et al. 2018); and antimicrobial activities (Reygaert 2014; Taylor et al. 2005).

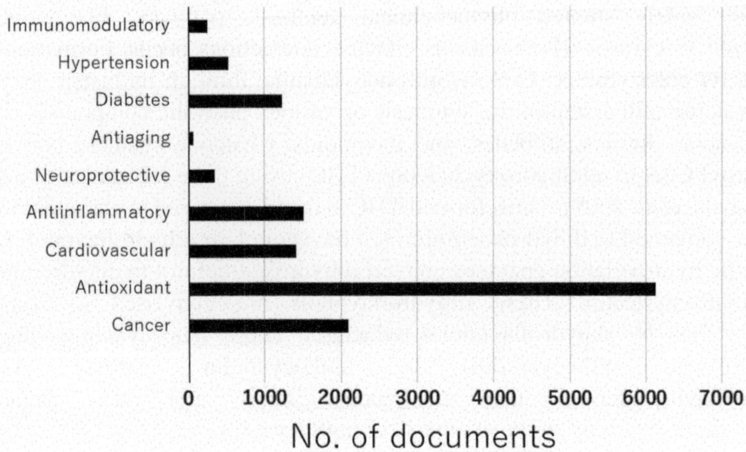

Fig. 1.22 No. of studies related to different biological activities of nutraceuticals

Naringenin, hesperidin, nobiletin, and other flavonoids obtained mainly from *Citrus* fruits are highly bioactive and multifunctional compounds with potent anti-inflammatory, antioxidant, anti-adipogenic, and cardioprotective activities (Salehi et al. 2019; Tapas et al. 2008; Zaidun et al. 2018; Yang et al. 2011; Adhikari-Devkota et al. 2019). Nobiletin and other polymethoxyflavonoids have received great attention as a potent antioxidanst (Wang et al. 2018), anticancer (Chen et al. 2014), and anti-inflammatory agents (Zhang et al. 2016) and as potential compounds to prevent neurodegenerative diseases (Nakajima et al. 2014). Many epidemiological studies have also reported the potential health beneficial and disease preventive and therapeutic effects of the dietary intake of these flavonoids (Pan et al. 2010). Some studies have also reported that the intake of flavonoids may help in delaying the progression of Alzheimer's disease (AD) and other neurodegenerative disorders (Williams and Spencer 2012; Ayaz et al. 2019).

Silibinin, a derivative of flavanonol moiety, is also known for its hepatoprotective activities. Mixture of silibinin with other stereoisomers (known as silymarin) has been widely used in treatment of liver diseases and also reported to be a potent cancer chemopreventive agent (Zhao and Agarwal 1999; Bosch-Barrera and Menendez 2015; Tyagi et al. 2002; Lu et al. 2012).

Flavonols such as quercetin, kaempferol, and fisetin found in many fruits and vegetables are reported as potent anti-inflammatory, antioxidant, cancer chemopreventive, and neuroprotective agents (Li et al. 2016; Wang et al. 2016; Boots et al. 2008; Khan et al. 2013; Donado et al. 2011; Park et al. 2007; Ahmad et al. 2017).

Genistein, a isoflavone, is linked to various pharmacological activities, and the high consumption of genistein-rich soy products by Asians has been linked with lower incidences of diabetes mellitus (Odegaard et al. 2011).

Resveratrol, a stilbenoid, is well known for its antioxidant, anti-inflammatory, and anticancer activities (Yeung et al. 2020b) along with potential benefits in modulation of lung inflammatory diseases (Vargas et al. 2016).

Similarly, curcumin and other curcuminoids are reported as potent antioxidant, anti-inflammatory, anti-cancer, antimicrobial, hepatoprotective, and neuroprotective agent (Li et al. 2019; Prasad et al. 2014; Anand et al. 2008; Gupta et al. 2012; Lee et al. 2013; Maheshwari et al. 2006; Singh and Khar 2008; Ohori et al. 2006; Weber et al. 2005; Basnet and Skalko-Basnet 2011).

Berberine, an alkaloid found in various plant species, is reported for its potent anti-inflammatory activity, anti-diabetic activity, and cardioprotective and immunomodulatory activities (Belwal et al. 2020a; Neag et al. 2018).

1.6 Pharmacokinetics and Advanced Drug Delivery Systems Related to Nutraceuticals

Plant-derived phytonutrients have shown various pharmacological activities in in vitro systems; however, the poor water solubility, low bioavailability, and extensive metabolism have been one of the main limiting factors for the therapeutic use of these phytochemicals (Selby-Pham et al. 2017; Aqil et al. 2013; Karaś et al. 2017). Studies in animals and humans have shown that the most of the ingested polyphenols are absorbed in the intestines in very low quantity (about 5%), and the remaining portion is passed unchanged to the large intestine.

For example, although EGCG has shown promising bioactivities in in vitro systems, poor bioavailability is of great concern for clinical effectiveness (Cai et al. 2018) as less than 1% of EGCG was measured in blood after oral administration in rats (Chen et al. 1997). Similar reports of low bioavailability and extensive metabolism of nutraceuticals are commonly reported for curcumin (Anand et al. 2008; Basnet and Skalko-Basnet 2011), resveratrol, and many other compounds.

In recent years, various novel drug delivery systems including nano-formulations such as liposomes, micelles, solid-liquid nanoparticles, nanoemulsions, etc. have been designed, developed, and evaluated for the effective delivery of these phytochemicals (Wang et al. 2014; Li et al. 2015a,b; Mangal et al. 2017; Bonferoni et al. 2017). These nanoparticles and advanced drug delivery formulations are aimed to increase the pharmaceutical properties of phytochemicals such as solubility and improve the bioavailability and targeted delivery (Davatgaran-Taghipour et al. 2017). High-quality research in the coming years in these areas can greatly improve the issues related to the bioavailability and bioactivity of nutraceuticals.

1.7 Research Trends, Natural Abundance, Sustainable Utilization, and New Technologies for the Production of Nutraceuticals

In recent years, there is a growing interest in the area of nutraceuticals research due to high market and industrial demand which need development of novel technologies in identification, estimation, and bioactivity evaluation of nutraceuticals. A simple Scopus search (www.Scopus.com, December 25, 2020) with the keyword "nutraceutical OR nutraceuticals" resulted into a total of 17,189 documents (search filed:

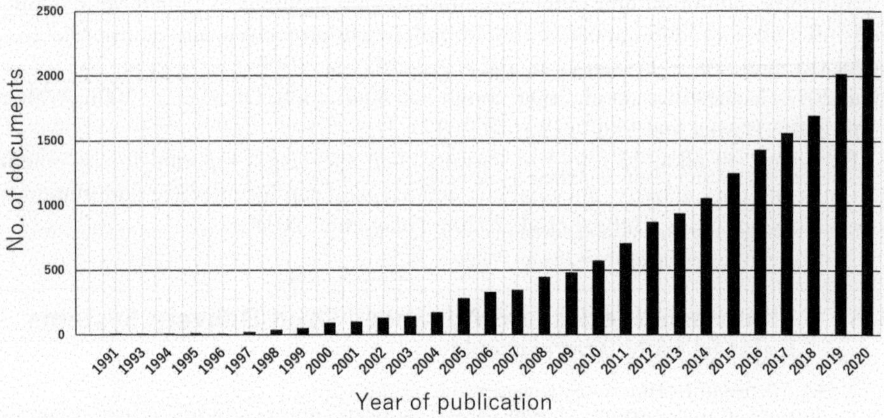

Fig. 1.23 Number of publications related to nutraceuticals

Abstract, Title, and Keywords). Based on author's affiliations, the United States, Italy, India, China, Canada, Spain, South Korea, the United Kingdom, Brazil, and Mexico were the leading countries on nutraceutical research. More than half of these studies are reported after 2010 (Fig. 1.23).

In recent years, there are many advanced techniques for the extraction and purification of phytochemicals that can be used as nutraceuticals (Belwal et al. 2018, 2020b). Similarly, many techniques related to harvesting and their processing have been developed and introduced in various fields. However, current cultivation practices may not fulfill the increasing demand of plant-based nutraceuticals, and extensive use of natural resources may result in extension of important plant species. Thus, sustainable methods for the natural resources use are very important. Tissue culture and cell culture methods have high potential to fulfill these demands as these methods can be easily manipulated to biosynthesize and produce target nutraceutical.

1.8 Conclusions and Future Recommendations

Nutraceuticals are the products that are consumed for their potential health beneficial effects which makes them different from normal food products. The market of nutraceuticals and related functional foods is growing worldwide, and there is an increasing demand. Most of these nutraceuticals are derived from plants such as polyphenols, alkaloids, carotenoids, etc. and are widely studied for their chemistry and pharmacological activities. However, in many cases, there are not much strong scientific evidences, and some reports also raise the concern of no benefits from the consumption of these products (Aronson 2017). Polypharmacy, drug interaction, side effects, and adulteration are also some other challenges in nutraceutical products. Future research studies should try to fulfill these research gaps by providing sufficient sources for production of these nutraceuticals. Furthermore, extensive studies related to their clinical efficacy and possible toxicity are necessary.

References

Abuajah CI, Ogbonna AC, Osuji CM (2015) Functional components and medicinal properties of food: a review. J Food Sci Technol 52:2522–2529. https://doi.org/10.1007/s13197-014-1396-5

Adhikari-Devkota A, Kurauchi Y, Yamada T et al (2019) Anti-neuroinflammatory activities of extract and polymethoxyflavonoids from immature fruit peels of Citrus 'Hebesu'. J Food Biochem. https://doi.org/10.1111/jfbc.12813

Agati G, Azzarello E, Pollastri S, Tattini M (2012) Flavonoids as antioxidants in plants: location and functional significance. Plant Sci 196:67–76. https://doi.org/10.1016/j.plantsci.2012.07.014

Ahmad A, Ali T, Park HY et al (2017) Neuroprotective effect of fisetin against amyloid-beta-induced cognitive/synaptic dysfunction, neuroinflammation, and neurodegeneration in adult mice. Mol Neurobiol 54:2269–2285. https://doi.org/10.1007/s12035-016-9795-4

Anand P, Thomas SG, Kunnumakkara AB et al (2008) Biological activities of curcumin and its analogues (Congeners) made by man and Mother Nature. Biochem Pharmacol 76:1590–1611. https://doi.org/10.1016/j.bcp.2008.08.008

Andrew R, Izzo AA (2017) Principles of pharmacological research of nutraceuticals. Br J Pharmacol 174:1177–1194. https://doi.org/10.1111/bph.13779

Aqil F, Munagala R, Jeyabalan J, Vadhanam MV (2013) Bioavailability of phytochemicals and its enhancement by drug delivery systems. Cancer Lett 334:133–141. https://doi.org/10.1016/j.canlet.2013.02.032

Aronson JK (2017) Defining 'nutraceuticals': neither nutritious nor pharmaceutical. Br J Clin Pharmacol 83:8–19. https://doi.org/10.1111/bcp.12935

Atanasov AG, Waltenberger B, Pferschy-Wenzig EM et al (2015) Discovery and resupply of pharmacologically active plant-derived natural products: a review. Biotechnol Adv 33:1582–1614. https://doi.org/10.1016/j.biotechadv.2015.08.001

Ayaz M, Sadiq A, Junaid M et al (2019) Flavonoids as prospective neuroprotectants and their therapeutic propensity in aging associated neurological disorders. Front Aging Neurosci 11:1–20

Barros L, Dueñas M, Ferreira ICFR et al (2009) Phenolic acids determination by HPLC-DAD-ESI/MS in sixteen different Portuguese wild mushrooms species. Food Chem Toxicol 47:1076–1079. https://doi.org/10.1016/j.fct.2009.01.039

Basnet P, Skalko-Basnet N (2011) Curcumin: an anti-inflammatory molecule from a curry spice on the path to cancer treatment. Molecules 16:4567–4598

Batiha GES, Beshbishy AM, Wasef L et al (2020) Uncaria tomentosa (Willd. ex Schult.) DC.: a review on chemical constituents and biological activities. Appl Sci 10:82668. https://doi.org/10.3390/APP10082668

Belwal T, Ezzat SM, Rastrelli L et al (2018) A critical analysis of extraction techniques used for botanicals: trends, priorities, industrial uses and optimization strategies. TrAC Trends Anal Chem 100:82–102. https://doi.org/10.1016/j.trac.2017.12.018

Belwal T, Bisht A, Devkota HP et al (2020a) Phytopharmacology and clinical updates of berberis species against diabetes and other metabolic diseases. Front Pharmacol 11:41

Belwal T, Chemat F, Venskutonis PR et al (2020b) Recent advances in scaling-up of non-conventional extraction techniques: learning from successes and failures. TrAC Trends Anal Chem 127:115895. https://doi.org/10.1016/j.trac.2020.115895

Bigliardi B, Galati F (2013) Innovation trends in the food industry: the case of functional foods. Trends Food Sci Technol 31:118–129. https://doi.org/10.1016/j.tifs.2013.03.006

Boehm K, Borrelli F, Ernst E et al (2009) Green tea (Camellia sinensis) for the prevention of cancer. Cochrane Database Syst Rev 3:CD005004

Bonferoni MC, Rossi S, Sandri G, Ferrari F (2017) Nanoparticle formulations to enhance tumor targeting of poorly soluble polyphenols with potential anticancer properties. Semin Cancer Biol 46:205–214. https://doi.org/10.1016/j.semcancer.2017.06.010

Boots AW, Haenen GRMM, Bast A (2008) Health effects of quercetin: from antioxidant to nutraceutical. Eur J Pharmacol 585:325–337

Bosch-Barrera J, Menendez JA (2015) Silibinin and STAT3: a natural way of targeting transcription factors for cancer therapy. Cancer Treat Rev 41:540–546

Brown DE, Rashotte AM, Murphy AS et al (2001) Flavonoids act as negative regulators of auxin transport in vivo in arabidopsis. Plant Physiol 126:524–535. https://doi.org/10.1104/pp.126.2.524

Cai ZY, Li XM, Liang JP et al (2018) Bioavailability of tea catechins and its improvement. Molecules 23:2346

Carl HB (2000) Phenolic-storing cells: keys to programmed cell death and periderm formation in wilt disease resistance and in general defence responses in plants? Physiol Mol Plant Pathol 57:101–110. https://doi.org/10.1006/pmpp.2000.0287

Carloni P, Tiano L, Padella L et al (2013) Antioxidant activity of white, green and black tea obtained from the same tea cultivar. Food Res Int 53:900–908. https://doi.org/10.1016/j.foodres.2012.07.057

Chen L, Lee MJ, Li H, Yang CS (1997) Absorption, distribution, and elimination of tea polyphenols in rats. Drug Metab Dispos 25:1045–1050

Chen C, Ono M, Takeshima M, Nakano S (2014) Antiproliferative and apoptosis-inducing activity of nobiletin against three subtypes of human breast cancer cell lines. Anticancer Res 34:1785–1792

Da Silva PM (2013) Tea: A new perspective on health benefits. Food Res Int 53:558–567

Davatgaran-Taghipour Y, Masoomzadeh S, Farzaei MH et al (2017) Polyphenol nanoformulations for cancer therapy: experimental evidence and clinical perspective. Int J Nanomedicine 12:2689–2702. https://doi.org/10.2147/IJN.S131973

Day L, Seymour RB, Pitts KF et al (2009) Incorporation of functional ingredients into foods. Trends Food Sci Technol 20:388–395. https://doi.org/10.1016/j.tifs.2008.05.002

Devkota HP, Watanabe M (2020) Role of medicinal plant gardens in pharmaceutical science education and research: an overview of medicinal plant garden at Kumamoto University, Japan. J Asian Assoc Sch Pharm 9:44–52

Dietrich H (2004) Bioactive compounds in fruit and juice. Fruit Process 2004:50–55

Donado F, Sandoval U, Carrillo JL (2011) Viscosity enhancement in dilute magnetorheological fluids through magnetic perturbations. Rev Mex Fis 57:426–434. https://doi.org/10.1016/j.intimp.2011.03.012

Fitzgerald M, Heinrich M, Booker A (2019) Medicinal plant analysis: a historical and regional discussion of emergent complex techniques. Front Pharmacol 10. https://doi.org/10.3389/fphar.2019.01480

Forkmann G (1991) Flavonoids as flower pigments: the formation of the natural spectrum and its extension by genetic engineering. Plant Breed 106:1–26. https://doi.org/10.1111/j.1439-0523.1991.tb00474.x

Gramza A, Korczak J (2005) Tea constituents (Camellia sinensis L.) as antioxidants in lipid systems. Trends Food Sci Technol 16:351–358. https://doi.org/10.1016/j.tifs.2005.02.004

Gul K, Singh AK, Jabeen R (2016) Nutraceuticals and functional foods: the foods for the future world. Crit Rev Food Sci Nutr 56:2617–2627

Gupta SC, Patchva S, Koh W, Aggarwal BB (2012) Discovery of curcumin, a component of golden spice, and its miraculous biological activities. Clin Exp Pharmacol Physiol. https://doi.org/10.1111/j.1440-1681.2011.05648.x

Hara Y (2011) Tea catechins and their applications as supplements and pharmaceutics. Pharmacol Res 64:100–104. https://doi.org/10.1016/j.phrs.2011.03.018

Harborne JB, Williams CA (2000) Advances in flavonoid research since 1992. Phytochemistry 55: 481–504. https://doi.org/10.1016/S0031-9422(00)00235-1
Higdon JV, Frei B (2003) Tea catechins and polyphenols: health effects, metabolism, and antioxidant functions. Crit Rev Food Sci Nutr 43:89–143
Hodgson JM, Croft KD (2010) Tea flavonoids and cardiovascular health. Mol Asp Med 31:495–502
Iwashina T (2015) Contribution to flower colors of flavonoids including anthocyanins: a review. Nat Prod Commun 10(3):529–544
Kao YH, Chang HH, Lee MJ, Chen CL (2006) Tea, obesity, and diabetes. Mol Nutr Food Res 50: 188–210
Karaś M, Jakubczyk A, Szymanowska U et al (2017) Digestion and bioavailability of bioactive phytochemicals. Int J Food Sci Technol 52:291–305. https://doi.org/10.1111/ijfs.13323
Khalid M, Saeed-ur-Rahman, Bilal M, Feng HD (2019) Role of flavonoids in plant interactions with the environment and against human pathogens—a review. J Integr Agric 18:211–230. https://doi.org/10.1016/S2095-3119(19)62555-4
Khan N, Mukhtar H (2010) Cancer and metastasis: prevention and treatment by green tea. Cancer Metastasis Rev 29:435–445
Khan N, Syed DN, Ahmad N, Mukhtar H (2013) Fisetin: a dietary antioxidant for health promotion. Antioxid Redox Signal 19:151–162. https://doi.org/10.1089/ars.2012.4901
Khan H, Sureda A, Belwal T et al (2019) Polyphenols in the treatment of autoimmune diseases. Autoimmun Rev 18:647–657
Khoo HE, Prasad KN, Kong KW et al (2011) Carotenoids and their isomers: color pigments in fruits and vegetables. Molecules 16:1710–1738
Kurauchi Y, Devkota HP, Hori K et al (2019) Anxiolytic activities of Matcha tea powder, extracts, and fractions in mice: contribution of dopamine D1 receptor- and serotonin 5-HT1A receptor-mediated mechanisms. J Funct Foods 59:301–308. https://doi.org/10.1016/j.jff.2019.05.046
Lee W-H, Loo C-Y, Bebawy M et al (2013) Curcumin and its derivatives: their application in neuropharmacology and neuroscience in the 21st century. Curr Neuropharmacol. https://doi.org/10.2174/1570159x11311040002
Legeay S, Rodier M, Fillon L et al (2015) Epigallocatechin gallate: a review of its beneficial properties to prevent metabolic syndrome. Nutrients 7:5443–5468
Li C, Zhang J, Zu YJ et al (2015a) Biocompatible and biodegradable nanoparticles for enhancement of anti-cancer activities of phytochemicals. Chin J Nat Med 13:641–652. https://doi.org/10.1016/S1875-5364(15)30061-3
Li Z, Jiang H, Xu C, Gu L (2015b) A review: using nanoparticles to enhance absorption and bioavailability of phenolic phytochemicals. Food Hydrocoll 43:153–164. https://doi.org/10.1016/j.foodhyd.2014.05.010
Li Y, Yao J, Han C et al (2016) Quercetin, inflammation and immunity. Nutrients. https://doi.org/10.3390/nu8030167
Li H, Sureda A, Devkota HP et al (2019) Curcumin, the golden spice in treating cardiovascular diseases. Biotechnol Adv 38:107343
Lin M, Zhang J, Chen X (2018) Bioactive flavonoids in Moringa oleifera and their health-promoting properties. J Funct Foods 47:469–479
Liu CW, Murray JD (2016) The role of flavonoids in nodulation host-range specificity: an update. Plants 5:4045–4053. https://doi.org/10.3390/plants5030033
Lu W, Lin C, King TD et al (2012) Silibinin inhibits Wnt/β-catenin signaling by suppressing Wnt co-receptor LRP6 expression in human prostate and breast cancer cells. Cell Signal 24:2291–2296. https://doi.org/10.1016/j.cellsig.2012.07.009
Maheshwari RK, Singh AK, Gaddipati J, Srimal RC (2006) Multiple biological activities of curcumin: a short review. Life Sci 78:2081–2087
Mandal SK, Maji AK, Mishra SK et al (2020) Goldenseal (Hydrastis canadensis L.) and its active constituents: a critical review of their efficacy and toxicological issues. Pharmacol Res 160: 105085. https://doi.org/10.1016/j.phrs.2020.105085

Mangal S, Gao W, Li T, Zhou QT (2017) Pulmonary delivery of nanoparticle chemotherapy for the treatment of lung cancers: challenges and opportunities. Acta Pharmacol Sin 38:782–797. https://doi.org/10.1038/aps.2017.34

Market Research (2020) Dried fruit market size, share & trends analysis report by product (apricots, dates, raisins, figs, berries), by distribution channel (supermarkets & hypermarkets, online), by region, and segment forecasts, 2020–2027

Martin C, Butelli E, Petroni K, Tonelli C (2011) How can research on plants contribute to promoting human health? Plant Cell 23:1685–1699

Masterjohn C, Bruno RS (2012) Therapeutic potential of green tea in nonalcoholic fatty liver disease. Nutr Rev 70:41–56. https://doi.org/10.1111/j.1753-4887.2011.00440.x

Mattila P, Hellström J, Törrönen R (2006) Phenolic acids in berries, fruits, and beverages. J Agric Food Chem 54:7193–7199. https://doi.org/10.1021/jf0615247

Mozos I, Stoian D, Caraba A et al (2018) Lycopene and vascular health. Front Pharmacol 9:521

Nakabayashi R, Saito K (2015) Integrated metabolomics for abiotic stress responses in plants. Curr Opin Plant Biol 24:10–16. https://doi.org/10.1016/j.pbi.2015.01.003

Nakajima A, Ohizumi Y, Yamada K (2014) Anti-dementia activity of nobiletin, a citrus flavonoid: a review of animal studies. Clin Psychopharmacol Neurosci 12:75–82

Namal Senanayake SPJ (2013) Green tea extract: chemistry, antioxidant properties and food applications—a review. J Funct Foods 5:1529–1541

Neag MA, Mocan A, Echeverría J et al (2018) Berberine: botanical occurrence, traditional uses, extraction methods, and relevance in cardiovascular, metabolic, hepatic, and renal disorders. Front Pharmacol 9:557. https://doi.org/10.3389/fphar.2018.00557

Negi PS (2012) Plant extracts for the control of bacterial growth: efficacy, stability and safety issues for food application. Int J Food Microbiol 156:7–17. https://doi.org/10.1016/j.ijfoodmicro.2012.03.006

Newman DJ, Cragg GM (2020) Natural products as sources of new drugs over the nearly four decades from 01/1981 to 09/2019. J Nat Prod 83:770–803. https://doi.org/10.1021/acs.jnatprod.9b01285

Nile SH, Park SW (2014) Edible berries: Bioactive components and their effect on human health. Nutrition 30:134–144

Nutraceuticals Market (2019) Nutraceuticals market to 2027—global analysis and forecasts by type; application. https://www.researchandmarkets.com/reports/4987763/nutraceuticals-market-to-2027-global-analysis?utm_source=dynamic&utm_medium=BW&utm_code=8swn8c&utm_campaign=1383035+-+Insights+into+the+Worldwide+Nutraceuticals+Industry+to+2027+-+Increase+in+Demand+for+N

Odegaard AO, Koh WP, Gross MD et al (2011) Combined lifestyle factors and cardiovascular disease mortality in Chinese men and women: the Singapore Chinese health study. Circulation 124:2847–2854. https://doi.org/10.1161/CIRCULATIONAHA.111.048843

Ohori H, Yamakoshi H, Tomizawa M et al (2006) Synthesis and biological analysis of new curcumin analogues bearing an enhanced potential for the medicinal treatment of cancer. Mol Cancer Ther. https://doi.org/10.1158/1535-7163.mct-06-0174

Pan MH, Lai CS, Ho CT (2010) Anti-inflammatory activity of natural dietary flavonoids. Food Funct 1:15–31. https://doi.org/10.1039/c0fo00103a

Park HH, Lee S, Oh JM et al (2007) Anti-inflammatory activity of fisetin in human mast cells (HMC-1). Pharmacol Res 55:31–37. https://doi.org/10.1016/j.phrs.2006.10.002

Park JH, Sung HY, Song DK (2009) Green tea and type 2 diabetes. In: Handbook of green tea and health research. Nova Science, Hauppauge, pp 413–420

Pawar RS, Krynitsky AJ, Rader JI (2013) Sweeteners from plants-with emphasis on Stevia rebaudiana (Bertoni) and Siraitia grosvenorii (Swingle). Anal Bioanal Chem 405:4397–4407. https://doi.org/10.1007/s00216-012-6693-0

Pervin M, Unno K, Ohishi T et al (2018) Beneficial effects of green tea catechins on neurodegenerative diseases. Molecules 23:61297. https://doi.org/10.3390/molecules23061297

Piironen V, Toivo J, Puupponen-Pimiä R, Lampi AM (2003) Plant sterols in vegetables, fruits and berries. J Sci Food Agric 83:330–337. https://doi.org/10.1002/jsfa.1316

Prasad S, Gupta SC, Tyagi AK, Aggarwal BB (2014) Curcumin, a component of golden spice: from bedside to bench and back. Biotechnol Adv 32:1053–1064

Rasul A, Millimouno FM, Ali Eltayb W et al (2013) Pinocembrin: a novel natural compound with versatile pharmacological and biological activities. Biomed Res Int 2013:379850. https://doi.org/10.1155/2013/379850

Research and Markets (2018) Global nutritional supplement market (2018–2023). https://www.researchandmarkets.com/reports/4667650/global-nutritional-supplement-market-2018-2023

Reygaert WC (2014) The antimicrobial possibilities of green tea. Front Microbiol 5:434

Salehi B, Fokou PVT, Sharifi-Rad M et al (2019) The therapeutic potential of naringenin: a review of clinical trials. Pharmaceuticals 12:11

Sanna V, Lubinu G, Madau P et al (2015) Polymeric nanoparticles encapsulating white tea extract for nutraceutical application. J Agric Food Chem 63:2026–2032. https://doi.org/10.1021/jf505850q

Sarkic A, Stappen I (2018) Essential oils and their single compounds in cosmetics-a critical review. Cosmetics 5:10011. https://doi.org/10.3390/cosmetics5010011

Selby-Pham SNB, Miller RB, Howell K et al (2017) Physicochemical properties of dietary phytochemicals can predict their passive absorption in the human small intestine. Sci Rep 7:01888. https://doi.org/10.1038/s41598-017-01888-w

Sereno AB, Bampi M, dos Santos IE et al (2018) Mineral profile, carotenoids and composition of cocona (Solanum sessiliflorum Dunal), a wild Brazilian fruit. J Food Compos Anal 72:32–38

Shahidi F (2004) Functional foods: their role in health promotion and disease prevention. J Food Sci 69:R146–R149. https://doi.org/10.1111/j.1365-2621.2004.tb10727.x

Shahidi F, Ambigaipalan P (2015) Phenolics and polyphenolics in foods, beverages and spices: Antioxidant activity and health effects—a review. J Funct Foods 18:820–897

Singh S, Khar A (2008) Biological effects of curcumin and its role in cancer chemoprevention and therapy. Anti Cancer Agents Med Chem. https://doi.org/10.2174/187152006776930918

Siró I, Kápolna E, Kápolna B, Lugasi A (2008) Functional food. Product development, marketing and consumer acceptance-a review. Appetite 51:456–467. https://doi.org/10.1016/j.appet.2008.05.060

Tanaka Y, Sasaki N, Ohmiya A (2008) Biosynthesis of plant pigments: anthocyanins, betalains and carotenoids. Plant J 54:733–749. https://doi.org/10.1111/j.1365-313X.2008.03447.x

Tapas A, Sakarkar D, Kakde R (2008) Flavonoids as nutraceuticals: a review. Trop J Pharm Res 7:14693. https://doi.org/10.4314/tjpr.v7i3.14693

Taylor PW, Hamilton-Miller JMT, Stapleton PD (2005) Antimicrobial properties of green tea catechins. Food Sci Technol Bull Funct Foods 2:71–81. https://doi.org/10.1616/1476-2137.14184

The Ministry of Health Labor and Welfare of Japan (2016) The Japanese pharmacopoeia, 17th edn. The Ministry of Health Labor and Welfare of Japan, Tokyo

Tsao R (2010) Chemistry and biochemistry of dietary polyphenols. Nutrients 2:1231–1246

Tyagi AK, Singh RP, Agarwal C et al (2002) Silibinin strongly synergizes human prostate carcinoma DU145 cells to doxorubicin-induced growth inhibition, G2-M arrest, and apoptosis. Clin Cancer Res 8:3512–3519. https://doi.org/10.1186/1471-2407-12-117

Vargas JE, Souto AA, Pitrez PMC et al (2016) Modulatory potential of resveratrol during lung inflammatory disease. Med Hypotheses 96:61–65. https://doi.org/10.1016/j.mehy.2016.09.023

Voon HC, Bhat R, Rusul G (2012) Flower extracts and their essential oils as potential antimicrobial agents for food uses and pharmaceutical applications. Compr Rev Food Sci Food Saf 11:34–55. https://doi.org/10.1111/j.1541-4337.2011.00169.x

Wai A, Yeung K, Aggarwal BB et al (2018) Dietary natural products and their potential to influence health and disease including animal model studies. Anim Sci Paper Rep 36:345–358

Wang S, Su R, Nie S et al (2014) Application of nanotechnology in improving bioavailability and bioactivity of diet-derived phytochemicals. J Nutr Biochem 25:363–376. https://doi.org/10.1016/j.jnutbio.2013.10.002

Wang W, Sun C, Mao L et al (2016) The biological activities, chemical stability, metabolism and delivery systems of quercetin: a review. Trends Food Sci Technol 56:21–38. https://doi.org/10.1016/j.tifs.2016.07.004

Wang M, Meng D, Zhang P et al (2018) Antioxidant protection of nobiletin, 5-demethylnobiletin, tangeretin, and 5-demethyltangeretin from citrus peel in saccharomyces cerevisiae. J Agric Food Chem 66:3155–3160. https://doi.org/10.1021/acs.jafc.8b00509

Weber WM, Hunsaker LA, Abcouwer SF et al (2005) Anti-oxidant activities of curcumin and related enones. Bioorg Med Chem. https://doi.org/10.1016/j.bmc.2005.03.035

Weinreb O, Mandel S, Amit T, Youdim MBH (2004) Neurological mechanisms of green tea polyphenols in Alzheimer's and Parkinson's diseases. J Nutr Biochem 15:506–516

Weston LA, Mathesius U (2013) Flavonoids: their structure, biosynthesis and role in the rhizosphere, including allelopathy. J Chem Ecol 39:283–297. https://doi.org/10.1007/s10886-013-0248-5

Williams RJ, Spencer JPE (2012) Flavonoids, cognition, and dementia: actions, mechanisms, and potential therapeutic utility for Alzheimer disease. Free Radic Biol Med 52:35–45

Yang CS, Wang H (2016) Cancer preventive activities of tea catechins. Molecules 21:1679

Yang J, Li Q, Zhou XD et al (2011) Naringenin attenuates mucous hypersecretion by modulating reactive oxygen species production and inhibiting NF-κB activity via EGFR-PI3K-Akt/ERK MAPKinase signaling in human airway epithelial cells. Mol Cell Biochem 351:29–40. https://doi.org/10.1007/s11010-010-0708-y

Yeung AWK, Aggarwal BB, Barreca D et al (2019) Dietary natural products and their potential to influence health and disease including animal model studies. Anim Sci Paper Rep 36:345–358

Yeung AWK, Orhan IE, Aggarwal BB et al (2020a) Berberine, a popular dietary supplement for human and animal health: quantitative research literature analysis-a review. Animal Science Papers and Reports 38:15

Yeung AWK, Orhan IE, Aggarwal BB et al (2020b) Berberine, a popular dietary supplement for human and animal health: quantitative research literature analysis—a review. Anim Sci Paper Rep 38:5–19

Zadernowski R, Naczk M, Nesterowicz J (2005) Phenolic acid profiles in some small berries. J Agric Food Chem 53:2118–2124. https://doi.org/10.1021/jf040411p

Zaidun NH, Thent ZC, Latiff AA (2018) Combating oxidative stress disorders with citrus flavonoid: naringenin. Life Sci 208:111–122. https://doi.org/10.1016/j.lfs.2018.07.017

Zaveri NT (2006) Green tea and its polyphenolic catechins: medicinal uses in cancer and noncancer applications. Life Sci 78:2073–2080

Zhang L, Zhang X, Zhang C et al (2016) Nobiletin promotes antioxidant and anti-inflammatory responses and elicits protection against ischemic stroke in vivo. Brain Res 1636:130–141. https://doi.org/10.1016/j.brainres.2016.02.013

Zhao J, Agarwal R (1999) Tissue distribution of silibinin, the major active constituent of silymarin, in mice and its association with enhancement of phase II enzymes: implications in cancer chemoprevention. Carcinogenesis 20:2101–2108. https://doi.org/10.1093/carcin/20.11.2101

In Vitro Production of Bioactive Compounds from Plant Cell Culture

2

Vasantha Veerappa Lakshmaiah, Akshatha Banadka, Gopishankar Thirumoorthy, Poornananda Madhava Naik, Jameel Mohammed Al-Khayri, and Praveen Nagella

Abstract

Secondary metabolites (SMs) are bioactive compounds widely used in various industries as pharmaceutical agents and food additives and serve as precursor substances for the synthesis of commercially important products. These natural bioactive metabolites are quickly replacing chemicals as efficient coloring, flavoring, texturizing, and preservative agents. Productions of these SMs are hampered due to physiological and technological parameters. Although SMs do not have any significant role in the growth and development of the organisms where they are found, they have commercial importance. Humankind has harnessed its application in every walk of their life. In the medical field, SMs are used as antibiotics, antifungal, antiviral, metabolic inhibitors, anticancer agents, and many more. The biological and pharmacological benefits of medicinal plants are attributed to SM produced by subsidiary pathways that are highly specific to target molecules. Most pharmaceuticals are either directly or indirectly derived from plant sources. Production of SMs from field plants suffers from various limitations like seasonal production, choosing specific plant organs for specific metabolites, low yield, cost of purification, and seasonal variations. Biotechnological approaches such as plant cell, tissue, and organ cultures are

V. V. Lakshmaiah · A. Banadka · G. Thirumoorthy · P. Nagella (✉)
Department of Life Sciences, CHRIST (Deemed to be University), Bengaluru, Karnataka, India
e-mail: vasantha.vl@christuniversity.in; akshatha.b@res.christuniversity.in; gopi.t@science.christuniversity.in; praveen.n@christuniversity.in

P. M. Naik
Department of Botany, Karnatak University, Dharwad, Karnataka, India

J. M. Al-Khayri
Department of Agricultural Biotechnology, College of Agriculture and Food Sciences, King Faisal University, Al-Ahsa, Saudi Arabia
e-mail: jkhayri@kfu.edu.sa

T. Belwal et al. (eds.), *Nutraceuticals Production from Plant Cell Factory*,
https://doi.org/10.1007/978-981-16-8858-4_2

the best alternative methods for commercial production. The current chapter focuses on establishment of plant cell culture system for the production of SMs, strategies to improve biomass yield and metabolite content, and biosynthetic pathways. The chapter also emphasizes elicitation strategies, application of CRISPR Cas9 in metabolite synthesis, large-scale production, and commercial aspects of SMs.

Keywords

Plant cell culture · Plant secondary metabolites · CRISPR Cas9 · Elicitation · Metabolic engineering · Biosynthesis

2.1 Introduction

Since time immemorial, plants have served as the primary source of carbohydrates, fats, and proteins for humankind. Furthermore, plants produce a broad range of valuable phytochemicals such as primary and secondary metabolites (SMs) (Hassanpour et al. 2011). Primary metabolites are involved in the growth, development, and reproduction of the plant. On the contrary, SMs do not take part in the growth and development of plants. However, they serve as a defense against biotic and abiotic stress as the plants are sessile organisms. The change in climatic conditions such as carbon dioxide and global warming influence SM production. The carbon dioxide increases the phenolic compounds and decreases terpenoids in foliage. On the contrary, global warming reduces phenolic compounds and increases terpenoids (Holopainen et al. 2018). Besides this, the SMs attract pollinators, aids in establishing symbiosis, and provide structural components (Guerriero et al. 2018). The study of SMs has gained a special interest in recent years as they possess various biological activities known to alleviate several human ailments. Albrecht Kossel, the winner of the Nobel Prize in physiology or medicine in 1910, first described the notion of the SMs. The biosynthesis of secondary metabolites is a multistep reaction that occurs through amino acid, acetate-malonate, glucose, 2-C-methyl-d-erythritol-4-phosphate, mevalonate pathway, phenylpropanoid pathways, or combined pathways (Gonçalves and Romano 2018). More than 2,140,000 secondary metabolites are known so far (Thirumurugan et al. 2018). Plant secondary metabolites are broadly classified as alkaloids, phenolic compounds, sulfur-containing compounds, and terpenoids based on structure function and its biosynthetic pathway (Guerriero et al. 2018).

Secondary metabolites have been extensively used in cancer treatment and modern medicine. SMs such as alkaloids, phenolic acids, flavonoids, and tocopherols significantly impact the global market (Marchev and Georgiev 2020). Over 60% of anticancer drugs and about 25–28% of modern medicines are derived from plants. The global herbal medicine market is expected to reach USD 129.7 billion by 2023. BBC reports that by 2022, plant-derived drugs will reach USD 39.2 billion (Gonçalves and Romano 2018). The increasing market demand for plant secondary metabolites has resulted in the overharvesting of these medicinal plants. This extensive harvesting of these medicinal plants has led to serious threats like habitat loss and habitat degradation. The chemical synthesis of some simple SMs is possible. However, the chemical synthesis method is time-consuming and economically infeasible. Besides this, the complex SM synthesis is either tedious or impossible. Thus, an alternative strategy for SM production has to be adopted. Plant tissue culture offers an alternative method for the conservation of plant species and also for production of bioactive molecules. Research work on different plant tissue culture techniques such as micropropagation, indirect and direct organogenesis, and hairy root culture has been adopted to enhance SM production. Over the past decade, the structure, biological significance, and biosynthesis of the secondary metabolites are extensively studied (Bourgaud et al. 2001). Additionally, the genes involved in the biosynthesis of plant secondary metabolites are manipulated through genetic engineering for enhanced SM production (Rastegari et al. 2019).

Plant tissue culture is an in vitro culture of cells, tissues, organs, or whole plants under controlled aseptic conditions. Gottlieb Haberlandt first established plant tissue culture in 1902, and it was patented in 1950 by Pfizer Inc. (Smetanska 2008). Plant cell and tissue culture is an eco-friendly technique for the enhanced production of secondary metabolites in a controlled environment. Moreover, the plant tissue culture helps in the mass propagation of plants in a short period independent of the seasonal constraints and place of cultivation. It aids in easier isolation or extraction of secondary metabolites in a short duration. Another advantage of plant tissue culture is the production of SMs not produced in the native plant (Gonçalves and Romano 2018).

Undifferentiated calli, or differentiated structures like shoots, roots, or somatic embryos, can be used as explants to produce SMs through in vitro culture. In some instances, SM production demands differentiated organ culture as specific organs produce particular SMs. Plant tissue culture techniques are of different types, which include seed culture where the seed is used as an explant; embryo culture, which involves the use of mature or immature embryo as an explant; organ culture, which uses shoot, root, or leaf as the explant; and callus culture, which is the development of an undifferentiated mass of cells from differentiated plant material. The callus then undergoes organ differentiation (Gonçalves and Romano 2018). Cell suspension culture is also one of the tissue culture techniques involving liquid medium where cells or tissues are suspended and is extensively used for SMs production (Bourgaud et al. 2001; Naik and Al-Khayri 2020).

Plant cell culture can enhance SM production through various strategies. The first strategy deals with selecting a high metabolite-producing plant genotype and high SM-producing cell line. However, after a few subcultures, they might lose their producing ability due to somaclonal variations. Thus, alternative strategies have to be adopted. The other approaches involve optimizing culture conditions, such as the temperature, pH, light intensity, agitation speed, aeration, selection and optimization of different culture media, inoculum density, medium strength, carbon source, and optimization of plant growth regulators (PGR) by subjecting the explant to various combinations of PGRs in varying concentrations. It is known that SMs are produced as a defense mechanism against stress conditions. In this regard, the nutrient medium is supplied with stress-inducing compounds known as elicitors, such as jasmonic acid and salicylic acid, which induce the upregulation of genes that are involved in secondary metabolite production. The nutrient medium is also fed with precursors utilized by the enzyme systems to produce the desired SM products. Immobilization of plant cell tissue is another strategy that protects the tissue from shear stress (Gonçalves and Romano 2018). The present chapter deals with establishing a plant cell culture system for the production of SMs, strategies employed for improving the yield of biomass and metabolite content, and biosynthetic pathway leading to metabolite production. Also, the chapter discusses different elicitation strategies used in SM production, application of CRISPR Cas9 in metabolite synthesis, large-scale production of SMs using bioreactors, and commercial aspects.

2.2 Production and Optimization of Bioactive Molecules Using Cell Cultures

Plants are the primary source of naturally derived modern drugs and anticancer agents. The SMs provide the scientific foundation for the pharmaceutical properties of these medicinal herbs. The demand for these natural plant-derived products has spiked in recent times due to their safety and reasonable cost. Thus, the medicinal plants are overexploited to extract SM, which has resulted in their depletion. On the other hand, the chemical synthesis of complex SMs is tedious. Growing medicinal plants in their natural habitat is time-consuming, and some medicinal plants do not extend outside their local habitat. This has led biotechnologists to develop and adopt a new strategy by culturing plant cells and tissues in vitro under controlled conditions for bioactive compound production (Bourgaud et al. 2001; Naik and Al-Khayri 2018, 2020). Some of the most important strategies that ameliorate bioactive compounds in cell cultures have been respresented in Table 2.1 and discussed below.

2.2.1 Selection of Cell Lines

The in vitro SMs production occurs in two stages: accumulation of biomass and biosynthesis of SMs—both of these need to be optimized by employing multiple tissue culture strategies (Chandran et al. 2020). The primary strategy for enhancing the SM is screening high SM yielding cultivar and selecting high-yielding cell lines. The genetic makeup of the parent plant is one of the significant determinants of the SM production potential. Different members of the same species show varied genetic potential for secondary metabolite production (Smetanska 2008). The variation in glucosinolate profiles has been reported in the members of the family Brassicaceae. After selecting the plant genotype, a callus is induced in different parts of the parent plant, such as leaves, roots, flowers, seeds, and other plants. This is followed by

Table 2.1 Enhanced secondary metabolite by optimizing the media and culture conditions

Culture condition	Plant name	Metabolite	Optimal culture condition	Metabolite yield	Reference
Carbon source	*Artemisia annua* L.	Artemisinin	10% fructose	Two-fold increase	Wang and Weathers 2007
	Gymnema sylvestre R. Br.	Gymnemic acid	4% sucrose	10.1 mg/g DW	Nagella et al. 2011a
	Ginkgo biloba	Bilobalide	5% sucrose	5 mg/g	Park et al. 2004
	Ginkgo biloba	Ginkgolide	7% sucrose	20 mg/g	Park et al. 2004
	Camptotheca acuminata Decne.	Camptothecin	6% sucrose	0.0029 mg/L	Kim et al. 1999
	Melissa officinalis L.	Rutin	300 mM sucrose	–	Kim et al. 2020
Nitrogen source	*Lactuca sativa* L.	Caffeic	0% nitrogen	0.64 mg/100 g FW	Galieni et al. 2015
	Lactuca sativa L.	Caftaric	0% nitrogen	300.85 mg/100 g FW	Galieni et al. 2015
	Lactuca sativa L.	Chicoric acids	0% nitrogen	67.38 mg/100 g FW	Galieni et al. 2015
	Panax quinquefolium L.	Ginsenoside	0.5 mM ammonium	12.45 mg/g dw	Kochan et al. 2016
	Panax quinquefolium L.	Saponin	0.5 mM ammonium	35.11 mg/g dw	Kochan et al. 2016
	Camptotheca acuminata Decne.	Camptothecin	40 mM, 5:1 NH4+/NO3	6.3 mg/L	Kim et al. 1999
Phosphate level	*Digitalis purpurea* L.	Digitoxin	–	180 mg/kg f.w	Hagimori et al. 1982
	Panax quinquefolium L.	Saponin	1.25 mM	960 mg/L	Liu and Zhong 1998
	Panax ginseng C.A.Mey.	Saponin	1.04 mM	643 mg/L	Liu and Zhong 1998
	Coffea arabica	Caffeine	0.0 mM	0.7 mg/g	Bramble et al. 1991
	Vitis vinifera L.	Anthocyanin	0.008 mM	350 ng/mol	Dedaldechamp et al. 1995
	Panax quinquefolium L.	Saponin	0.83 mM	35.11 mg/g dw	Kochan et al. 2016
Plant growth regulators	*Oxalis linearis* var. β. latior	Anthocyanin	2 μM 2,4-Dichlorophenoxyacetic acid		Meyer and Van Staden 1995

(continued)

Table 2.1 (continued)

Culture condition	Plant name	Metabolite	Optimal culture condition	Metabolite yield	Reference
	Thalictrum minus L.	Berberine	1-Naphthaleneacetic acid	0.66 mg/mL	Suzuki et al. 1987
	Haplopappus gracilis (Nutt.) A. Gray	Anthocyanin	Benzyl adenine (BA) and kinetin		Constabel et al. 1971
	Populus hybrids (P. maximowiczii X P. nigra)	Anthocyanin	0 mg/L kinetin	0.3 mg/L	Matsumoto et al. 1973
	Artemisia annua L.	Artemisinin	5 mg/L gibberellic acid	550 mg/L	Liu et al. 1997
	Hyoscyamus muticus L.	Hyoscyamine	Gibberellic acid	–	Vanhala et al. 1998
	Camptotheca acuminata Decne.	Camptothecin	0.5 mg/L 2,4-D and 0.1 mg/L kinetin	0.98 mg/L	Van Hengel et al. 1992
	Nothapodytes nimmoniana (J. Grah.) D.J. Mabberley	Camptothecin	4.52 μM 2,4D and 2.22 μM BA	1.30%	Isah and Mujib 2015
Temperature	*Catharanthus roseus* (L.) G.Don	Ajmalicine	16 °C	Tenfold increase	Morris 1986
	Catharanthus roseus (L.) G.Don	Serpentine	25 °C	11 fold increase	Morris 1986
	Camptotheca acuminata Decne.	Camptothecin	30 °C	–	Pan et al. 2004
pH	*Camptotheca acuminata* Decne.	Camptothecin	4.3	–	Pan et al. 2004
Light	*Perilla frutescens* (L.) Britton	Anthocyanin	27.2 W/m^2 light irradiation	3.0 g/L	Zhong et al. 1991
	Melastoma malabathricum L.	Anthocyanin	Light for ten days	–	Chan et al. 2010

developing fast-growing, high-producing cell lines by suspending the callus in a suitable liquid media under controlled conditions (Schreiner 2005).

The parent plant used as the explant would show heterogeneous metabolic productivity. This heterogeneity would increase due to genetic variations when the plants are grown in controlled culture conditions. These genetic variations, in turn, would result in decreased SMs production. These undesirable variations can be prevented or reduced by selecting cell populations with high-yielding potential from heterogeneous cultures (Smetanska 2008). Cell cloning of high-metabolite yielding cell lines is a beneficial technique for enhancing SMs (Bhojwani and Razdan 1996). In the study conducted by Yamamoto et al. (1982), the *Euphorbia milii* Des Moul. plant leaves were used as explants from which the callus was developed and subcultured in regular intervals. It has been reported that the high-yielding cell lines of *E. milii* selected after 24 rounds of subculturing produced sevenfold levels of anthocyanin. In *Coptis japonica* (Thunb.) Makino, it has been observed that there was sixfold higher production of berberine after subjecting the leaf and petiole segments to callus culture for four weeks (Yamada and Sato 1981). The cell lines of *Vitis vinifera* L. derived from the callus cultures derived from Dr. Francois Cormier's group cell lines have reported a 2.3 to 4-fold increase in anthocyanin production (Curtin et al. 2003). In suspension cultures of high-yielding cell lines of *Lithospermum erythrorhizon* Siebold & Zucc., a 13 to 20-fold increase in shikonin production was observed after repeated subculturing at intervals (Kim and Chang 1990). *Lavandula vera* L. cells cultured under light conditions showed a ninefold accumulation of free biotin (Watanabe et al. 1982). Cell lines of *Nicotiana tabacum* L. BY-2, showed 15-fold increase in ubiquinone-10 content after subjecting the parent cell line to 13 rounds of subculturing (Matsumoto et al. 1971). The increased production of capsaicin has been reported in PEP cell lines of *Capsicum annuum* L. developed from callus culture, derived from hypocotyl explants (Salgado-Garciglia and Ochoa-Alejo 1990). In *Phoenix dactylifera* L. cell suspension culture, results showed that cell line 2 induced a higher quantity of polyphenols (catechin, caffeic acid, kaempferol, and apigenin) when compared to cell line 1, and high-yielding cell line is used for further polyphenol production (Naik and Al-Khayri 2018).

Screening methods of high-yielding cell lines are different in different plant species depending upon the type of SM produced. Cell lines producing colored SMs such as berberine, betanin, and shikonin can be easily selected from the cell mass by separating the colored areas from the rest of the cell mass. Cell lines producing fluorescent SMs can be identified by fluorescent microscope or flow cytometry, thus helping in its rapid identification and separation. In some other cases, the colorless high-yielding cell lines can be screened by immunological methods. The next challenge is the relative stability of the screened high-yielding cell line, as they are susceptible to genetic variations over generations. These altered variations might affect the SM biosynthesis. Thus, it is essential to maintain stable cell lines, which can be achieved by repeated selection of stable cell lines. A study on the choice of stable cell lines has been conducted in *E. milii*, by subjecting the cell lines to 30 successive subcloning (Yamamoto et al. 1982). Sato and Yamada (1984), in another study on cell lines of *Cjaponica*, have subjected the cell lines to recurrent subcloning for maintaining the stability and for enhanced production of berberine. With the recent advancements in plant biotechnology, cell lines are now treated with mutagens to overproduce SMs. These mutagens alter the cell permeability and help release the intracellular accumulated SMs to the exterior. Studies on induction of a mutagen p-fluorophenylalanine in *N. tabacum* have been studied by Berlin et al. (1981). It has been reported that p-fluorophenylalanine-resistant cell lines of *Nicotiana tabacum* produced six- to tenfolds higher cinnamoyl putrescine. Despite selecting and maintaining highly productive cell lines, the production efficiency will be lost after long periods of culture. Thus, alternative strategies, including traditional and metabolic engineering strategies, are employed for enhanced SM production (Gonçalves and Romano 2018).

2.2.2 Optimization of Culture Media and Culture Conditions

The plant's growth, biomass accumulation, and SM production depend on the physicochemical factors. The composition of media in which the explant is cultured influences the physiology and metabolism of the explant. Gamborg's (B5) (Gamborg et al. 1968) media, Murashige and Skoog media (Murashige and Skoog 1962), Linsmaier and Skoog media (Linsmaier and Skoog 1965), and Schenk and Hildebrandt media (Schenk and Hildebrandt 1972) are some of the most commonly used media in plant tissue culture. Selecting a suitable culture medium is critical in enhancing SMs as the plant growth and development vary in different plants based on their nutritional requirements. The carbon, nitrogen, and phosphate levels and the concentration of plant growth regulators in the culture medium can influence biomass accumulation. Thus, optimization of culture medium plays a significant role in SM enhancement (Dörnenburg and Knorr 1995; Nagella and Murthy 2010, 2011; Naik and Al-Khayri 2020).

The growing demand for SMs in today's market world has paved the way for employing traditional and metabolic strategies for consistent high-yield production of SMs (Chandran et al. 2020; Hussain et al. 2012). SMs are biosynthesized from substrates of primary metabolism that are deflected to secondary biosynthetic pathways. The biosynthesis of SMs depends on both abiotic and biotic factors like temperature, pH, humidity, and light intensity. The optimization for SM production in in vitro cultures can be done in two stages: biomass accumulation stage and SM biosynthesis stage. Optimization of culture media components like carbohydrate, nitrogen, phosphate, and various combinations of plant growth regulators and optimization of culture conditions like pH, temperature, light intensity, agitation, and aeration can be done during the biomass accumulation stage. In the biosynthetic stage, elicitation, precursor feeding, permeabilization, and immobilization can be employed for enhancing the secondary metabolite production (Murthy et al. 2014).

2.2.2.1 Different Media

Bioactive compounds accumulation in the cell culture may vary dramatically. Different media compositions/nutrients are responsible for such variation in the SMs accumulation in cell cultures (Rao and Ravishankar 2002). In cell culture of *Withania somnifera* L. dunal among the various media studied (MS, B5, NN, and N6), MS medium emerged as a promising medium composition for the production of withanolide A (Nagella and Murthy 2010). From the various media tested (MS, B5, N6, and NN) in cell suspension culture of *P. dactylifera*, MS medium-treated culture showed optimum accumulation of phenolic and flavonoid content (Naik and Al-Khayri 2020).

2.2.2.2 Medium Strength

The choice of suitable media and the appropriate strength of media are vital for the growth of in vitro cultures. In cell suspension cultures of *W. somnifera*, full strength (1.0X) MS nutrient medium showed the highest biomass accumulation and withanolide A production (Nagella and Murthy 2010). Similarly, among various strengths of MS media ranging from 0.25 to 2.0X, the highest biomass and gymnemic acid production were achieved with full strength (1.0X) MS media (Nagella et al. 2011a). Naik and Al-Khayri (2020) reported that in cell suspension culture of *P. dactylifera*, MS media with 2.0X induced maximum phenolic content compared to other media strengths.

2.2.2.3 Inoculum Density

The inoculum density is also an important parameter affecting plant cells, tissues, and organ cultures. Plant cell suspensions are initiated from callus using relatively high cell concentrations as there is a minimum inoculation density below which growth does not occur or is preceded by a long lag phase. Medium conditioning can be used to reduce the minimum inoculum density; however, the chemical basis of the conditioning effect has not been fully defined, and it is primarily empirical (Lee and Shuler 2000). Nagella and Murthy (2010) reported that 10 g/L inoculum density induced optimum level of production of withanolide A in cell culture of *W. somnifera*. Similarly, in the cell culture of *Gymnema sylvestre* R. Br., gymnemic acid production was achieved maximum at culture treated with 10 g/L inoculum density (Nagella et al. 2011a). Among the various cell inoculum density tested (1, 2, 5, 10, and 20 g/L) in cell suspension culture of *P. dactylifera*, MS medium with 20 g/L-treated culture yielded maximum phenolic content (Naik and Al-Khayri 2020).

2.2.2.4 Carbon Source

The in vitro plant cell, tissue, and organ cultures are not fully autotrophic. Hence, they require carbon sources for energy production, maintenance of osmotic potential, and the plant's growth and development. Carbon sources take part in cell growth, root and shoot induction, embryogenesis, and organogenesis. The effect of a wide range of carbon sources, which include sugars (reducing and non-reducing) and sugar alcohol, has been extensively studied in different plants (Yaseen et al. 2013). One of the most used carbon sources is sucrose, at a concentration of 2–5%.

However, various other carbon sources such as glucose, lactose, galactose, maltose, and starch are used. Glucose is also found to have the same effect as sucrose. Fructose is also an effective carbon source. However, it is less effective than glucose (Saad and Elshahed 2012). Among the tested carbon sources (sucrose, glucose, fructose, and maltose) in the cell suspension culture of *P. dactylifera*, glucose treatment emerged as best over other carbon sources in the accumulation of phenolic content flavonoids (Naik and Al-Khayri 2020). The effect of different sucrose concentrations ranging from 1 to 8% on biomass accumulation and gymnemic acid production has been studied in *G. sylvestre*; 3% and 4% of sucrose are optimal for enhanced biomass and gymnemic acid (10.1 mg g^{-1} DW) production, respectively (Nagella et al. 2011a). A similar report was observed in cell suspension culture of *W. somnifera*, where, among the tested carbohydrates, 3% sucrose produced an optimum level of withanolide A content (Nagella and Murthy 2014). Studies on sucrose concentration on biomass accumulation and metabolite production in *Ginkgo biloba* L. have been conducted by Park et al. (2004). It has been found that 3%, 5%, and 7% sucrose are the optimal concentrations for biomass accumulation, ginkgolide, and bilobalide production, respectively. Sucrose, a carbon source, is an effective osmoprotectant in *V. vinifera* (Do and Cormier 1990). In *Panax ginseng* C.A.Meyer. cell suspension cultures, 3% (w/v) sucrose favored the production of ginsenosides when compared with other ranges of 1 to 7% (w/v) (Lian et al. 2002). Ali et al. (2016) investigated various sucrose concentrations in cell suspension culture of *Artemisia absinthium* L. and found that 5% sucrose concentration favored the accumulation of phenolic content and 7% sucrose induced flavonoids. In cell suspension culture of *P. dactylifera*, the accumulation of optimum flavonoid and phenolic content was observed in 2% and 4% sucrose-treated media, respectively (Naik and Al-Khayri 2020).

2.2.2.5 Nitrogen Source

Nitrogen is a vital element in synthesizing amino acids, nitrogenous bases, lipids, and organic compounds. It helps in growth, vegetative development, reproduction, and, most importantly, its involvement in metabolite production. Thus, it is an essential element in biomass accumulation and biosynthesis stages of the in vitro cultures. The selection of an appropriate nitrogen source with a proper concentration is another key for SM enhancement. Nitrogen is supplemented as an organic or inorganic source (Radušienė et al. 2019). The effect of decreased levels and deficiency of nitrogen on metabolite production has been studied. Ravishankar et al. (1988) reported the increased capsaicin levels were achieved in the cell cultures of *C. annum*. In another investigation, the complete elimination of nitrogen enhanced the pyrethrin production by two-fold in *Chrysanthemum cinerariaefolium* (Rajasekaran et al. 1991). In cell suspension culture of *W. somnifera*, the optimum level of biomass was accumulated at NH_4^+/NO_3^- ratio of 7.19/18.80 and withanolide A content at NH_4^+/NO_3^- ratio of 14.38/37.60 mM (Nagella and Murthy 2011). Similarly, in cell culture of *G. sylvestre*, cell accumulation and gymnemic acid content were influenced by NH_4^+/NO_3^- ratio; optimum biomass growth and gymnemic acid was found at NH_4^+/NO_3^- ratio of 7.19/18.80 (Nagella et al. 2011b).

2.2.2.6 Phosphate Levels

Phosphorus is one of the components that are required in large amounts for plant metabolism. The phosphorus forms the principal constituent of nucleic acids and phospholipids. They play an essential role in primary and SM production and take part in energy metabolism. Phosphate levels in the media have shown varied responses by the in vitro culture of different plant species. In some plants, the increased phosphate levels have shown better SM yield, while in some other cases, the limitation of phosphate has shown better yield. Enhanced digitoxin production, i.e., 180 mg/kg FW, has been observed in the callus cultures of *Digitalis purpurea* L. when treated with increased phosphate levels (Hagimori et al. 1982). Studies on the effect of different concentrations of phosphate on *Panax quinquefolius* L. (960 mg/L) and *P. ginseng* (643 mg/L) have shown the highest saponin production when supplied with 1.25 mM and 1.04 mM of phosphate, respectively (Liu and Zhong 1998). On the contrary, SM enhancement has been observed in a culture medium deprived of phosphate; the callus cultures of *Coffea arabica* L. were subjected to phosphate concentrations of 0.0, 0.35, and 0.70 mM; the highest caffeine content of 0.7 mg/g was achieved in the media without phosphate (Bramble et al. 1991). In *Vitis* hybrid grapes, the anthocyanin production was highest (350 ng/mol) in media supplemented with the least phosphate level of 0.008 mM (Dedaldechamp et al. 1995).

2.2.2.7 Plant Growth Regulators

Growth regulators are the phytohormones that regulate the growth, proliferation of biomass, and metabolites accumulation in a cell, adventitious shoot, and hairy root cultures. They are the crucial factors in the induction of callus and roots in non-transformed cells. Auxins, cytokinins, gibberellins, and abscisic acid are some of the most commonly used phytohormones in plant tissue/cell culture (Murthy et al. 2014). The concentration and the combination of auxin and cytokinins regulate the product formation type (callus/cell/adventitious root/shoot). Thus, choosing the appropriate growth regulators in optimal concentrations is very important (Murthy et al. 2014). Auxins are growth regulators that induce callus formation at higher concentrations. They induce rooting and somatic embryogenesis and, in some cases, enhance SM production at lower concentrations. Indole-3-acetic acid (IAA), Indole-3-butyric acid (IBA), 2,4-Dichlorophenoxyacetic acid (2,4-D), and Napthaleneacetic acid (NAA) are the most commonly used auxins (Saad and Elshahed 2012). Among 2,4-D ranging from 0.5 to 32 mM, the highest anthocyanin production was achieved in the callus cultures of *Oxalis linearis* var. β. latior supplemented with 2 mM 2,4-D (Meyer and Van Staden 1995). On the contrary, the limitation of 2,4-D and its replacement with NAA triggered berberine's enhanced production (0.66 mg/mL) in *Thalictrum minus* L. (Suzuki et al. 1987). In *P. ginseng* cell suspension culture, the cytokinin did not affect the biomass accumulation. However, a lower concentration of benzyl adenine (BA) or Kinetin (KN) and higher concentrations of auxin (IBA) showed the highest production of ginsenosides (Lian et al. 2002). Among the tested auxins, NAA was found more effective than 2,4-D for the growth of *Papaver bracteatum* Lindl. cell suspension cultures (Farjaminezhad et al. 2013). In the suspension culture of *W. somnifera*, the highest content of withanolide A (2.26 mg/g DW) has been observed with the cultures supplemented with 2.0 mg/L 2,4-D + 0.5 mg/L KN (Nagella and Murthy 2010).

Cytokinins are phytohormones that stimulate cell division and induce shoot formation and its proliferation. BA, 6-(γ,γ-dimethylallylamino)purine (2iP), kinetin, zeatin, and Thidiazuron (TDZ) are the most commonly used cytokinins. Each cytokinin has a different effect on the type of plant species. In the study conducted by Constabel et al. (1971), cytokinins like BA and KN have improved the production of anthocyanin accumulation in cell cultures of *Haplopappus gracilis* (Nutt.) A. Gray However, in another study by Matsumoto et al. (1973), anthocyanin production decreased in *Populus* cell suspension cultures when treated with KN. Gibberellins are growth regulators that enhance the callus growth and take part in shoot elongation. The most commonly used gibberellin is gibberellic acid. It has been reported by Fett-Neto and DiCosmo (1996) that the addition of gibberellic acid enhanced the callus growth of *Taxus cuspidata* Siebold & Zucc. Gibberellin was found to enhance artemisinin production in *Artemisia annua* L. (Liu et al. 1997). On the contrary, the application of gibberellin inhibited the production of hyoscyamine in *Hyoscyamus muticus* L. (Vanhala et al. 1998).

2.2.2.8 Temperature

Temperature is one of the physical factors that influence plant growth. Temperature ranging from 17 to 25 °C is optimal for callus and suspension cultures. However, the temperature is plant-specific; it varies from species to species. The effect of temperature on biomass accumulation and metabolite production has been studied by Morris (1986), and it has been found that maximum biomass yield (0.47 g/g) and maximum production of ajmalicine and serpentine have been reported at 35 °C, 16 °C, and 25 °C, respectively.

2.2.2.9 pH

The hydrogen ion concentration in the media influences plant growth. The pH concentration changes during nutrient uptake and biomass accumulation. Several studies on the effect of pH on biomass and metabolite accumulation have been carried out. *Fragaria ananassa* Duchesne cell suspension cultures grew well. Apparently, the biomass was not affected by the pH levels of 3.7 to 8.7, but the highest anthocyanin production was observed at the high pH 8.7 compared to the normal pH 5.7 and lower than this (Zhang and Furusaki 1997). In cell suspension culture of *W. somnifera* among the various pH studied (4–6.5), MS medium with pH 5.8 influenced the accumulation of cell biomass and withanolide A (Nagella and Murthy 2010). Naik and Al-Khayri (2020) showed that in cell suspension culture of *P. dactylifera*, MS media at pH 6.0 favored the accumulation of phenolic content and at pH 5.0 induced the accumulation of flavonoids.

2.2.2.10 Light Irradiation

Light is a predominant regulatory factor in the biosynthesis of SMs. Light influences the expression of genes that take part in biosynthetic pathways of alkaloids, flavonoids, cardenolides, and betacyanins [14]. Light quality, intensity, and exposure time are factors that elicit SM production. The effect of light intensity, quality, and duration of irradiation on anthocyanin production by *Perilla frutescens* (L.) Britton has been studied by Zhong et al. (1991). It was observed that the highest anthocyanin production of 3.0 g/L was observed in the suspended culture when irradiated with 27.2 W/m^2 light irradiation. The effect of different light intensities and irradiance on anthocyanin pigment production has been studied in *Melastoma malabathricum* L. by Chan et al. (2010). The highest anthocyanin production was observed in cultures exposed to continuous light for ten days, and the least was observed in cultures exposed to 10 days of constant darkness. Various light sources like red, blue, blue with red, fluorescent, and metal halide affect biomass accumulation. Inhibitory effects of light are also reported for shikonin production in *L. erythrorhizon* (Tabata et al. 1974).

2.2.3 Precursor Feeding

The biosynthesis of SMs is multiple reactions that involve the conversion of primary to SMs. Precursors are chemicals that serve as the starting material or intermediate for the biosynthetic pathway of SMs. The purpose of using precursors in a plant tissue culture media is to facilitate the induction and enhancement of SMs synthesis by utilizing subsisting enzyme systems. Thus, precursor feeding is a promising approach for enhancing SM production (Arias et al. 2021). Experiments on SM induction by feeding with various precursors have been conducted. The type, concentration, and time of addition of the precursor should be taken into consideration. The vanillin accumulation increased by 1.7-fold in *Vanilla planifolia* Jacks. ex Andrews by addition of precursor ferulic acid at the concentration of 1 mM (Romagnoli and Knorr 1988). The use of amino acids as inexpensive precursors has been reported. Amino acids like phenylalanine have been fed as a precursor in the cultures of *Taxus cuspidata*, and they have shown improved paclitaxel accumulation (Fett-Neto and DiCosmo 1996). The effect of isocaproic acid, a near precursor, and phenylalanine, a distant precursor, on capsaicin enhancement in cell cultures of *Capsicum frutescens* L. has been reported (Sudhakar and Ravishankar 1996). Plant cells are known to produce SMs in the stationary phase towards the late log phase. Thus, the treatment of elicitors and precursors in later log phases increases biomass and SM production. During the early log phase, the therapy might just enhance SM production but not the biomass, thus causing overall low productivity (Shilpa et al. 2010).

2.2.4 Permeabilization

The secondary plant metabolites formed are generally stored in vacuole, which might inhibit the further accumulation of SMs. Thus, the cell membranes have to be made permeable to exudate the metabolites synthesized. This can be done by membrane permeabilization by the formation of pores in the plant membrane. This promotes the production of secondary metabolites, and most importantly, it helps in continuous recovery and easier purification process. Permeabilization, however, causes the loss of cell viability due to pore formation. The product recovery by permeabilization can be achieved by chemical method and physical method. The chemical treatments like high salt concentration solutions, organic solvents like DMSO, chitosan, and physical treatment like ultrasonic vibrations, high pressure, and high electric fields can be applied for membrane permeabilization. 0.7 mM Triton X-100 when used as a chemical treatment for permeabilization in *Beta vulgaris* L. cell culture for 15 minutes, 30% release of betacyanins was reported (Trejo-Tapia et al. 2007). The cells of *Chenopodium rubrum* were subjected to chitosan permeabilization for a long duration in the study by Dörnenburg and Knorr (1997). It was subjected to 250 MPa pressure in the study by Knorr et al. (1993). Both the treatments enhanced the release of amaranthine.

2.3 Biosynthesis of Secondary Metabolites

The biosynthesis of secondary metabolites is a multistep reaction that involves converting primary metabolites to the end product by a series of enzymes. Understanding the biosynthetic pathway of secondary metabolites is very important for its enhancement. The study on biosynthetic pathways deals with identifying the primary metabolites' elementary units, identifying and elucidating intermediates, and recognizing the genes that take part in biosynthesis, regulatory factors, and enzymes that influence the biosynthetic pathway. The biosynthetic reactions are categorized into three classes:

- Class I: Biosynthetic reactions involve the transformation of primary metabolites to intermediates.
- Class II: Biosynthetic reactions involve the polymerization of similar small molecules to large molecules.
- Class III: Biosynthetic reactions involve the modification of the essential assembled molecule.

The biosynthesis of secondary metabolites is carried out through the acetate-malonate pathway for alkaloids and phenol; pentose for glycosides and polysaccharides; shikimic acid for tannins, aromatic alkaloids, and phenols; and mevalonic acid for terpenes, steroids, and alkaloids (Dewick 2009) (Fig. 2.1).

Phenolic compounds are a diverse group of biosynthesized compounds, and having two basic pathways: the shikimate pathway and the malonate pathway. The shikimate pathway is a seven-step metabolic pathway that serves as an intermediate for the synthesis of phenylpropanoids from carbohydrates. The simple carbohydrates through glycolysis or pentose phosphate pathway form carbohydrate precursors, entering the shikimic acid pathway to form aromatic amino acids such as phenylalanine, tyrosine, and tryptophan. These aromatic amino acids serve as precursors for

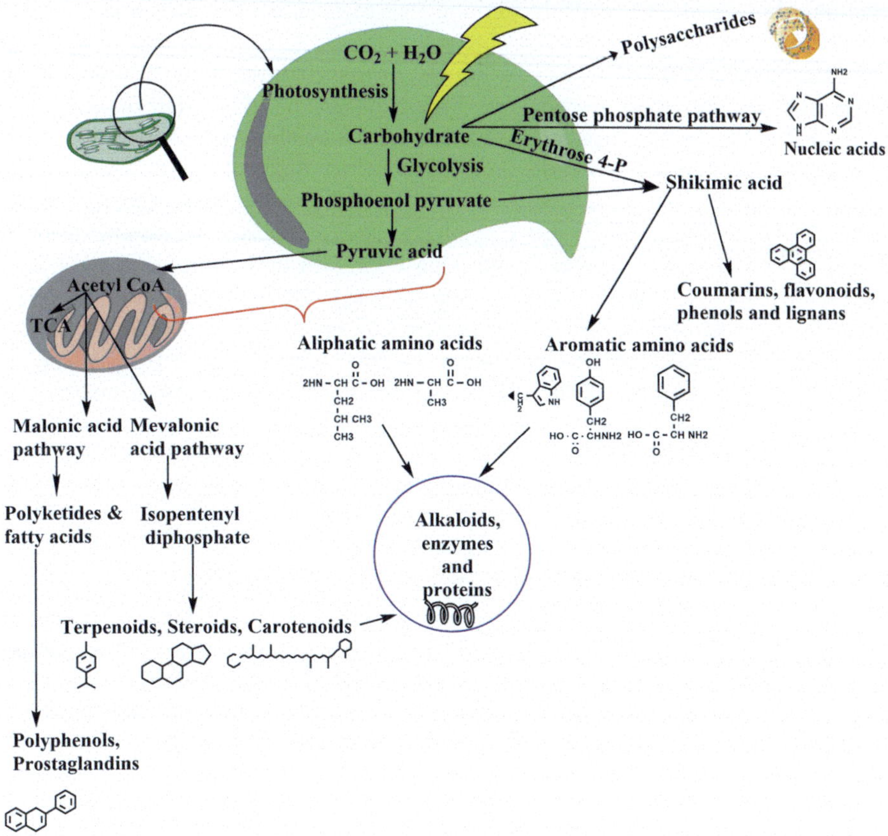

Fig. 2.1 General biosynthetic pathway of secondary metabolites (polyphenols, prostaglandins, macrocyclic antibiotics) (Modified from Parsaeimehr et al. 2011)

the biosynthesis of phenylpropanoids catalyzed by phenylalanine ammonia lyase (PAL). The phenylpropanoids, in turn, are used to produce phenolic compounds such as coumarins, flavonoids, etc. The malonate pathway serves as an intermediate pathway for simple phenols and is less significant in higher plants (Lincoln et al. 2015).

Alkaloids are a large family of nitrogen-containing compounds generally derived from amino acids such as lysine, tyrosine, and tryptophan. Alkaloids are biosynthesized in four steps. The first step involves the accumulation of an amine precursor. The amines such as lysine, arginine, and ornithine and aromatic amines such as tryptophan and tyrosine serve as amine precursors. The amine accumulation is followed by aldehyde accumulation as alkaloid biosynthesis requires aldehyde precursors. The aldehyde and amine precursors condense to form iminium intermediates which then enable general alkaloid biosynthesis by Mannich-like reaction. This serves as a scaffold forming step to establish the heterocyclic structure, the basic structure of alkaloids (Lichman 2021).

Terpenes, also referred to as isoprenoids, are the largest group of compounds that are the constituents of essential oils. They are made up of isoprene units that are the building blocks of terpenes. Terpenes are biosynthesized primarily by the mevalonic acid pathway wherein three acetyl CoA molecules are combined to form a six-carbon mevalonic acid in a sequential manner catalyzed by acetyl CoA acetyltransferase, HMG CoA synthase, and reductase. Mevalonic acid is phosphorylated by phosphomevalonate kinase to form mevalonic acid pyrophosphate (MVA-PP), which then undergoes decarboxylation and dehydration to form isopentenyl pyrophosphate (IPP) in the presence of diphosphomevalonate decarboxylase. The IPP undergoes isomerization to form dimethylallyl pyrophosphate (DMAPP) in the presence of the IPP isomerase enzyme. Both these isomers condense to form 10-C geranyl pyrophosphate (GPP), which is then converted to 15-C farnesyl pyrophosphate (FPP). The IPP and derivatives of IPP like GGPP and FPP polymerize to form polyterpenes (Manfredi 2007).

2.4 Metabolomics of Plant Secondary Metabolites

For the increase in the production of the phytocompounds, exploiting the plants in vitro machinery has significantly increased during these decades. Metabolic pathways have been targeted for direct manipulation at the genome level by modifying the cell's enzymatic and regulatory functions (Sinha et al. 2019). Metabolic engineering (ME) facilitates the overexpression or upregulation of the genes, which intensifies the metabolic reflux in specific pathways and produces novel compounds. ME has proven to be an essential tool for scalable, selective, and cost-effective secondary metabolite development (Alper and Avalos 2018). ME was stringently used to increase the development of pharmaceutically useful SMs in plants and in vitro systems by transformation with *Agrobacterium* species accompanied by modulation of the plant metabolic signaling pathway (Hidalgo et al. 2017). This approach uses *Agrobacterium's* natural ability to infect and transfer genes to the cells of plants. The effectiveness of this approach depends on several variables, such as the nature of target cells or tissues and their ability to regenerate and transform the efficacy of DNA distribution, the strictness of the system used to pick transformed cells, and the time to retrieve fertile transgenic plants (Hwang et al. 2017). Secondary metabolite stilbene was produced by *Agrobacterium* transformation in the hairy root of tobacco after inserting stilbene synthase (STS) encoding gene in tobacco (Hidalgo et al. 2017). Polyacetylenes, an alkaloid, were produced in hairy root cultures of *Campanula medium* L. by Agrobacterium transformation (Tada et al. 1996). Furthermore, glycoside, iridoids, polyphenols, naphthoquinones, anthraquinones, and several essential oils can be yielded with this metabolic engineering method. Secondary metabolites can also be produced by transferring the complete metabolic pathway into heterologous hosts to create a novel string of metabolic circuits. This can overcome several endogenous barriers and controls of the native host. Additionally, it is cheap and cost-effective and requires less time. Two primary hosts used for SM production are *E. coli* and *S. cerevisiae* (Razdan 2018). One such potent anticancer compound, paclitaxel (taxol), was produced in *E. coli*, an appropriate host for producing necessary precursor, i.e., taxadiene; its hydroxylation leads to the formation of paclitaxel (Ajikumar et al. 2010). A group of SM phenylpropanoids such as naringenin and resveratrol were yielded up to 7 mg/L and 300 μg/L, respectively, in *S. cerevisiae* as a host (Razdan 2018). Another way of gene modification at its expression level is RNA interference (RNAi), which is done by silencing the gene at the post-transcriptional level in plants. Thus, this modification technique targets and specifically degrades mRNA associated with double-stranded DNA (Sinha et al. 2019). However, being a potent tool for gene silencing is not much used to produce secondary metabolites in plants (Ossowski et al. 2008). This method can be used to silence one pathway of metabolism to increase the level of required metabolite. One such approach was made to produce non-narcotic alkaloids named reticuline in opium poppy. Here in this plant, the biosynthetic pathway for morphine production was silenced by targeting the enzyme Codeinone Reductase (COR) encoding for morphine. Thus, increased reticuline levels were observed due to feedback mechanisms (Wijekoon and Facchini 2011). Another example of the downregulation of the cinnamate-4-hydroxylase gene increased the biosynthesis of artemisinin in *A. annua*, a SM used to treat fever (Kumar et al. 2016). Using this modification method, many genetic manipulations were done in several plants to enhance metabolite production or reduce degrading pathways. Another way of increasing SM production is the "push-pull" effect involving overexpression of several genes concurrently. One such example is by co-introduction of 2 genes in *A. acutangulus* hairy roots that leads to an increase in alkaloid accumulation up to 8.7-fold higher than usual (Kai et al. 2011). This approach of overexpressing the targeted genes has led to an increase in the yield of SMs. Another method would be downregulating genes that take part in competitive pathways.

2.4.1 Upregulation (Overexpression)

The expression of genes that take part in the biosynthetic pathway of secondary metabolite production can be upregulated utilizing transcription factors that bind to the gene's promoter or enhancer regions. The overexpression of transcription elements ORCA2 or ORCA3 in *C. roseus* cell suspension culture has enhanced the production of metabolites like ajmalicine, serpentine, tryptamine, and catharanthine (Sun et al. 2017. The overexpression of transcription factor proteins (AaERF1 and AaERF2) has improved artemisinin and artemisinic acid accumulation (Yu et al. 2012). The other strategy to enhance secondary metabolite production is the upregulation of precursor genes. Overexpression of the GrDXS precursor gene has enhanced terpenoid secondary metabolites' accumulation (Jadaun et al. 2017).

2.4.2 Downregulation (Silencing)

The genes that code for enzymes that take part in competitive pathways have to be silenced to impede unwanted metabolites' enzymes. These undesirable metabolites can be ruled out through antisense RNA and RNA interference (RNAi). The enhanced production of codeine and morphine in the opium poppy by manipulating the phenylpropanoid pathway using RNAi has been reported by Wagner and Kroumova 2008.

2.5 Role of CRISPR-Cas9 in Secondary Metabolite Synthesis

In recent years, genome editing has become a defined tailored strategy for altering an organism's genetic composition. The primary tool utilized (since 2001 and 2011, respectively) in the editing technology is the site-specific nuclease (SSN), a multi-function nuclease that concisely targets specific gene sequences. In order to precisely remove, insert, and substitute a particular gene sequence, the use of these engineered nucleases highlights the boundaries of the site-oriented mutagenesis method. It, therefore, favors random mutagenesis techniques (Zaidi and Mansoor 2017). Some of the genomic approaches include customized homing nucleases (meganucleases), zinc-finger nucleases, transcriptions like activators, and the effectors (TALEN), which depend on protein-based systems with tailored DNA binding specifications to target the gene series (Sander and Joung 2014; Voytas and Gao 2014). Later on, GE's development began in 2013 when CRISPR (clustered frequently interspersed short palindromic repeats)/Cas system editing technology was introduced. In general, the CRISPR/Cas method is one of the most flexible, less expensive, and easy-to-use methods for plant genome editing since it relies on DNA-RNA interaction instead of DNA-protein interaction for target DNA sequence recognition (Miao et al. 2013; Schaeffer and Nakata 2015). The main idea behind genetic modification is the location, identification, and division of foreign DNA sequences into small guide RNA (gRNA) through associated DNA endonuclease (Cas9) (Fig. 2.2). There are mainly three classes of CRISPR/Cas systems Types I, II, and III; Cas9 belongs to the Type II class because the Cas9 system is characterized by comparative construction simplicity and relative high efficacy in human, animal, and plant cells. The genome editing using Cas9 is majorly determined by two types of gRNAs, defined as CRISPR RNA (crRNA or the "protospacer") and trans-activating crispr RNA (tracrRNA) (Hanania et al. 2017; Nemudryi et al. 2014). The two different gRNA sequences are combined into one single guide RNA molecule (sgRNA) to improve this method's functionality, particularly with eukaryotic genome variants. The sgRNA loop structure binds to the target sequence and constructs a Cas9 complex which fixes the double-stranded DNA and forms a double-strand DNA (DSB) break at the destination locus that is further restructured by the host cell repair system (non-homologous end junction (NHEJ) or homology-directed repair (HDR)) (Nishimasu et al. 2018). Following the successful demonstration as a programmable RNA-driven gene-editing method, CRISPR-Cas9 was used by the scientist in plant

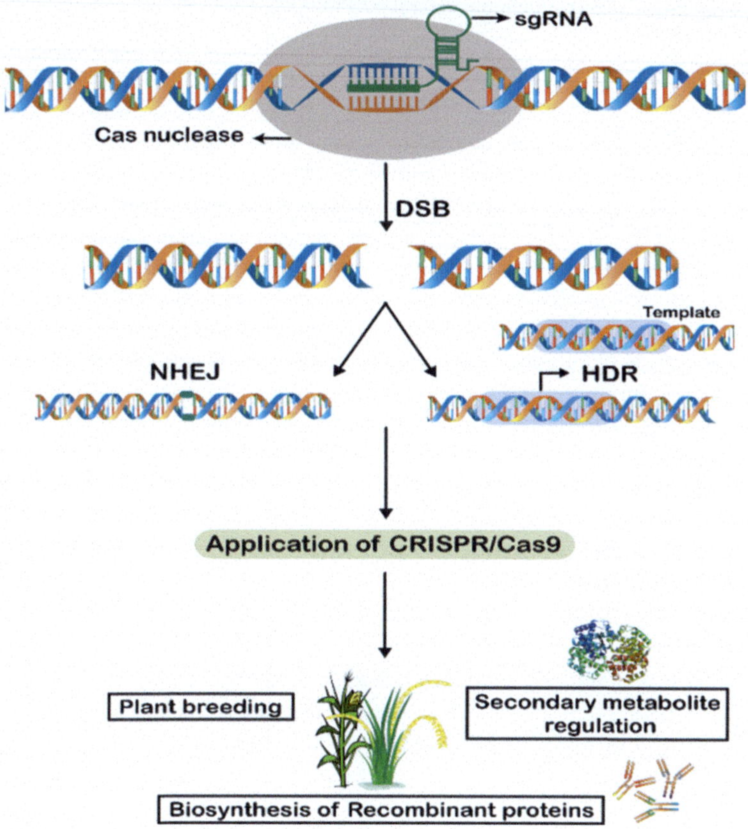

Fig. 2.2 Plant genome editing by CRISPR/Cas9 system (Modified from Marchev et al. 2020)

breeding, manipulation of secondary metabolism (SM), and the development of recombinant protein (Jinek et al. 2012). However, the use of CRISPR/Cas9 for the modulation of SM biosynthesis is still at an early stage. It has recently been successfully used to knock out the SmCPS1 gene in *S. miltiorrhiza*, resulting in the blocking of metabolic flux by geranylgeranyl diphosphate (GGPP) and disruption of tanshinone biosynthesis. This plant is commonly used for different pharmacological properties, such as vasorelaxation, antiarrhythmic, and ischemia prevention, in the traditional Chinese medicine method. In theory, GGPP is considered to serve as a source for another useful biosynthesis of diterpenes, i.e., taxol, since tanshinone and taxol have a common precursor (GGPP) (Li et al. 2017a). Furthermore, the knockout of 4'OMT2 in the opium poppy, a gene that controls the biosynthesis of benzylisoquinoline alkaloids (BIAs), showed a substantial decrease in BIA biosynthesis (e.g., morphine, thebaine), suggesting that the use of CRISPR/Cas9 in *Papaver* species, opium poppy, and related species can simply be transformed into biofactories for the mass development of BIAs through introducing breaks in related gene sequences (Alagoz et al. 2016). The active mutation of the fatty acid desaturase 2 (FAD2) genes, OsFAD2–1 and FAD2, respectively, through the CRISPR-Cas9 system, has also resulted in a significant increase in the oleic acid content of rice seeds (Abe et al. 2018) and oilseed rape (*Brassica napus* L.) Okuzaki et al. 2018). This genome editing platform was also used to develop lycopene-enriched tomato (*Solanum lycopersicum*) by simultaneously altering the genes involved in lycopene biosynthesis (Li et al. 2018).

In the same way, a pYLCRISPR/Cas9 system with one or two single-site guide RNAs targeted the slyPDS gene involved in GABA shunt in tomato cultivar *Solanum lycopersicum*, which resulted in 19-fold higher accumulation of y-aminobutyric acid (GABA) than in wild plants (Li et al. 2017b). Despite the promising advances made by CRISPR/Cas9 genome editing, there are still many discrepancies in its plant genome off-target effects and how the mutation transmission rate can be eliminated and successful in future generations. Apart from all these concerns, the CRISPR framework benefits from conventional methods of genetic engineering.

2.6 Large-Scale Production of Plant Secondary Metabolites (Bioreactor Scale)

A diverse group of plant-derived products can be produced at a low cost using in vitro systems. It is initially produced at flask scale (small scale). However, to meet the increased market demands of these biologically active metabolites, they have to be produced in large scale using bioreactors. The large-scale production of these SMs is challenged with barriers like low metabolite yield, cell heterogeneity, genetic instability, sensitivity to shear stress, exquisite mixing, aeration, cell aggregation, foaming, and adhesion of the cells. To promote cell growth and enhance SM production cultivation, the culture parameters include the inoculum concentration, media composition, aeration, agitation speed, temperature, light irradiation, and bioreactor design. The design of the bioreactor in industries depends on the type and the biological significance of the SMs that have to be produced. To ensure high SM yield, the selection of economically feasible bioreactors with optimal design and operating conditions is essential. The optimization of media composition enhanced anthocyanins production (387 mg/L) in *V. vinifera* cultured in a 2-L Rushton turbine stirred bioreactor at 75 rpm (Aumont et al. 2004). In the study conducted by Rani et al. (2018), the production of squalene (5.5 mg/g DW) in *Santalum album* L. cultured in a 7-L stirred tank bioreactor was enhanced by optimizing the nutrient medium with 5% sucrose and 1.5 mg/L 2.4-D. An enhanced accumulation of 53.80 mg/L of Resveratrol was achieved in cell suspension cultures of *Vitis amurensis* Rupr. cultured in a 3-L balloon-type bioreactor optimized with 30 g/L sucrose and 500 mg/L casein (Sun et al. 2016). Rosmarinic acid production of 17.90 mg/g DW in *Dracocephalum forrestii* W.W.Sm. shoots was enhanced when cultured in a 10/L nutrient sprinkle bioreactor with 25 s pump operating time (Weremczuk-Jeżyna et al. 2019). About 30 mg/g DW of rosmarinic acid was accumulated in the cell suspension cultures of *Ocimum basilicum* L. cultured in stirred tank bioreactor operated at 100 rpm (Pandey et al. 2019). In large-scale bioreactor cultures, the supply of gases (carbon dioxide and oxygen) is an essential criterion in producing secondary metabolites. The process of aeration maintains aerobic conditions, prevents excessive heating, and helps release volatile compounds. Studies on different concentrations of oxygen ranging from 20% to 50% and carbon dioxide ranging from 0.03 to 5.0% on biomass accumulation in *P. ginseng* cultures have been conducted, and 40% oxygen and 1% carbon dioxide were found to be optimal for biomass accumulation (Thanh et al. 2006a, b). The supply of oxygen has enhanced serpentine production in *C. roseus* in the study conducted by Leckie et al. (1991). In shake flask and large-scale bioreactors of plant cell cultures, agitation is an essential factor that must be maintained for the uniform distribution of nutrients and dispersion of bubbles for aeration. Agitation helps in the uptake of nutrients and promotes plant growth. Agitation is usually held at low speeds in plant cell cultures. Studies on agitation speed on growth and metabolism of *Thevetia peruviana* (L.) Lippold have been studied, and the highest biomass accumulation of 11.92 ± 0.25 g DW/L was achieved at 550 rpm (Arias et al. 2021).

2.7 Elicitation for Production of Secondary Metabolites

Plants produce SMs when exposed to biotic and abiotic stress factors like infection by pathogens, attack by herbivorous, high salinity, temperature, drought, radiations, and so on. In the course of their defense and adaptation to these stress agents, plants produce novel SMs that accumulate in cells. Thus, SM production in the in vitro cell or tissue culture can be enhanced by mimicking the stress by elicitation. Elicitors are substances that elicit a morphological or physiological response in plants or improve SMs biosynthesis. The eliciting stress factors can be categorized into biotic and abiotic elicitors. Biotic elicitors include enzymes, cell wall fragments of bacteria, virus, fungus, phytochemicals produced in response to damage, and products of a microbial attack like jasmonic acid and salicylic acid; and abiotic elicitors include chemicals such as heavy metals and inorganic salts, physical factors like UV radiation, high temperature, and mechanical wounding (Al-Khayri and Naik 2020; Naik and Al-Khayri 2016a, b). The choice of elicitor, the concentration, the exposure time, and the age of the culture are some of the factors that influence the elicitation.

2.7.1 Abiotic Elicitor

Substances that are not of biological origin, such as physical, chemical, and hormonal factors, are termed abiotic elicitors. Different environmental stresses like high, low temperature, salinity, drought, and radiation have a prominent impact on plants' bioactive metabolite production (Naik and Al-Khayri 2016a, b).

Table 2.2 Light as an elicitor in metabolite production

Elicitor	Plant species	Type of culture	Compound	Reference
Blue and red light	*Stevia rebaudiana* (bertoni)	Callus	Phenolic and flavonoids	Ahmad et al. 2016
Blue light	*Eclipta alba* (L.) Hassk.	Callus	Stigmasterol	Khurshid et al. 2020
Light irradiation	*Perilla frutescens* (L.) Britton	Cell suspension culture	Anthocyanin	Zhong et al. 1991
Light	*Artemisia absinthium* L.	Cell suspension culture	Total phenol	Ali and Abbasi 2014
Red light	*Eclipta alba* (L.) Hassk.	Callus	Eclalbatin, wedelolactone, and dimethyl wedelolactone	Khurshid et al. 2020
UV-B radiation	*Camellia sinensis* (L.) Kuntze	Callus	Phenolic and flavans	Zagoskina et al. 2003
UV rays	*Passiflora quadrangularis* L.	Callus	Isoorientin, orientin, isovitexin, and vitexin	Antognoni et al. 2007
UV C	*Vitis vinifera* L. Öküzgözü	Callus	α-Tocopherol, catechin	Cetin 2014
White and green light	*Artemisia absinthium* L.	Callus	Chlorophyll and carotenoid	Tariq et.al. 2014
Yellow light	*Artemisia absinthium* L.	Callus	MDA	Tariq et.al. 2014

2.7.1.1 Light as Physical Elicitor

Studies have proved that by choosing the optimal light intensity and a specific wavelength of light for the plant in vitro culture, cells had significantly improved their ability to produce metabolites. Thus, standardizing the quality and quantity of light exposure can be a promising means to elicit plants for SM's commercial production. Light can have both beneficial and inhibitory effects on plant cells for the production of SMs. The blue light had a strong inhibitory effect on cell suspension cultures of *L. erythrorhizon* for shikonin production (Yazaki 2017). Light serves as one of the prime factors influencing plants' primary and secondary metabolism (Tariq et al. 2014). Table 2.2 presents few examples that emphasize the significance of light as an elicitor in metabolite production.

2.7.1.2 Salinity Stress as Elicitor

Studies have emphasized environmental factors to greatly influence plant SMs, such as temperature, water availability, soil fertility, and salinity. Human activities have resulted in soil salinization due to improper irrigation practices and extensive use of pesticides and fertilizers. This kind of increase in high salinized landmass has induced stress response in plants resulting in a change in their metabolic pathway that facilitates their survival in these harsh conditions with increased production and accumulation of secondary metabolites (Yang et al. 2018). Salinity induced stress results in variation in physiological, biochemical processes in plant cells. Plant cells lose their membrane stability and homeostasis and disrupt nutrition balance in cells, affecting primary metabolism. Plants produce precursors for SM production under this kind of stress conditions (Gupta and Huang 2014). Safflower callus culture showed a better response to two elicitors, salicylic acid (SA) and chitosan (CHT), with increasing salinity stress (1.5%, W/V) of sodium chloride. The culture produced elevated levels of total phenolics, total flavonoids, total flavonols, anthocyanin, and antioxidant activity (Golkar et al. 2019). *Solanum nigrum* L. callus, when cultured in media with 150 mM NaCl, produced an enhanced yield of solasodine, a steroidal alkaloid, and an alternative to diosgenin, a precursor for steroid drugs (Bhat et al. 2008). *Carthamus tinctorius* L. hypocotyl-derived callus was subjected to varying concentrations of NaCl (0–300 mM) proline, MDA, TPC, and TFD content increased with an increase in salinity stress (Golkar and Taghizadeh 2018).

2.7.1.3 Heavy Metal as Elicitors

Generally, plants need essential nutrients such as Cu^{2+}, Ca^{2+}, Mn^{2+}, Fe^{2+} in trace amount as a cofactor for enzyme activity, nucleic acid metabolism, growth, and differentiation of plant cells, but these metal ions, when present at higher concentration in the medium, have an inhibitor effect on plant cells and serve as stress factor inducing SM production (Saba et al. 2000). Certain unusual earth element ions such as cerium (as CeO_2), yttrium (as Y_2O_3), neodymium (as $NdCl_3$), and lanthanum (as La_2O_3) can also be supplemented in media, which can induce secondary production and accumulation of secondary metabolites. Furthermore, heavy metals such as Co^{2+}, Zn^{2+}, and Fe^{2+}, when added to media, can be a potent elicitor for SM production (Chen et al. 2000; Ouyang et al. 2003; Zhao et al. 2000) (Table 2.3).

2.7.1.4 Thermal Stress as Elicitor

The biosynthesis of SM is highly regulated and exhibits tissue-specific expression. Expressions of genes associated with specific biosynthetic pathways for SM are modulated based on various environmental stimuli such as temperature (Rai et al.

Table 2.3 Heavy metal as an elicitor for metabolite production

Elicitor	Plant species	Type of culture	Compound	Reference
Cadmium chloride	*Gymnema sylvestre* R. Br.	Callus	Gymnemic acids	Ch et al. 2011
Copper ions	*Agave amanuensis*	Cell suspension	Hecogenin	Kartosentono et al. 2002
NaCl and $MgSO_4$	*Carthamus tinctorius* L.	Cell suspension	α-Tocopherol	Chavan et al. 2010
La	*Cistanche deserticola* Ma	Cell suspension culture	Phenylethanoid glycosides	Ouyang et al. 2003
Zinc and copper ions	*Lepidium sativum* L.	Callus	Lepidine	Saba et al. 2000

Table 2.4 Temperature as an elicitor for metabolite production

Elicitor/temperature	Plant species	Type of culture	Compound	Reference
4 °C	*Galega officinalis*	Callus	Luteolin, genistein, p-coumaric, naringenin	Karakas and Bozat 2020
13 °C	*Cyclopia subternata* Vogel	Callus	Bioflavonoid	Kokotkiewicz et al. 2014
29 °C	*Taxus chinensis* (Rehder & E.H.Wilson) Rehder	Cell suspension	Paclitaxel	Choi et al. 2000
30 °C	*Fragaria ananassa* Duchesne	Cell suspension	Anthocyanin	Zhang et al. 1997

2017). Plants, when grown at extreme temperatures, have an adverse effect on plant physiology and metabolism. In response to this thermal shift, plants produce certain SMs that can protect the plants from these unusual conditions (Naik and Al-Khayri 2016a) (Table 2.4).

2.7.1.5 Signaling Molecules as Elicitors

Plant hormones can serve as signal molecules that can bind to specific receptors on the plant cell surface, leading to cascade reactions in the signal transduction pathway, ultimately expressing the genes involved in the biosynthesis of SM (Table 2.5).

2.7.2 Biotic Elicitors

Biotic elicitors are mostly biomolecules derived from plant pathogens and the plant itself (Radman et al. 2003). Mostly, complex biological preparations such as proteins, glycoproteins, oligosaccharides, and fatty acids from pathogenic organisms are widely used as elicitors. Different types of biotic elicitors are tabulated as an example (Table 2.6).

2.8 Commercial Aspects

Consumers are increasingly conscious of lifestyle choices that have boosted demand for organic and natural products. Due to their health benefits in terms of protection and cost relative to conventional medicines, interest in medicinal plants has grown in the past two decades (Anand et al. 2019; Thomford et al. 2018). Currently, many companies are working to propagate plant cell cultures to develop bioprocesses delivering SMs for the pharmaceutical, food, and cosmetics industries more economically and sustainably. Different commercial SMs (e.g., arbutin, vanillin, artemisinin, rishitin, pyrethrins, and vincristine) which are manufactured based on plant cell cultures have been established and have a remarkable place in the global market (Deno et al. 1987; Hibino and Ushiyama 1999; Sato and Yamada 1984). To name a few, Mibelle Biochemistry in 2008 laid the foundation for the successful

Table 2.5 Signaling molecules as an elicitor for metabolite production

Elicitor	Plant species	Type of culture	Compound	Reference
Methyl jasmonate	*Nicotiana tabacum* L.	Cell suspension	Scopoletin and scopolin	Sharan et al. 1998
Methyl jasmonate	*Medicago truncatula* Gaertn.	Cell suspension	Triterpene	Suzuki et al. 2004
Methyl jasmonate and jasmonic acid	*Mentha piperita* L.	Cell suspension	Rosmarinic acid	Krzyzanowska et al. 2011
Methyl jasmonate	*Solenostemon scutellarioides* (L.) Benth.	Cell suspension	Rosmarinic acid	Sahu et al. 2012
Methyl jasmonate	*V. vinifera* cv. Barbera	Cell suspension	Trans- and cis-resveratrol	Tassoni et al. 2005
Methyl jasmonate	*Gymnema sylvestre* R. Br.	Cell suspension	Gymnemic acid	Chodisetti et al. 2014
Salicylic acid	*Fagonia indica* L.	Callus	Phenolic and flavonoid	Khan et al. 2019
Salicylic acid	*Taxus chinensis* var. mairei	Cell suspension	Taxol	Wang et al. 2004
Salicylic acid	*Garcinia brasiliensis* Mart.	Callus	Phenolic content	Teixeira et al. 2019
Salicylic acid and jasmonic acid	*Azadirachta indica* A. Juss.	Cell suspension	Azadirachtin	Prakash and Srivastava 2008
Salicylic acid	*Taxus baccata* L.	Cell suspension	Taxol	Rezaei et al. 2011
Salicylic acid and methyl jasmonate	*Vitis vinifera* cv. Cabernet sauvignon	Cell suspension	Stilbene	Xu et al. 2015

course of plant stem cell culture extracts into the cosmetics industry, which is used by leading cosmetics brands such as Dior, Lancôme, Guerlain, and La Prairie in their cosmetic formulations (Schmid et al. 2013; Schürch et al. 2007). A further milestone in plant cell culture technology is represented by the FDA (Food and Drug Administration) approval of the anticancer compound paclitaxel in early 2000 (Imseng et al. 2014). In Europe, Toeside 10 was the first plant cell culture-based food supplement approved officially for commercial production (Eibl et al. 2018). Table 2.7 contains a selection of plant cell culture-based products that are commercially produced and approved.

2.9 Conclusion and Recommendations

The plant tissue culture technique is a potent and promising tool for mass production of industrial, medically, and commercially significant SMs over decades. Optimizing the culture medium, optimizing the culture conditions, feeding the culture with

Table 2.6 Biotic elicitors for metabolite production in cell suspension culture

Elicitor	Plant species	Type of culture	Compound	Reference
Aspergillus Niger	*Psoralea corylifolia* L.	Cell suspension	Psoralen	Ahmed and Baig 2014
Aspergillus flavus (mycelial extract)	*Hypericum perforatum* L.	Cell suspension	Phenylpropanoids Hypericin	Gadzovska-Simic et al. 2012
Aspergillus Niger and *Rhizopus oryzae*	*Plumbago rosea* L.	Cell suspension	Plumbagin	Komaraiah et al. 2002
Aspergillus Niger and Penicillium expansum	*Andrographis paniculata* (Burm.f.) Nees	Cell suspension	Andrographolide	Vakil and Mendhulkar 2013
Aspergillus Niger	*Gymnema sylvestre* R. Br.	Cell suspension	Gymnemic acid	Chodisetti et al. 2013
Aspergillus Niger	*Hypericum triquetrifolium* Turra	Cell suspension	Rutin, hypersoid and quercetin	Azeez and Kadhim 2013
Aspergillus flavus	*Daucus carota* L	Callus	Anthocyanin	Rajendran et al. 1994
Aspergillus Niger	*Psoralea corylifolia* L.	Cell suspension	Psoralen	Ahmed and Baig 2014
A. niger	*Coleus forskohlii* (Andrews) Benth. ex G.Don	Cell suspension	Forskolin	Swaroopa et al. 2013
Bacillus subtilis	*Gymnema sylvestre* R. Br.	Cell suspension	Gymnemic acids	Chodisetti et al. 2013
Candida albicans KCTC 7121	*Gingko biloba* L.	Cell suspension	Bilobalide, ginkgolide A and ginkgolide B	Kang et al. 2009
Chitosan	*Fagonia indica* L.	Callus	Phenolic and flavonoid content	Khan et al. 2019
Chitosan	*Carthamus tinctorius* L.	Callus	Total Phenolics, flavonoids, anthocyanin	Golkar et al. 2019
Chitosan	*Cistanche deserticola* Ma	Cell suspension	Phenylethanoid glycosides	Cheng et al. 2006
Chitosan	*Vitis vinifera* L.	Cell suspension	Anthocyanin and phenolic acid	Cai et al. 2011
Chitosan	*Withania somnifera* L. dunal	Callus	Withanolides	Sivanandhan et al. 2012
Chitosan	*Rubia tinctorum* L.	Cell suspension	Anthraquinone	Vasconsuelo 2003
Chitosan	*Taxus chinensis* (Rehder & E.H.Wilson) Rehder	Cell suspension	Paclitaxel	Zhang et al. 2007

(continued)

Table 2.6 (continued)

Elicitor	Plant species	Type of culture	Compound	Reference
Chitosan	*Scrophularia striata* L.	Cell suspension	Aromatic amino acids Echinacoside	Kamalipour Azad et al. 2016
Chitosan	*Linum album* L.	Cell suspension	Podophyllotoxin, lariciresinol	Esmaeilzadeh Bahabadi et al. 2014
Chitosan	*Betula platyphylla* Sukaczev	Cell suspension	Triterpenoid	Fan et al. 2010
Chitosan	*Petroselinum crispum* (Mill.) Nym	Cell suspension	Callose Coumarin	Conrath et al. 1989
Chitosan	*Calendula officinalis* L.	Cell suspension	Oleanolic acid	Wiktorowska et al. 2010
Escherichia coli	*Gymnema sylvestre* R. Br.	Cell suspension	Gymnemic acids	Chodisetti et al. 2013
Penicillium Citrinum	*Catharanthus roseus* (L.) G.Don	Suspension	Catharanthine	Xu and Dong 2004
Trichoderma Viride	*Catharanthus roseus* (L.) G.Don	Cell suspension	Ajmalicine	Namdeo et al. 2002
Yeast extract	*Gymnema sylvestre* R. Br.	Cell suspension	Gymnemic acid	Veerashree et al. 2011

precursors for the production of desired metabolites is a common strategy employed for an enhanced product yield. Elicitation is another means of stimulating the culture for increased yield of metabolites. Just choosing the biotic or abiotic elicitors does not serve the purpose; it is necessary to standardize other parameters associated with elicitation and precursor feeding for best performance, such as the dose, duration of treatment, age, and type of culture. Extensive research is carried out in this field as one can easily and economically scale up for mass production of a wide range of SMs. Further, the SM production can be enhanced by scaling up by means of bioreactors which serve as biotechnological tools for large-scale metabolite production. Optimization of culture conditions and the designing of the bioreactors are essential for successful large-scale production. Metabolomics is an emerging field of science mainly focusing on the metabolic pathways in which the SMs are biosynthesized. Metabolic engineering deals with the study of biosynthetic processes at the gene, transcriptomic, and protein levels and allows gene manipulation by upregulating the genes involved in biosynthetic pathways or by downregulating the genes that are involved in the competitive pathway of SMs to enhance the SM production.

Table 2.7 Products manufactured from plant cell cultures of commercial importance (Modified from Eibl et al. 2018)

Product	Species	Manufacturer	Use and notes	Reference
Acteos 10P	*Lippia citriodora* Kunth	Active Botanicals Research (ABR), Brendola, Italy	Food supplementary ingredient	Fremont 2017
Anthocyanins	*Aralia cordata* Thunb.	Nippon Paint Co., Ltd., Osaka, Japan	Textile dyes	Kolewe et al. 2008, Obembe et al. 2011
Arbutin	*Catharanthus roseus* (L.) G.Don	Mitsui Chemicals, Inc., Tokyo, Japan	Pigment antiseptic	
Berberines	*Coptis japonica* Kunth *Thalictrum minus* L.	Mitsui Chemicals, Inc., Tokyo, Japan	Anticancer, antibiotic, anti-inflammatory	
Betacyanins	*Beta vulgaris* L.	Nippon Shinyaku Co., Kyoto, Japan	Red to red-violet pigment	
Celtosome	*Crithmum maritimum* L.	BiotechMarine by Seppic, Pontrieux France	Cosmetics for skin rejuvenation And care	SEPPIC n.d.
Cocovanol	*Theobroma cacao* L.	Diana Plant Sciences, Portland, USA	Food supplementary ingredient	Linda 2021
Ginseng	*Panax ginseng* C.A. Mey.	Nitto Denko Corporation, Osaki, Tokyo	Dietary supplement	Kolewe et al. 2008, Obembe et al. 2011
Paclitaxel	*Taxus* spp.	Phyton Biotech, Inc. Genexol—Samyang Genex, Ahrensburg, Germany	Anticancer FDA approved for the treatment of ovarian, breast, and lung cancers	
RootBioTec HO	*Ocimum basilicum* L.	Mibelle Biochemistry Group, Bolimattstrasse, Buchs	Treatment of hair loss	Belser 2015
Scopolamine	*Duboisia* spp.	Sumitomo Chemical Co., Tokyo, Japan	Anticholinergic	Kolewe et al. 2008, Obembe et al. 2011
Shikonin	*Lithospermum erythrorhizon* Siebold & Zucc.	Mitsui Chemicals, Tokyo, Japan	Red pigment, antibiotic	

(continued)

Table 2.7 (continued)

Product	Species	Manufacturer	Use and notes	Reference
Stem cell culture extracts: BerryFlux Vita, cell integrity, bionymph Peptide, cell pulse, Daphne VitaSense, FicuCell Vita, Hibiskin Vita, Lykosin defense, Vita freeze Mythus Vita, Vita Nova VitaShape, VitaLight	*Rubus idaeus* L, *Nicotiana sylvestris* Speg. & Comes *Psilanthus bengalensis* (Roxb. ex Schult.) J.-F.Leroy, *Daphne odora* Thunb., *Opuntia ficus-indica* (L.) Mill., *Hibiscus syriacus* L., *Solanum lycopersicum* L. *Actinidia deliciosa* (A. Chev.) C.F.Liang & A. R.Ferguson, *Lotus japonicus* L. *Coleus forskohlii* (Andrews) Benth. ex G.Don, *Cirsium eriophorum* (L.) Scop.	Vitalab, McDonough, USA	Skin rejuvenation and care, skin brightening, and Firming	Apone et al. 2010, Barbulova et al. 2010, Bimonte et al. 2011, Tito et al. 2011, Tito et al. 2015, Di Martino et al. 2017

References

Abe K, Araki E, Suzuki Y et al (2018) Production of high oleic/low linoleic rice by genome editing. Plant Physiol Biochem 131:58–62. https://doi.org/10.1016/j.plaphy.2018.04.033

Ahmad N, Rab A, Ahmad N (2016) Light-induced biochemical variations in secondary metabolite production and antioxidant activity in callus cultures of *Stevia rebaudiana* (Bert). J Photochem Photobiol B 154:51–56

Ahmed SA, Baig MMV (2014) Biotic elicitor enhanced production of psoralen in suspension cultures of *Psoralea corylifolia* L. Saudi J Biol Sci 21(5):499–504. https://doi.org/10.1016/j.sjbs.2013.12.008

Ajikumar PK, Xiao WH, Tyo KEJ et al (2010) Isoprenoid pathway optimization for taxol precursor overproduction in *Escherichia coli*. Science 330(6000):70–74. https://doi.org/10.1126/science.1191652

Alagoz Y, Gurkok T, Zhang B et al (2016) Manipulating the biosynthesis of bioactive compound alkaloids for next-generation metabolic engineering in *Opium poppy* using CRISPR-Cas 9 genome editing technology. Sci Rep 6(1):30901. https://doi.org/10.1038/srep30910

Ali M, Abbasi BH (2014) Light-induced fluctuations in biomass accumulation, secondary metabolites production and antioxidant activity in cell suspension cultures of *Artemisia absinthium* L. J Photochem Photobiol B 140:223–227. https://doi.org/10.1016/j.jphotobiol.2014.08.008

Ali M, Abbasi BH, Ahmad N et al (2016) Sucrose enhanced biosynthesis of medicinally important antioxidant secondary metabolites in cell suspension cultures of *Artemisia absinthium* L. Bioprocess Biosyst Eng 39:1945–1954

Al-Khayri JM, Naik PM (2020) Elicitor-induced production of biomass and pharmaceutical phenolic compounds in cell suspension culture of date palm (*Phoenix dactylifera* L.). Molecules 25(20):4669

Alper HS, Avalos JL (2018) Metabolic pathway engineering. Synth Syst Biotechnol 3(1):1–2. https://doi.org/10.1016/j.synbio.2018.01.002

Anand U, Jacobo-Herrera N, Altemimi A et al (2019) A comprehensive review on medicinal plants as antimicrobial therapeutics: potential avenues of biocompatible drug discovery. Metabolites 9(11):258. https://doi.org/10.3390/metabo9110258

Antognoni F, Zheng S, Pagnucco C et al (2007) Induction of flavonoid production by UV-B radiation in *Passiflora quadrangularis* callus cultures. Fitoterapia 78(5):345–352. https://doi.org/10.1016/j.fitote.2007.02.001

Apone F, Tito A, Carola A et al (2010) A mixture of peptides and sugars derived from plant cell walls increases plant defense responses to stress and attenuates ageing-associated molecular changes in cultured skin cells. J Biotechnol 145:367–376. https://doi.org/10.1016/j.jbiotec.2009.11.0

Arias JP, Mendoza D, Arias M (2021) Agitation effect on growth and metabolic behavior of plant cell suspension cultures of *Thevetia peruviana* at bench scale reactor. Plant Cell Tiss Org Cult 145(2):307–319. https://doi.org/10.1007/s11240-021-02009-2

Aumont V, Larronde F, Richard T et al (2004) Production of highly 13C-labeled polyphenols in *Vitis vinifera* cell bioreactor cultures. J Biotechnol 109(3):287–294

Azeez HA, Kadhim MI (2013) Effect of biotic elicitors on secondary metabolite production in cell suspensions of *Hypericum triquetrifolium* Turra. Bull UASVM Horticult 70(1):26–33

Barbulova A, Tito A, Carola A, Bimonte M, de Laurentis F, Ambrosio DP, Apone F, Colucci G (2010) Raspberry stem cell extract to protect skin from inflammation and oxidative stress. Cosmet Toilet 125:38–47

Belser E (2015) Anti-ageing hair care: preventing hair loss and ensuring fuller hair. COSMA 1: 10–11

Berlin J, Kukoschke KG, Knobloch KH (1981) Selection of tobacco cell lines with high yields of *Cinnamoyl putrescines*. Planta Med 42(6):173–180

Bhat MA, Ahmad S, Aslam J et al (2008) Salinity stress enhances production of solasodine in *Solanum nigrum* L. Chem Pharm Bull 56(1):17–21. https://doi.org/10.1248/cpb.56.17

Bhojwani SS, Razdan MK (1996) Production of secondary metabolites. In: Studies in plant science, vol 5. Elsevier, London, pp 537–562

Bimonte M, Tito A, Carola A, Barbulova A, Monoli I, Cucchiara M, Hill J, Colucci G, Apone F (2011) A new active extract obtained from *Coffea bengalensis* stem cells acts on different skin cell types attenuating signs of ageing. Cosmet Toilet 126:644–650

Bourgaud F, Gravot A, Milesi S et al (2001) Production of plant secondary metabolites: a historical perspective. Plant Sci 161(5):839–851

Bramble JL, Graves DJ, Brodelius P (1991) Calcium and phosphate effects on growth and alkaloid production in *Coffea arabica*: experimental results and mathematical model. Biotechnol Bioeng 37(9):859–868

Cai Z, Kastell A, Mewis I et al (2011) Polysaccharide elicitors enhance anthocyanin and phenolic acid accumulation in cell suspension cultures of *Vitis vinifera*. Plant Cell Tissue Organ Cult 108(3):401–409. https://doi.org/10.1007/s11240-011-0051-3

Cetin E (2014) Induction of secondary metabolite production by UV-C radiation in *Vitis vinifera* L. Öküzgözü callus cultures. Biol Res 47(1):37. https://doi.org/10.1186/0717-6287-47-37

Ch B, Rao K, Gandi S, Giri A (2011) Abiotic elicitation of gymnemic acid in the suspension cultures of *Gymnema sylvestre*. World J Microbiol Biotechnol 28(2):741–747. https://doi.org/10.1007/s11274-011-0870-8

Chan LK, Koay SS, Boey PL, Bhatt A (2010) Effects of abiotic stress on biomass and anthocyanin production in cell cultures of *Melastoma malabathricum*. Biol Res 43(1):127–135

Chandran H, Meena M, Barupal T et al (2020) Plant tissue culture as a perpetual source for production of industrially important bioactive compounds. Biotechnol Rep 26:e00450

Chavan SP, Lokhande VH, Nitnaware KM et al (2010) Influence of growth regulators and elicitors on cell growth and α-tocopherol and pigment productions in cell cultures of *Carthamus tinctorius* L. Appl Microbiol Biotechnol 89(6):1701–1707. https://doi.org/10.1007/s00253-010-3014-4

Chen WJ, Gu YH, Zhao GW et al (2000) Effects of rare earth ions on activity of RuBPcase in tobacco. Plant Sci 152:145–151

Cheng XY, Zhou HY, Cui X et al (2006) Improvement of phenylethanoid glycosides biosynthesis in *Cistanche deserticola* cell suspension cultures by chitosan elicitor. J Biotechnol 121(2): 253–260. https://doi.org/10.1016/j.jbiotec.2005.07.012

Chodisetti B, Rao K, Gandi S et al (2013) Improved gymnemic acid production in the suspension cultures of *Gymnema sylvestre* through biotic elicitation. Plant Biotechnol Rep 7:519–525. https://doi.org/10.1007/s11816-013-0290-3

Chodisetti B, Rao K, Gandi S, Giri A (2014) Gymnemic acid enhancement in the suspension cultures of *Gymnema sylvestre* by using the signaling molecules—methyl jasmonate and salicylic acid. In Vitro Cell Dev Plant 51(1):88–92. https://doi.org/10.1007/s11627-014-9655-8

Choi HK, Kim SI, Son JS et al (2000) Enhancement of paclitaxel production by temperature shift in suspension culture of *Taxus chinensis*. Enzym Microb Technol 27(8):593–598. https://doi.org/10.1016/s0141-0229(00)00255-6

Conrath U, Domard A, Kauss H (1989) Chitosan-elicited synthesis of callose and of coumarin derivatives in parsley cell suspension cultures. Plant Cell Rep 8(3):152–155. https://doi.org/10.1007/bf00716829

Constabel F, Shyluk JP, Gamborg OL (1971) The effect of hormones on anthocyanin accumulation in cell cultures of *Haplopappus gracilis*. Planta 96(4):306–316

Curtin C, Zhang W, Franco C (2003) Manipulating anthocyanin composition in *Vitis vinifera* suspension cultures by elicitation with jasmonic acid and light irradiation. Biotechnol Lett 25(14):1131–1135

Dal Toso R, Melandri F (2011) *Echinacea angustifolia* cell culture extract. Nutrafoods 10:19–24. https://doi.org/10.1007/BF03223351

Dedaldechamp F, Uhel C, Macheix JJ (1995) Enhancement of anthocyanin synthesis and dihydroflavonol reductase (DFR) activity in response to phosphate deprivation in grape cell suspensions. Phytochemistry 40(5):1357–1360

Deno H, Suga C, Morimoto T et al (1987) Production of shikonin derivatives by cell suspension cultures of *Lithospermum erythrorhizon* VI: production of shikonin derivatives by a two-layer culture containing an organic solvent. Plant Cell Rep 6:197–199

Dewick PM (2009) Secondary metabolism: the building blocks and construction mechanisms. In: Medicinal natural products: a biosynthetic approach. Wiley, Hoboken, pp 7–38

Di Martino O, Tito A, de Lucia A et al (2017) *Hibiscus syriacus* extract from an established cell culture stimulates skin wound healing. Biomed Res Int 2017:7932019. https://doi.org/10.1155/2017/7932019

Do CB, Cormier F (1990) Accumulation of anthocyanins enhanced by a high osmotic potential in grape (*Vitis vinifera* L.) cell suspensions. Plant Cell Rep 9(3):143–146

Dörnenburg H, Knorr D (1995) Strategies for the improvement of secondary metabolite production in plant cell cultures. Enzyme Microb Technol 17(8):674–684

Dörnenburg H, Knorr D (1997) Challenges and opportunities for metabolite production from plant cell and tissue cultures. Food Technol 51(11):47–54

Eibl R, Meier P, Stutz I, Schildberger D, Hühn T, Eibl D (2018) Plant cell culture technology in the cosmetics and food industries: current state and future trends. Appl Microbiol Biotechnol 102(20):8661–8675. https://doi.org/10.1007/s00253-018-9279-8

Esmaeilzadeh Bahabadi S, Sharifi M, Murata J et al (2014) The effect of chitosan and chitin oligomers on gene expression and lignans production in *Linum album* cell cultures. J Med Plants 13(49):46–53

Fan G, Li X, Wang X, Zhai Q et al (2010) Chitosan activates defense responses and triterpenoid production in cell suspension cultures of *Betula platyphylla* Suk. Afr J Biotechnol 9(19): 2816–2820

Farjaminezhad R, Zare N, Asghari-Zakaria R, Farjaminezhad M (2013) Establishment and optimization of cell growth in suspension culture of *Papaver bracteatum*: a biotechnology approach for thebaine production. Turkish J Biol 37:689–697

Fett-Neto AG, DiCosmo F (1996) Production of paclitaxel and related toxoids in cell cultures of *Taxus cuspidata*: perspectives for industrial applications. In: DiCosmo F (ed) Plant cell culture: secondary metabolism toward industrial application. CRC Press, Boca Raton, pp 139–166

Fremont M (2017) Cell culture: an innovative approach for production of plant actives. New Food Magazine. https://www.newfoodmagazine.com/article/33682/cell-culture-plant-actives/. Accessed 23 January 2021

Gadzovska-Simic S, Tusevski O, Antevski S, Atanasova-Pancevska N, Petreska J, Stefova M, Kungulovski D, Spasenoski M (2012) Secondary metabolite production in *Hypericum perforatum* L. cell suspensions upon elicitation with fungal mycelia from *Aspergillus flavus*. Arhiv Biol Sci 64(1):113–121. https://doi.org/10.2298/abs1201113g

Galieni A, Di Mattia C, Gregorio D et al (2015) Effects of nutrient deficiency and abiotic environmental stresses on yield, phenolic compounds and antiradical activity in lettuce (*Lactuca sativa* L.). Sci Hortic 187:93–101

Gamborg OL, Miller RA, Ojima K (1968) Nutrient requirements of suspension cultures of soybean root cells. Exp Cell Res 50(1):151–158

Golkar P, Taghizadeh M (2018) In vitro evaluation of phenolic and osmolite compounds, ionic content, and antioxidant activity in safflower (*Carthamus tinctorius* L.) under salinity stress. Plant Cell Tissue Organ Cult 134(3):357–368. https://doi.org/10.1007/s11240-018-1427-4

Golkar P, Taghizadeh M, Yousefian Z (2019) The effects of chitosan and salicylic acid on elicitation of secondary metabolites and antioxidant activity of safflower under in vitro salinity stress. Plant Cell Tissue Organ Cult. https://doi.org/10.1007/s11240-019-01592-9

Gonçalves S, Romano A (2018) Production of plant secondary metabolites by using biotechnological tools. In: Vijayakumar R, Raja S (eds) Secondary metabolites. IntechOpen

Guerriero G, Berni R, Muñoz-Sanchez JA et al (2018) Production of plant secondary metabolites: Examples, tips and suggestions for biotechnologists. Genes 9(6):309

Gupta B, Huang B (2014) Mechanism of salinity tolerance in plants: physiological, biochemical, and molecular characterization. Int J Genomics 2014:1–8. https://doi.org/10.1155/2014/701596

Hagimori M, Matsumoto T, Obi Y (1982) Studies on the production of *Digitalis cardenolides* by plant tissue culture: II. Effect of light and plant growth substances on digitoxin formation by undifferentiated cells and shoot-forming cultures of *Digitalis purpurea* L. grown in liquid media. Plant Physiol 69(3):653–656

Hanania U, Ariel T, Tekoah Y et al (2017) Establishment of a tobacco BY2 cell line devoid of plant-specific xylose and fucose as a platform for the production of biotherapeutic proteins. Plant Biotechnol J 15(9):1120–1129. https://doi.org/10.1111/pbi.12702

Hassanpour S, Maherisis N, Eshratkhah B et al (2011) Plants and secondary metabolites (Tannins): a review. Int For Soil Eros 1(1):47–53

Hibino K, Ushiyama K (1999) Commercial production of ginseng by plant tissue culture technology. In: Fu TJ, Singh G, Curtis WR (eds) Plant cell and tissue culture for the production of food ingredients. Springer, Boston, pp 215–224

Hidalgo D, Georgiev M, Marchev A et al (2017) Tailoring tobacco hairy root metabolism for the production of stilbenes. Sci Rep 7(1):18330. https://doi.org/10.1038/s41598-017-18330-w

Holopainen JK, Virjamo V, Ghimire RP et al (2018) Climate change effects on secondary compounds of forest trees in the northern hemisphere. Front Plant Sci 9:1445

Hussain MS, Fareed S, Ansari S et al (2012) Current approaches toward production of secondary plant metabolites. J Pharm Bioallied Sci 4(1):10–20

Hwang HH, Yu M, Lai EM (2017) Agrobacterium-mediated plant transformation: biology and applications. Arabidopsis Book 15:e0186. https://doi.org/10.1199/tab.0186

Imseng N, Schillberg S, Schürch C et al (2014) Suspension culture of plant cells under heterotrophic conditions. In: Hans-Peter M, Schmidhalter DR (eds) Industrial scale suspension culture of living cells. Wiley, Blackwell, pp 224–258. https://doi.org/10.1002/9783527683321.ch07

Isah T, Mujib A (2015) Camptothecin from *Nothapodytes nimmoniana*: review on biotechnology applications. Acta Physiol Plant 37(6):106

Jadaun JS, Sangwan NS, Narnoliya LK et al (2017) Over-expression of DXS gene enhances terpenoidal secondary metabolite accumulation in rose-scented geranium and *Withania somnifera*: active involvement of plastid isoprenogenic pathway in their biosynthesis. Physiol Plant 159(4):381–400

Jinek M, Chylinski K, Fonfara I et al (2012) A programmable dual-RNA-guided DNA endonuclease in adaptive bacterial immunity. Science 337(6096):816–821. https://doi.org/10.1126/science.1225829

Kai G, Yang S, Luo X et al (2011) Co-expression of AaPMT and AaTRI effectively enhances the yields of tropane alkaloids in *Anisodus acutangulus* hairy roots. BMC Biotechnol 11:43

Kamalipour Azad M, Sharifi M, Maivan HZ et al (2016) Induction of aromatic amino acids and phenylpropanoid compounds in *Scrophularia striata* Boiss. Cell culture in response to chitosan-induced oxidative stress. Plant Physiol Biochem 107:374–384. https://doi.org/10.1016/j.plaphy.2016.06.034

Kang SM, Min JY, Kim YD et al (2009) Effect of biotic elicitors on the accumulation of bilobalide and ginkgolides in *Ginkgo biloba* cell cultures. J Biotechnol 139(1):84–88. https://doi.org/10.1016/j.jbiotec.2008.09.007

Karakas FP, Bozat BG (2020) Fluctuation in secondary metabolite production and antioxidant defense enzymes in *in vitro* callus cultures of goat's rue (*Galega officinalis*) under different abiotic stress treatments. Plant Cell Tissue Organ Cult. https://doi.org/10.1007/s11240-020-01870-x

Kartosentono S, Indrayanto G, Zaini NC (2002) The uptake of copper ions by cell suspension cultures of *Agave amaniensis*, and its effect on the growth, amino acids and hecogenin content. Plant Cell Tissue Organ Cult 68(3):287–292. https://doi.org/10.1023/a:1013920919889

Khan T, Khan T, Hano C et al (2019) Effects of chitosan and salicylic acid on the production of pharmacologically attractive secondary metabolites in callus cultures of *Fagonia indica*. Ind Crop Prod 129:525–535. https://doi.org/10.1016/j.indcrop.2018.12.048

Khurshid R, Ullah MA, Tungmunnithum D et al (2020) Lights triggered differential accumulation of antioxidant and antidiabetic secondary metabolites in callus culture of *Eclipta alba* L. PLoS One 15(6):e0233963. https://doi.org/10.1371/journal.pone.0233963

Kim DJ, Chang HN (1990) Enhanced shikonin production from *Lithospermum erythrorhizon* by in situ extraction and calcium alginate immobilization. Biotechnol Bioeng 36(5):460–466

Kim SH, Hur BK, Byun SY (1999) Effect of sugar concentration on camptothecin production in cell suspension cultures of *Camptotheca acuminata*. Biotechnol Bioproc E 4(4):277–280

Kim S et al (2020) Metabolomic elucidation of the effect of sucrose on the secondary metabolite profiles in *Melissa officinalis* by ultraperformance liquid chromatography-mass spectrometry. ACS Omega 5(51):33186–33195

Knorr D, Caster C, Dörneburg H, Dorn R et al (1993) Biosynthesis and yield improvement of food ingredients from plant cell and tissue cultures. Food Technol 47(12):57–63

Kochan E, Szymczyk P, Kuźma Ł et al (2016) Nitrogen and phosphorus as the factors affecting ginsenoside production in hairy root cultures of *Panax quinquefolium* cultivated in shake flasks and nutrient sprinkle bioreactor. Acta Physiol Plant 38(6):149

Kokotkiewicz A, Bucinski A, Luczkiewicz M (2014) Light and temperature conditions affect bioflavonoid accumulation in callus cultures of *Cyclopia subternata* Vogel (honeybush). Plant Cell Tissue Organ Cult 118(3):589–593. https://doi.org/10.1007/s11240-014-0502-8

Kolewe ME, Gaurav V, Roberts SC (2008) Pharmaceutically active natural product synthesis and supply via plant cell culture technology. Mol Pharm 5(2):243–256. https://doi.org/10.1021/mp7001494

Komaraiah P, Naga Amrutha R, Kavi Kishor P, Ramakrishna S (2002) Elicitor enhanced production of plumbagin in suspension cultures of *Plumbago rosea* L. Enzym Microb Technol 31(5): 634–639. https://doi.org/10.1016/s0141-0229(02)00159-x

Krzyzanowska J, Czubacka A, Pecio L et al (2011) The effects of jasmonic acid and methyl jasmonate on rosmarinic acid production in *Mentha piperita* cell suspension cultures. Plant Cell Tiss Org 108(1):73–81. https://doi.org/10.1007/s11240-011-0014-8

Kumar R, Vashisth D, Misra A et al (2016) RNAi down-regulation of cinnamate-4-hydroxylase increases artemisinin biosynthesis in *Artemisia annua*. Sci Rep 6(1). https://doi.org/10.1038/srep26458

Leckie F, Scragg AH, Cliffe KC (1991) Effect of bioreactor design and agitator speed on the growth and alkaloid accumulation by cultures of *Catharanthus roseus*. Enzyme Microb Technol 13(4): 296–305

Lee CWT, Shuler ML (2000) The effect of inoculum density and conditioned medium on the production of ajmalicine and catharanthine from immobilized *Catharanthus roseus* cells. Biotechnol Bioeng 67:61–67

Li R, Li R, Li X et al (2017a) Multiplexed CRISPR/Cas9-mediated metabolic engineering of γ-aminobutyric acid levels in *Solanum lycopersicum*. Plant Biotechol J 16(2):415–427. https://doi.org/10.1111/pbi.12781

Li B, Cui G, Shen G et al (2017b) Targeted mutagenesis in the medicinal plant *Salvia miltiorrhiza*. Sci Rep 7(1). https://doi.org/10.1038/srep43320

Li X, Wang Y, Chen S et al (2018) Lycopene is enriched in tomato fruit by CRISPR/Cas9-mediated multiplex genome editing. Front Plant Sci 9:559. https://doi.org/10.3389/fpls.2018.00559

Lian ML, Chakrabarty D, Paek KY (2002) Effect of plant growth regulators and medium composition on cell growth and saponin production during cell-suspension culture of mountain ginseng (*Panax ginseng* C.A. Mayer). J Plant Biol 45:201–206

Lichman BR (2021) The scaffold-forming steps of plant alkaloid biosynthesis. Nat Prod Rep 38(1): 103–129

Lincoln T, Eudardo Z, Ian Max M et al (2015) Secondary metabolites. In: Plant physiology and development, 6th edn. Oxford University Press, Oxford, pp A41–A48

Linda BS (2021) Oregon bioscience. Diana plant sciences creates natural, plant-based products. https://www.barneyassoc.com/documents/dianaplantsciences_lbarney_D2_6_18_13_final.pdf. Accessed 21 Jan 2021

Linsmaier EM, Skoog F (1965) Organic growth factor requirements of tobacco tissue cultures. Physiol Plant 18(1):100–127

Liu S, Zhong JJ (1998) Phosphate effect on production of ginseng saponin and polysaccharide by cell suspension cultures of *Panax ginseng* and *Panax quinquefolium*. Process Biochem 33(1): 69–74

Liu CZ, Wang YC, Ouyang F, Ye et al (1997) Production of artemisinin by hairy root cultures of *Artemisia annua* L. Biotechnol Lett 19(9):927–929

Manfredi KP (2007) Terpenes. Flavors, fragrances, pharmaca, pheromones by Eberhard Breitmaier (University of Bonn). Wiley-VCH, Weinheim. 2006. ix + 214 pp. ISBN 3-527-31786-4. J Nat Prod 70(4):711–711

Marchev AS, Georgiev MI (2020) Plant in vitro systems as a sustainable souce of active ingredients for cosmeceutical application. Molecules 25(9):2006

Marchev AS, Yordanova ZP, Georgiev MI (2020) Green (cell) factories for advanced production of plant secondary metabolites. Crit Rev Biotechnol 40(4):443–458

Matsumoto T, Okunishi K, Nishida K et al (1971) Studies on the culture conditions of higher plant cells in suspension culture. Agric Biol Chem 35(4):543–551

Matsumoto T, Nishida K, Noguchi M et al (1973) Some factors affecting the anthocyanin formation by Populus cells in suspension culture. Agric Biol Chem 37(3):561–567

Meyer HJ, Van Staden J (1995) The in vitro production of an anthocyanin from callus cultures of *Oxalis linearis*. Plant Cell Tissue Organ Cult 40(1):55–58

Miao J, Guo D, Zhang J et al (2013) Targeted mutagenesis in rice using CRISPR-Cas system. Cell Res 23(10):1233–1236. https://doi.org/10.1038/cr.2013.123

Morris P (1986) Regulation of product synthesis in cell cultures of *Catharanthus roseus*. Effect of culture temperature. Plant Cell Rep 5(6):427–429.

Murashige T, Skoog F (1962) A revised medium for rapid growth and bioassays with tobacco tissue cultures. Physiol Plant 15(3):473–497

Murthy HN, Dandin VS, Zhong JJ et al (2014) Strategies for enhanced production of plant secondary metabolites from cell and organ cultures. In: Paek KY, Murthy HN, Zhong JJ (eds) Production of biomass and bioactive compounds using bioreactor technology. Springer, Netherlands, pp 471–508

Nagella P, Murthy HN (2010) Establishment of cell suspension cultures of *Withania somnifera* for the production of withanolide A. Bioresour Technol 10:6735–6739

Nagella P, Murthy HN (2011) Effects of macro elements and nitrogen source on cell suspension growth and withanolide-A production in *Withania somnifera* (L.) Dunal. Plant Cell Tissue Organ Cult 104:119–124

Nagella P, Murthy HN (2014) Production of withanolides from cell and organ cultures of *Withania somnifera* (L.) Dunal. In: Paek KY, Murthy HN, Zhong JJ (eds) Production of biomass and bioactive compounds using bioreactor technology. Springer, Dordrecht, pp 285–315

Nagella P, Chung IM, Murthy HN (2011a) In vitro production of gymnemic acid from cell suspension cultures of *Gymnema sylvestre* R. Br. Eng Life Sci 11(5):537–540

Nagella P, Chung IM, Murthy HN (2011b) Improvement of growth and gymnemic acid production by altering the macro elements concentration and nitrogen source supply in cell suspension cultures of *Gymnema sylvestre* R. Br. Ind Crops Prod 33:282–286

Naik PM, Al-Khayri JM (2016a) Abiotic and biotic elicitors-role in secondary metabolites production through in vitro culture of medicinal plants. In: Shanker AK, Shankar C (eds) Abiotic and biotic stress in plants—recent advances and future perspectives. Rijeka, Croatia, pp 247–277

Naik PM, Al-Khayri JM (2016b) Impact of abiotic elicitors on in vitro production of plant secondary metabolites: a review. J Adv Res 1(2):7

Naik PM, Al-Khayri JM (2018) Cell suspension culture as a means to produce polyphenols from date palm (*Phoenix dactylifera* L.). Cienc Agrotec 42:464–473

Naik PM, Al-Khayri JM (2020) Influence of culture parameters on phenolics, flavonoids and antioxidant activity in cell culture extracts of date palm (*Phoenix dactylifera* L.). Erwerbs Obstbau 62:181–188

Namdeo A, Patil S, Fulzele DP (2002) Influence of fungal elicitors on production of ajmalicine by cell cultures of *Catharanthus roseus*. Biotechnol Prog 18(1):159–162. https://doi.org/10.1021/bp0101280

Nemudryi AA, Valetdinova KR, Medvedev SP et al (2014) TALEN and CRISPR/Cas genome editing systems: tools of discovery. Acta Nat 6(3):19–40

Nishimasu H, Shi X, Ishiguro S et al (2018) Engineered CRISPR-Cas9 nuclease with expanded targeting space. Science 361:1259–1262. https://doi.org/10.1126/science.aas9129

Obembe OO, Popoola JO, Leelavathi S et al (2011) Advances in plant molecular farming. Biotechnol Adv 29(2):210–222. https://doi.org/10.1016/j.biotechadv.2010.11.004

Okuzaki A, Ogawa T, Koizuka C et al (2018) CRISPR/Cas9-mediated genome editing of the fatty acid desaturase 2 gene in *Brassica napus*. Plant Physiol Biochem 131:63–69. https://doi.org/10.1016/j.plaphy.2018.04.025

Ossowski S, Schwab R, Weigel D (2008) Gene silencing in plants using artificial microRNAs and other small RNAs. Plant J 53(4):674–690. https://doi.org/10.1111/j.1365-313x.2007.03328.x

Ouyang J, Wang X, Zhao B, Yuan X et al (2003) Effects of rare earth elements on the growth of *Cistanche deserticola* cells and the production of phenylethanoid glycosides. J Biotechnol 102(2):129–134. https://doi.org/10.1016/s0168-1656(03)00019-1

Pan XW, Xu HH, Liu X et al (2004) Improvement of growth and camptothecin yield by altering nitrogen source supply in cell suspension cultures of *Camptotheca acuminata*. Biotechnol Lett 26(22):1745–1748

Pandey P, Singh S, Banerjee S (2019) *Ocimum basilicum* suspension culture as resource for bioactive triterpenoids: yield enrichment by elicitation and bioreactor cultivation. Plant Cell Tiss Org Cult 137(1):65–75

Park YG, Kim SJ, Kang YM et al (2004) Production of ginkgolides and bilobalide from optimized the *Ginkgo biloba* cell culture. Biotechnol Bioprocess E 9(1):41–46

Parsaeimehr A, Sargsyan E, Vardanyan A (2011) Expression of secondary metabolites in plants and their useful perspective in animal health. https://papers.ssrn.com/abstract=2986314

Prakash G, Srivastava AK (2008) Statistical elicitor optimization studies for the enhancement of azadirachtin production in bioreactor *Azadirachta indica* cell cultivation. Biochem Eng J 40(2): 218–226. https://doi.org/10.1016/j.bej.2007.12.017

Radman R, Saez T, Bucke C, Keshavarz T (2003) Elicitation of plants and microbial cell systems. Biotechnol Appl Biochem 37(Pt. 1):91–102. https://doi.org/10.1042/ba20020118. PMID: 12578556

Radušienė J, Marksa M, Ivanauskas L et al (2019) Effect of nitrogen on herb production, secondary metabolites and antioxidant activities of *Hypericum pruinatum* under nitrogen application. Ind Crops Prod 139:111519

Rai M, Rai A, Kawano N et al (2017) De novo RNA sequencing and expression analysis of *Aconitum carmichaelii* to analyze key genes involved in the biosynthesis of diterpene alkaloids. Molecules 22(12):2155. https://doi.org/10.3390/molecules22122155

Rajasekaran T, Rajendran L, Ravishankar GA et al (1991) Influence of nutrient stress on pyrethrin production by cultured cells of pyrethrum (*Chrysanthemum cinerariaefolium*). Curr Sci 60(12): 705–707

Rajendran L, Suvarnalatha G, Ravishankar GA et al (1994) Enhancement of anthocyanin production in callus cultures of *Daucus carota* L. under the influence of fungal elicitors. Appl Microbiol Biotechnol 42(2–3):227–231. https://doi.org/10.1007/bf00902721

Rani A, Meghana R, Kush A (2018) Squalene production in the cell suspension cultures of Indian sandalwood (*Santalum album* L.) in shake flasks and air lift bioreactor. Plant Cell Tiss Org Cult 135(1):155–167

Rao SR, Ravishankar GA (2002) Plant cell cultures: chemical factories of secondary metabolites. Biotechnol Adv 20:10–153

Rastegari AA, Yadav AN, Yadav N (2019) Genetic manipulation of secondary metabolites producers. In: Gupta VK, Pandey A (eds) New and future developments in microbial biotechnology and bioengineering. Elsevier, pp 13–29

Ravishankar GA, Sarma KS, Venkataraman LV, Kadyan AK (1988) Effect of nutritional stress on capsaicin production in immobilized cell cultures of *Capsicum annuum*. Curr Sci 57(7): 381–383

Razdan S (2018) An update towards the production of plant secondary metabolites. In: Yadav SK et al (eds) Recent trends and techniques in plant metabolic engineering. https://doi.org/10.1007/978-981-13-2251-8_1

Rezaei A, Ghanati F, Behmanesh M et al (2011) Ultrasound-potentiated salicylic acid-induced physiological effects and production of taxol in hazelnut (*Corylus avellana* L.) cell culture. Ultrasound Med Biol 37(11):1938–1947. https://doi.org/10.1016/j.ultrasmedbio.2011.06.013

Romagnoli LG, Knorr D (1988) Effects of ferulic acid treatment on growth and flavor development of cultured *Vanilla planifolia* cells. Food Biotechnol 2(1):93–104

Saad AIM, Elshahed AM (2012) Plant tissue culture media. In: Leva A, Rinaldi LMR (eds) Recent advances in plant in vitro culture. IntechOpen. https://doi.org/10.5772/50569

Saba PD, Iqbal M et al (2000) Effect of ZnSO4 and CuSO4 on regeneration and lepidine content in *Lepidium Sativum* L. Biol Plant 43:253–256. https://doi.org/10.1023/A:1002708427984

Sahu R, Gangopadhyay M, Dewanjee S (2012) Elicitor-induced rosmarinic acid accumulation and secondary metabolism enzyme activities in *Solenostemon scutellarioides*. Acta Physiol Plant 35(5):1473–1481. https://doi.org/10.1007/s11738-012-1188-3

Salgado-Garciglia R, Ochoa-Alejo N (1990) Increased capsaicin content in PFP-resistant cells of chili pepper (*Capsicum annuum* L.). Plant Cell Rep 8(10):617–620

Sander JD, Joung JK (2014) CRISPR-Cas systems for editing, regulating and targeting genomes. Nat Biotechnol 32(4):347–355. https://doi.org/10.1038/nbt.2842
Sato F, Yamada Y (1984) High berberine-producing cultures of *Coptis japonica* cells. Phytochemistry 23(2):281–285
Schaeffer SM, Nakata PA (2015) CRISPR/Cas9-mediated genome editing and gene replacement in plants: transitioning from lab to field. Plant Sci 240:130–142. Suzuki. https://doi.org/10.1016/j.plantsci.2015.09.011
Schenk RU, Hildebrandt AC (1972) Medium and techniques for induction and growth of monocotyledonous and dicotyledonous plant cell cultures. Can J Bot 50(1):199–204. https://doi.org/10.1139/b72-026
Schmid D, Zülli F, Belser E (2013) Stem cell activation for smoother, more even skin., https://www.personalcaremagazine.com/story/11280/stem-cell-activation-for-smoother-and-more-even-skin. Accessed July 2013
Schreiner M (2005) Vegetable crop management strategies to increase the quantity of phytochemicals. Eur J Nutr 44(2):85–94
Schürch C, Blum P, Zülli F (2007) Potential of plant cells in culture for cosmetic application. Phytochem Rev 7(3):599–605. https://doi.org/10.1007/s11101-007-9082-0
SEPPIC (n.d.) Crithmum maritimum. SEPPIC. https://www.seppic.com/en/celtosome-crithmum-maritimum. Accessed 12 February 2021
Sharan M, Taguchi G, Gonda K, Jouke T, Shimosaka M, Hayashida N, Okazaki M (1998) Effects of methyl jasmonate and elicitor on the activation of phenylalanine ammonia-lyase and the accumulation of scopoletin and scopolin in tobacco cell cultures. Plant Sci 132(1):13–19. https://doi.org/10.1016/s0168-9452(97)00260-4
Shilpa K, Varun K, Lakshmi BS (2010) An alternate method of natural drug production: eliciting secondary metabolite production using plant cell culture. J Plant Sci 5(3):222–247
Sinha S, Sandhu K, Bisht N, Naliwal T, Saini I, Kaushik P (2019) Ascertaining the paradigm of secondary metabolism enhancement through gene level modification in therapeutic plants. J Young Pharm 11(4):337–343
Sivanandhan G, Arun M, Mayavan S et al (2012) Chitosan enhances withanolides production in adventitious root cultures of *Withania somnifera* (L.) Dunal. Ind Crops Prod 37(1):124–129. https://doi.org/10.1016/j.indcrop.2011.11.022
Smetanska I (2008) Production of secondary metabolites using plant cell cultures. Adv Biochem Eng Biotechnol 111:187–228
Sudhakar JT, Ravishankar GA (1996) Precursor biotransformation in immobilized placental tissues of *Capsicum frutescens* Mill: I. influence of feeding intermediate metabolites of the capsaicinoid pathway on capsaicin and dihydrocapsaicin accumulation. J Plant Physiol 147(5):481–485
Sun D, Li C, Qin H et al (2016) Somatic embryos cultures of *Vitis amurensis* Rupr. in air-lift bioreactors for the production of biomass and resveratrol. J Plant Biol 59(5):427–434
Sun Y, Abdukader A, Lu D, Zhang H, Liu C (2017) Synthesis of (E)-β-iodo vinylsulfones via iodine-promoted iodosulfonylation of alkynes with sodium sulfinates in an aqueous medium at room temperature. Green Chem 19(5):1255–1258. https://doi.org/10.1039/c6gc03387c
Suzuki M, Nakagawa K, Fukui H, Tabata M (1987) Relationship of berberine-producing capability between Thalictrum plants and their tissue cultures. Plant Cell Rep 6(3):260–263
Suzuki H, Reddy MSS, Naoumkina M et al (2004) Methyl jasmonate and yeast elicitor induce differential transcriptional and metabolic re-programming in cell suspension cultures of the model legume *Medicago truncatula*. Planta 220(5):696–707. https://doi.org/10.1007/s00425-004-1387-2
Swaroopa G, Anuradha M, Pullaiah T (2013) Elicitation of forskolin in suspension cultures of *coleus forskohlii* (Willd.) Briq. using elicitors of fungal origin. Curr Trends Biotechnol Pharm 7(3):755–762
Tabata M, Mizukami H, Hiraoka N, Konoshima M (1974) Pigment formation in callus cultures of *Lithospermum erythrorhizon*. Phytochemistry 13(6):927–932. https://doi.org/10.1016/s0031-9422(00)91425-0

Tada H, Nakashima T, Kunitake H et al (1996) Polyacetylenes in hairy root cultures of *Campanula medium* L. J Plant Physiol 147:617–619

Tariq U, Ali M, Abbasi BH (2014) Morphogenic and biochemical variations under different spectral lights in callus cultures of *Artemisia absinthium* L. J Photochem Photobiol B 130:264–271. https://doi.org/10.1016/j.jphotobiol.2013.11.026

Tassoni A, Fornalè S, Franceschetti M et al (2005) Jasmonates and Na-orthovanadate promote resveratrol production in *Vitis vinifera* cv. Barbera cell cultures. New Phytol 166(3):895–905. https://doi.org/10.1111/j.1469-8137.2005.01383.x

Teixeira MG, Carvalho M, Leite MA et al (2019) Effect of salicylic acid, 2,4-D and 2i-P on the production of secondary metabolites in *Garcinia brasiliensis* Mart. Callus. Braz Arch Biol Technol 62:e19170303

Thanh NT, Murthy HN, Pandey DM, Yu KW, Hahn EJ, Paek KY (2006a) Effect of carbon dioxide on cell growth and saponin production in suspension cultures of *Panax ginseng*. Biol Plant 50(4):752–754

Thanh NT, Murthy HN, Yu KW et al (2006b) Effect of oxygen supply on cell growth and saponin production in bioreactor cultures of *Panax ginseng*. J Plant Physiol 163(12):1337–1341

Thirumurugan D, Cholarajan A, Raja S et al (2018) An introductory chapter: secondary metabolites. In: Vijayakumar R, Raja S (eds) Secondary metabolites. IntechOpen

Thomford NE, Senthebane DA, Rowe A et al (2018) Natural products for drug discovery in the 21st century: innovations for novel drug discovery. Int J Mol Sci 19(6):1578. https://doi.org/10.3390/ijms19061578

Tito A, Carola A, Bimonte M, Barbulova A, Arciello S, de Laurentiis F, Monoli I, Hill J, Gibertoni S, Colucci G, Apone F (2011) A tomato stem cell extract, containing antioxidant compounds and metal chelating factors, protects skin cells from heavy metal-induced damages. Int J Cosmet Sci 33:543–552. https://doi.org/10.1111/j.1468-2494.2011.00668.x

Tito A, Bimonte M, Carola A et al (2015) An oil-soluble extract of *Rubus idaeus* cells enhances hydration and water homeostasis in skin cells. Int J Cosmet Sci 37:588–594. https://doi.org/10.1111/ics.12236

Trejo-Tapia G, Cuevas-Celis J, Salcedo-Morales G et al (2007) *Beta vulgaris* L. suspension cultures permeabilized with triton X-100 retain cell viability and betacyanines production ability: a digital image analysis study. Biotechnol Prog 23(2):359–363

Vakil MMA, Mendhulkar VD (2013) Enhanced synthesis of andrographolide by *Aspergillus niger* and *Penicillium expansum* elicitors in cell suspension culture of *Andrographis paniculata* (Burm. f.) Nees. Bot Stud 54(1):49. https://doi.org/10.1186/1999-3110-54-49

Van Hengel AJ, Harkes MP, Wichers HJ et al (1992) Characterization of callus formation and camptothecin production by cell lines o*f Camptotheca acuminata*. Plant Cell Tissue Organ Cult 28(1):11–18

Vanhala L, Eeva M, Lapinjoki S et al (1998) Effect of growth regulators on transformed root cultures of *Hyoscyamus muticus*. J Plant Physiol 153(3):475–481

Vasconsuelo A (2003) Involvement of the PLC/PKC pathway in chitosan-induced anthraquinone production by *Rubia tinctorum* L. cell cultures. Plant Sci 165(2):429–436. https://doi.org/10.1016/s0168-9452(03)00208-5

Veerashree V, Anuradha CM, Kumar V (2011) Elicitor-enhanced production of gymnemic acid in cell suspension cultures of *Gymnema sylvestre* R. Br. Plant Cell Tiss Org Cult 108(1):27–35. https://doi.org/10.1007/s11240-011-0008-6

Voytas DF, Gao C (2014) Precision genome engineering and agriculture: opportunities and regulatory challenges. PLoS Biol 12(6):e1001877. https://doi.org/10.1371/journal.pbio.1001877

Wagner GJ, Kroumova AB (2008) The use of RNAi to elucidate and manipulate secondary metabolite synthesis in plants. In: Ying SY (ed) Current perspectives in microRNAs (miRNA). Springer, Netherlands, pp 431–459

Wang Y, Weathers PJ (2007) Sugars proportionately affect artemisinin production. Plant Cell Rep 26(7):1073–1081

Wang YD, Yuan YJ, Wu JC (2004) Induction studies of methyl jasmonate and salicylic acid on taxane production in suspension cultures of *Taxus chinensis* var. mairei. Biochem Eng J 19(3): 259–265. https://doi.org/10.1016/j.bej.2004.02.006

Watanabe K, Yano SI, Yamada Y (1982) The selection of cultured plant cell lines producing high levels of biotin. Phytochemistry 21(3):513–516

Weremczuk-Jeżyna I, Kochan E, Szymczyk P et al (2019) The antioxidant and antimicrobial properties of phenol-rich extracts of *Dracocephalum forrestii* W. W. Smith shoot cultures grown in the nutrient sprinkle bioreactor. Phytochem Lett 30:254–260

Wijekoon CP, Facchini PJ (2011) Systematic knockdown of morphine pathway enzymes in opium poppy using virus-induced gene silencing. Plant J 69(6):1052–1063. https://doi.org/10.1111/j.1365-313x.2011.04855.x

Wiktorowska E, Długosz M, Janiszowska W (2010) Significant enhancement of oleanolic acid accumulation by biotic elicitors in cell suspension cultures of *Calendula officinalis* L. Enzym Microb Technol 46(1):14–20. https://doi.org/10.1016/j.enzmictec.2009.09.002

Xu M, Dong J (2004) Elicitor-induced nitric oxide burst is essential for triggering catharanthine synthesis in *Catharanthus roseus* suspension cells. Appl Microbiol Biotechnol 67(1):40–44. https://doi.org/10.1007/s00253-004-1737-9

Xu A, Zhan JC, Huang WD (2015) Effects of ultraviolet C, methyl jasmonate and salicylic acid, alone or in combination, on stilbene biosynthesis in cell suspension cultures of *Vitis vinifera* L. cv. Cabernet Sauvignon. Plant Cell Tiss Org Cult 122(1):197–211. https://doi.org/10.1007/s11240-015-0761-z

Yamada Y, Sato F (1981) Production of berberine in cultured cells of *Coptis japonica*. Phytochemistry 20(3):545–547

Yamamoto Y, Mizuguchi R, Yamada Y (1982) Selection of a high and stable pigment-producing strain in cultured *Euphorbia millii* cells. Theor Appl Genet 61(2):113–116

Yang L, Wen KS, Ruan X et al (2018) Response of plant secondary metabolites to environmental factors. Molecules 23(4):762. https://doi.org/10.3390/molecules23040762

Yaseen M, Ahmad T, Sablok G et al (2013) Review: role of carbon sources for in vitro plant growth and development. Mol Biol Rep 40(4):2837–2849

Yazaki K (2017) *Lithospermum erythrorhizon* cell cultures: present and future aspects. Plant Biotechnol 34(3):131–142. https://doi.org/10.5511/plantbiotechnology.17.0823a

Yu ZX, Li JX, Yang CQ et al (2012) The jasmonate-responsive AP2/ERF transcription factors AaERF1 and AaERF2 positively regulate artemisinin biosynthesis in *Artemisia annua* L. Mol Plant 5(2):353–365

Zagoskina NV, Dubravina GA, Alyavina AK et al (2003) Effect of ultraviolet (UV-B) radiation on the formation and localization of phenolic compounds in tea plant callus cultures. Russ J Plant Physiol 50(2):270–275. https://doi.org/10.1023/a:1022945819389

Zaidi SSA, Mansoor S (2017) Viral vectors for plant genome engineering. Front Plant Sci 8. https://doi.org/10.3389/fpls.2017.00539

Zhang W, Furusaki S (1997) Regulation of anthocyanin synthesis in suspension cultures of strawberry cell by pH. Biotechnol Lett 19:1057–1061

Zhang W, Seki M, Furusaki S (1997) Effect of temperature and its shift on growth and anthocyanin production in suspension cultures of strawberry cells. Plant Sci 127(2):207–214. https://doi.org/10.1016/s0168-9452(97)00124-6

Zhang CH, Fevereiro PS, He G et al (2007) Enhanced paclitaxel productivity and release capacity of *Taxus chinensis* cell suspension cultures adapted to chitosan. Plant Sci 172(1):158–163. https://doi.org/10.1016/j.plantsci.2006.08.002

Zhao J, Zhu WH, Hu Q (2000) Promotion of indole alkaloid production in *Catharanthus roseus* cell cultures by rare earth elements. Biotechnol Lett 22:825–828

Zhong JJ, Seki T, Kinoshita S, Yoshida T (1991) Effect of light irradiation on anthocyanin production by suspended culture of *Perilla frutescens*. Biotechnol Bioeng 38(6):653–658

Scale-Up Production of Bioactive Compounds Using Bioreactors

3

M. R. Rohini and P. E. Rajasekharan

Abstract

The increased demand for plant secondary metabolites by the pharmaceutical, food, flavor, beverage, and cosmetic industries has necessitated the rapid and mass production of these metabolites using in vitro plant culture systems. Bioreactors provided a suitable alternative to conventional plant culture by facilitating large-scale propagation of plants and production of secondary metabolites. Bioreactors proved to be effective plant culture systems which are genetically stable, low cost, easy to operate, and fully automated. Bioreactors play a very important role in medicinal plant industry and have evolved over time. At present, a variety of bioreactor configurations are available each customized for specific plant cell/tissue so that a stable optimum yield of bioactives is obtained. The chapter discusses briefly about the use of bioreactors in scaling up the production of secondary metabolites, different categories and designs of bioreactors available, factors on which bioreactor function depends, and the different crops in which bioreactor scaling up is attempted.

Keywords

Bioreactor · Secondary metabolites · Plant cell cultures

3.1 Introduction

Plants have been the only source of medicine available to human beings from time immemorial until the modern medical research and drug synthesis emerged. With the realization of the innumerable side effects that modern medicines impose on the

M. R. Rohini · P. E. Rajasekharan (✉)
Indian Institute of Horticultural Research, Bengaluru, India

T. Belwal et al. (eds.), *Nutraceuticals Production from Plant Cell Factory*,
https://doi.org/10.1007/978-981-16-8858-4_3

health conditions, there has been a gradual readoption of the plant-based medicines. As per the reports of WHO, 80% of the global population relies on medicinal herbs for their healthcare needs. Many of the modern-day drugs contain plant extracts or active ingredients as their integral part. The very famous analgesic drug, aspirin, was derived from the plant species *Salix* and *Spiraea*, and anticancer drugs like paclitaxel is derived from *Taxus* species, while antimalarial drug quinine is obtained from *Cinchona* species. It is roughly estimated that one-fourth of the modern-day drugs contain active ingredients or bioactive compounds from plants. The active ingredients or bioactive compounds are the high-value compounds which are responsible for the pharmacological activities of the plant. Plant extracts provide a vast scope of applications by virtue of the chemical diversity present in them. Bioactives have numerous industrial applications in different sectors like pharmaceutical industry, cosmetic industry, food and beverage industry, etc. The global plant extract market for phytomedicines and herbal extracts segment is the largest as well as fastest-growing and is experiencing high demand from functional food and beverages which require ingredients for preventing chronic diseases. The plant extract market is estimated to be valued at USD 23.7 billion in 2019 and is projected to reach USD 59.4 billion by 2025 (http://www.marketsandmarkets.com, 2019).

Owing to all these factors, the demand for plant extracts has increased, but the supply is scanty; the reason is because nearly 72% of the species in high demand is continued to be sourced from the wild. The unscrupulous collection from the forest resources has threatened the survival of many important medicinal species. Apart from this, the destruction of forest resources due to increased anthropogenic activities and climate change scenario has also been responsible for the dwindling of populations of many economically important species. In such a scenario, it becomes essential to find the alternate sources for the supply of plant extracts or bioactives on a large scale for meeting the commercial demands. Traditionally, secondary metabolites were obtained through different extraction procedure using whole plants and tissues which again depend on the type of species, age of plant, plant part used, etc. Later on, with the advent of biotechnological methods like tissue culture techniques, large-scale in vitro plant cultures were used for the production of secondary metabolites. Findings from a study that anthraquinone synthesis occurred from undifferentiated cells of *Morinda citrifolia*, it was proved that both differentiated as well as undifferentiated cells can be used for production of secondary metabolites in plant cell cultures. Thus, the in vitro cultures for bioactive synthesis may be either callus cultures, cell suspension cultures, and/or organ cultures. Organ cultures such as root and shoot cultures were found to be better for the production of secondary metabolites as compared to undifferentiated cultures like callus and suspension cultures. In vitro culture of whole plant organs, viz., root and shoot cultures, was attempted in many medicinal plants for the production of secondary metabolites (Biondi et al. 2002). Analysis of secondary metabolites in organ cultures and intact plants showed that there was no difference with respect to quality and quantity of metabolites obtained. The added advantage of organ cultures is that they are less sensitive to shear stress and relatively more stable in the production of bioactives as compared to undifferentiated cells (Bais et al. 2002).

Plant roots are also active sites of secondary metabolite synthesis and hence contribute as a potent explant for bioactive production in tissue culture. Hairy roots induced by *Agrobacterium rhizogenes* shows a high productivity of secondary metabolites and are yet another valuable source of phytochemicals used in pharmaceuticals.

Plant cell cultures are routinely used for the production of valuable bioactives by the pharmaceutical, flavoring, as well as fragrance industry. Around 20,000 chemical metabolites are produced in plants including both primary and secondary chemicals, and it is reported that every year about 1600 new plant chemicals are discovered (Sajc et al. 2000). Paek et al. (2005) described that in vitro cultures for secondary metabolite production were advantageous in terms of different factors like:

- Mass production of metabolites in short period of time.
- Ensures regular availability of bioactives independent of season and plant availability.
- Controlled and sterile environments with well-defined production system ensuring high yield with consistent quality.
- Prevention of abiotic and biotic stresses.

Considering the benefits of in vitro cultures, researchers implemented the automation of in vitro propagation using bioreactor system, thereby reducing the cost and labor for mass production of secondary metabolites. Bioreactor systems are particularly useful for species:

- Which is difficult to cultivate.
- Which has a long cultivation period.
- Produces commercially important metabolite which cannot be synthesized chemically.
- Which naturally produces only small quantity of metabolite of interest.

Bioreactor system is more advanced and efficient as compared to conventional tissue culture system as the bioreactor process is completely automated and allows for the manipulation of culture condition such as temperature, pH and dissolved oxygen, carbon dioxide, and nutrients at any point of time. The process of continuous aeration, mixing, and circulation of media also enhances the nutrient uptake by the cells and increases their proliferation (Ibrahim 2015).

3.2 Bioreactors

Bioreactors refer to large vessels with controlled and sterile environments containing liquid medium for culturing cells, tissues, or organs intensively with the purpose of mass multiplication or mass production of certain bioactives produced by these cells. Bioreactor systems mostly use liquid culture mediums for mass production of various plant cells and tissues. These systems were developed initially for mass multiplication of microbes, but now it's applied to plants for rapid mass propagation of plant species which are difficult to propagate by conventional means and also for production of secondary metabolites. The objective of bioreactor system is to provide the optimum physical and chemical conditions for obtaining maximum yield and quality of explants and at the same time to keep minimum production cost by integrating automated facilities and simple low-cost devices (Preil 2005).

Three types of cultures are used in the bioreactor system as below:

- Cultures for mass propagation of planting material.
- Cultures for mass production of secondary metabolites and enzymes.
- Cultures used for biotransformation of exogenously added metabolites (which may be precursors in a metabolic pathway).

Bioreactor systems ensure optimal growth and biochemical functions of plant cell to synthesize bioactive compounds. The advantages of bioreactor system involve:

- It allows scaling up of bioactive production in suspension cultures under controlled conditions.
- It reduces manual handling of cultures, thus saving labor, time, and energy.
- Regulation and manipulation of culture conditions can be done at different stages of bioreactor.
- The system is easy for accessibility like inoculation or harvesting of cultures.
- Submerged culture conditions will enhance the nutrient absorption by the cells and thereby increases the secondary metabolite production. Large numbers of plantlets are easily produced and can be scaled up.

Bioreactor technology is already in place for mass propagation of many medicinal species like *Rauwolfia serpentina*, *Chlorophytum borivilianum*, *Bacopa monnieri*, *Swertia chirayita*, etc. However, with respect to bioactive production from medicinal plants, in spite of the enormous advantages, commercial scale production is achieved only in Japan in few species like *Lithospermum erythrorhizon* (for shikonin), *Panax ginseng* (for ginsenosides), and *Berberis aristata* (for berberine) (Bourgaud et al. 2001).

3.3 Bioreactor Design

First bioreactor system for plant propagation was developed for in vitro mass propagation of *Begonia* × *hiemalis* (Takayama and Misawa 1981). Since then lot of bioreactor systems have been developed for different types of plant species. Standardizing a bioreactor system for plant propagation or secondary metabolite production is an interdisciplinary research which includes study of plant biotechnology, plant physiology, biochemistry, molecular biology, pharmacology, and engineering. Understanding all the complex processes involved is important to assess the tissue morphology, bioreactor conditions in terms of oxygen availability, heat transfusion, agitation, mixing, kinetics of cell growth and metabolite production, control of culture environment, and the potential for process scale-up. The overall objective of bioreactor system is to develop an optimum protocol to achieve high productivity. The design and development of bioreactors depend on the purpose as well as the different engineering parameters like oxygen requirement, shear resistance, controlled physical and chemical environment, and amenability for scaling up. The factors affecting the metabolite production in bioreactors include: the gaseous atmosphere, oxygen supply and CO2 exchange, pH, minerals, carbohydrates, growth regulators, the liquid medium rheology, and cell density.

The basic requirements of bioreactor system are as follows:

- Provision for cell-to-cell contact.
- Provision for homogenous mixing with minimal shearing and sufficient aeration.
- Provision for effective heat transfer.
- Provision for adequate dispersion of air and gas.
- Should avoid segregation of substrates.
- Provision for measuring various growth and yield parameters.
- Provision for scaling up of bioactive production.
- Provision for sterile and stable environment.
- Provision for easy handling and maintenance.
- Should minimize all possibilities of contamination.

Bioreactors are classified into different categories based on certain criteria; one among them is based on agitation method. Based on agitation, bioreactors are fundamentally classified as mechanically agitated, pneumatically agitated, and non-agitated types. Mechanically agitated bioreactors include stirred tank bioreactor, rotating drum bioreactor, and spin filter bioreactor. Pneumatically agitated bioreactors comprise of simple aeration bioreactor, bubble column bioreactor, airlift bioreactor, and balloon-type bioreactor. Non-agitated bioreactors include gaseous phase bioreactors, oxygen permeable membrane aerator bioreactors, and overlay aeration bioreactors.

Mechanically agitated bioreactors use impellers or magnetic stirrers or vibrating perforated boards for aeration, mixing, and circulation; on the other hand, in pneumatically agitated bioreactors like airlift bioreactors, aeration and mixing are done by the inserted gas which is entering though the side or basal inlets via airlift pumps. The entering air will lift the plant biomass and provide sufficient oxygen (Ziv 2005). Mechanically agitated bioreactors have higher energy consumption, and tall bioreactors have sealing issues and are more prone to cell damage and shearing, whereas airlift bioreactors are advantageous in terms of its simple design, low shearing, high gas and nutrient transfer rates, low electricity consumption, and relatively high yields (Denchev et al. 1992). Since airlift bioreactors don't have any moving parts, the maintenance time and cost is also less. The disadvantages of this system include excessive foam formation and growth of cells at the top which can be avoided by using antifoams and wide top bioreactors or bubble bioreactors (Paek et al. 2001).

Recently, novel bioreactor systems based on the extent and period of immersion of explants or cells in culture media have emerged which have been categorized as continuous immersion and temporary immersion bioreactors. In continuous immersion system (CIS), the explant material is continuously in contact with the liquid medium. Stirred tank, airlift, and balloon bioreactors are examples of continuous immersion system. The limitations associated with this system are that continuous immersion causes hyperhydricity and malformations because of insufficient oxygen availability for the submerged tissues. The deficiency in oxygen availability induces oxidative stress producing reactive oxygen species causing injury to the plant tissue. To avoid this problem, oxygen should be provided either by agitation or by aeration or by exposing a part of the explant to air. This has led to the development of temporary immersion system (TIS) in which there is only temporary contact between the plants and the liquid medium, thus avoiding continuous immersion and providing adequate oxygen transfer to the cultures. The working principle of TIS system is to prevent complete immersion using different ways like using separate sides of culture vessels or keeping the explant above a platform. The medium comes in contact with the plant material for specific period of time and then the medium gets back to the storage tank. The whole process is controlled by electronic system. Bioreactors employing TIS system involve different types like:

- RITA®, Plantform™, and Rocker bioreactors: Explant material is kept separate from the liquid medium in the same culture vessel but in different zone or different compartments. Explant material is kept above the medium using different support materials like nets, glass beads, etc. The medium comes in contact with the material by the force of filtered air pumped at specific intervals in RITA® and Plantform™ systems and by mechanical movement of the vessel in the case of rocker bioreactors. Plantform™ systems are especially useful for plants that are more prone to hyperhydricity. The optimization of protocols in terms of immersion duration and frequency needs to be standardized for the species of interest.
- Two flask system: Explant material is kept in one vessel and medium is present in a separate vessel. Here also the explant material comes in contact with the medium by the driven force of filtered air pumped at specific intervals.

Plants cultured by TIS generally show increased vigor and better quality than those grown completely submerged in liquid medium or conventionally in semisolid medium.

Based on the nature of continuous phase used, bioreactors are classified as liquid-phase bioreactors, gas-phase bioreactors, and hybrid bioreactors.

Liquid-Phase Bioreactors In this bioreactor system, plant cells will be immersed in the liquid medium continuously, and oxygen will be given by passing air through the culture medium. Mechanically driven bioreactor, pneumatically driven bioreactor, and hydraulically driven bioreactor belong to this category. The limitation of liquid-phase bioreactors includes low solubility of gases, reduced availability of nutrients, growth inhibition, etc.

Gas-Phase Bioreactors Gas-phase bioreactors were introduced for very sensitive plant tissues like hairy roots which get damaged by other bioreactor system like stirred tank system. Gas-phase bioreactors include mist or spray technology in which the roots are exposed to nutrient mist/spray as in aeroponics. In these bioreactors, oxygen availability is not limited even at high culture density and also resistant to shearing. Studies also showed that the production of secondary metabolites is often greater in mist bioreactors than in liquid-phase reactors.

Hybrid Bioreactors These bioreactors are a combination of liquid-phase and gas-phase bioreactors. In this system, there occurs a shift from liquid to gaseous phase after the inoculation, distribution, attachment to immobilization points, and short growth phase of the cells. An example is the Wilson bioreactor for hairy root cultures.

Table 3.1 and 3.2 depicts the different types of bioreactors used in different medicinal species for secondary metabolite production and hairy root cultures.

3.4 Factors Affecting Bioactive Production in Bioreactors

3.4.1 Ventilation and Availability of Dissolved Oxygen

Regeneration capacity as well as viability of plant tissues in the culture mostly depends on the ventilation as well as availability of oxygen in the liquid medium. Bioreactor conditions have to be optimized with respect to agitation speed and aeration rate, gas mixing, etc. so that effective oxygen diffusion occurs from the gaseous to the liquid phase. The presence of sufficient oxygen in the liquid media is important for the growth as well as mass production of cells or tissues.

3.4.2 Mixing of Liquid Medium

In order to facilitate uniform distribution of nutrients to all the cells in the liquid media, there should be proper mixing of the medium. The mixing can be done either mechanically or pneumatically depending on the type of the bioreactor, but care should be taken to ensure that there is no cell or tissue damage happening during mixing and at the same time cells are getting enough nutrients.

3.4.3 Medium pH

The pH of the liquid media determines the mass production of cells, tissues, and thereby bioactive production in bioreactor systems. Studies have shown that liquid medium of pH 5 inhibits embryogenesis of cultures. pH of the medium also affects the availability of nutrients to the plant tissues.

Table 3.1 Different types of bioreactors used in different plant species for secondary metabolite production

Bioreactor type	Secondary metabolite	Plant species	Explant type	References
Shake flask system	Hypericin	*Hypericum perforatum* L.	Adventitious root	Cui et al. 2011
Airlift bioreactor system	Caftaric acid, chlorogenic acid, cichoric acid	*Echinacea purpurea* (L.) Moench.	Adventitious roots	Jeong et al. 2009
Airlift bioreactor system	Anthraquinones, phenolics, flavonoids	*Morinda citrifolia* L.	Leaf cells	Ahmed et al. 2008
Shake flask system	Anthocyanin dyes	*Melastoma malabathricum* L.	Callus cell cultures	Chan et al. 2010
Balloon-type bubble bioreactors	Ginsenosides	*Panax ginseng* Meyer	Adventitious roots	Smolenskaya et al. 2007
Shake flask system	Human serum albumin	*Oryza sativa* L.	Transgenic rice seeds	He et al. 2011
Shake flask system	Ginsenoside	*Panax japonicus* C.A. Mey. Var. repens	Callus cell cultures	Smolenskaya et al. 2007
Nutrient sprinkle bioreactor	Rosmarinic acid	*Dracocephalum forrestii* W. W. Smith	Shoot cultures	Weremczuk-Jeźyna et al. 2019
Stirred tank bioreactor	Resveratrol	*Vitis labrusca L*	Suspension culture	Chastang et al. 2018
Stirred tank bioreactor	Anthocyanin	*Vitis vinifera (L.) cv Gamay Fréaux var. Teinturier*	Suspension culture	Aumont et al. 2004
Nutrient sprinkle bioreactor	Caffeic acid	*Dracocephalum forrestii W.W. smith*	Shoot cultures	Weremczuk-Jeźyna et al. 2019
Nutrient sprinkle Bioreactor	Verbascoside	*Scutellaria alpina*	Shoots	Grzegorczyk-Karolak et al. 2017

3.4.4 Availability of Nutrients

Nutrients are the major chemical factors which determine the productivity of tissue culture systems in bioreactors. Nutrients have to be provided regularly at periodic intervals based on the amount of nutrient removed by the cultures in order to get sustained yields from bioreactors.

Table 3.2 Bioreactors used for hairy root cultures

Bioreactor type	Volume (l)	Plant species	Secondary metabolite	References
Stirred tank bioreactor	12	*Datura stramonium*	Tropane alkaloids	Hilton and Rhodes 1990
Stirred tank bioreactor	25	*Atropa belladonna*	Tropane alkaloids	Lee et al. 1999
Stirred tank bioreactor	2.0	*Panax ginseng*	Saponins	Inomata et al. 1993
Bubble column bioreactor	2.0	*Hyoscyamus muticus*	Tropane alkaloids	Cuello et al. 1991
Airlift bioreactor	1.0	*Catharanthus roseus*	Indole alkaloids	Toivonen et al. 1989
Nutrient sprinkle bioreactor		*Salvia officinalis* L	Rosmarinic acid	Grzegorczyk and Wysokinska 2011
Airlift bioreactor	2.0	*Duboisia leichhardtii*	Scopolamine	Muranaka et al. 1992
Airlift bioreactor	1	*Coleus blumei* L.	Rosmarinic acid	Bauer et al. 2015
Airlift bioreactor	1.5	*Nicotiana rustica*	Nicotine	Rhodes et al. 1986
Trickle bed bioreactor	2.0	*Hyoscyamus muticus*	Tropane alkaloids	Flores and Curtis 1992
Nutrient mist bioreactor	9	*Mucuna pruriens*	L-DOPA	Huang et al. 2004
Nutrient sprinkle bioreactor	5	*Leonurus sibiricus* L.	Caffeic acid	Sitarek et al. 2018

3.5 Case Studies of Medicinal Plants

3.5.1 *Catharanthus roseus*

This plant produces a very important compound—ajmalicine—in its leaves which has got antihypertensive property and also is used in the treatment of circulatory diseases. The low availability of the compound in the plant necessitated alternative production systems for the rapid and mass production of ajmalicine. The bacterium-induced hairy roots were considered as a better alternative for in vitro production of ajmalicine. Mass cultivation of hairy roots was carried out in different bioreactor systems like bubble column, rotating drum, modified bubble column, etc. Bioreactor conditions in terms of inoculum size and density, aeration rate, agitation speed, etc. were standardized. The yield of ajmalicine was only 4.6 $\pm$ 0.4 mg/l in rotating drum bioreactor, whereas it gave 34 $\pm$ 2.3 mg/l ajmalicine in shake flask bioreactor. The yield was still higher in a modified bioreactor where the hairy roots were anchored to a polyurethane foam (Thakore et al. 2017).

3.5.2 *Panax ginseng*

Panax ginseng is valued for the saponins—ginsenosides—present in its roots which is responsible for the pharmacological activity. Cultivation of this plant being time consuming and laborious needed some alternate sources of ginsenosides. Large-scale biomass production of ginseng was achieved on industrial scale by Nitto Denko Corporation (Ibaraki, Osaka, Japan) in big dimension stirred tank bioreactors. Stirred tank and airlift bioreactors were commonly used for the suspension cultures. In stirred tank bioreactor system for ginseng, it was found that cell growth and ginsenoside production was affected by agitator design and the agitation rate (Furuya et al. 1984). Hairy root cultures were also tried by many workers for rapid saponin (Yoshikawa and Furuya 1987; Yu et al. 2000), but these hairy roots were observed to be lethal to mammalian cells. Therefore, adventitious root production was attempted for the production of secondary metabolite, and it was found successful (Yu et al. 2001), and ultimately balloon bubble bioreactor system was used for the scaling up of adventitious root production (Yu et al. 2000, 2001).

3.5.3 *Artemisia annua* L

Artemisinin obtained from the leaves of *Artemisia annua* is used for the treatment of cerebral malaria along with quinine and chloroquine. The availability of this bioactive is very limited in the plant accounting to only 0.01–0.5%; as a result only 6 kg of artemisinin can be obtained from 1000 kg dry plant leaves. This necessitated the in vitro production of artemisinin using plant cell and hairy root cultures. Different bioreactor systems have been tried in *Artemisia annua* for artemisinin production. It was found that artemisinin production was highest in mist bioreactors as compared to other systems (Patra and Srivastava 2014). The yield of artemisinin was 0.025 mg/l (Kim et al. 2003) and 0.088 mg/l (Xie et al. 2000), respectively, in bubble column type and shake flask bioreactors, whereas a higher artemisinin content of 0.031 mg/l was observed in nutrient mist bioreactors. It was also found that the genes responsible for artemisinin synthesis are overexpressed during scale-up of the cultures in bioreactor (Souret et al. 2003).

3.5.4 *Bacopa monnieri*

Bacopa monnieri yields the saponin bacoside which has got the pharmacological memory boosting property. In vitro production of bacoside has been reported by several workers from differentiated as well as undifferentiated tissues. Liquid medium for culturing *B. monnieri* is standardized by Ahuja et al. (2016). Shake flask bioreactor, Growtek®, and modified bench top air agitated bioreactors were used for shoot cultures of *Bacopa monnieri*. Air agitated bioreactor produced more biomass (169.67 g) as compared to Growtek® (23.17 g) and shake flask (15.70 g) bioreactors. Bacoside production was three times higher in the air agitated bioreactor

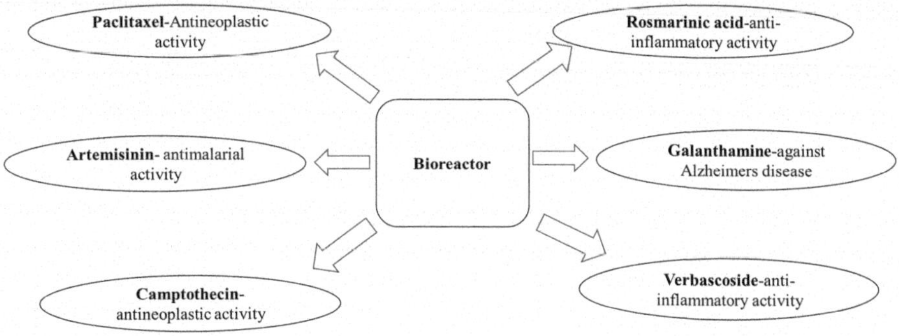

Fig. 3.1 Selected examples of high-value bioactives produced by plant cell/organ culture in bioreactors

than shake flask system. These findings suggested that bioreactor system is advantageous for *B. monnieri* shoot proliferation and bacoside production. Figure 3.1 shows high value compound which are commercially produced from few of the medicinal crops.

3.6 Conclusions and Recommendations

Plant bioreactors have turned out to be the most suitable system for the industrial production of plant secondary metabolites on a large scale. Most of the bioreactors utilize liquid culture media for the optimum growth of cultures and production of secondary metabolites. An efficient bioreactor system should provide a better control of the contact of the plant tissue with the culture medium and optimal nutrient and growth regulator supply, as well as aeration, medium circulation, the filtration of the medium, and the scaling up of the cultures. Better understanding of different metabolic pathways governing the biosynthesis of bioactive compounds is essential for developing a successful bioreactor system. Efforts should be made in the direction of shifting the bioreactor research from the laboratories to the commercial level so as to meet the trade demands of plant extracts in the international market worldwide.

References

Ahmed S, Hahn EJ, Paek KY (2008) Aeration volume and photosynthetic photon flux affect cell growth and secondary metabolite contents in bioreactor cultures of Morindacitrifolia. J Plant Biol 51(3):209–212

Ahuja A, Tripathi MK, Singh SP (2016) Plant cell cultures-an efficient resource for the production of biologically important metabolites: recent developments—a review. Progressive Res 11(1): 1–8

Aumont V, Larronde F, Richard T, Budzinski H, Decendit A, Deeux G, Krisa S, Mérillon J-M (2004) Production of highly 13C-labeled polyphenols in Vitis vinifera cell bioreactor cultures. J Biotechnol 109:287–294

Bais HP, Suresh B, Ramachandra Rao S, Raghavarao KSMS, Ravishankar GA (2002) Performance of Cichorium intybus hairy root cultures in various bioreactor configurations. In Vitro Cell Dev Biol Plant 38:573–580

Bauer N, Vuković R, Likić S et al (2015) Potential of different Coleus blumei tissues for rosmarinic acid production. Food Technol Biotechnol 53(1):3–10

Biondi S, Scaramagli S, Oksman-Caldentey KM, Poli F (2002) Secondary metabolism in root and caallus cultures of Hyoscyamus muticus L: the relationship between morphological organization and response to methyl jasmonate. Plant Sci 163:563–569

Bourgaud F, Gravot A, Milesi S, Gonteir E (2001) Production of plant secondary metabolites: a historical perspective. Plant Sci 161:839–851

Chan LK, Koay SS, Boey PL, Bhatt A (2010) Effects of abiotic stress on biomass and anthocyanin production in cell cultures of Melastoma malabathricum. Biol Res 43:127–135

Chastang T, Pozzobon V, Taidi B, Courot E, Clément C, Pareau D (2018) Resveratrol production by grapevine cells in fed-batch bioreactor: experiments and modelling. Biochem Eng J 131:9–16

Cuello JL, Walker PN, Curtis WR (1991) Am. Soc. Agric. Eng. Winter Meet., Chicago, pp 17–20

Cui XH, Chakrabarty D, Lee EJ, Paek KY (2011) Production of adventitious roots and secondary metabolites by Hypericum perforatum L. in a bioreactor. Bioresour Technol 101(12): 4708–4716

Denchev PD, Kuklin AI, Scragg AH (1992) Somatic embryo production in bioreactors. J Biotechnol 26:99–109

Flores HE, Curtis WR (1992) Approaches to understanding and manipulating the biosynthetic potential of plant roots. Ann N Y Acad Sci 665(1):188–209

Furuya T, Yoshikawa T, Orihara Y, Oda H (1984) Studies of the culture conditions for Panax ginseng cells in jar fermentors. J Nat Prod 47(1):70–75

Grzegorczyk I, Wysokinska H (2011) Antioxidant compounds in Salvia officinalis L. shoot and hairy root cultures in the nutrient sprinkle bioreactor. Acta Soc Bot Pol 79(1):7–10

Grzegorczyk-Karolak I, Rytczak P, Bielecki S, Wysokinska H (2017) The influence of liquid systems for shoot multiplication, secondary metabolite production and plant regeneration of Scutellariaalpina. Plant Cell Tissue Organ Cult 128:479–486

He Y, Ning T, Xie T, Qiu Q, Zhang L, Sun Y, Jiang D, Fu K, Yin F, Zhang W, Shen L, Wang H, Li J, Lin Q, Sun Y, Li H, Zhu Y, Yanga D (2011) Large-scale production of functional human serum albumin from transgenic rice seeds. Proc Natl Acad Sci 108(47):19078–19083

Hilton MG, Rhodes MJC (1990) Growth and hyoscyamine production of 'hairy root' cultures of Datura stramonium in a modified stirred tank reactor. Appl Microbiol Biotechnol 33(2): 132–138

Huang SY, Hung CH, Chou SN (2004) Innovative strategies for operation of mist trickling reactors for enhanced hairy root proliferation and secondary metabolite productivity. Enzym Microb Technol 35(1):22–32

Ibrahim R (2015) The potential of bioreactor technology for large-scale plant micropropagation. In: VI International symposium on production and establishment of micropropagated plants, vol 1155, pp 573–584

Inomata S, Yokoyama M, Gozu Y, Shimizu T, Yanagi M (1993) Growth pattern and ginsenoside production of Agrobacterium-transformed Panax ginseng roots. Plant Cell Rep 12(12):681–686

Jeong JA, Wu CH, Murthy HN, Hahn EJ, Paek KY (2009) Application of an airlift bioreactor system for the production of adventitious root biomass and caffeic acid derivatives of Echinacea purpurea. Biotechnol Bioprocess Eng 14(1):91–98

Kim YJ, Weathers PJ, Wyslouzil BE (2003) Growth dynamics of Artemisia annua hairy roots in three culture systems. Biotechnol Bioeng 83:428–443

Lee KT, Suzuki T, Yamakawa T, Kodama T, Igarashi Y, Shimomura K (1999) Production of tropane alkaloids by transformed root cultures of Atropa belladonna in stirred bioreactors with a stainless steel net. Plant Cell Rep 18(7–8):567–571

Muranaka T, Ohkawa H, Yamada Y (1992) Scopolamine release into media by Duboisialeichhardtii hairy root clones. Appl Microbiol Biotechnol 37(5):554–559

Paek KY, Hahn EJ, Son SH (2001) Application of bioreactors of large scale micropropagation systems of plants. In Vitro Cell Dev Biol Plant 37:149–157

Paek KY, Chakrabarty D, Hahn EJ (2005) Application of bioreactor systems for large scale production of horticultural and medicinal plants. In: Liquid culture systems for in vitro plant propagation. Springer, Dordrecht, pp 95–116

Patra N, Srivastava AK (2014) Enhanced production of Artemisinin by hairy root cultivation of Artemisia annua in a modified stirred tank reactor. Appl Biochem Biotechnol 174:2209–2222

Preil W (2005) General introduction: a personal reflection on the use of liquid media for in vitro culture. In: Liquid culture systems for in vitro plant propagation. Springer, Dordrecht, pp 1–18

Rhodes MJC, Hilton M, Parr AJ, Hamill JD, Robins RJ (1986) Nicotine production by "hairy root" cultures of Nicotiana rustica: fermentation and product recovery. Biotechnol Lett 8(6):415–420

Sajc L, Grubisic D, Vunjak-Novakovic G (2000) Bioreactors for plant engineering: an outlook for further research. Biochem Eng J 4(2):89–99

Sitarek P, Kowalczyk T, Picot L, Michalska-Hejduk D, Bijak M, Białas A, Wielanek M, Śliwiński T, Skała E (2018) Growth of Leonurus sibiricus L. roots with over-expression of AtPAP1 transcriptional factor inclosed bioreactor, production of bioactive phenolic compounds and evaluation of their biological activity. Ind Crop Prod 122:732–739

Smolenskaya I, Reshetnyak O, Nosov A, Zoriniants S, Chaiko A, Smirnova Y (2007) Ginsenoside production, growth and cytogenetic characteristics of sustained Panax japonicus var. repens cell suspension culture. Biol Plant 51:235–241

Souret FF, Kim Y, Wyslouzil BE, Wobbe KK, Weathers PJ (2003) Scale-up of Artemisia annua L. hairy root cultures produces complex patterns of terpenoid gene expression. Biotechnol Bioeng 83:653–667

Takayama S, Misawa M (1981) Mass propagation of begonia x hiemalis plantlet by shake culture. Plant Cell Physiol 22:461–467

Thakore D, Srivastava AK, Sinha AK (2017) Mass production of Ajmalicine by bioreactor cultivation of hairy roots of Catharanthus roseus. Biochem Eng J 119:84–91

Toivonen L, Balsevich J, Kurz WGW (1989) Indole alkaloid production by hairy root cultures of Catharanthus roseus. Plant Cell Tissue Organ Cult 18:79–93

Weremczuk-Jeżyna I, Kochan E, Szymczyk P et al (2019) The antioxidant and antimicrobial properties of phenol rich extracts of Dracocephalum forrestii W. W. Smith shoot cultures grown in the nutrient sprinkle bioreactor. Phytochem Lett 30:254–260

Xie D, Wang L, Ye H, Li G (2000) Isolation and production of artemisinin and stigmasterol in hairy root cultures of Artemisia annua. Plant Cell Tissue Organ Cult 63:161–166

Yoshikawa T, Furuya T (1987) Saponin production by cultures of Panax ginseng transformed with Agrobacterium rhizogenes. Plant Cell Rep 6(6):449–453

Yu KW, Gao WY, Son SH, Paek KY (2000) Improvement of ginsenoside production by jasmonic acid and some other elicitors in hairy root culture of ginseng (Panax ginseng CA Meyer). In Vitro Cell Dev Biol Plant 36(5):424–428

Yu KW, Gao WY, Hahn EJ, Paek KY (2001) Effects of macro elements and nitrogen source on adventitious root growth and ginsenoside production in ginseng (Panax ginseng CA Meyer). J Plant Biol 44(4):179–184

Ziv M (2005) Simple bioreactors for mass propagation of plants. In: Liquid culture systems for in vitro plant propagation. Springer, Dordrecht, pp 79–93

Factors Affecting In Vitro Production of Nutraceuticals

4

Lalit Giri, Laxman Singh, Kuldeep Joshi, Arti Bisht, and Indra D. Bhatt

Abstract

In vitro culture techniques have emerged as a promising source for production of various nutraceutical compounds and are employed in a variety of industries, including agrochemicals, pharmaceuticals, flavours, fragrances, food colouring, etc. Recognizing the importance of nutraceutical compounds, different strategies have been extensively studied on in vitro cultures with the objective of improving the production of nutraceutical compound in the plants. However, success for obtaining the desired output is yet to be achieved. There are many factors which are responsible in hindering the in vitro production of nutraceuticals. The review thus focuses on the study of various biotic and abiotic factors and their impact on three prime groups of nutraceutical compounds, i.e. phenolics, terpenes and nitrogen-containing compounds. Based on the review of literature, various strategies for improving yield of nutraceuticals in in vitro conditions are discussed, and factors responsible for in vitro production of nutraceuticals are highlighted.

Keywords

Alkaloid · In vitro culture · Nutraceuticals · Phenolic · Terpenes

L. Giri
Ladakh Regional Center, G. B. Pant National Institute of Himalayan Environment, Leh, Ladakh, India

L. Singh · K. Joshi · A. Bisht · I. D. Bhatt (✉)
G. B. Pant National Institute of Himalayan Environment, Kosi-Katarmal, Almora, Uttarakhand, India

T. Belwal et al. (eds.), *Nutraceuticals Production from Plant Cell Factory*,
https://doi.org/10.1007/978-981-16-8858-4_4

4.1 Introduction

Nutraceutical compounds are naturally derived bioactive compounds and are being used in different traditional medicine by various cultures around the world (Yuan et al. 2016); the therapeutical effects of these compounds are well recognised since ancient times. Plants are rich source of three prime groups of nutraceutical compounds, i.e. phenolics, terpenes and nitrogen- and sulphur-containing compounds. Although phenolics and terpenes are primarily generated by the shikimic and mevalonate system, respectively, the third major group of nitrogen- and sulphur-containing compounds are generally generated by common amino acids (Van Etten et al. 2001). These compounds are often present in plant species or a taxonomically related group of species whose principal ecological activity is to biosynthesise molecules to strengthen the plant's ability to adapt to the wild environment, significantly in relation of defence and signalling (Wink 2003). As a result, these molecules have a variety of biological properties, including antioxidant, antibacterial, anti-inflammatory, hypocholesterolemic, anti-diabetic and cytotoxic capabilities. Considering these facts, the demands of nutraceutical compounds are increasing many folds. This can be judged from the fact that the business of nutraceutical compounds is estimated to be around $373 billion in 2025 at the rate of 7.5% compound annual growth rate (PMMI Business Intelligence on Nutraceutical Market Assessment 2019). The high demand for nutraceutical compounds directly depends upon the large-scale exploitation/extraction from the selected wild plant species. This has a negative influence on the availability of raw materials and, as a result, on the supply chain. In this context, there is a need to develop some alternative approaches, which may help in addressing the increasing demands of natural nutraceuticals.

In recent years, besides production of large-scale quality plant material, in vitro culture methods are rapidly being employed for the synthesis of diverse nutraceutical components since they offer a great way to control nutraceutical synthesis under diverse culture conditions including nutrient medium, plant growth regulators (PGRs) and their concentration, elicitor types, pH variations, etc. Hairy root culture technology, in addition to in vitro cultures, is regarded as one of the most promising methods for large-scale synthesis of nutraceutical chemicals (Halder et al. 2018). However, all these approaches are affected by various biotic and abiotic factors. This review focuses on understanding the role of different factors and their effect on in vitro production of nutraceuticals.

4.2 Factors Affecting the Production of Nutraceutical Compounds

Available literatures on indicate that there are a number of factors responsible for altering the in vitro production of nutraceutical compounds (Fig. 4.1). However, their specific roles and mechanism are poorly known. Considering the above, the present review focused on understanding the possible factors that affect the in vitro

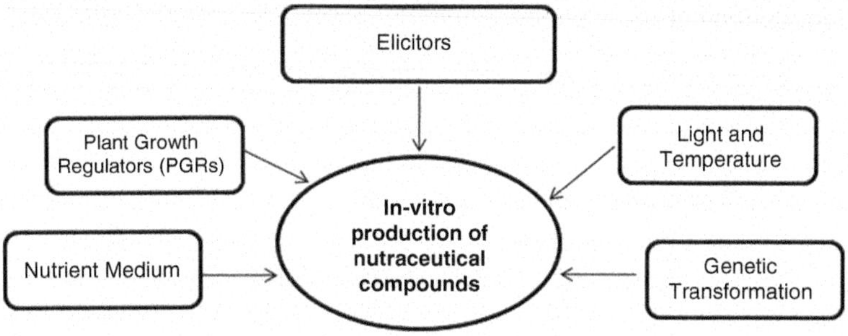

Fig. 4.1 Factors affecting in vitro production of nutraceutical compounds

production of three most important groups of nutraceutical compounds, i.e. phenolics, terpenes and nitrogen- and sulphur-containing compounds. Details of each factors is described in following sections:

4.2.1 Optimization of Cultural Conditions

Culture medium is required for the in vitro cell or tissue cultures of plants, as well as the subsequent production of nutraceutical compounds. Gamborg (B5 1968), Linsmaier and Skoog (LS 1965), Murashige and Skoog (MS 1962), and Schenk and Hildebrandt are just few of the many mediums that have been used to establish plant cell and organ cultures. The most fundamental technique for optimizing the synthesis of specific nutraceuticals in in vitro culture is to vary the quantity of micro- and macronutrients with a supply of nitrogen, sugar and plant growth regulators (PGRs). Various studies have been carried out on the accumulation of nutraceutical compounds like phenol, flavonoid, anthocyanin, etc., considering nutritional stress by changing the concentrations of nutrients in the medium (Schiozer and Barata 2007). For example, in *Arabidopsis thaliana*, *Cleome rosea* and *Rosa hybrida*, changing the ratio of $NH4^+$ and $NO3^-$ in the MS medium greatly enhanced anthocyanin and phenol contents (Simões et al. 2009; Shi and Xie 2010; Ram et al. 2011; Liu et al. 2014). Furthermore, anthocyanin synthesis in *Vitis vinifera* cell suspension cultures was reported to be regulated by osmotic stress caused by sucrose and other osmotic agents (Do and Cormier 1990). Reports indicate that sugar is a signalling molecule that affects the growth, development and metabolism of cultured cells (Wang and Weathers 2007; Praveen and Murthy 2012). Therefore, optimization of culture medium is essential whilst designing experiments for in vitro production of nutraceutical compounds.

The condition of solid and liquid medium is also a key role in increasing the concentration of various nutraceutical substances in vitro. For example, liquid media clearly enhanced alkaloid content in *Brugmansia suaveolens* (Zayed and Wink 2004), *Ophiorrhiza pumila* (Saito et al. 2001) and *Nothapodytes foetida* (Fulzele et al. 2001). Meanwhile, literatures also indicate solid media also enhanced alkaloid production in many plants like *Hyoscyamus muticus* (Dehghan et al. 2012), *Leucojum aestivum* (Ivanov et al. 2010) and *Narcissus pseudonarcissus* (Ferdausi et al. 2020).

4.2.2 Plant Growth Regulators (PGRs)

Among plant growth regulators, auxins and cytokinins are major plant growth regulators used individually and in combinations for producing nutraceutical compounds (Pullman et al. 2005). Cell division and growth are promoted by auxins and cytokinins. The most common auxin is 2,4-dichlorophenoxyacetic acid (2,4-D), but auxins like 2,4-5-trichlorophenoxyacetic acid (2,4,5-T) and indoleacetic acid (IAA) are also commonly used in in vitro production of nutraceutical compounds; however, in case of cytokinin, zeatin and benzylaminopurine (BAP) are commonly used (Hiraoka et al. 2004; Slater et al. 2008). The auxin to cytokinins ratio has a significant impact on plant development, since a high auxin to cytokinins ratio promotes shoot production, whilst a higher cytokinins to auxins ratio promotes root growth (Slater et al. 2008). A number of studies have found that PGRs have a direct influence on the production of nutraceutical substances in in vitro cells (Table 4.1). Auxin or cytokinin concentrations, or the ratio of auxin and cytokinin in cultured cells of various plant species, affected biomass growth and nutraceutical compound production (Mantell and Smith 1984). In case of anthocyanin, exogenous cytokinins stimulate the transcription level of key regulatory genes (PAL1, CHS, CHI, DFR), which is responsible for the biosynthesis of anthocyanin compounds (Deikman and Hammer 1995).

4.2.3 Light and Temperature

Light is a physical factor which influences plant nutraceutical production (Table 4.1). Different light intensity influence the production of phenol, flavonoid and anthocyanin in in vitro culture *of Cleome rosea*, *Centella asiatica* and *Vitis vinifera* (Simões et al. 2009; Rao et al. 2015; Zhang et al. 2002; Kapoor et al. 2018). Illumination was used for anthocyanin production in many publications on cell culture systems (Blando et al. 2005); however, anthocyanin synthesis has also been reported in the dark (Konczak-Islam et al. 2000). According to studies, the intensity of light is directly proportional to anthocyanin synthesis. In *Vitis vinifera* cells, for example, maximum anthocyanin synthesis was attained with 8000 lux light (Zhang et al. 2002), but *Cleome rosea* cells required 3478 lux light (Simões et al. 2009). The light increases the expression of regulatory factors and structural genes involved in nutraceutical chemical biosynthesis (Zhang et al. 2018). Temperature is one of the factors which influence the production of plant nutraceutical compounds in various plant species in in vitro culture. Generally, extreme temperature causes the adverse impact on plant growth including synthesis and degradation of primary metabolites (Al-Khayri and Ibraheem 2014). For the development of callus tissues

Table 4.1 Nutraceutical compounds produced by in vitro culture techniques

SN	Plant species	Nutraceutical compounds	Explant used/culture medium/ conditions	References
A. Phenolics				
1	*Arabidopsis thaliana*	Anthocyanin	In vitro germinated seed-derived leaves/MS (1/2 strength of NH_4NO_3 and KN03)/ 50μmol m^{-2} s^{-1} of light intensity	Shi and Xie 2010
2	*Arabidopsis thaliana*	Anthocyanin	In vitro germinated seed-derived leaves/modified MS (removed NH_4NO_3 and reduced concentration of KNO_3 to 9.4 mM/L)/9μM NAA	Liu et al. 2014
3	*Artemisia absinthium*	Phenol and flavonoid	Leaf/MS (1.0 mg/L MeJA, jasmonic acid and GA_3) liquid/ gyratory shaker 25 °C, 120 rpm, 25 ± 1 °C	Ali et al. 2015
4	*Azadirachta indica*	Anthocyanin	Leaf/MS/TDZ (0.2 mg/L)	Ashokhan et al. 2020
5	*Beta vulgaris*	Phenol	Roots/*A. rhizogenes* strain (ATCC 15834)/liquid MS medium supplemented with 30 g/L sucrose/25 °C (light/dark cycle of 16/8 h) on a shaker at 11 rad/s	Georgiev et al. 2010
6	*Brassica rapa ssp. rapa*	Phenol	Leaf explants/*A. rhizogenes* strain (KCTC 2703)/MS medium with 4% sucrose/25 ± 1 °C in dark condition	Chung et al. 2016
7	*Cajanus cajan*	Flavonoid	Leaf/*A. rhizogenes* strain (LBA9402)/MS liquid supplemented with 3.0% sucrose/25 °C in the dark condition	Jiao et al. 2020
8	*Camellia sinensis*	Phenol	Shoot/cobalt metal ions (0.006 ppm)/MS/24 ± 2 °C dark condition	Sutini et al. 2019
9	*Centella asiatica*	Phenol and flavonoid	Leaf, stem/MeJA (100μM), salicylic acid (100μM)/MS/light intensity of 3000 lux	Rao et al. 2015
10	*Cleome rosea*	Anthocyanin	Stem/1/2 MS (1:4 ratio of NH_4^+ to NO_3^-)/2,4-D (0.9μM)/sucrose (70 g/L)/light intensity (2478 lux)	Simões et al. 2009
11	*Cleome rosea*	Anthocyanin	Seedling/MS/IAA (2 mg/L)/kin (0.2 mg/L)	Sudha and Ravishankar 2003
12	*Cleome rosea*	Anthocyanin	Leaves/MS/11.41μM IAA and 0.93μM kin	Saad et al. 2018

(continued)

Table 4.1 (continued)

SN	Plant species	Nutraceutical compounds	Explant used/culture medium/ conditions	References
13	*Daucus carota L. ssp. sativus var. atrorubens*	Anthocyanins	Taproot/*A. rhizogenes* strain (A4)/1/2 MS + 200 mg L^{-1} ethephon/20 °C/18 °C (light/ dark cycle of 12/12 h) on an oscillatory shaker at 100 rpm	Barba-Espín et al. 2020
14	*Echinacea purpurea*	Phenol	Leaf/*A. rhizogenes* strain (ATCC 43,057)/MS + 50 g/l sucrose +1.0 mg/l (24-eBL) 24-Epibrassinolide/25 °C in a growth chamber under the dark condition	Demirci et al. 2020
15	*Fagopyrum tataricum*	Anthocyanins and phenolic compounds	Seeds/*A. rhizogenes* strain (R1000)/1/2 MS/25 ± 2 °C incubated under dark conditions	Thwe et al. 2016
16	*Ficus carica*	Phenol and flavonoid	Shoot/*A. rhizogenes* strain (A7)/ 1/2 MS liquid +100μmol L^{-1} MeJA elicitation/25 ± 2 °C (light/dark cycle of 16/8 h)	Amani et al. 2020
17	*Habenaria edgeworthii*	Phenolic	Seed/MS/BA (0.0–3.0μM) methyl jasmonate (10,000 μM/) liquid/16 h 40 E m^{-2} s^{-1} light cool white fluorescent tubes	Giri et al. 2012
18	*Hypericum perforatum L.*	Flavonoid	Shoot, root/BA (1.2 mg/L), NAA (0.02 mg/L)/liquid MS/25 ± 2 °C, photoperiod 16/8 h light/dark photoperiod	Mohammed et al. 2019
19	*Knautia sarajevensis*	Flavonoid	Shoot/MS/salicylic acid (100 mm)/liquid/dark, humidity (70%), photoperiod 50 mmol/m^2 s; light/dark 16:8	Karalija et al. 2020
20	*Melastomam alabathricum*	Anthocyanin	Cells (suspension culture)/MS/ 0.25 mg/L BA and 0.5 mg/L NAA/45 gL^{-1} sucrose/ 3.5 mgL^{-1} MeJA	Suan-See et al. 2011
21	*Melastomam alabathricum*	Anthocyanin	Cells (suspension culture)/MS/ 0.25 mg/L BA and 0.5 mg/L NAA/45 g/L sucrose/3.5 mg/L MeJA/pH 5.75/light intensity (301–600 lux), temperature (20 ± 2 °C)	Chan et al. 2010
22	*Melissa officinalis*	Anthocyanin	Apical part/MS/0.5 mg/L BA/O_3 fumigation (200 ppb, 3 h)	Tonelli et al. 2015
23	*Ocimum basilicum*	Phenol	Shoots/*A. rhizogenes* strain (R1000)/1/2 MS liquid medium/ 25 ± 2 °C in dark conditions for 2 days	Kwon et al. 2020

(continued)

Table 4.1 (continued)

SN	Plant species	Nutraceutical compounds	Explant used/culture medium/ conditions	References
24	*Panax sikkimensis*	Anthocyanin	Root/MS/5.4µM NAA	Biswas et al. 2015
25	*Panax sikkimensis*	Anthocyanin	Root/MS/4.5µM 2,4-D and 1.2µM kin with 3.0% sucrose 0.01% myo-inositol/0.33µM thiamine hydrochloride, 2.5µM Pyridoxine hydrochloride 4.0µM nicotinic acid	Mathur et al. 2010
26	*Phoenix dactylifera*	Phenol and flavonoid	Leaf/salicylic acid (50 mg/L), cadmium chloride and silver nitrate (50–200 mg/L)/solid/ rotary shaker (150 rpm) light intensity (40 mol/m^2/s) 16 h, temperature (25 ± 2 °C)	Al-Khayri and Naik 2020
27	*Plumbago europaea*	Phenol	Stem/*A. rhizogenes* (MSU440)/ 1/2 MS-B5 liquid medium containing 30 g L^{-1} sucrose/ 25 ± 1 °C in dark condition for 48 h	Beigmohamadi et al. 2020
28	*Polygonum multiflorum*	Phenol	Leaf/*A. rhizogenes* strain (KCTC 2703)/MS liquid medium supplemented with 30 g/l sucrose/25 ± 2 °C (light/dark cycle of 16/8 h) on shaker with the speed of 100 rpm	Thiruvengadam et al. 2014
29	*Raphanus sativus*	Anthocyanin	Root tip/1/2 MS/0.5 mg/L IBA/25 °C under the 14 h/day light at 100	Betsui et al. 2004
30	*Rhodiola imbricata*	Phenol and flavonoid	Leaves/MS/5 mg/L NAA and 5 mg/L BAP/blue light	Kapoor et al. 2018
31	*Rosa hybrida*	Anthocyanin	Leaf/modified EM/2.45µM IBA and 2.33µM kin/0.5µM MeJA	Ram et al. 2013
32	*Rosa hybrida* cv. *'Pusa Ajay'*	Anthocyanin	Leaf disc/EM/4.0 mg/L 2,4-D/70 g/L sucrose/NH_4^+ and NO_3^+ (low: High)	Ram et al. 2011
33	*Salvia bulleyana*	Phenolic compounds	Leaves/*A. rhizogenes* strain (A4)/MS + 0.1 mg/l IAA + 0.5 mg/l BAP/26 °C in dark condition	Wojciechowska et al. 2020
34	*Salvia miltiorrhiza*	Flavonoid	Leaves/*A. tumefaciens* strain C58C1 (pRiA4)/1/ 2 MS/25 ± 2 °C in dark condition on a shaker with the speed of 100 rpm for 60 days	Shi et al. 2020a, b
35	*Salvia officinalis*	Flavonoid	Shoot/sodium salicylate (500µM)/solid/25 ± 2 °C, photoperiod 16 h light/8 dark h cycle	Kolarevic et al. 2015

(continued)

Table 4.1 (continued)

SN	Plant species	Nutraceutical compounds	Explant used/culture medium/ conditions	References
36	*Scutellaria lateriflora*	Phenol	Internode/*A. rhizogenes* strain (ATCC 15834)/MS medium +0.5 mg/l IBA/25 °C in the dark condition	Marsh et al. 2014
37	*Silybum marianum*	Flavonoid	Cotyledons/*A. rhizogenes* strain (A4 strain)/MS medium + 0.2 mg/l NAA/25 ± 2 °C (light/ dark cycle of 16/8 h)	Gabr et al. 2016
38	*Thevetia peruviana*	Phenol and flavonoid	Leaf/salicylic acid (300μM) methyl jasmonate (3μM)/MS liquid/25 °C, photoperiod 12 h light/12 h darkness	Mendozaa et al. 2018
39	*Vitis vinifera*	Anthocyanin	Cell suspension line/B5/ 0.1 mg L^{-1} NAA and 0.2 mg L^{-1} kin/30 g L^{-1} sucrose, 250 mg L^{-1} casein hydrolysate/20μM MeJA and 8000 lux light	Zhang et al. 2002
40	*Vitis vinifera*	Anthocyanin	Berries/B5/0.1 mg L^{-1} NAA and 0.2 mg L^{-1} kin/20 g L^{-1} sucrose, 250 mg^{-}L^{1} casein hydrolysate, 10–4 M ABA	Gagné et al. 2011
41	*Vitis vinifera*	Anthocyanin	Pulp fragments/B5/0.54μM NAA/0.93μM kin/88 mM sucrose combined with 165 mM mannitol/250 mg L^{-1}	Do and Cormier 1990
42	*Vitis vinifera L. cv. Gamay Fréaux*	Anthocyanin	Berries/B5/0.1 mg L^{-1} NAA, 0.2 mg L^{-1} kin/30 g L^{-1} sucrose, 250 mg^{-}L^{1} casein hydrolysate/50 mg L^{-1} MeJA and 1 mg L^{-1} dextran	Qu et al. 2011
B. Terpenes				
43	*Artemisia annua*	Monoterpenes	Plant cells/MS (liquid medium)/ 3% (w/v) sucrose/0.1 mg/l NAA and 0.1 mg/l Kn/20 mg/l acetyl salicylic acid	Baldi and Dixit 2008
44	*Artemisia annua*	Monoterpenes	Leaves/*A. rhizogenes* strain (LBA9402)/MS/liquid supplemented with 3.0% sucrose/25 °C in the dark condition	Rao et al. 1998
45	*Ambrosia tenuifolia*	Monoterpenes	Calls/MS/10μM kn and 1μM 2–4 D/10μM ascorbic acid and cysteine	Goleniowski and Tirppi 1999
46	*Taxus baccata*	Monoterpenes	Callus line/B5 medium/0.05 mM VSO4	Cusidó et al. 1999

(continued)

Table 4.1 (continued)

SN	Plant species	Nutraceutical compounds	Explant used/culture medium/ conditions	References
47	*Coleus forskohlii*	Diterpenes	Leaves/*A. rhizogenes* strain (MAFF 03-01724)/B5 medium/ liquid supplemented with 3.0% sucrose/100 rpm/25 °C in the dark condition	Sasaki et al. 1998
48	*Citrullus colocynthis*	Diterpenes	Stem-derived callus/MS/2 mg/l 2,4-D and 4 mg/l kin	Hegazy et al. 2010
49	*Azadirachta indica*	Diterpenes	Calls/liquid MS medium with altered nitrate: Ammonium ratios (4:1)/sucrose 15 g/l	Sujanya et al. 2008
50	*Astragalus mongholicus*	Diterpenes	Sterile grown seedlings/ *A. rhizogenes* strain (LBA9402/ MS without nitrate/the cultures were kept in the dark at 25 °C	Ionkova et al. 1997
51	*Panax ginseng*	Diterpenes	Callus/MS liquid/2.0 mg l^{-1} NAA, 0.1 mg l^{-1} kin/105 rpm/ dark at 25 °C	Thanh et al. 2005;
52	*Centella asiatica*	Diterpenes	Calls/MS liquid medium/ supplemented with 2 mg l^{-1} 2,4-D and 0.1 mg l^{-1} BA/100 rpm/dark at 25 °C	Bonfill et al. 2010
C. Alkaloids				
53	*Ophiorrhiza pumila*	Alkaloids	Stem/*A. tumefaciens* strain (C58C1)/B5 medium	Shi et al. 2020a, b
54	*Brugmansia suaveolens*	Alkaloids	Root tips/WP liquid/200μm methyl jasmonate + quercetin	Zayed and Wink 2004
55	*Bupleurum falcatum*	Alkaloids	Root/B5/8 mgL^{-1} IBA	Kusakari et al. 2000
56	*Catharanthus roseus*	Alkaloids	Seeds/half strength MS supplemented with 0.5 mg/L 2, 4-D + 1.0 mgL^{-1} BA	Verma et al. 2012
57	*Capsicum annuum*	Alkaloids	Fruits/MS medium supplemented with 9.04μM 2,4-D + 2.32μM kinetin (17.11 mM NaCl and 14.71μM AgNO3)	Gammoudi et al. 2019
58	*Cassia acutifolia*	Alkaloids	Root/MS liquid medium supplemented with 1.0 mg/L 2,4-D + 0.1 mg/L kinetin	Nazif et al. 2000
59	*Cereus peruvianus*	Alkaloids	Shoots/MS/4.0 mgL^{-1} 2,4-D, 4.0 mgL^{-1} kinetin and 125 mgL^{-1} tyrosine	De Oliveira and da Silva Machado 2003
60	*Ephedra foliata*	Alkaloids	Axillary buds/MS/0.5 mgL^{-1} 2, 4-D and kinetin, 100 mgL^{-1}, L-phenylalanine and 5 mgL^{-1} IBA	Lodha et al. 2014

(continued)

Table 4.1 (continued)

SN	Plant species	Nutraceutical compounds	Explant used/culture medium/ conditions	References
61	*Eriobotrya japonica*	Alkaloids	Leaf/LS medium/10μM BA	Taniguchi et al. 2002
62	*Erythrina americana*	Alkaloids	Seeds/MS liquid/3 mgL^{-1} NAA and 2 mgL^{-1} kin	Miguel-Chávez et al. 2003
63	*Ginkgo biloba*	Alkaloids	Leaf/MS liquid medium/ 1 mgL^{-1} NAA and 0.1 mgL^{-1} kin	Carrier et al. 1990
64	*Leucojum aestivum*	Alkaloids	Shoot/MS/1.15 mgL^{-1} NAA and 2.0 mgL^{-1} BA	Ivanov et al. 2010
65	*Nothapodytes foetida*	Alkaloids	Stem/MS liquid/10.74 mM 2,4-D and 10.74 NAA and 2.22 mM BA	Fulzele et al. 2001
66	*Stemona* sp.	Alkaloids	Shoot tips/MS liquid/100μM SA	Chaichana et al. 2012
67	*Torreya nucifera*	Alkaloids	Leaves/MS liquid medium/ 10 mgL^{-1} 2,4-D and 1 gL^{-1} casamino acid	Orihara et al. 2002
68	*Withania somnifera*	Alkaloids	Shoot/MS/1 mgL^{-1} BA/sucrose (3%)	Ray and Jha 2001
69	*Withania somnifera*	Alkaloids	Shoot tips/MS/0.4μM IAA and 0.4μM BAP	Sharada et al. 2007

and the proliferation of cultured cells, temperatures of 17–25 °C are widely adopted (Rao and Ravishankar 2002). However, the phenol, flavonoid and anthocyanin yields are maximised at the optimum temperature (25°C2oC) (Table 4.1).

4.2.4 Elicitors

In response to numerous biotic (e.g. pathogens or insects) and abiotic (e.g. temperature, salinity, water, radiation, heavy metals and mineral) stresses, plant neutralities develop in plant cells (Ramakrishna and Ravishankar 2011). These biotic and abiotic stress conditions have been named as 'elicitors' (Dornenburg and Knorr 1995). This elicitation method is widely utilised in in vitro plant culture to produce a variety of plant nutraceutical compounds (Dornenburg and Knorr 1995; Ramakrishna and Ravishankar 2011). Fungal, bacterial and yeast elicitors (e.g. polysaccharides, glycoproteins, inactivated enzymes, purified xanthan, crudlan and chitosan) have been reported to induce the overproduction of various nutraceutical compounds, whereas abiotic elicitors (e.g. heavy metal salts) have been reported to induce the overproduction of various nutraceutical compounds. Some chemical elicitors commonly used in the in vitro production of nutraceuticals include jasmonic acid (JA), methyl jasmonate (MeJA), 2-hydroxyl ethyl jasmonate, salicylic acid (SA), acetyl salicylic acid (ASA), trifluoromethylsalicylic acid, ethylene (ET), nitric oxide (NO) and sodium nitroprusside (SNP) (Giri and Zaheer 2016). SA regulates resistance to pathogens like bacterial, fungal and viral. On the other hand MeJA regulates the production of proteins by the octadecanoid pathway. Therefore, hormonal MeJA and SA have been widely used to increase the yield of different nutraceutical compounds in the cell cultures. For example, methyl jasmonate (Ali et al. 2015; Rao et al. 2015; Giri et al. 2012; Ram et al. 2013; Mendozaa et al. 2018; Zhang et al. 2002; Suan-See et al. 2011) and salicylic acid (Karalija et al. 2020; Kolarevic et al. 2015; Al-Khayri and Naik 2020) have been efficiently used for the production of phenol, flavonoid and anthocyanin in in vitro cell or tissue culture of various plant species (Table 4.1). The concentration and duration of exposure of elicitor concentrations are also important factors that influence the production of biomass and nutraceutical compounds. For example, Giri et al. (2012) studied the effect of methyl jasmonate (i.e. at concentrations of 0, 1.0, 2.0, 5.0 and 10.0 mg L^{-1}) on callus culture of *Habenaria edgeworthii*. Results indicated that an increase in concentration resulted in a decrease in both fresh and dry biomass. However, phenolic compounds were reported significantly higher than control. Further, the time duration of subculture also affects the production of nutraceutical compounds. For example, Wang et al. (2015) obtained higher amounts of flavonoids in 15–20 exposure days after the use of MJ elicitor in the *Hypericum perforatum*. Similar findings were found in suspension cultures of *Bacopa monnieri*, with salicylic elicitor-treated cultures showing an increase in biomass and bacoside production around the sixth–ninth day (Koul and Mallubhotla 2020). Bota and Deliu (2011) recorded a considerable enhancement in the production of flavonoids on *Digitalis lanata* by the use of copper sulphate about 24 hrs. JA and SA along with light irradiation in cell suspension cultures resulted the significant increase in phenol, flavonoid and anthocyanin concentration, showing the synergistic effect of integrated processes under in vitro conditions (Zhang et al. 2002; Rao et al. 2015). Reports indicates that suspension culture is the best method for the elicitation and production of nutraceuticals in in vitro (Gadzovska et al. 2013; Ramírez-Mosqueda and Iglesias-Andreu 2017).

4.2.5 Genetic Transformation or Hairy Root Culture

Genetic information or hairy root culture is one of the advanced technologies of plant biotechnology for producing various nutraceutical compounds (Hansen and Wright 1999; Sheludko 2010). The use of transformed bacterial strain of *Agrobacterium tumefaciens* and *Agrobacterium rhizogenes* are reported to have its effect on in vitro production of nutraceutical compounds (Halder et al. 2018). These bacterial strains have been shown to be efficient and promising vehicles for introducing genes into plant genomes, resulting in the transfer and integration of genes from plasmids from bacteria into plant DNA, transformed neoplastic tissues, crown galls and hairy roots (Saito 1993). Hairy root culture techniques emerged as an ideal biotechnological tool for the uniform production of bioactive compounds, independent of geographical, climatic or environmental constraints (Chung et al. 2016) It has several advantages over field-raised plants like high yield, fast growth cycles, high genetic uniformity and stability in quality of bioactive molecule production (Boobalan and Kamalanathan 2020). Several strategies, such as medium optimization and selection of high-yielding lines, precursor feeding, growth simulation and modelling, upscaling in bioreactors, elicitation and metabolic engineering of metabolite biosynthetic pathways, have been developed to achieve continuous extraction of bioactive molecules (Mehrotra et al. 2015; Srivastava et al. 2017). Hairy root culture is used for more than just plant-based bioactive compounds; it is also used for biotransformation, phytoremediation, molecular pharming and transgenic plant development (Georgiev et al. 2012). To date more than 50 families representing over 150 genera of angiosperms have been exploited for their hairy roots and associated secondary metabolite production (Mehrotra et al. 2015). Metabolites of the class alkaloids, phenolics, carbonyl compounds, coumarins and terpenes and several subclasses are currently exploited for scale-up production using this technique (Table 4.1). Some examples of specialised metabolites already produced are rosmarinic acid (Kwon et al. 2020), azadirachtin (Thakore and Srivastava 2017), camptothecin (Shi et al. 2020a, b; Xu et al. 2020), betalain (Pavlov et al. 2005), umckalin (Yousefian et al. 2020), ginsenoside (Zhang et al. 2015), taxol (Exposito et al. 2010), nicotine (Zhao et al. 2013), plumbagin (Beigmohamadi et al. 2020), morphine (Hashemi and Naghavi 2016), daidzin (Zaheer et al. 2016), withanolide A (Shajahan et al. 2017) and catharanthine (Hu and Du 2006). Available literatures indicate that *A. rhizogenes* strains are most suitable for large-scale in vitro production of nutraceutical compounds as compared to *A. tumefaciens* (Table 4.1).

4.3 Challenges Faced in in Vitro Culture Technique

The increasing demands of nutraceutical compounds, mainly due to increasing consumer preferences towards natural food additives and nutraceutical and healthy foods, is one of the challenges to fulfil the demand. Further, nutraceutical compounds are also popular due to their safety and potential applications in medicine, cosmetics and food products, and therefore the wide use of new technologies is the need of the hour. For instance, the extraction of nutraceutical compounds through in vitro culture is more preferable than direct extraction because it is sustainable and effective ways of commercial and large-scale production. However, various challenges and limitations of these technologies are noticed. These include:

- The basic approaches that should be defined with reference to individual plant species in the first stage of the in vitro cultivation process include improving medium type and chemical ingredients, as well as numerous environmental conditions.
- For large-scale production, parameters such as inoculum density, elicitation/PGR type and concentration, agitation/aeration, nutrition feeding precursor feeding and so on should be determined.
- The cost-effectiveness for industrial production of plant nutraceutical compounds is a big concern, and adopting some of the recommendations that can minimize

the cost, such as (1) thorough knowledge of the regulatory mechanisms that influence the commencement and flow of the pathways during in vitro production of plant nutraceutical compounds, (2) using metabolic engineering approaches to manipulate regulatory enzymes to increase cellular activities, (3) designing of low-cost techniques with minimum resources, and (4) the advancement of bioprocessing technology for continuous metabolite accumulation and release.

4.4 Conclusion

Nutraceutical compounds have been extracted from diverse medicinal and wild edible plant species across the globe for medical purposes since the ancient time. The demands are gradually escalating which can be fulfilled by using the advanced techniques such as in vitro culture. The current state of knowledge on in vitro synthesis, problems and prospects for these nutraceutical chemicals could aid in the development of a novel method for investigating high-value medicinal plant species. The scientific community will benefit greatly from comprehensive information on these compounds in order to conduct intensive research in this field.

Acknowledgements The authors are thankful to the Director of the Institute for providing facilities during the work. Partial financial support for thisresearch work provided by in-house project 4 (GBPNIHE) is greatly acknowledged.

References

Ali M, Abbasi BH, Ali GS (2015) Elicitation of antioxidant secondary metabolites with jasmonates and gibberellic acid in cell suspension cultures of *Artemisia absinthium* L. Plant Cell Tissue Organ Cult 120(3):1099–1106

Al-Khayri JM, Ibraheem Y (2014) In vitro selection of abiotic stress tolerant date palm (*Phoenix dactylifera* L.): a review. Emirates J Food Agric 26(11):921–933

Al-Khayri JM, Naik PM (2020) Elicitor-induced production of viomass and pharmaceutical phenolic compounds in cell suspension culture of date palm (*Phoenix dactylifera L.*). Mol 25(20):4669

Amani S, Mohebodini M, Khademvatan S, Jafari M (2020) *Agrobacterium rhizogenes* mediated transformation of *Ficus carica* L. for the efficient production of secondary metabolites. J Sci Food Agric 100:2185–2197. https://doi.org/10.1002/jsfa.10243

Ashokhan S, Othman R, Abd Rahim MH, Karsani SA et al (2020) Effect of plant growth regulators on coloured callus formation and accumulation of azadirachtin, an essential biopesticide in *Azadirachta indica*. Plan Theory 9(3):e352

Baldi A, Dixit VK (2008) Yield enhancement strategies for artemisinin production by suspension cultures of *Artemisia annua*. Bioresour Technol 99(11):4609–4614

Barba-Espín G, Chen ST, Agnolet S, Hegelund JN et al (2020) Ethephon-induced changes in antioxidants and phenolic compounds in anthocyanins producing black carrot hairy root cultures. J Exp Bot 71(22):7030–7045

Beigmohamadi M, Movafeghi A, Jafari S, Sharafi A (2020) Potential of the genetically transformed root cultures of *Plumbago europaea* for biomass and plumbagin production. Biotechnol Prog 36:e2905. https://doi.org/10.1002/btpr.2905

Betsui F, Tanaka-Nishikawa N, Shimomura K (2004) Anthocyanin production in adventitious root cultures of *Raphanus sativus* L. cv. Peking Koushin. Plant Biotechnol 21(5):387–391

Biswas T, Singh M, Mathur AK, Mathur A (2015) A dual purpose cell line of an Indian congener of ginseng-*Panax sikkimensis* with distinct ginsenoside and anthocyanins production profiles. Protoplasma 252(2):697–703

Blando F, Scardino AP, De Bellis L, Nicoletti I et al (2005) Characterization of in vitro anthocyanin-producing sour cherry (*Prunus cerasus* L.) callus cultures. Food Res Int 38(8–9): 937–942

Boobalan S, Kamalanathan D (2020) Tailoring enhanced production of aervine in Aerva lanata (L.) Juss. Ex Schult by Agrobacterium rhizogenes- mediated hairy root cultures. Ind Crop Prod 155, 112814

Bonfill M, Mangas S, Moyano E, Cusido RM et al (2010) Production of centellosides and phytosterols in cell suspension cultures of *Centella asiatica*. Plant Cell Tissue Organ Cult 104:61–67

Bota C, Deliu C (2011) The effect of copper sulphate on the production of flavonoids in *Digitalis lanata* cell cultures. Farmacia 59(1):113–118

Carrier DJ, Cosentino G, Neufeld R, Rho D, Weber M, Archambault J (1990) Nutritional and hormonal requirements of Ginkgo biloba embryo-derived callus and suspension cell culture. Plant Cell Rep 8(11):635–638

Chaichana N, Dheeranupattana S, Jatisatienr A, Wangkarn S, Pyne SG, Mungkornnasawakul P, Sangthong P, Sastraruji T (2012) Response of Stemona alkaloid production in Stemona sp. to chitosan and yeast extract elicitors. Curr Res J Biol Sci 4(4):449–454

Chan LK, Koay SS, Boey PL, Bhatt A (2010) Effects of abiotic stress on biomass and anthocyanin production in cell cultures of *Melastoma malabathricum*. Biol Res 43(1):127–135

Chung IM, Rekha K, Rajakumar G, Thiruvengadam M (2016) Production of glucosinolates, phenolic compounds and associated gene expression profiles of hairy root cultures in turnip (*Brassica rapa* ssp. rapa). 3 Biotech 6:175. https://doi.org/10.1007/s13205-016-0492-9

Cusidó RM, Palazón J, Navia-Osorio A, Mallol A et al (1999) Production of Taxol® and baccatin III by a selected *Taxus baccata* callus line and its derived cell suspension culture. Plant Sci 146(2):101–107. https://doi.org/10.1016/s0168-9452(99)00093-x

De Oliveira AJB, da Silva Machado MDFP (2003) Alkaloid production by callous tissue cultures of Cereus peruvianus (Cactaceae). Appl Biochem Biotechnol 104(2):149–155

Dehghan E, Häkkinen ST, Oksman-Caldentey KM et al (2012) Production of tropane alkaloids in diploid and tetraploid plants and in vitro hairy root cultures of Egyptian henbane *(Hyoscyamus muticus L.)*. Plant Cell Tissue Organ Cult 110(1):35–44

Deikman J, Hammer PE (1995) Induction of anthocyanin accumulation by cytokinins in *Arabidopsis thaliana*. Plant Physiol 108(1):47–57

Demirci T, Akçay UC, Baydar NG (2020) Effects of 24-epibrassinolide and l-phenylalanine on growth and caffeic acid derivative production in hairy root culture of *Echinacea purpurea* L. Moench. Acta Physiol Plant 42:66. https://doi.org/10.1007/s11738-020-03055-7

Do C, Cormier F (1990) Accumulation of anthocyanins enhanced by a high osmotic potential in grape *(Vitis vinifera* L.*)* cell suspensions. Plant Cell Rep 9(3):143–146. https://doi.org/10.1007/bf00232091

Dornenburg H, Knorr D (1995) Strategies for improvement of secondary metabolite production in plant cell cultures. Enzym Microb Technol 17:674–684

Exposito O, Syklowska-Baranek K, Moyano E, Onrubia M, Bonfill M, Palazon J, Cusido RM (2010) Metabolic responses of Taxus media transformed cell cultures to the addition of methyl jasmonate. Biotechnol Prog 26(4):1145–1153

Ferdausi A, Chang X, Hall A et al (2020) Galanthamine production in tissue culture and metabolomic study on amaryllidaceae alkaloids in *Narcissus pseudonarcissus* cv. Carlton. Ind Crops Prod 144:112058

Fulzele DP, Satdive RK, Pol BB (2001) Growth and production of camptothecin by cell suspension cultures of *Nothapodytes foetida*. Planta Med 67(2):150–152

Gabr AMM, Ghareeb H, Shabrawi HME, Smetanska I et al (2016) Enhancement of silymarin and phenolic compound accumulation in tissue culture of Milk thistle using elicitor feeding and hairy root cultures. J Genet Eng Biotechnol 14(2):327–333. https://doi.org/10.1016/j.jgeb.2016.10.003

Gadzovska S, Maury S, Delaunay A, Spasenoski M, Hagège D, Courtois D, Joseph C (2013) The influence of salicylic acid elicitation of shoots, callus, and cell suspension cultures on production of naphtodianthrones and phenylpropanoids in Hypericum perforatum L. Plant Cell Tissue Org Cult 113(1):25–39

Gagné S, Cluzet S, Mérillon JM, Gény L (2011) ABA initiates anthocyanin production in grape cell cultures. J Plant Growth Regul 30(1):1–10

Gammoudi N, Zerria K, Nagaz K, Ferchichi A (2019) Enhancement of capsaicinoids in vitro production by abiotic elicitors in placenta-derived callus of Capsicum annuum L. Tunisian var.'Baklouti Medenine. Biol 74(6):725–732

Georgiev VG, Weber J, Kneschke EM, Denev PN et al (2010) Antioxidant activity and phenolic content of betalain extracts from intact plants and hairy root cultures of the red beetroot *Beta vulgaris* cv. Detroit dark red. Plant Foods Hum Nutr 65:105–111. https://doi.org/10.1007/s11130-010-0156-6

Georgiev MI, Agostini E, Ludwig-Müller J, Xu J (2012) Genetically transformed roots: from plant disease to biotechnological resource. Trends Biotechnol 30(10):528–537

Giri L, Dhyani P, Rawat S, Bhatt ID et al (2012) In vitro production of phenolic compounds and antioxidant activity in callus suspension cultures of *Habenaria edgeworthii*: a rare Himalayan medicinal orchid. Ind Crop Prod 39:1–6

Giri CC, Zaheer M (2016) Chemical elicitors versus secondary metabolite production in vitro using plant cell, tissue and organ cultures: recent trends and a sky eye view appraisal. Plant Cell Tissue Org Cult 126(1):1–18

Goleniowski M, Tirppi VS (1999) Effect of growth medium composition on psilostachyinoides and altamisine production. Plant Cell Tissue Organ Cult 56:215–218

Halder M, Roychowdhury D, Jha S (2018) A critical review on biotechnological interventions for production and yield enhancement of secondary metabolites in hairy root cultures. Hairy Roots 2018:21–44

Hansen G, Wright MS (1999) Recent advances in the transformation of plants. Trends Plant Sci 4(6):226–231

Hashemi SM, Naghavi MR (2016) Production and gene expression of morphinan alkaloids in hairy root culture of Papaver orientale L. using abiotic elicitors. Plant Cell Tiss Org Cult 125:31–41

Hegazy AK, Mohamed AA, Ali SI, Saker MM (2010) Enhancement of callus induction and cucurbitacin content in *Citrullus colocynthis* L. (Schrad) using plant growth regulators. J Ala Acad Sci 81(1):23–36

Hiraoka N, Bhatt ID, Sakurai Y et al (2004) Alkaloid production by somatic embryo cultures of *Corydalis ambigua*. Plant Biotechnol 21(5):361–366

Hu ZB, Du M (2006) Hairy root and its application in plant genetic engineering. J Integr Plant Biol 48(2):121–127. https://doi.org/10.1111/j.1744-7909.2006.00121.x

Ionkova I, Karting T, Alfermann W (1997) Cycloartane saponin production from hairy root cultures of *Astragalus mongholicus*. Phytochemistry 45:1597–1600

Ivanov I, Georgiev V, Georgiev M et al (2010) Galanthamine and related alkaloids production by *Leucojum aestivum* L. shoot culture using a temporary immersion technology. Appl Biochem Biotechnol 163(2):268–277

Jiao J, Gai QJ, Wang X, Liu J et al (2020) Effective production of phenolic compounds with health benefit in pigeon pea [*Cajanus cajan* (L.) Millsp.] hairy root cultures. J Agric Food Chem 68(31):8350–8361. https://doi.org/10.1021/acs.jafc.0c02600

Kapoor S, Raghuvanshi R, Bhardwaj P, Sood H et al (2018) Influence of light quality on growth, secondary metabolites production and antioxidant activity in callus culture of *Rhodiola imbricata* Edgew. J Photochem Photobiol B 183:258–265

Karalija E, Zeljković SĆ, Parić A (2020) Harvest time–related changes in biomass, phenolics and antioxidant potential in Knautia sarajevensis shoot cultures after elicitation with salicylic acid and yeast. In Vitro Cell Dev Biol Plant 56:177–183. https://doi.org/10.1007/s11627-019-10028-0

Kolarevic MK, Dmitrovic S, Filipovic B, Peric M et al (2015) Influence of sodium salicylate on rosmarinic acid, carnosol and carnosic acid accumulation by *Salvia officinalis* L. shoots grown in vitro. Biotechnol Lett 37(8):1693–1701

Konczak-Islam I, Yoshinaga M, Nakatani M, Terahara N et al (2000) Establishment and characteristics of an anthocyanin-producing cell line from sweet potato storage root. Plant Cell Rep 19(5):472–477

Koul A, Mallubhotla S (2020) Elicitation and enhancement of bacoside production using suspension cultures of Bacopa monnieri (L.) Wettst. 3 Biotech 10:1–4

Kusakari K, Yokoyama M, Inomata S (2000) Enhanced production of saikosaponins by root culture of Bupleurum falcatum L. using two-step control of sugar concentration. Plant Cell Rep 19(11): 1115–1120

Kwon DY, Kim YB, Kim JK, Park SU (2020) Production of rosmarinic acid and correlated gene expression in hairy root cultures of green and purple basil (*Ocimum basilicum* L.). Prep Biochem Biotechnol 51(1):35–43. https://doi.org/10.1080/10826068.2020.1789990

Liu Z, Shi MZ, Xie DY (2014) Regulation of anthocyanin biosynthesis in Arabidopsis thaliana red pap1-D cells metabolically programmed by auxins. Planta 239(4):765–781

Lodha D, Patel AK, Rai MK, Shekhawat NS (2014) In vitro plantlet regeneration and assessment of alkaloid contents from callus cultures of Ephedra foliata (Unth phog), a source of anti-asthmatic drugs. Acta Physiol Plant 36(11):3071–3079

Mantell SH, Smith H (1984) Culture factors that influence secondary metabolite accumulation in plant cell and tissue cultures. In: Mantell SH, Smith H (eds) Plant biotechnology. Cambridge University Press, Cambridge, pp 75–108

Marsh Z, Yang T, Nopo-Olazabal L, Wu S et al (2014) Effect of light, methyl jasmonate and cyclodextrin on production of phenolic compounds in hairy root cultures of *Scutellaria lateriflora*. Phytochemistry 107:50–60. https://doi.org/10.1016/j.phytochem.2014.08.020

Mathur A, Mathur AK, Gangwar A, Yadav S et al (2010) Anthocyanin production in a callus line of *Panax sikkimensis* ban. In Vitro Cell Dev Biol Plant 46(1):13–21

Mehrotra S, Srivastava V, Rahman LU, Kukreja AK (2015) Hairy root biotechnology—indicative timeline to understand missing links and future outlook. Protoplasma 252:1189–1201

Mendozaa D, Cuaspuda O, Ariasa JP, Ruizc O et al (2018) Effect of salicylic acid and methyl jasmonate in the production of phenolic compounds in plant cell suspension cultures of *Thevetia peruviana*. Biotechnol Rep 19:e00273

Miguel-Chávez S, Soto-Hernández M, Ramos-Valdivia AC, Kite G, Martínez-Vázquez M, García M, Terrazas TS (2003) Production of alkaloids by in vitro culture of Erythrina americana Miller. Biotechnol Lett 25(13):1055–1059

Mohammed NR, Dalar A, Ozdemir FA, Turker M (2019) In vitro propagation and secondary metabolites investigation of *Hypericum perforatum* L. Fresenius Environ Bull 28(7):5569–5576

Nazif NM, Rady MR, El-Nasr MS (2000) Stimulation of anthraquinone production in suspension cultures of Cassia acutifolia by salt stress. Fitoterapia 71(1):34–40

Orihara Y, Yang JW, Komiya N, Koge K, Yoshikawa T (2002) Abietane diterpenoids from suspension cultured cells of Torreya nucifera var. radicans. Phytochemistry 59(4):385–389

Pavlov A, Georgiev V, Ilieva M (2005) Betalain biosynthesis by red beet (Beta vulgaris L.) hairy root culture. Process Biochem 40:1531–1533

Praveen N, Murthy HN (2012) Synthesis of withanolide a depends on carbon source and medium pH in hairy root cultures of *Withania somnifera*. Ind Crop Prod 35:241–243

Pullman GS, Johnson S, Tassel SV, Zhang Y (2005) Somatic embryogenesis in loblolly pine (Pinus taeda) and Douglas fir (Pseudotsuga menziesii): improving culture initiation and growth with MES pH buffer, biotin, and folic acid. Plant Cell Tissue Org Cult 80(1):91–103

Qu J, Zhang W, Yu X (2011) A combination of elicitation and precursor feeding leads to increased anthocyanin synthesis in cell suspension cultures of *Vitis vinifera*. Plant Cell Tissue Organ Cult 107(2):261–269

Ram M, Prasad KV, Kaur C, Singh SK et al (2011) Induction of anthocyanin pigments in callus cultures of *Rosa hybrida* L. in response to sucrose and ammonical nitrogen levels. Plant Cell Tissue Organ Cult 104(2):171–179

Ram M, Prasad KV, Singh SK et al (2013) Influence of salicylic acid and methyl jasmonate elicitation on anthocyanin production in callus cultures of *Rosa hybrida* L. Plant Cell Tissue Organ Cult 113(3):459–467

Ramírez-Mosqueda MA, Iglesias-Andreu LG (2017) Vanilla (Vanilla planifolia Jacks.) cell suspension cultures: establishment, characterization, and applications. 3. Biotech 7(4):242. https://doi.org/10.1007/s13205-017-0871-x

Ramakrishna A, Ravishankar GA (2011) Influence of abiotic stress singles on secondary metabolites in plants. Plant Signal Behav 6:1720–1731

Rao SR, Ravishankar GA (2002) Plant cell cultures: chemical factories of secondary metabolites. Biotechnol Adv 20(2):101–153

Rao KV, Venkanna N, Narasu ML (1998) *Agrobacterium rhizogenes* mediated transformation of *Artemisia annua*. J Sci Ind Res 57:773–776

Rao S, Usha K, Arjun (2015) Production of secondary metabolites from callus cultures of *Centella asiatica* (L.) Urban. Ann Phytomed 4(1):4–78

Ray S, Jha S (2001) Production of withaferin A in shoot cultures of Withania somnifera. Planta Med 67(5):432–436

Saad KR, Parvatam G, Shetty NP (2018) Medium composition potentially regulates the anthocyanin production from suspension culture of *Daucus carota*. 3 Biotech 8(3):134

Saito K (1993) Genetic engineering in tissue culture of medicinal plants. Plant Tissue Cult Lett 10: 1–8

Saito K, Sudo H, YamazakiM K-NM et al (2001) Feasible production of camptothecin by hairy root culture of *Ophiorrhiza pumila*. Plant Cell Rep 20:267–271

Sasaki K, Udagawa A, Ishimaru H, Hayashi T et al (1998) High forskolin production in hairy roots of *coleus forskohlii*. Plant Cell Rep 17(6):457–459

Schiozer AL, Barata LES (2007) Stability of natural pigments and dyes. Rev Fitos 3:6–23

Shajahan A, Thilip C, Faizal K, Mehaboob VM, Raja P, Aslam A, Kathiravan K (2017) An efficient hairy root system for withanolide production in Withania somnifera (L.) Dunal. In: S. Malik (ed.), Production of plant derived natural compounds through hairy root culture, pp 133–143. https://doi.org/10.1007/978-3-319-69769-7_7

Sharada M, Ahuja A, Suri KA, Vij SP, Khajuria RK, Verma V, Kumar A (2007) Withanolide production by in vitro cultures of Withania somnifera and its association with differentiation. Biologia Plantarum 51(1):161–164

Sheludko YV (2010) Recent advances in plant biotechnology and genetic engineering for production of secondary metabolites. Cytol Genet 44(1):52–60

Shi MZ, Xie DY (2010) Features of anthocyanin biosynthesis in pap1-D and wild-type *Arabidopsis thaliana* plants grown in different light intensity and culture media conditions. Planta 231(6): 1385–1400

Shi M, Gong H, Cui L, Wang Q et al (2020a) Targeted metabolic engineering of committed steps improves anti-cancer drug camptothecin production in *Ophiorrhiza pumila* hairy roots. Ind Crop Prod 148:112277. https://doi.org/10.1016/j.indcrop.2020.112277

Shi M, Wang Y, Wang X, Deng C et al (2020b) Simultaneous promotion of tanshinone and phenolic acid biosynthesis in *salvia miltiorrhiza* hairy roots by overexpressing Arabidopsis MYC2. Ind Crop Prod 155:112826. https://doi.org/10.1016/j.indcrop.2020.112826

Simões C, Bizarri CHB, da Silva CL et al (2009) Anthocyanin production in callus cultures of *Cleome rosea*: modulation by culture conditions and characterization of pigments by means of HPLC-DAD/ESIMS. Plant Physiol Biochem 47(10):895–903

Slater A, Scott N, Fowler M (2008) Plant biotechnology: the genetic manipulation of plants. Oxford University Press, pp 35–53

Srivastava V, Mehrotra S, Verma PK (2017) Biotechnological interventions for production of therapeutic secondary metabolites using hairy root cultures of medicinal plants. In: Dubey SK, Pandey A, Sangwan RS (eds) Current developments in biotechnology and bioengineering. Elsevier, pp 259–282. https://doi.org/10.1016/B978-0-444-63661-4.00012-8

Suan-See K, Bhatt A, Keng CL (2011) Effect of sucrose and methyl jasmonate on biomass and anthocyanin production in cell suspension culture of *Melastoma malabathricum* (Melastomaceae). Rev Biol Trop 59(2):597–606

Sudha G, Ravishankar GA (2003) Elicitation of anthocyanin production in callus cultures of *Daucus carota* and the involvement of methyl jasmonate and salicylic acid. Acta Physiol Plant 25(3):249–256

Sujanya S, Devi BP, Sai I (2008) In vitro production of azadirachtin from cell suspension cultures of *Azadirachta indica*. J Biosci 33(1):113–120

Sutini W, Augustien N, Purwanto DA et al (2019) The production of cinnamic acid secondary metabolites through in vitro culture of callus *Camellia sinensis* L with the elicitor of cobalt metal ions. AIP Conf Proc 2120(1):030028

Taniguchi S, Imayoshi Y, Kobayashi E, Takamatsu Y, Ito H, Hatano T, Sakagami H, Tokuda H, Nishino H, Sugita D, Shimura S (2002) Production of bioactive triterpenes by Eriobotrya japonica calli. Phytochemistry 59(3):315–323

Thanh NT, Murthy HN, Yu KW, Hahn EJ et al (2005) Methyl jasmonate elicitation enhanced synthesis of ginsenoside by cell suspension cultures of *Panax ginseng* in 5-l balloon type bubble bioreactors. Appl Microbiol Biotechnol 67:197–201

Thakore D, Srivastava AK (2017) Production of biopesticide azadirachtin using plant cell and hairy root cultures. Eng Life Sci 17:997–1005

Thiruvengadam M, Praveen N, Kim EH, Kim SH et al (2014) Production of anthraquinones, phenolic compounds and biological activities from hairy root cultures of *Polygonum multiflorum* Thunb. Protoplasma 251:555–566. https://doi.org/10.1007/s00709-013-0554-3

Thwe A, Valan Arasu M, Li X, Park CH et al (2016) Effect of different *agrobacterium rhizogenes* strains on hairy root induction and phenylpropanoid biosynthesis in tartary buckwheat (*Fagopyrum tataricum* Gaertn). Front Microbiol 7:318. https://doi.org/10.3389/fmicb.2016.00318

Tonelli M, Pellegrini E, D'Angiolillo F et al (2015) Ozone-elicited secondary metabolites in shoot cultures of *Melissa officinalis* L. Plant Cell Tissue Organ Cult 120(2):617–629

Van Etten H, Temporini E, Wasmann C (2001) Phytoalexin (and phytoanticipin) tolerance as a virulence trait: why is it not required by all pathogens? Physiol Mol Plant Pathol 59(2):83–93

Verma AK, Singh RR, Singh S (2012) Improved alkaloid content in callus culture of Catharanthus roseus. Bot Serbica 36(2):123–30

Wang Y, Weathers PJ (2007) Sugars proportionately affect artemisinin production. Plant Cell Rep 26:1073–1081

Wang J, Qian J, Yao L, Lu Y (2015) Enhanced production of flavonoids by methyl jasmonate elicitation in cell suspension culture of *Hypericum perforatum*. Bioresour Bioprocess 2(1):5

Wink M (2003) Evolution of secondary metabolites from an ecological and molecular phylogenetic perspective. Phytochemistry 64:3–19

Wojciechowska M, Owczarek A, Kiss AK, Grąbkowska R et al (2020) Establishment of hairy root cultures of *salvia bulleyana* Diels for production of polyphenolic compounds. J Biotechnol 318: 10–19

Xu M, Wu C, Zhao L, Wang Y, Wang C, Zhou W, Ming Y, Kai G (2020) WRKY transcription factor OpWRKY1 acts as a negative regulator of camptothecin biosynthesis in Ophiorrhiza pumila hairy roots. Plant Cell Tissue Organ Cult 142:69–78

Yousefian Z, Golkar P, Mirjalili MH (2020) Production Enhancement of Medicinally Active Coumarin and Phenolic Compounds in Hairy Root Cultures of Pelargonium sidoides: The Effect of Elicitation and Sucrose. J Plant Growth Regul 40(2):628–641

Yuan H, Ma Q, Ye L, Piao G (2016) The traditional medicine and modern medicine from natural products. Molecules 21(5):559. https://doi.org/10.3390/molecules21050559

Zaheer M, Reddy VD, Giri CC (2016) Enhanced daidzin production from jasmonic and acetyl salicylic acid elicited hairy root cultures of Psoralea corylifolia L. (Fabaceae). Nat Prod Res 30(13):1542–1547

Zayed R, Wink M (2004) Induction of tropane alkaloid formation in transformed root cultures of *Brugmansia suaveolens* (Solanaceae). Z Naturforsch C 59(11–12):863–867

Zhang R, Zhang BL, Li GC, Xie T, Hu T, Luo ZY (2015) Enhancement of ginsenoside Rg1 in Panax ginseng hairy root by overexpressing the α-l-rhamnosidase gene from Bifidobacterium breve. Biol Lett 37(10):2091–2096

Zhang Y, Jiang L, Li Y et al (2018) Effect of red and blue light on anthocyanin accumulation and differential gene expression in strawberry (Fragaria× ananassa). Molecules 23(4):820

Zhang W, Curtin C, Kikuchi M, Franco C (2002) Integration of jasmonic acid and light irradiation for enhancement of anthocyanin biosynthesis in Vitis vinifera suspension cultures. Plant Sci 162(3):459–468

Zhao B, Agblevor FA, KC R, Jelesko JG (2013) Enhanced production of the alkaloid nicotine in hairy root cultures of Nicotiana tabacum L. Plant Cell Tissue Organ Cult 113:121–129

Part II

In Vitro Production of Nutraceutical Compounds

In Vitro Production of Phenolic Compound 5

Lalit Giri, Laxman Singh, and Indra D. Bhatt

Abstract

The in vitro production of phenolic compounds in different plant culture techniques offers an attractive substitute rather than separation from plant materials. The variability and instability of phenolic compound composition hinders the standardization and quality assurance of crude materials. Therefore, in vitro studies offer several allied advantages, such as little or no seasonal variability, the use of high metabolite yielding cell lines for scaling-up production, and reduced production time. Despite the fact that various strategies have been used to increase phenolic compound content, elicitation, along with hairy root culture, is one of the most feasible strategies currently in use. The present work intends to provide an illustrative understanding of the factors having conspicuous impacts on phenolic compound production in in vitro culture conditions. In addition, emphasis was placed on highlighting important phenolic compounds with broad implications for human health and contrasting metabolic pathways involved. In conclusion, we further postulate in-depth research on specific compound/compounds and their optimal extraction techniques rather than studies encircling the phenolic extract as a whole and their corresponding bioactivities. Such a strategy finds wide acceptability and application, considering the needs of the pharmaceutical industry or for encapsulation in functional food.

L. Giri (✉)
G. B. Pant National Institute of Himalayan Environment, Ladakh Regional Center, Leh, Ladakh, India

L. Singh · I. D. Bhatt
G. B. Pant National Institute of Himalayan Environment, Kosi-Katarmal, Almora, Uttarakhand, India

T. Belwal et al. (eds.), *Nutraceuticals Production from Plant Cell Factory*,
https://doi.org/10.1007/978-981-16-8858-4_5

Keywords

Hairy root culture · In vitro culture · Phenolic compounds · Secondary metabolite

5.1 Introduction

Plant secondary metabolites are a group of organic compounds produced to facilitate interaction with the biotic environment and the establishment of a defense mechanism against various stresses (Wink 1988; Verpoorte et al. 2000). Secondary metabolites have a far lower occurrence in plants than primary metabolites; a compound may only be present in a few species or perhaps a few variants within a species. The synthesis of these compounds is often minimal (less than 1% DW) and is highly dependent on plant species, physiological and developmental stages, and plant stage (Namdeo 2007). Secondary metabolites are frequently found in specific cells or organs of plants. Secondary metabolites like phenolics, alkaloids, terpenes, etc. are classified by their biosynthetic origin, have a variety of biological functions, and are employed as medications, agrochemicals, tastes, perfumes, colors, biopesticides, and food additives. In traditional medicine, several plants with large amounts of certain secondary metabolites, such as alkaloids or phenolics, are utilized. About 75% of the world's population, particularly in underdeveloped nations, relies on herbal medicine to prevent and cure ailments, while medicinal plants account for about 25% of synthetic pharmaceuticals (Raskin et al. 2002; Alfermann and Petersen 1995).

Among the secondary metabolites of plants, phenolic compounds are an essential components of the human diet due to their high antioxidant activity (Vamanu and Nita 2013) and ability to reduce oxidative stress and thereby prevent the development of chronic diseases (Bulgakov et al. 2018) or cancer (Raskin et al. 2002). The addition of phenolic compounds into food material is an essential job for food producers due to increasing consumer demand for food supplemented with natural health-promoting ingredients. Environmental factors, political and labor instability in the producing countries, unpredictable changes in crop quality, authorities' failure to prevent crop adulteration, and losses in storage and handling are all issues in extracting various phenolic chemicals from plants (Georgiev et al. 2009). Chemical production from some plants is generally either highly difficult or economically unfeasible (Namdeo 2007). Plant in vitro cultures can be employed as natural sources of phenolic compounds in the long run. In contrast to plant tissues, plant in vitro cells may manage the concentration and individual profile of phenolic compounds, resulting in a uniform product (Murthy et al. 2014; Moreno et al. 1995).

The aim of this review article is to highlight the key advantages of adopting plant tissue culture to produce high-value bioactive compounds, with an emphasis on phenolic compounds and their bioactive properties. As a result, the production of these secondary metabolites in plant tissues, as well as the biochemical pathways involved, will be explained.

5.2 Biological Activities (Major) and Biosynthesis Pathway of Phenolic Compounds

Phenolic compounds are secondary metabolites produced by phenylpropanoid metabolization in plant shikimic acid and pentose phosphate pathway (Randhir et al. 2004). From simple phenolic molecules to highly polymerized compounds, they all comprise benzene rings with one or more hydroxyl substituents (Velderrain-Rodríguez et al. 2014) (Fig. 5.1). The commitment of glucose to the pentose phosphate pathway (PPP) and irreversible transformation of glucose-6-phosphate to ribulose-5-phosphate is the initial step in the synthesis of phenolic chemicals. Glucose-6-phosphate dehydrogenase initiates the first committed process in the conversion to ribulose-5-phosphate (G6PDH). On the one hand, converting ribulose-5-phosphate to reducing equivalents of nicotinamide adenine dinucleotide phosphate (NADPH) for cellular anabolic activities results in reducing equivalents of NADPH. PPP, on the other hand, produces erythrose-4-phosphate and phospho-enolpyruvate from glycolysis, which is then used to make phenolic compounds via the phenylpropanoid route before being directed to the shikimic acid pathway to produce phenylalanine (Vattem et al. 2005; Lin et al. 2010) (Fig. 5.2). Phenolics are the most prominent secondary metabolites in plants, and their distribution can be seen throughout the metabolic process. Polyphenols, or phenolic compounds, contain a wide variety of molecules, including simple flavonoids, phenolic acids, complex flavonoids, and pigmented anthocyanins (Babbar et al. 2014) (Fig. 5.1).

These phenolic chemicals are frequently linked to plant defensive responses. Other mechanisms involving phenolic metabolites include adding appealing chemicals to speed pollination, coloration for camouflage and protection against herbivores, and antibacterial and antifungal actions (Alasalvar et al. 2001; Acamovic and Brooker 2005; Edreva et al. 2008). Phenolic substances, especially stress-related phytochemicals, have been linked to positive effects induced by fruit and vegetable

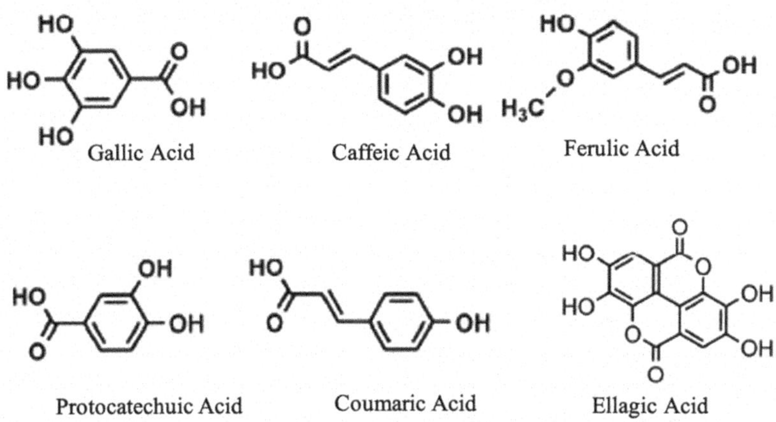

Fig. 5.1 Examples of some individual phenolic compounds produced by in vitro culture techniques

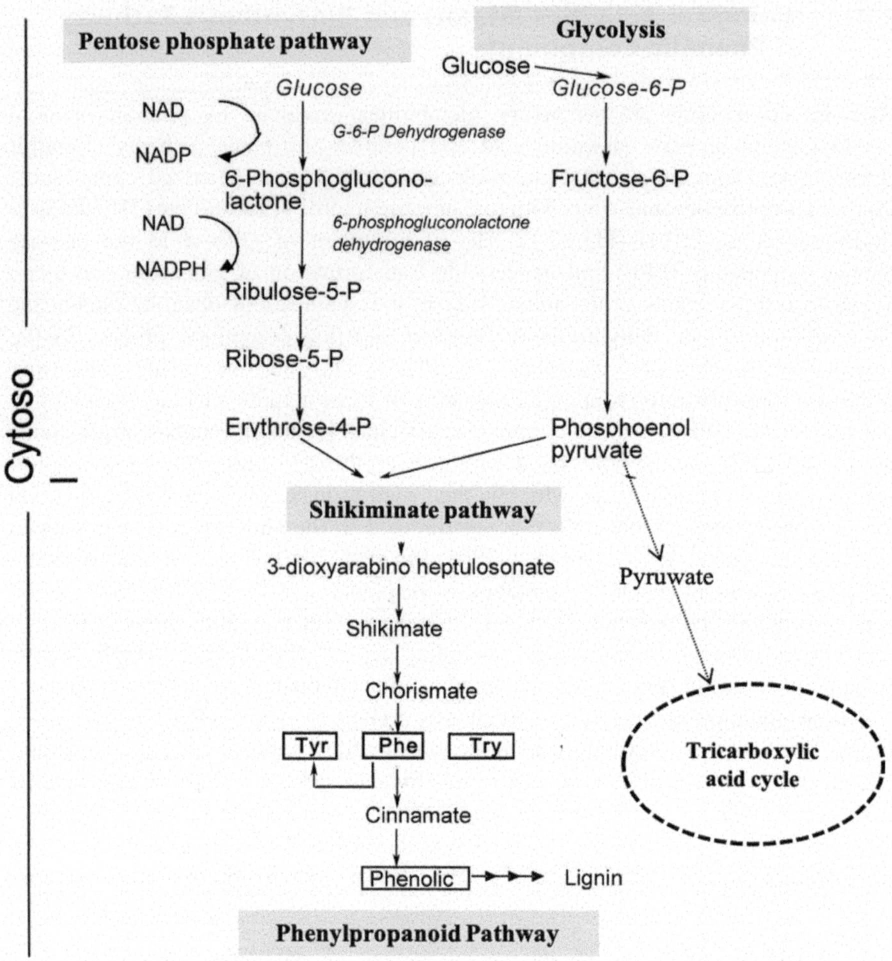

Fig. 5.2 Biosynthesis of phenol compounds in the pentose phosphate, shikimate, and phenylpropanoid pathways in plants (modified from Vattem et al. 2005; Lin et al. 2010)

diet, owing to their antioxidant activity (Heim et al. 2002). The antioxidant activity, occurrence, and possible uses of phenolic compounds in plants and agri-industrial by-products were reviewed by Balasundram et al. (2006). According to the studies, the main sources of phenolic compounds in the human diet are fruits, vegetables, and beverages. Plant polyphenols as dietary antioxidants in human health and disease could provide some protection from oxidative stress. In the prevention of carbohydrate absorption, such as diabetes, chemical compounds taken from plants, phenolic compounds, can inhibit amylase absorption (Sales et al. 2012). Anti-aging, anti-inflammatory, antioxidant, and antiproliferative properties of phenolic compounds have been demonstrated in numerous research. In addition to the aforementioned adjustments, antioxidant enzymes are important in the fight against oxidants

(Shukitt-Hale et al. 2008; Moo-Huchin et al. 2015) Polyphenols, particularly flavonoids, phenolic acids, and tannins, have the crucial virtue of suppressing glucosidase and amylase, which are key enzymes involved in the digestion of carbohydrates to glucose in the diet.

5.3 In Vitro Cultures as Source for Phenolic Compounds

Plant in vitro culture techniques facilitate scientists to employ cells, tissues, or organs from commercially important plants to grow them in vitro and/or genetically alter them to produce high-value compounds. As a result, a corresponded plant cell culture provides an attractive alternative source of secondary metabolites including phenolic compounds that can overcome the limitations of extracting useful metabolites from limited natural resources. There are various studies on the production of phenolic compounds by using in vitro culture (Table 5.1). The production of polyphenol is the main goal of the vast majority of those studies; however, maximum of those studies are also focused specifically on phenolic compounds, i.e., gallic acid, chlorogenic acid, caffeic acid, and rosmarinic acid. The in vitro production of phenolic compounds and other compounds requires a number of systematic steps, including the development of explants, the formation of calli and cell suspensions, and their induction for metabolite biosynthesis. To ensure the optimum production of phenolic compounds, many abiotic parameters (such as medium composition, growth hormones, precursors, elicitors, pH, light, and temperature) are studied. These abiotic variables or factors have an impact on the in vitro production of cell biomass and, as a result, phenolic compounds, which have been thoroughly examined in order to build present and future research strategies. Details of each factor are explained in the following sections:

5.3.1 Medium Optimization

A number of chemical and physical factors have been identified as having an impact on biomass growth and secondary metabolite synthesis in plant cell and organ cultures. The type of culture medium, the appropriate salt strength of the medium, the types and levels of carbohydrates, nitrate levels, phosphate levels, and growth regulator levels are all important components (Dörnenburg and Knorr 1995; Rao and Ravishankar 2002; Stafford et al. 1986).

The media formulations of Gamborg (B5—1968), Linsmaier and Skoog (LS—1965), Murashige and Skoog (MS—1962), and Schenk and Hildebrandt (SH—1972) have widely been used for establishing plant cell and organ cultures. Establishing in vitro culture requires the selection of an appropriate medium and concentration. Furthermore, the amounts of certain salts are important in both plant culture in vitro and phenolic chemical formation. For instance, among 1/2 and full-strength MS media tested, the 1/2 strength medium for *Habenaria edgeworthii* cell suspension cultures was found to yield greater biomass accumulation and phenolic

Table 5.1 In vitro production of phenolic compounds by plant cell and tissue cultures

Plant species	Culture conditions	Solvent type, concentration, and extraction techniques	Detection technique	Phenolic compounds	References
Agastache rugosa	MS + 1 mg/l L-phenylalanine	Methanol (80%) + acidified with 0.025% formic acid (v/v),ultrasonic bath for 15 min	UPLC-ESI-qTOF-MS	4-*O*-caffeoylquinic acid, feruloylquinic acid, ferulic acid, rosmarinic acid, rosmarinic acid methyl ester, apigenin derivatives	Zielińska et al. 2019
Allamanda cathartica	MS + 0.3 μM TDZ + 2.5 μM BA	Methanol, ultrasound-assisted extraction	UV-visible spectrophotometer, HPLC-MS	Total flavonoids (172.90 mg QE/g), quercetin (51.39 mg/g)	Khanam and Anis 2018
Aster scaber	MS + 100 μM MeJA	Methanol	UHPLC	Myricetin (2073.40 ± 4.0 μg/g), quercetin (689.50 ± 3.5 μg/g), kaempferol (255.75 ± 2 μg/g), rutin (192.25 ± 2.0 μg/g), naringenin (50.50 ± 1.0 μg/g), biochanin A (25.00 ± 1.0 μg/g), formononetin (34.25 ± 1.0 μg/g), caffeic acid (225.50 ± 2.0 μg/g), p-coumaric acid (155.00 ± 2.0 μg/g), ferulic acid (175.10 ± 1.5 μg/g), m-coumaric acid (25.00 ± 1.0 μg/g), o-coumaric acid (25.00 ± 1.0 μg/g), chlorogenic acid (211.20 ± 2.0 μg/g), gallic acid (165.75 ± 1.5 μg/g), protocatechuic	Ghimire et al. 2019

				(125.25 ± 1.5 μg/g), β-resorcylic (51.60 ± 1.0 μg/g), vanillic acid (85.50 ± 1.0 μg/g), syringic acid (175.50 ± 1.0 μg/g), gentisic acid (255.05 ± 2.0 μg/g), salicylic acid (444.25 ± 2.5 μg/g), vanillin (35.50 ± 1.0 μg/g), homogentisic (135.25 ± 2.0 μg/g), resveratrol (52.00 ± 1.0 μg/g)	
Azadirachta indica	MS + 1 mg/L picloram and 2 mg/L Kn	UAE solvent concentration (50% ethanol), temperature (35–45 °C), ultrasonication time (10–30 min)	HPLC-DAD	Azadirachtin (86.44 mg/g), mevalonic acid (33.67 mg/g), squalene (8.28 mg/g)	Farjaminezhad and Garoosi 2020
Bellis perennis L.	MSMO +0.5 mg/l TDZ + 0.5 mg/l IAA	Dichloromethane (at 55–60 °C), Soxhlet apparatus	LC-MS/MS	Vanillic acid (129.1 ± 0.05 μg/g), procyanidin C1 (3.1 ± 0.001 μg/g), p-coumaric acid (30.4 ± 0.007 μg/g), luteolin-7-O-β-D-glucoside (0.9 ± 0.001 μg/g), rutin hydrate (2.7 ± 0.002 μg/g), myricetin (4.0 ± 0.01 μg/g), quercetin (1.0 ± 0.002 μg/g), genistein (10.9 ± 0.001 μg/g), apigenin (77.9 ± 0.02 μg/g)	Karakas and Turker 2013

(continued)

Table 5.1 (continued)

Plant species	Culture conditions	Solvent type, concentration, and extraction techniques	Detection technique	Phenolic compounds	References
Cullen corylifolium	Gamborg B5 medium +5 μM BA +10 μM IBA+ 5 μm salicylic acid +50 mg/ L chitosan	Methanol Rotary shaker (48 h)	TLC, HPLC-PDA, NMR (^{1}H &^{13}C) and HR-MS	Daidzein (17159.73 μg/ g ± 8.81), psoralen (5337.83 ± 6.33), and genistein (2438.74 ± 3.93)	Singh et al. 2020
Cynara scolymus	MS + 2 mg/l NAA + 5 mg/l BAP	Ethanol (70%) + acidified with 37% HCl (pH 2.2 ± 0.1), for 1 h	UV–visible spectrophotometer, HPLC	3-*O*-Caffeoylquinic acid (0.62 g/kg), 5-*O*-caffeoylquinic acid (2.51 ± 0.1 g/kg), 3,5-*O*-dicaffeoylquinic acid (0.40 g/ kg), 1,5-*O*-dicaffeoylquinic acid (21.18 ± 1.1 g/kg), 4,5-*O*-dicaffeoylquinic acid (1.86 ± 0.1 g/kg), total caffeoylquinic acids (26.6 g/ kg), total polyphenols (28.9 g/ kg), cyanidin-3-glucoside (0.21 ± 0.01 g/kg), cyanidin malonyl glucoside (0.87 ± 0.01 g/kg), cyanidin (0.13 g/kg), total measured anthocyanins (1.21 g/kg), total anthocyanins (2.04 g/kg)	Pandino et al. 2016
Dracocephalum forrestii	MS medium + BPA 5 mg/l and IAA 0.2 mg/l	Hydromethanolic extract	UPLC-PDA-ESI-MS	Chlorogenic acid, dicaffeoylquinic acid, acacetin rhamnosyl-trihexoside, rosmarinic acid, apigenin caffeoyl-rhamnoside, apigenin *p*-coumaroyl-rhamnoside (I), acacetin acetyl-rhamnosyl-trihexoside, apigenin *p*-coumaroyl-rhamnoside (II)	Weremczuk-Jeżyna et al. 2019

Dracocephalum kotschyi	1/2 MS + cefotaxime (500 mg/L) and (IBA) (2 mg/L) + SiO2 NPs (100 mg/L)	Methanol (80%) Ultrasonic at 30 °C and 30 min	HPLC–DAD	Rosmarinic acid (1016.6 μg/g), gallic acid (10.40 μg/g), caffeic acid (40.33 μg/g), chlorogenic acid (2.58 μg/g), cinnamic acid (0.39 μg/g), coumaric acid (8.11 μg/g), rutin (1.81 μg/g), apigenin (27.35 μg/g), cirsimaritin (0.2 μg/g), isokaempferid (0.2 μg/g), penduletin (0.1 μg/g), xanthomicrol (0.21 μg/g)	Nourozi et al. 2019
Dracocephalum polychaetum Bornm	Gamborg B5 media +1.0 mg/L NAA +2.5 mg/L BAP	Methanol for 20 min with ultrasonic bath	HPLC-DAD	Gallic acid (24.45), 4-hydroxybenzoic acid (11.7), chlorogenic acid (35.1), epicatechin (143.2), syringic acid (35.7), benzoic acid (39.6), p-coumaric acid (17.41), naringin (11.8) Rosmarinic acid (68.64), rutin (9.74), quercetin (10.61), apigenin (30.92), thymol (30.45), carvacrol (4.75)	Taghizadeh et al. 2020
Eryngium alpinum L	MS + BAP (1.0 mg/l) + IAA (1.0 mg/l) + and GA3 (1.0 mg/l)	Methanol Sonication 30 min	HPLC–DAD	3,4-dihydroxyphenylacetic acid (52.15 ± 3.43 mg/100 g), chlorogenic acid (94.26 ± 1.64 mg/100 g), isochlorogenic acid (20.50 ± 0.81 mg/100 g), neochlorogenic acid (26.49 ± 0.96 mg/100 g), caffeic acid (3.22 ± 0.14 mg/100 g), caftaric acid	Kikowska et al. 2020

(continued)

Table 5.1 (continued)

Plant species	Culture conditions	Solvent type, concentration, and extraction techniques	Detection technique	Phenolic compounds	References
				(17.57 ± 0.05 mg/100 g), rosmarinic acid (179.20 ± 5.50 mg/100 g), isoquercetin (29.26 ± 1.50 mg/100 g), quercitrin (41.85 ± 1.78 mg/100 g), robinin (73.63 ± 7.21 mg/100 g)	
Erythrina lysistemon	MS + 1 mg/l or 2 mg/l Kn + 2,4D + 1 mM/l MeJA	Methanol (80%), for 10 h using an orbital shaker in the dark	UPLC-PDA, UPLC-ESI-MS	Dihydroxybenzoic acid, pentosylhexoside, N-caffeoyl aspartic acid, N-p-coumaroyl aspartic acid, N-(hydroxycinnamoyl) tyraminehexoside, N-feruloylaspartic acid, dihydroxyflavone hexoside, apigeninhexosylmalonate, Demethylmedicarpin hexosyl malonate, dihydroxyisoflavone, sandwicensin, demethylmedicarpin, erythromycin B, eryvarin D, apigenin, Voegelin A, dimethoxyisoflavone	Farag et al. 2016

Fagonia indica	MS0 + 3% (w/v) sucrose +65.50 μM chitosan	Methanol (80%), sonication (10 min)	UV/VIS-DAD spectrophotometer HPLC-PDA	TPC (13.87 ± 0.09 μg/g), TFC (1.83 ± 0.06 μg/g) Gallic acid (0.205 ± 0.001 μg/mg), caffeic acid (0.189 ± 0.003 μg/g), myricetin (0.649 ± 0.007 μg/g), catechin (1.129 ± 0.003 μg/g), kaempferol (0.862 ± 0.002 μg/g), isorhamnetin (0.524 ± 0.005 μg/g), apigenin (0.987 ± 0.009 μg/g), nahagenin (0.114 ± 0.0034 μg/g), hederagenin (0.104 ± 0.004 μg/g), ursolic acid (0.311 ± 0.003 μg/g), betulinic acid (0.433 ± 0.004 μg/g)	Khan et al. 2019
Ficus deltoidea var. *kunstleri*	MS + 2.75 g/L gelrite +2 mg/L picloram +1 mg/L Kn	Distilled water	UV–visible spectrophotometer	Flavonoid (3.3 mg RE/g DW)	Haida et al. 2019
Fragaria vesca	MS + 1.0 mg/l TDZ +0.5 mg/l IBA +0.1% pluronic F-68	Methanol at 45 °C in a water bath for 24 h	LC–ESI-MS/MS	Gallic acid monohydrate (26.2 ± 0.0003 μg/g), pyrocatechol (≤0.0005), procyanidin B1 (≤0.001), (−) epigallocatechin (0.5 ± 0.000), (+) catechin (16.0 ± 0.0000), procyanidin B2 (≤0.001), vanillic acid (111.5 ± 0.0048), caffeic acid	Yildirim and Turker 2014

(continued)

Table 5.1 (continued)

Plant species	Culture conditions	Solvent type, concentration, and extraction techniques	Detection technique	Phenolic compounds	References
				(0.6 ± 0.0000), procyanidin C1 (≤0.001), (−) epicatechin (≤0.0002), p-coumaric acid (86.2 ± 0.0000), (±) taxifolin hydrate (5.1 ± 0.006), coumarin (≤0.0005), luteolin-7-O-β-D glucoside (5.7 ± 0.0022), rutin hydrate (1.3 ± 0.0000), resveratrol (≤0.0002), myricetin (8.5 ± 0.0009), kaempferol-3-β-D-glucopyranoside (≤0.001), daidzein (≤0.0002), quercetin (14.6 ± 0.0006), genistein (≤0.0002), apigenin (4.3 ± 0.0001)	
Habenaria edgeworthii	1/2 MS + 3.0 μM BA	Methanol (80%); rotatory shaker (12 h) + sonication (10 min)	HPLC-UV–VIS detector	Gallic acid (143.63 ± 0.93 mg/ 100 g), catechin 2.98 ± 0.05 mg/ 100 g), hydroxy benzoic acid (5.50 ± 0.15 mg/100 g)	Giri et al. 2012a
Habenaria edgeworthii	1/2 MS + 0.05 μM NAA + 0.05 μM IBA + 100 μM MeJA	Methanol (80%), 12 h with continuous shaking and ultrasonicated for 10 min	UV–VIS spectra, HPLC-UV–VIS	Total phenol (10.05 ± 0.03 mg /g), hydroxybenzoic acid (20.86 ± 0.07 mg/100 g), gallic acid (5.35 ± 0.10 mg/ 100 g)	Giri et al. 2012b

Helicteres angustifolia	MS + 3.0 mg L^{-1} NAA, 0.4 mg L^{-1} ascorbic acid and 3% (w/v) sucrose	Ethanol (80%), hot reflux extraction for 2 h	UV–VIS detector, HPLC-UV–VIS detector	TPC (72.74 ± 1.13 mg/g), TFC (230.13 ± 0.83 mg/g) Caffeic acid (1.02 ± 0.02 μg/mg), catechol (1.45 ± 0.04 μg/mg), p-coumaric acid (0.62 ± 0.02 μg/mg), gallic acid (6.96 ± 0.36 μg/mg), quercetin (1.25 ± 0.03 μg/mg), rosmarinic acid (0.98 ± 0.02 μg/mg), rutin (0.98 ± 0.01 μg/mg), vanillin (0.05 ± 0.00 μg/mg)	Yang et al. 2019a, b
Hypericum undulatum	MS + sucrose (2%)	Deionized water, ultrasonic bath	HPLC–UV, LC–UV–ESI-MS	Chlorogenic acid (15.39 ± 0.71 mg/g), epicatechin (4.23 ± 0.54 mg/g), hyperoside (0.88 ± 0.03 mg/g), isoquercitrin (2.31 ± 0.24 mg/g), quercitrin (5.72 ± 0.05 mg/g) 3-O-caffeoylquinic acid, 3-p-coumaroylquinic acid, 4-O-caffeoylquinic acid, o-feruloylquinic acid, catechin, mangiferin, catechin, epicatechin, rutin, quercetin sulfate, hyperoside, isoquercitrin, luteolin hexoside, quercetin-pentoside, quercitrin, apigenin-7-O-glucoside, flavonoid aglycone	Rainha et al. 2012

(continued)

Table 5.1 (continued)

Plant species	Culture conditions	Solvent type, concentration, and extraction techniques	Detection technique	Phenolic compounds	References
Knautia arvensis	MS + 100 mm salicylic acid	Methanol (80%), centrifugation at 2000 rpm for 15 min	UV-VIS, HPLC-UV	Total phenols (16.118b ± 0.775 mg/mL), total flavonoids (15.415a ± 1.27 mg/mL), total flavanols (0.725c ± 0.015 mg/mL) Gallic acid (38.32 nmol/mL), 4-hydroxybenzoic acid (86.71 nmol/mL), chlorogenic acid (426.84 nmol/mL), vanillic acid (106.73 nmol/mL), caffeic acid (5.80 nmol/mL), syringic acid (42.70 nmol/mL), sinapic acid (16.68 nmol/mL), ferulic acid (415.70 nmol/mL), salicylic acid (443.41 nmol/mL), rosmarinic acid (1387.48 nmol/mL), apigenin (6.59 nmol/mL)	Karalija et al. 2019
Linum album	MS + 200 mg/L chitosan	Ethanol (80%)	HPLC	Cinnamic acid (14.66 ± 0.69 μg/g), coumaric acid (0.67 ± 0.08 μg/g), caffeic acid (6.23 ± 0.45 μg/g), ferulic acid (2.51 ± 0.37 μg/g), salicylic acid (4.81 ± 0.36 μg/g), total phenolic acids (64.82 ± 1.49 μg/g), total	Samari et al. 2019

				flavonoids (63.38 ± 6.65 μg/g), catechin (85.12 ± 6.00 μg/g), kaempferol (64.86 ± 2.90 μg/g), myricetin (1.71 ± 0.11 μg/g), vitexin (44.30 ± 2.20 μg/g), naringenin (1.16 ± 0.07 μg/g), diosmin (1.29 ± 0.17 μg/g), resveratrol (0.60 ± 0.09 μg/g), daidzein (23.76 ± 1.46 μg/g), apigenin (0.94 ± 0.08 μg/g), lariciresinol (21.38 ± 0.64 μg/g), podophyllotoxin (146.20 ± 2.27 μg/g), 6-methoxypodophyllotoxin (39.04 ± 0.73 mg/g)	
Lonicera japonica	MS + BA 1.5 mg L^{-1}, NAA 0.2 mg L^{-1} and 2,4D 0.1 mg L^{-1}	Ethanol (50%) and sonicated for 30 min at 40 °C using a bath ultrasonic	HPLC-PDA, UPLC-Q-TOF-MS	3,5-Di-O-caffeoylquinic acid (8.1 mg/g), 3-O-caffeoylquinic acid (7.4 mg/g), 4,5-di-O-caffeoylquinic acid (4.5 mg/g), and 3,4-di-O-caffeoylquinic acid (2.7 mg/g)	Hu et al. 2019
Lycium barbarum	MS + 0.25 mg/l TDZ	Methanol, ultrasonic water bath at 37 ± 2 °C, 30 min	LC–ESI–MS/MS	Gallic acid (28.36 ± 0.0001 μg/g), chlorogenic acid (9724.23 ± 0.2693 μg/g), caffeic acid (177.86 ± 0.0001 μg/g), rutin hydrate (37.42 ± 0.0031 μg/g), myricetin (≤0.01 μg/g), quercetin (9.77 ± 0.0002 μg/	Karakas 2020

(continued)

Table 5.1 (continued)

Plant species	Culture conditions	Solvent type, concentration, and extraction techniques	Detection technique	Phenolic compounds	References
				g), vanillic acid (142.75 ± 0.0003 μg/g), luteolin (≤0.01 μg/g), isorhamnetin (≤0.01 μg/g), apigenin (≤0.01 μg/g), total phenol (10120.39 μg/g)	
Mammillaria candida	MS medium with 13.57 μM 2,4-D + 1.25 mg L^{-1} chitosan	Ethanol (60%) and stirred for 3 h at room temperature	UV–VIS, HPLC–DAD, LC/MS-TOF	Phenolics (176 ± 1.12 μmol/g), flavonoids (106 ± 0.25 μmol/g), betalains (11.9 ± 1.03 μmol/g) Quinic acid (69.91 ± 20.72 mg/100 g), aconitic acid (60.78 ± 5.39 mg/100 g), diallyl rhamnoside (490.45 ± 36.81 mg/100 g), epicatechingallate (322.03 ± 68.48 mg/100 g), methyl-(epi) gallocatechin derivative (64.82 ± 4.50 mg/100 g), decarboxy-dehydro-neobetanin (1.183 ± 0.08 mg/100 g), bidecarboxy-dehydro-neobetanin (1.733 ± 0.08 mg/100 g)	Reyes-Martínez et al. 2019

Medicago truncatula	MSO/2 medium +2 mg/l (2, 4-D) + 0.25 mg/l (kinetin) +250 mg/l (casein hydrolysate)	Methanol (80%, ultrasonic bath for 30 min	LC/ESI/MS/MS	Daidzein, biochanin A, afrormosin, irisolidone, formononetin, medicarpine, genistein, naringenin, liquiritigenin, chrysoeriol	Staszków et al. 2011
Morinda coreia	½ MS + 1.0 mg/l IBA + 0.4 mg/ml chitosan	Ethyl alcohol (80%) at 80 °C for 2 h in water bath	UV–VIS spectrophotometer	Anthraquinones (292.038 ± 2.782 mg/g), phenolic compounds (86.8 ± 4.0 mg/g)	Kannan et al. 2020
Moringa oleifera	MS + 8.88 μM BA and 9.05 μM 2,4-D	Methanol	LC MS	Rutin, isorhamentin, quercetin, luteolin, kaempferol, apigenin, p-coumaric acid, gallic acid, cholorogenic acid	Mustafa et al. 2020
Nardostachys jatamansi	1/2 MS + 23.1 μM NAA + 4 μM BA +4.16 μMKn +6.0 μM MeJA	Methanol (80%) + rotatory shaker (24 h)	UV–VIS Spectrophotometer	Phenolics (0.097 ± 0.1), flavonoids (0.12 ± 0.1), tannin (0.0 17 ± 0.1)	Rawat et al. 2019
Ocimum basilicum	MS + 0.5 mg/L Kn +2.5 mg/L NAA +50 mg/L sorbitol	Methanol, ultrasonic-assisted bath (35 kHz, power 180 W)	RP-HPLC-PDA	Rosmarinic acid (11.78 ± 0.63 mg/g), chicoric acid (3.98 ± 0.64), rutin (7.23 ± 0.60), and isoquercetin (4.68 ± 0.30)	Açıkgöz 2020
Ocimum basilicum	MS+ 100 μM melatonin	Methanol	UV–visible spectrophotometer, HPLC-DAD	Total phenolic acid (784.6 μg / g), rosmarinic acid (754.2 ± 6.35 μg/g), caffeic acid (16.6 ± 0.32 μg/g), cinnamic acid (11.5 ± 0.14 μg/g), p-coumaric (1.5 ± 0.5 μg/g) and vanillin (0.8 ± 0.04 μg/g)	Duran et al. 2019

(continued)

Table 5.1 (continued)

Plant species	Culture conditions	Solvent type, concentration, and extraction techniques	Detection technique	Phenolic compounds	References
Origanum vulgare	½ MS + 50 μM IBA	Methanol (80%) and homogenized under ultrasonicator for 5 min	HPLC-DAD	3-Hydroxy benzoic acid (0.96 ± 0.16 mg/g),3-hydroxy cinnamic acid (0.05 ± 0.004 mg/g), catechin (2.48 ± 0.164 mg/g), chlorogenic acid (1.93 ± 0.08 mg/g), ellagic acid (0.10 ± 0.0002 mg/g), ferulic acid (0.22 ± 0.017 mg/g), gallic acid (0.16 ± 0.02 mg/g), p-coumaric acid (0.08 ± 0.026 mg/g), phloridzin (1.03 ± 0.17 mg/g), vanillic acid (0.17 ± 0.02 mg/g)	Pandey et al. 2019
Rhodiola imbricata	MS + 0.5 mg/L TDZ and 1 mg/L NAA	Methanol (70%) and sonicated for 30 min	UPLC-PDA	Salidroside (3.96 ± 0.03 mg/g), tyrosol (0.99 ± 0.00 mg/g), rosavin (1.62 ± 0.00 mg/g), rosarin (0.92 ± 0.00 mg/g), p-coumaric acid (0.88 ± 0.00 mg/g)	Rattan et al. 2020
Rosmarinus officinalis	WPM + 1 mg/L NAA	Methanol, ultrasonic sonicated at 35 kHz, 45 min with (50 °C)	UV–visible spectrophotometer, HPLC-DAD	Total phenol content (115.6 mg/g), rosmarinic acid (34.4 mg/g)	Yesil-Celiktasa et al. 2007
Rosmarinus officinalis	MS + 1.0 mg/l BAP + 2.0 mg/l NAA+ 200 μM melatonin	Methanol, homogenate was centrifuged at 4000 rpm for 5 min	HPLC-DAD	Rosmarinic acid (680 μg/g), caffeic acid yield (19 μg/g), p-Coumaric acid (0.8 μg/g)	Coskun et al. 2019

Salvia nemorosa	MS + 16 μM 2,4-D + 8 μM BA +70 gamma irradiation +100 mg/l MWCNT-COOH	Methanol sonicated for 30 min	HPLC-PDA	Rosmarinic acid (82.71 mg/g), salvianolic acid B (28.31 mg/g), cinnamic acid (23.48 mg/g)	Heydari et al. 2020
Sophora flavescens	MS +1.0 mg/L BAP + 2.0 mg/L picloram +10 μM MeJA	Methanol (100%); sonication	HPLC, NMR(^{1}H and ^{13}C)	Trifolirhizin (10.81 ± 4.35 mg/g), trifolirhizin malonate (26.16 ± 6.86), maackiain (3.69 ± 0.42)	Kim et al. 2020
Sphaeralcea angustifolia	MS + 2 μM copper +2.74 mM total nitrate	Dichloromethane: Methanol (9:1) three times by maceration (24 h for each procedure)	HPLC–DAD	Coumarins (4137.00 μg/L), sphaeralcic acid (1441.00 μg/L), and scopoletin (999.00 μg / L)	Perez-Hernandez et al. 2019
Thevetia peruviana	MS + 2 mg/L 2,4-D + 0.5 mg/L Kn, +1 g/L myoinositol +3 μM MeJA	Ethanol (50%); ultrasonic bath (40 kHz)	TLC, HPLC-DAD	Chlorogenic acid, (±), hesperetin, trans-sinapic acid, quercetin, kaempferol	Mendoza et al. 2018
Thymus lotocephalus	MS medium with 2% (w/v) sucrose and 0.7% (w/v) agar+ YE (500 mg/L),	Methanol, ultrasound bath (37 kHz) for 30 min	UV–VIS spectrometer, HPLC-PDA	TPC (118.34 ± 6.24 mg/g), TFC (54.92 ± 2.62 mg/g), rosmarinic acid (78.57 ± 0.99 mg/g)	Goncalves et al. 2019
Tilia americana var. *mexicana*	MS + 0.005 mg/l TDZ + 0.1 mg/l IBA	Methanol	HPLC-PDA	3-*O*-*β*-d-glucoside (26.43 ± 2.26 μg/g), scopoletin (10.89 ± 0.76 μg/g)	Cisneros-Torres et al. 2019
Turbinicarpus laui	MS medium with 5.71 μM IAA + 1.25 mg L^{-1} chitosan	Ethanol 60% and stirred for 3 h at room temperature	UV–VIS, HPLC–DAD, LC/MS-TOF	Phenolics (97.1 ± 5.02 μmol/g), flavonoids (39.7 ± 1.03), betalains (14.2 ± 0.49)	Reyes-Martínez et al. 2019

(continued)

Table 5.1 (continued)

Plant species	Culture conditions	Solvent type, concentration, and extraction techniques	Detection technique	Phenolic compounds	References
				Caffeic acid derivative (77.93 ± 46.07 mg/100 g), Di-feruloylquinic acid (3.50 ± 0.08 mg/100 g), acacetin-hexoside (9.45 ± 1.70 mg/100 g), diosmetin-glucuronide (3.23 ± 0.12 mg/100 g), hydroxy gallic acid (16.88 ± 3.77 mg/100 g), luteolin-acetyl-hexoside (4.95 ± 0.27 mg/100 g), luteolin glucuronide (4.18 ± 0.17 mg/100 g), acacetin-malonyl-hexoside (5.18 ± 0.45 mg/100 g), acacetin (3.77 ± 0.31 mg/100 g), eriodictyol (3.41 ± 0.04 mg/100 g), diosmetin (4.32 ± 0.05 mg/100 g), 17-decarboxy neobetanin (8.76 ± 4.60 mg/100 g), chrysosplentin (3.24 ± 0.03 mg/100 g)	

compound production (Giri et al. 2012a, b). Plant cell cultures are usually grown by using a single simple sugar or a combination of simple sugars such as glucose, fructose, maltose, and sucrose. In case of phenolic compounds, maximum reports indicate that sucrose was used as a main carbon source. Nitrogen concentration was found to affect biomass growth and metabolite accumulation in cell and organ suspension cultures. The plant-tissue culture media such as MS, LS, SH, and B5 contain both nitrate and ammonium as nitrogen sources. The results showed that the highest production of coumarins (4137.00 μg/L), sphaeralcic acid (1441.00 μg/L), and scopoletin (999.00 μg/L) was obtained in *Sphaeralcea angustifolia* with 2.74 mM total nitrate (Perez-Hernandez et al. 2019). Many plant cell cultures have also been used to convert precursors into products by utilizing preexisting enzyme systems. For example, the addition of L-phenylalanine in the media enhances the production of 4-*O*-caffeoylquinic acid and feruloylquinic acid in *Agastache rugosa* (Zielińska et al. 2019).

5.3.2 Plant Growth Regulators

Exogenous plant growth regulators (PGRs) are required for the growth and proliferation of biomass and metabolite accumulation in in vitro cell, adventitious root, or shoot cultures. In general, PGRs play a significant role in cell and organ growth, proliferation, and metabolite production (DiCosmo and Towers 1984). Among the PGRs, auxins, i.e., indole acetic acid (IAA), and naphthalene acetic acid (NAA) have been shown to enhance the production of hydroxybenzoic acid, gallic acid, rosmarinic acid, betalains, caffeic acid, catechol, and anthraquinones in various plant species (Yang et al. 2019a, b; Yesil-Celiktasa et al. 2007; Reyes-Martínez et al. 2019; Pandey et al. 2019; Kannan et al. 2020; Reyes-Martínez et al. 2019). In addition, 2,4-dichlorophenoxyacetic acid (2,4-D) has also been shown to exhibit a stimulatory effect on the accumulation of betalains, quinic acid, aconitic acid and isorhamentin in *Mammillaria candida* (Reyes-Martínez et al. 2019), and rutin and quercetin in *Moringa oleifera* (Mustafa et al. 2020). Among the cytokinins and auxin concentration, benzyladenine (BA) with a combination of auxin has been shown to improve the production of 3-*O*-caffeoylquinic acid and caffeoylquinic acid in *Cynara scolymus*, chlorogenic acid in *Dracocephalum forrestii* (Weremczuk-Jeżyna et al. 2019), and gallic acid, 4-hydroxybenzoic acid, and chlorogenic acid in *Dracocephalum polychaetum* (Taghizadeh et al. 2020).

5.3.3 Permeabilization

Plant secondary metabolites are normally stored in the vacuoles; thus, extracting the products into the culture medium in a way that makes the purifying step easier is ideal. Releasing secondary metabolites from the cell's vacuoles would also help to reduce product inhibition and boost productivity. Many attempts have been made to use organic solvents to permeabilize plant cell membranes in a reversible manner.

Several studies have employed organic solvents such as isopropanol, dimethylsulfoxide (DMSO), and polysaccharides (e.g., chitosan) as permeabilizing agents (Beaumont and Knorr 1987; Knorr and Teutonico 1986; Brodelius et al. 1988). In case of phenolic compounds, the chitosan is used for daidzein, psoralen, and genistein permeabilization in *Cullen corylifolium* (Singh et al. 2020). Moreover, other phenolic compounds like gallic acid, caffeic acid, myricetin, catechin, betalains, quinic acid and anthraquinones work similarly in other species like *Fagonia indica (*Khan et al. 2019), *Turbinicarpus laui* (Reyes-Martínez et al. 2019), *Morinda coreia* (Kannan et al. 2020), *Mammillaria candida* (Reyes-Martínez et al. 2019), and *Linum album* (Samari et al. 2019).

5.3.4 Elicitation

In in vitro plant culture, the elicitation method is commonly utilized to produce a variety of secondary metabolites (Dörnenburg and Knorr 1995; Akula and Ravishankar 2011). Abiotic elicitors (e.g., heavy metal salts) have been reported to induce the overproduction of various nutraceutical compounds, whereas fungal, bacterial, and yeast elicitors (e.g., polysaccharides, glycoproteins, inactivated enzymes, purified xanthan, and crudlan) have been reported to induce the overproduction of various nutraceutical compounds. Various chemical elicitors like jasmonic acid (JA), methyl jasmonate (MeJA), 2-hydroxyl ethyl jasmonate, salicylic acid (SA), acetyl salicylic acid (ASA), trifluoromethyl salicylic acid, ethylene (ET), nitric oxide (NO), and sodium nitroprusside (SNP) are being used (Giri and Zaheer 2016). Among the elicitors, MeJA (Ghimire et al. 2019; Farag et al. 2016; Giri et al. 2012b; Rawat et al. 2019; Kim et al. 2020; Mendoza et al. 2018) and SA (Singh et al. 2020; Karalija et al. 2019) have been efficiently used for the production of phenol, flavonoid, and anthocyanin in in vitro cell or tissue culture of various plant species (Table 5.1). The exposure of elicitor concentrations, as well as duration, are also important factors that influence the production of biomass and nutraceutical compounds. For example, Giri et al. (2012a, b) studied the effect of methyl jasmonate (i.e., at concentrations of 0, 1.0, 2.0, 5.0, and 10.0 mg L^{-1}) on callus culture of *Habenaria edgeworthii*. Results indicated that an increase in concentration resulted in a decrease in both fresh and dry biomass. However, phenolic compounds were reported significantly higher than control. Further, the time duration of subculture also has effects on the production of nutraceutical compounds; for example, Wang et al. (2015) obtained higher amounts of flavonoids in 15–20 exposure days after the use of MJ elicitor in the *Hypericum perforatum*.

5.3.5 Hairy Root Culture

The lack of cell differentiation can be attributed for the low output of secondary metabolites in cell cultures. The structured culture of roots is an alternative to cell culture. The lack of cell differentiation can be attributed to the low output of secondary metabolites in cell cultures. The structured culture of roots or shoots is an alternative to cell culture (Verpoorte et al. 2000; Kolewe et al. 2008). Hairy root culture is an excellent demonstration; it is caused by infecting the roots with *Agrobacterium rhizogenes* and then transferring the Ri plasmid, which results in abundant growth of neoplastic roots that may be maintained in vitro (Ron et al. 2014). Hairy root cultures provide a number of advantages over cell suspension cultures, including rapid growth rates without the use of plant growth regulators, genetically and biochemically stable cultures, and a similar capability for secondary metabolite production (Guillon et al. 2006; Georgiev et al. 2012; Mora-Pale et al. 2014). Table 5.2 lists phenolic extracts and compounds produced by hairy root culture in various plant species. The hairy roots had a high production capacity, and withanolide A levels were 2.7 times greater than in nontransformed roots. Many medicinal plants have had natural adventitious roots produced in flask scale to bioreactor growth for the synthesis of diverse bioactive chemicals (Baque et al. 2012; Murthy et al. 2014). When compared to field-grown or greenhouse-grown plants, adventitious root cultures of *Morinda citrifolia* produced in bioreactors revealed a several-fold increase in anthraquinone concentration (Baque et al. 2012).

5.4 Extraction and Detection Techniques

Bioactive molecules from plant extract are an attractive value-added product with wide bioactivity spectrum. The distributions of plant's secondary metabolites inside specialized cells or organs are known to affect the extraction process. Selection of suitable extraction procedure/tools for the recovery of bioactive compounds mostly depends on efficiency of the extraction methods. In fact, it is a critical process used for isolation and quantification of compounds of interest at industrial level. Composition of the bioactive compounds isolated by these methods constitutes a reliable source that has a profound effect on various biological activities (Viera et al. 2017). Various factors such as solvent type (organic and inorganic), sample to solvent ratio (g/ml), extraction temperature (°C), time (min), power (W), solvent concentrations, pH, etc. are known to influence the extraction process of polyphenols (Pandey et al. 2018; Chakraborty et al. 2020). Traditionally, methods like percolation, maceration decoctions or infusion, hydrodistillation, heated reflux, and Soxhlet extraction under solid–liquid extraction technique was quite familiar that involves solvent application and leaching out of bioactive constituent (Belwal et al. 2018 and 2020; Singh et al. 2022). These techniques are categorized as conventional extraction flaws and have several drawbacks at industrial level being cost-ineffective, labor and time intense, high solvent consumption, and low yield (Wen et al. 2019). These fundamental drawbacks have triggered for more efficient, cost-effective, sustainable, and greener techniques as preferred in recent years. Considering these various sophisticated analytical techniques, i.e., microwave-assisted extraction, ultrasound/ultrasound enzyme-assisted extraction, supercritical fluids extraction, etc. are being actively utilized (Belwal et al. 2018). Owing to their fast and efficient extraction; least solvent consumption, or no solvent; and higher yield with leaves, physicochemical properties intact is the most preferred scale-up tools.

Table 5.2 In vitro production of phenolic compounds by hairy root cultures

Plant source(s)	Explant	Strain	Culture medium	Solvent type, concentration, and extraction techniques	Detection technique	Bioactive compounds (quantity)	Reference(s)
Ficus carica	Shoot	A. Rhizogenes strains 7 (A7)	1/2 MS liquid +100 μmol L – 1 MeJA elicitation	Methanol, diethyl ether +6 h of hydro-distillation with clevenger apparatus	HPLC-DAD detector GC-MS	Gallic acid (709.75 μg/g), caffeic acid (45.92 μg/g), chlorogenic acid (6499.93 μg/g), rutin (34.42 μg/g), p-coumaric acid (40.83 μg/g), rosmarinic acid (137.6 μg/g), quercetin (22.46 μg/g), cinnamic acid (20.87 μg/g), apigenin (31.58 μg/g) Thymol (p-cymen-3-ol) (27.24%), Z, 3-penten-2-one (3.89%), palmitic acid (16.25%), benzaldehyde (12.84%), aziridine (3.5%), octane (13.03%), oleic acid (5.25%), citric acid (3.89%), 2H-pyrrol-2-one, 1,5-dihydro-1-methyl (5.25%), 2,4-dichloro-6-nitrophenol (2,4-D) (3.11%), 2-nitro-4-(trifluoromethyl) phenol (4.47%)	Amani et al. 2020
Centella asiatica	Leaf and petiole	KCCM 11,879	MS + 2.5 mM squalene +400 μM MeJA	80% methanol + sonication for 2 h	HPLC-PDA	Madecassoside, asiaticoside, madecassic acid, and asiatic acid	Baek et al. 2020

Daucus carota L. ssp. *sativus* var. *atrorubens*	Taproot	*A. Rhizogenes* strain A4	½ MS + 200 mg L^{-1} ethephon	50% methanol	LC-MS-ESI	Cyanidin 3-xylosyl (glucosyl) galactoside, cyanidin 3-xylosylgalactoside, cyanidin 3-xylosyl (caffeoylglucosyl)galactoside, cyanidin 3-xylosyl (hydroxybenzoylglucosyl) galactoside, cyanidin 3-xylosyl (coumaroylglucosyl)galactoside, peonidin 3-xylosil (sinapoylglucosyl) galactoside,	Barba-Espín et al. 2020
Plumbago europaea	Stem	*A. Rhizogenes* MSU440 strain	½ MS-B5 liquid medium containing 30 g L^{-1} sucrose	Cold maceration using 100 mL chloroform for 72 h	HPLC	3.2 mg g − 1 DW plumbagin	Beigmohamadi et al. 2021
Aerva lanata	Leaf	MTCC 532 strain	1/2 MS medium + with sucrose (1.5% w/v)	Methanol	HPLC-UV	10-Hydroxycanthin-6-one (aervine) (83.40 ± 0.31 μg/ml)	Boobalan and Kamalanathan 2020
Brassica rapa ssp. *rapa*	Leaf explants	KCTC 2703	MS medium with 4% sucrose	EtOH (70%) in a boiling water bath (70 °C)	UHPLC-TQMS UPLC	Progoitrin, sinigrin, gluconapin, glucobrassicanapin, glucoallysin, glucobrassicin, 4-hydroxyglucobrassicin, 4-methoxyglucobrassicin, gluconasturtiin, neoglucobrassicin Major phenolic p-hydroxybenzoic acid (178.25 ± 1.04 μg/g), protocatechuic acid (17.27 ± 0.35), syringic acid (89.17 ± 0.62), gentisic acid	Chung et al. 2016

(continued)

Table 5.2 (continued)

Plant source(s)	Explant	Strain	Culture medium	Solvent type, concentration, and extraction techniques	Detection technique	Bioactive compounds (quantity)	Reference(s)
						(174.62 ± 1.10), p-coumaric acid (94.50 ± 1.29), ferulic acid (326.37 ± 1.10), chlorogenic acid (135.75 ± 1.04), m-coumaric acid (10.17 ± 0.62), t-cinnamic acid (15.30 ± 0.53), myricetin (115.25 ± 0.95), quercetin (110.17 ± 0.92), kaempferol (65.87 ± 0.85), catechin (451.50 ± 1.08), naringenin (12.40 ± 0.63), rutin (152.25 ± 1.04), biochanin A (7.22 ± 0.69), formononetin (2.53 ± 0.13), hesperidin (24.27 ± 0.60)	
Echinacea purpurea	Leaf	ATCC 43,057	MS medium +50 g/l, sucrose + 1.0 mg/l (24-eBL) 24-epibrassinolide	70% methanol added 0.1% phosphoric acid in an ultrasonic bath for 30 min	HPLC-DAD	Cichoric acid (24.05 ± 0.01 mg/g), chlorogenic acid (3.42 ± 0.10), caftaric acid (6.85 ± 0.06), echinacoside (4.34 ± 0.08), *p*-coumaric acid (0.49 ± 0.01)	Demirci et al. 2020
Silybum marianum	Cotyledons	A4 strain	MS medium +0.2 mg/l NAA	70% methanol + ultrasonic water bath	HPLC-PDA	Silydianin (4.33 ± 0.42 µg/g), silybine A&B (16.92 ± 1.20), total silymarin (21.26 ± 0.95)	Gabr et al. 2016

Linum usitatissimum	Roots	A4 strain	–	–	HPLC	Secoisolariciresinol diglucoside (0.130 µmol/g), secoisolariciresinol (2.107), matairesinol (0.236) 4-hydroxybenzoic acid (3.394 µmol/g), vanillic acid (2.898), chlorogenic acid (2.789), caffeic acid (0.390), *p*-coumaric acid (0.186), sinapic acid (21.786), ferulic acid (1.541), trans-3-hydroxy-4-methoxycinnamic acid (2.968), trans-*o*-hydroxycinnamic (4.239), *p*-anisic acid (0.068)	Gabr et al. 2018
Beta vulgaris	Roots	ATCC 15834	Liquid MS medium supplemented with 30 g/L sucrose	70% ethanol	HPLC	4-Hydroxybenzoic acid (0.396 mg/g), caffeic acid (0.203), catechin hydrate (0.372), epicatechin (0.857), rutin (1.096)	Georgiev et al. 2010
Cajanus cajan	Leaf	LBA9402	MS liquid supplemented with 3.0% sucrose	70% ethanol solution at 60 °C in an ultrasonic bath power of 300 W for 8 min	UPLC-MS/MS	Cajaninstilbene acid (2996.23 µg/g)	Jiao et al. 2020
Ocimum basilicum	Shoots	R1000 strain	1/2 MS liquid medium	80% MeOH and 0.1% acetic acid + sonicated for 40 min at room temperature	HPLC-UV	Rosmarinic acid (214.30 µg/mg)	Kwon et al. 2021
Scutellaria lateriflora	Internode	ATCC 15834	MS medium +0.5 mg/l IBA	Methanol +sonication	HPLC-UV	Verbascoside, scutellarin, baicalin, scutellarein, wogonoside, baicalein, wogonin	Marsh et al. 2014

(continued)

Table 5.2 (continued)

Plant source(s)	Explant	Strain	Culture medium	Solvent type, concentration, and extraction techniques	Detection technique	Bioactive compounds (quantity)	Reference(s)
Turbinicarpus lophophoroides	Root tip	–	MS + 30 g L^{-1}	MeOH	UPLC-ESI-MS NMR	Feruloyl-glucoside (2.7267 ± 0.041 mg/g)	Solis-Castañeda et al. 2020
Polygonum multiflorum	Leaf	A. rhizogenes KCTC 2703	MS liquid medium supplemented with 30 g/l sucrose	Methanol and 20 min sonication	HPLC–PDA, UPLC	Anthraquinones content (emodin 211.32 μg/g DW and physcion 353.23 μg/g DW), myricetin (408.50 ± 4.94 μg/g), quercetin (287.50 ± 16.26), kaempferol (108.00 ± 11.31), caffeic acid (257.50 ± 3.53), p-coumaric acid (91.00 ± 0.0), ferulic acid (281.00 ± 2.82), m-coumaric acid (1.50 ± 0.70), o-coumaric acid (137.50 ± 16.26), chlorogenic acid (65.00 ± 2.51), gallic acid (557.50 ± 4.94), protocatechuic acid (304.00 ± 15.55), β-resorcylic acid (141.00 ± 9.89), vanillic acid (255.00 ± 1.41), syringic acid (97.50 ± 2.12), pyrogallol (1368.00 ± 12.72), rutin (352.50 ± 14.84), vanillin (149.50 ± 2.12), veratric acid (135.50 ± 3.53), hesperidin (136.00 ± 8.48), resveratrol (168.50 ± 2.12), naringenin (86.50 ± 4.94), formononetin (113.50 ± 7.77), biochanin A (93.50 ± 4.94)	Thiruvengadam et al. 2014

Fagopyrum tataricum	Seeds	*A. Rhizogenes* R1000	1/2 MS	100% methanol and heated at 60 °C for 1 h in a sonicator	HPLC	Catechin (0.53 ± 0.02 μg/g), chlorogenic acid (0:50 ± 0:01), ferulic acid (0.30 ± 0.01), benzoic acid (0.16 ± 0.02), rutin (22:31 ± 0:54), quercetin (0.34 ± 0.03), anthocyanin, cyanidin 3-*O*-glucoside (800 ± 0:00), cyanidin 3-*O*-rutinoside (2400 ± 20:0)	Thwe et al. 2016
Salvia sclarea	Leaf	A. rhizogenes ATCC 15835	Liquid hormone-free MS medium	Acetone for 72 h at room temperature	HPLC-DAD	Carnosic acid (62.36 ± 3.89 mg/l), salvipisone (133.18 ± 7.80), aethiopinone (208.98 ± 2.18), 1-oxo-aethiopinone (13.38 ± 0.82), 1-oxo-ferruginol (20.34 ± 4.32), ferruginol (230.40 ± 3.74)	Vaccaro et al. 2020
Salvia bulleyana	Leaves	A. Rhizogenes strain A4	MS + 0.1 mg/l IAA + 0.5 mg/l BAP	Methanol: Water solution (4:1, v/v) using ultrasonic disintegrator at 40 °C for 15 min	UPLC-PDA-ESI-MS/MS	Caffeic acid, rosmarinic acid hexoside, salvianolic acid E, rosmarinic acid, salvianolic acid K, caffeic acid derivative I, methyl rosmarinate, salvianolic acid F isomer I, salvianolic acid F isomer II, caffeic acid derivative II	Wojciechowska et al. 2020

Bioactive constituent, once extracted, further needs to be identified and the amount present in defined concentration. For this, various strategies are adopted for preliminary identification of phenolics, the simplest one being UV–VIS spectrophotometry. In this two widely accepted and adopted methods to measure total phenolics are Folin-Ciocalteu and Folin-Denis methods. The methods described involve chemical reduction of molybdenum and tungsten having absorption spectrum at 760 nm (Stalikas 2007). Over the year's novel approaches, TLC, high-performance liquid chromatography (HPLC), gas chromatography (GC), ultra-high pressure chromatography (UHPLC), high-performance thin-layer chromatography (HPTLC), etc. have been used as separation, identification, and quantification of phenolic compounds (Table 5.1). These find wide applicability based on adsorption, partition; size exclusion, filtration rate, etc. are used, depending on the physicochemical nature of the compound to be purified. Other than this, several hyphenated tools and techniques are currently in use. Mass spectrometric (MS) detectors are attached to high-performance liquid chromatography (HPLC–MS), GC MS, electrospray ionization mass spectrometry (ESI-MS), matrix-assisted laser desorption/ionization mass spectrometry (MALDI–MS), LC-MS-ESI$^+$, LC/DAD/MS ESI$^+$, ^{1}HNMR, and ^{13}CNMR (Chun et al. 2020; Wen et al. 2019).

5.5 Commercial Utilization and Prospects

Plant in vitro culture is an important method for producing important metabolites efficiently. Several techniques, such as suspension cultures and hairy roots, currently allow for large-scale plant metabolite production (Xu et al. 2012). However, as previously stated, the high costs of this technology render it uncompetitive when compared to less priced but environmentally unsustainable procedures like wild plant gathering or chemical synthesis. Because these culture methods still require bioreactors, their scale-up potential is limited (Nogueira et al. 2018; Buyel et al. 2017). New molecular technologies have opened up new avenues for the production of important metabolites using plant systems. Gene editing could be used to introduce new alleles, replace promoters, or create new pathways, all of which could lead to the development of plant-based systems capable of novel expression of useful bioactive chemicals (Nogueira et al. 2018). Once the plant has been designed, it can be utilized to produce the desired metabolites on a massive scale, including plant-made medications. Some plants, such as medicinal plants, benefit from in vitro conditions (higher growth rates) and should be kept in those conditions (Rao and Ravishankar 2002). With this in mind, a variety of activities can be taken to boost the production of secondary metabolites by plant cells, some of which have already been highlighted, such as the usage of precursors and elicitors. The manipulation of environmental factors, including high/low temperature, drought, UV, alkalinity, salinity, exposure to heavy metals, and others, is now emphasized. In plant in vitro cultures, these stress conditions often improve the capacity for production or even trigger de novo synthesis of phenolic compounds (Korkina et al. 2017; Lajayer et al. 2017; Moon et al. 2017; Puente-Garza et al. 2017).

5.6 Conclusions

The use of in vitro culture for the production of secondary metabolites including phenolic compounds remains a viable method, especially if the plant source material is an overexploited (or endangered), slow-growing, or low-yielding plant. However, because of the higher expense, it is necessary to conduct a cost-benefit analysis of in vitro culture before implementing the approach. Similarly, the production of pharmaceuticals using plant culture systems can provide significant benefits, such as cost savings, speed, low human pathogen burden, and scalability; all of these benefits are plant product specific and are dependent on production efficiencies compared to those offered by alternative sources.

References

Acamovic T, Brooker JD (2005) Biochemistry of plant secondary metabolites and their effects in animals. Proc Nutr Soc 64(3):403–412

Açıkgöz MA (2020) Effects of sorbitol on the production of phenolic compounds and terpenoids in the cell suspension cultures of *Ocimum basilicum* L. Biologia 76(1):395–409. https://doi.org/10.2478/s11756-020-00581-0

Alasalvar C, Grigor JM, Zhang D, Quantick PC, Shahidi F (2001) Comparison of volatiles, phenolics, sugars, antioxidant vitamins, and sensory quality of different colored carrot varieties. J Agric Food Chem 49(3):1410–1416

Alfermann AW, Petersen M (1995) Natural product formation by plant cell biotechnology. Plant Cell Tissue Org Cult 43(2):199–205

Akula R, Ravishankar GA (2011) Influence of abiotic stress signals on secondary metabolites in plants. Plant Signal Behav 6(11):1720–1731

Amani S, Mohebodini M, Khademvatan S, Jafari M (2020) *Agrobacterium rhizogenes* mediated transformation of Ficuscarica L. for the efficient production of secondary metabolites. J Sci Food Agric 100(5):2185–2197

Babbar N, Oberoi HS, Sandhu SK, Bhargav VK (2014) Influence of different solvents in extraction of phenolic compounds from vegetable residues and their evaluation as natural sources of antioxidants. J Food Sci Technol 51(10):2568–2575

Balasundram N, Sundram K, Samman S (2006) Phenolic compounds in plants and Agri-industrial by-products: antioxidant activity, occurrence, and potential uses. Food Chem 99(1):191–203

Baek S, Ho TT, Lee H, Jung G, Kim YE, Jeong CS, Park SY (2020) Enhanced biosynthesis of triterpenoids in *Centella asiatica* hairy root culture by precursor feeding and elicitation. Plant Biotechnol Rep 14(1):45–53

Barba-Espín G, Chen ST, Agnolet S, Hegelund JN, Stanstrup J, Christensen JH, Lütken H (2020) Ethephoninduced changes in antioxidants and phenolic compounds in anthocyanin-producing black carrot hairy root cultures. J Exp Bot 71(22):7030–7045

Baque MD, Elgirban A, Lee EJ, Paek KY (2012) Sucrose regulated enhanced induction of anthraquinone, phenolics, flavonoids biosynthesis and activities of antioxidant enzymes in adventitious root suspension cultures of *Morinda citrifolia* (L.). Acta Physiol Plant 34 (2):405–415

Beaumont MD, Knorr D (1987) Effects of immobilizing agents and procedures on viability of cultured celery (*Pium graveolens*) cells. Biotechnol Lett 9(6):377–382

Belwal T, Ezzat SM, Rastrelli L, Bhatt ID, Daglia M, Baldi A, Devkota HP, Orhan IE, Patra JK, Das G, Anandharamakrishnan C (2018) A critical analysis of extraction techniques used for botanicals: trends, priorities, industrial uses and optimization strategies. TrAC Trends Anal Chem 100:82–102

Belwal T, Chemat F, Venskutonis PR, Cravotto G et al (2020) Recent advances in scaling-up of non-conventional extraction techniques: learning from successes and failures. Trends Anal Chem:115895. https://doi.org/10.1016/j.trac.2020.115895

Beigmohamadi M, Movafeghi A, Jafari S, Sharafi A (2021) Efficient in vitro organogenesis, micropropagation, and plumbagin production in *Plumbago europaea* L. In Vitro Cell Dev Biol Plant 57(5):820–830

Boobalan S, Kamalanathan D (2020) Tailoring enhanced production of aervine in *Aerva lanata* (L.) Juss. Ex Schult by *Agrobacterium rhizogenes*-mediated hairy root cultures. Ind Crop Prod 155:112814

Brodelius PE, Funk C, Shillito RD (1988) Permeabilization of cultivated plant cells by electroporation for release of intracellularly stored secondary products. Plant Cell Rep 7(3):186–188

Bulgakov VP, Vereshchagina YV, Veremeichik GN (2018) Anticancer polyphenols from cultured plant cells: production and new bioengineering strategies. Curr Med Chem 25(36):4671–4692. https://doi.org/10.2174/0929867324666170609080357

Buyel JF, Twyman RM, Fischer R (2017) Very-large-scale production of antibodies in plants: the biologization of manufacturing. Biotechnol Adv 35:458–465

Chakraborty S, Uppaluri R, Das C (2020) Optimization of ultrasound-assisted extraction (UAE) process for the recovery of bioactive compounds from bitter gourd using response surface methodology (RSM). Food Bioprod Process 120:114–122. https://doi.org/10.1016/j.fbp.2020.01.003

Chun SC, Gopal J, Iyyakannu S, Muthu M (2020) An analytical retrospection of mass spectrometric tools established for plant tissue culture: current endeavours and future perspectives. Trends Anal Chem 126:115843. https://doi.org/10.1016/j.trac.2020.115843

Chung IM, Rekha K, Rajakumar G, Thiruvengadam M (2016) Production of glucosinolates, phenolic compounds and associated gene expression profiles of hairy root cultures in turnip (*Brassica rapa* ssp. *rapa*). 3 Biotech 6(2):1–16

Cisneros-Torres D, Cruz-Sosa F, Gonzalez-Cortazar M, Martinez-Trujillo A et al (2019) Enhancing the production of scopoletin and quercetin 3-*O*-*β*-d-glucoside from cell suspension cultures of *Tilia americana* var. *mexicana* by modulating the copper and nitrate concentrations. Plant Cell Tissue Organ Cult 139:305–316. https://doi.org/10.1007/s11240-019-01683-7

Coskun Y, Duran RE, Kilic S (2019) Striking effects of melatonin on secondary metabolites produced by callus culture of rosemary (*Rosmarinus officinalis* L.). Plant Cell Tissue Organ Cult 138:89–95. https://doi.org/10.1007/s11240-019-01605-7

Demirci T, ÇelikkolAkçay U, GöktürkBaydar N (2020) Effects of 24-epibrassinolide and l-phenylalanine on growth and caffeic acid derivative production in hairy root culture of *Echinacea purpurea* L. Moench. Acta Physiol Plant 42(4):1–11

DiCosmo F, Towers GHN (1984) Stress and secondary metabolism in cultured plant cells. In: Timmermann BN, Steelink C, Loewus FA (eds) Phytochemical adaptations to stress. Recent advances in phytochemistry (proceedings of the Phytochemical Society of North America), vol 18. Springer, Boston, MA, pp 97–175. https://doi.org/10.1007/978-1-4684-1206-2_5

Dörnenburg H, Knorr D (1995) Strategies for the improvement of secondary metabolite production in plant cell cultures. Enzyme Microb Technol 17(8):674–684

Duran RE, Kilic S, Coskun Y (2019) Melatonin influence on in vitro callus induction and phenolic compound production in sweet basil (*Ocimum basilicum* L.). In Vitro Cell Dev Biol Plant 55(4):468–475. https://doi.org/10.1007/s11627-019-10006-6

Edreva A, Velikova V, Tsonev T, Dagnon S, Gürel A, Aktaş L, Gesheva E (2008) Stress-protective role of secondary metabolites: diversity of functions and mechanisms. Gen Appl Plant Physiol 34(1–2):67–78

Farag MA, Mekky H, El-Masry S (2016) Metabolomics driven analysis of *Erythrina lysistemon* cell suspension culture in response to methyl jasmonate elicitation. J Adv Res 7:681–689. https://doi.org/10.1016/j.jare.2016.07.002

Farjaminezhad R, Garoosi G (2020) Establishment of green analytical method for ultrasound-assisted extraction of azadirachtin, mevalonic acid and squalene from cell suspension culture of *Azadirachta indica* using response surface methodology. Ind Crop Prod 144:111946. https://doi.org/10.1016/j.indcrop.2019.111946

Gabr AM, Ghareeb H, El Shabrawi HM, Smetanska I, Bekheet SA (2016) Enhancement of silymarin and phenolic compound accumulation in tissue culture of Milk thistle using elicitor feeding and hairy root cultures. J Genet Eng Biotechnol 14(2):327–333

Gabr AM, Mabrok HB, Abdel-Rahim EA, El-Bahr MK, Smetanska I (2018) Determination of lignans, phenolic acids and antioxidant capacity in transformed hairy root culture of *Linum usitatissimum*. Nat Prod Res 32(15):1867–1871

Georgiev MI, Agostini E, Ludwig-Muller J, Xu J (2012) Genetically transformed roots: from plant disease to biotechnological resource. Trends Biotechnol 30:528–537. https://doi.org/10.1016/j.tibtech.2012.07.001

Georgiev MI, Weber J, Maciuk A (2009) Bioprocessing of plant cell cultures for mass production of targeted compounds. Appl Microbiol Biotechnol 83(5):809–823

Georgiev VG, Weber J, Kneschke EM, Denev PN, Bley T, Pavlov AI (2010) Antioxidant activity and phenolic content of betalain extracts from intact plants and hairy root cultures of the red beetroot *Beta vulgaris* cv. Detroit dark red. Plant Foods Hum Nutr 65(2):105–111

Ghimire BK, Thiruvengadam M, Chung IM (2019) Identification of elicitors enhances the polyphenolic compounds and pharmacological potential in hairy root cultures of *Aster scaber*. S Afr J Bot 125:92–101. https://doi.org/10.1016/j.sajb.2019.07.006

Giri L, Dhyani P, Rawat S, Bhatt ID et al (2012a) In vitro production of phenolic compounds and antioxidant activity in callus suspension cultures of *Habenaria edgeworthii*: a rare Himalayan medicinal orchid. Ind Crops Prod 39:1–6. https://doi.org/10.1016/j.indcrop.2012.01.024

Giri L, Jugran A, Rawat S, Dhyani P et al (2012b) In vitro propagation, genetic and phytochemical assessment of *Habenaria edgeworthii*: an important Astavarga plant. Acta Physiol Plant 34: 869–875. https://doi.org/10.1007/s11738-011-0884-8

Giri CC, Zaheer M (2016) Chemical elicitors versus secondary metabolite production in vitro using plant cell, tissue and organ cultures: recent trends and a sky eye view appraisal. Plant Cell Tissue Organ Cult 126(1):1–18

Goncalves S, Mansinhos I, Rodriguez-Solana R, Perez-Santin E et al (2019) Elicitation improves rosmarinic acid content and antioxidant activity in *Thymus lotocephalus* shoot cultures. Ind Crop Prod 137:214–220. https://doi.org/10.1016/j.indcrop.2019.04.071

Guillon S, Trémouillaux-Guiller J, Pati PK, Rideau M, Gantet P (2006) Hairy root research: recent scenario and exciting prospects. Curr Opin Plant Biol 9(3):341–346

Haida Z, Syahida A, Ariff SM, Maziah M et al (2019) Factors affecting cell biomass and flavonoid production of *Ficus deltoidea* var. *kunstleri* in cell suspension culture system. Sci Rep 9:9533. https://doi.org/10.1038/s41598-019-46042-w

Heim KE, Tagliaferro AR, Bobilya DJ (2002) Flavonoid antioxidants: chemistry, metabolism and structure-activity relationships. J Nutr Biochem 13(10):572–584

Heydari HR, Chamani E, Esmaielpour B (2020) Cell line selection through gamma irradiation combined with multi-walled carbon nanotubes elicitation enhanced phenolic compounds accumulation in *Salvia nemorosa* cell culture. Plant Cell Tissue Organ Cult 142:353–367. https://doi.org/10.1007/s11240-020-01867-6

Hu M, Hu Z, Du L, Du J et al (2019) Establishment of cell suspension culture of *Lonicera japonica* Thunb and analysis its major secondary metabolites. Ind Crop Prod 137:98–104. https://doi.org/10.1016/j.indcrop.2019.05.024

Jiao J, Gai QY, Wang X, Liu J, Lu Y, Wang ZY, Fu YJ (2020) Effective production of phenolic compounds with health benefits in pigeon pea [*Cajanuscajan* (L.) Millsp.] hairy root cultures. J Agric Food Chem 68(31):8350–8361

Kannan N, Manokari M, Shekhawat MS (2020) Enhanced production of anthraquinones and phenolic compounds using chitosan from the adventitious roots of *Morinda coreia* Buck. and Ham. Ind Crops Prod 148:112321. https://doi.org/10.1016/j.indcrop.2020.112321

Karakas FP (2020) Efficient plant regeneration and callus induction from nodal and hypocotyls explants of goji berry (*Lycium barbarum* L.) and comparison of phenolic profiles in calli formed under different combinations of plant growth regulators. Plant Physiol Biochem 146:384–391. https://doi.org/10.1016/j.plaphy.2019.11.009

Karakas FP, Turker AU (2013) An efficient in vitro regeneration system for *Bellis perennis L.* and comparison of phenolic contents of field-grown and in vitro-grown leaves by LC-MS/MS. Ind Crops Prod 48:162–170. https://doi.org/10.1016/j.indcrop.2013.04.008

Karalija E, Zeljković SC, Parić A (2019) Harvest time–related changes in biomass, phenolics and antioxidant potential in *Knautia sarajevensis* shoot cultures after elicitation with salicylic acid and yeast. In Vitro Cell Dev Biol Plant:1–7. https://doi.org/10.1007/s11627-019-10028-0

Khan T, Khan T, Hano C, Abbasi BH (2019) Effects of chitosan and salicylic acid on the production of pharmacologically attractive secondary metabolites in callus cultures of *Fagonia indica*. Ind Crop Prod 129:525–535. https://doi.org/10.1016/j.indcrop.2018.12.048

Khanam MN, Anis M (2018) Organogenesis and efficient in vitro plantlet regeneration from nodal segments of *Allamanda cathartica* L. using TDZ and ultrasound assisted extraction of quercetin. Plant Cell Tissue Organ Cult 134(2):241–250. https://doi.org/10.1007/s11240-018-1416-7

Kikowska M, Thiem B, Szopa A, Ekiert H (2020) Accumulation of valuable secondary metabolites: phenolic acids and flavonoids in different in vitro systems of shoot cultures of the endangered plant species—*Eryngium alpinum* L. Plant Cell Tissue Organ Cult 141:381–391. https://doi.org/10.1007/s11240-020-01795-5

Kim S, Jeong J, Park SH, Park SC et al (2020) The synergistic effect of co-treatment of methyl jasmonate and cyclodextrins on pterocarpan production in *Sophora flavescens* cell cultures. Int J Mol Sci 21:3944. https://doi.org/10.3390/ijms21113944

Knorr D, Teutonico RA (1986) Chitosan immobilization and permeabilization of Amaranthus tricolor cells. J Agric Food Chem 34(1):96–97

Kolewe ME, Gaurav V, Roberts SC (2008) Pharmaceutically active natural product synthesis and supply via plant cell culture technology. Mol Pharm 5:243–256

Korkina LG, Mayer W, de Luca C (2017) Meristem plant cells as a sustainable source of redox actives for skin rejuvenation. Biomol Ther 7:1–22

Kwon DY, Kim YB, Kim JK, Park SU (2021) Production of rosmarinic acid and correlated gene expression in hairy root cultures of green and purple basil (*Ocimum basilicum* L.). Prep Biochem Biotechnol 51(1):35–43

Lajayer BA, Ghorbanpour M, Nikabadi S (2017) Heavy metals in contaminated environment: destiny of secondary metabolite biosynthesis, oxidative status and phytoextraction in medicinal plants. Ecotoxicol Environ Saf 145:377–390

Lin D, Hu L, You H, Sarkar D, Xing B, Shetty K (2010) Initial screening studies on potential of high phenolic-linked plant clonal systems for nitrate removal in cold latitudes. J Soil Sediment 10(5):923–932

Marsh Z, Yang T, Nopo-Olazabal L, Wu S, Ingle T, Joshee N, Medina-Bolivar F (2014) Effect of light, methyl jasmonate and cyclodextrin on production of phenolic compounds in hairy root cultures of *Scutellaria lateriflora*. Phytochemistry 107:50–60

Mendoza D, Cuaspud O, Arias JP, Ruiz O et al (2018) Effect of salicylic acid and methyl jasmonate in the production of phenolic compounds in plant cell suspension cultures of *Thevetia peruviana*. Biotechnol Rep 19:e00273. https://doi.org/10.1016/j.btre.2018.e00273

Moo-Huchin VM, Moo-Huchin MI, Estrada-León RJ, Cuevas-Glory L, Estrada-Mota IA, Ortiz-Vázquez E, Betancur-Anconad D, Sauri-Duch E (2015) Antioxidant compounds, antioxidant activity and phenolic content in peel from three tropical fruits from Yucatan, Mexico. Food Chem 166:17–22

Moon SH, Mistry B, Kim HD, Pandurangan M (2017) Antioxidant and anticancer potential of bioactive compounds following UV-C light-induced plant cambium meristematic cell cultures. Ind Crop Prod 109:762–772

Mora-Pale M, Sanchez-Rodriguez SP, Linhardt RJ, Dordick JS et al (2014) Biochemical strategies for enhancing the in vitro production of natural products with pharmaceutical potential. Curr Opin Biotechnol 25:86–94

Moreno PR, van der Heijden R, Verpoorte R (1995) Cell and tissue cultures of *Catharanthus roseus*: a literature survey. Plant Cell Tissue Organ Cult 42(1):1–25

Murthy HN, Dandin VS, Zhong JJ, Paek KY (2014) Strategies for enhanced production of plant secondary metabolites from cell and organ cultures. In: Production of biomass and bioactive compounds using bioreactor technology. Springer, Dordrecht, pp 471–508. https://doi.org/10.1007/978-94-017-9223-3_2011

Mustafa R, El-Naggar EMB, Svajdlenka E, Omran G et al (2020) Enhancement of phenolic content, antioxidant and cytotoxic activities of *Moringa oleifera* leaf and seed by suspension culture. Nat Prod Res:1–5. https://doi.org/10.1080/14786419.2020.1744137

Namdeo AG (2007) Plant cell elicitation for production of secondary metabolites. Pharmacogn Rev 1(1):69–79

Nogueira M, Enfissi EM, Almeida J, Fraser PD (2018) Creating plant molecular factories for industrial and nutritional isoprenoid production. Curr Opin Biotechnol 40:80–87

Nourozi E, Hosseini B, Maleki R, Mandoulakani BA (2019) Pharmaceutical important phenolic compounds overproduction and gene expression analysis in *Dracocephalum kotschyi* hairy roots elicited by SiO2 nanoparticles. Ind Crop Prod 133:435–446. https://doi.org/10.1016/j.indcrop.2019.03.053

Pandey A, Belwal T, Sekar KC, Bhatt ID et al (2018) Optimization of ultrasonic-assisted extraction (UAE) of phenolics and antioxidant compounds from rhizomes of *Rheum moorcroftianum* using response surface methodology (RSM). Ind Crop Prod 119:218–225. https://doi.org/10.1016/j.indcrop.2018.04.019

Pandey A, Belwal T, Tamta S, Bhatt ID et al (2019) Phenolic compounds, antioxidant capacity and antimutagenic activity in different growth stages of in vitro raised plants of *Origanum vulgare* L. Mol Biol Rep 46:2231–2241. https://doi.org/10.1007/s11033-019-04678-x

Pandino G, Meneghini M, Tavazza R, Lombardo S et al (2016) Phytochemicals accumulation and antioxidant activity in callus and suspension cultures of *Cynara scolymus* L. Plant Cell Tissue Organ Cult 128(1):223–230. https://doi.org/10.1007/s11240-016-1102-6

Perez-Hernandez J, Martinez-Trujillo A, Nicasio-Torres P (2019) Optimization of active compounds production by interaction between nitrate and copper in *Sphaeralcea angustifolia* cell suspension using response surface methodology. Plant Cell Tissue Organ Cult 136:407–413. https://doi.org/10.1007/s11240-018-1516-4

Puente-Garza CA, Meza-Miranda C, Ochoa-Martínez D, García-Lara S (2017) Effect of in vitro drought stress on phenolic acids, flavonols, saponins and antioxidant activity in *Agave salmiana*. Plant Physiol Biochem 115:400–407

Rainha N, Koci K, Coelho AV, Lima E et al (2012) HPLC–UV–ESI-MS analysis of phenolic compounds and antioxidant properties of *Hypericum undulatum* shoot cultures and wild-growing plants. Phytochemistry 86:83–91. https://doi.org/10.1016/j.phytochem.2012.10.006

Randhir R, Lin YT, Shetty K (2004) Stimulation of phenolics, antioxidant and antimicrobial activities in dark germinated mung bean sprouts in response to peptide and phytochemical elicitors. Process Biochem 39(5):637–646

Rao SR, Ravishankar GA (2002) Plant cell cultures: chemical factories of secondary metabolites. Biotechnol Adv 20(2):101–153

Raskin I, Ribnicky DM, Komarnytsky S, Ilic N, Poulev A, Borisjuk N, Brinker A, Moreno DA, Ripoll C, Yakoby N, O'Neal JM (2002) Plants and human health in the twenty-first century. Trends Biotechnol 20(12):522–531

Rattan S, Sood A, Kumar P, Kumar A et al (2020) Phenylethanoids, phenylpropanoids, and phenolic acids quantification vis-à-vis gene expression profiling in leaf and root derived callus lines of *Rhodiola imbricata* (Edgew.). Ind Crops Prod 154:112708. https://doi.org/10.1016/j.indcrop.2020.112708

Rawat V, Ghildiyal A, Singh L, Jugran AK et al (2019) Methyl jasmonate induced polyphenols and antioxidant production in callus suspension culture of *Nardostachys jatamansi*. Plant Biosyst 154(6):851–859. https://doi.org/10.1080/11263504.2019.1701124

Reyes-Martínez A, Antunes-Ricardo M, Gutiérrez-Uribe J, Santos-Díaz MS (2019) Enhanced production and identification of antioxidants in in vitro cultures of the cacti *Mammillaria candida* and *Turbinicarpus laui*. Appl Microbiol Biotechnol 103(6):2583–2595. https://doi.org/10.1007/s00253-019-09656-8

Ron M, Kajala K, Pauluzzi G, Wang D et al (2014) Hary root transformation using *agrobacterium rhizogenes* as a tool for exploring cell type-specific gene expression and function using tomato as a model. Plant Physiol 166:455–469

Sales PM, Souza PM, Simeoni LA, Magalhães PO, Silveira D (2012) α-Amylase inhibitors: a review of raw material and isolated compounds from plant source. J Pharm Pharm Sci 15(1): 141–183

Samari E, Sharifi M, Ghanati F, Fuss E et al (2019) Chitosan-induced phenolics production is mediated by nitrogenous regulatory molecules: NO and PAs in *Linum album* hairy roots. Plant Cell Tissue Organ Cult 140:563–576. https://doi.org/10.1007/s11240-019-01753-w

Shukitt-Hale B, Lau FC, Joseph JA (2008) Berry fruit supplementation and the aging brain. J Agric Food Chem 56(3):636–641

Singh T, Yadav R, Agrawal V (2020) Effective protocol for isolation and marked enhancement of psoralen, daidzein and genistein in the cotyledon callus cultures of *Cullen corylifolium* (L.) Medik. Ind Crop Prod 143:111905

Singh L, Singh B, Kewlani P, BelwalT BID, Nandi SK, Bisht AK (2022) Process optimization and bioactive compounds quantification from *Dactylorhiza hatagirea* tuber for alleviating glycemic and oxidative stress. J Appl Res Med Aromat Plants 26:100352

Solis-Castañeda GJ, Zamilpa A, Cabañas-García E, Bahena SM, Pérez-Molphe-Balch E, Gómez-Aguirre YA (2020) Identification and quantitative determination of feruloyl-glucoside from hairy root cultures of *Turbinicarpus lophophoroides* (Werderm.) Buxb. & Backeb. (Cactaceae). In Vitro Cell Dev Biol Plant 56(1):8–17

Stafford A, Morris P, Fowler MW (1986) Plant cell biotechnology: a perspective. Enzym Microb Technol 8(10):578–587

Stalikas CD (2007) Review: extraction, separation, and detection methods for phenolic acids and flavonoids. J Sep Sci 30:3268–3295

Staszków A, Swarcewicz B, Banasiak J, Muth D, Jasiński M, Stobiecki M (2011) LC/MS profiling of flavonoid glycoconjugates isolated from hairy roots, suspension root cell cultures and seedling roots of *Medica gotruncatula*. Metabolomics 7(4):604–613

Taghizadeh M, Nasibi F, Kalantari KM, Benakashani F (2020) Callogenesis optimization and cell suspension culture establishment of *Dracocephalum polychaetum* Bornm. and *Dracocephalum kotschyi* Boiss.: an in vitro approach for secondary metabolite production. S Afr J Bot 132:79–86

Thiruvengadam M, Praveen N, Kim EH, Kim SH, Chung IM (2014) Production of anthraquinones, phenolic compounds and biological activities from hairy root cultures of *Polygonum multiflorum* Thunb. Protoplasma 251(3):555–566

Thwe A, ValanArasu M, Li X, Park CH, Kim SJ, Al-Dhabi NA, Park SU (2016) Effect of differentA *grobacterium rhizogenes* strains on hairy root induction and phenylpropanoid biosynthesis in tartary buckwheat *(Fagopyrum tataricum* Gaertn). Front Microbiol 7:318

Vamanu E, Nita S (2013) Antioxidant capacity and the correlation with major phenolic compounds, anthocyanin, and tocopherol content in various extracts from the wild edible boletus edulis mushroom. Biomed Res Int 2013:313905

Vaccaro MC, Alfieri M, De Tommasi N, Moses T, Goossens A, Leone A (2020) Boosting the synthesis of pharmaceutically active abietanediterpenes in *S. sclarea* hairy roots by engineering the GGPPS and CPPS genes. Front Plant Sci 11:924

Vattem DA, Randhir R, Shetty K (2005) Cranberry phenolics-mediated antioxidant enzyme response in oxidatively stressed porcine muscle. Process Biochem 40(6):2225–2238

Velderrain-Rodríguez GR, Palafox-Carlos H, Wall-Medrano A, Ayala-Zavala JF, Chen CO, Robles-Sánchez M, Astiazaran-García H, Alvarez-Parrilla E, González-Aguilar GA (2014) Phenolic compounds: their journey after intake. Food Funct 5(2):189–197

Verpoorte R, van der Heijden R, Memelink J (2000) Engineering the plant cell factory for secondary metabolite production. Transgenic Res 9:323–343

Viera VB, Piovesan N, Rodrigues JB, Mello R et al (2017) Extraction of phenolic compounds and evaluation of the antioxidant and antimicrobial capacity of red onion skin Allium cepa L. Int Food Res J 24(3):990–999

Wang J, Qian J, Yao L, Lu Y (2015) Enhanced production of flavonoids by methyl jasmonate elicitation in cell suspension culture of *Hypericum perforatum*. Bioresour Bioprocess 2(1):1–9

Wen L, Zhang Z, Sun DW, Sivagnanam SP et al (2019) Combination of emerging technologies for the extraction of bioactive compounds. Crit Rev Food Sci Nutr 60(11):1826–1841. https://doi.org/10.1080/10408398.2019.1602823

Weremczuk-Jeżyna I, Skała E, Kuźma L, Kiss AK et al (2019) The effect of purine-type cytokinin on the proliferation and production of phenolic compounds in transformed shoots of *Dracocephalum forrestii*. J Biotechnol 306:125–133. https://doi.org/10.1016/j.jbiotec.2019.09.014

Wink M (1988) Plant breeding: importance of plant secondary metabolites for protection against pathogens and herbivores. Theor Appl Genet 75(2):225–233

Wojciechowska M, Owczarek A, Kiss AK, Grąbkowska R, Olszewska MA, Grzegorczyk-Karolak I (2020) Establishment of hairy root cultures of *Salvia bulleyana* Diels for production of polyphenolic compounds. J Biotechnol 318:10–19

Xu JF, Dolan MC, Medrano G, Cramer CL et al (2012) Green factory: plants as bioproduction platforms for recombinant proteins. Biotechnol Adv 30:1171–1184

Yang X, Lei Z, Yu Y, Xiao L et al (2019a) Phytochemical characteristics of callus suspension culture of *Helicteres angustifolia* L. and its in vitro antioxidant, antidiabetic and immunomodulatory activities. S Afr J Bot 121:178–185. https://doi.org/10.1016/j.sajb.2018.11.006

Yang X, Zhang J, Lei Z, Yan X et al (2019b) Adventitious root cultures from leaf explants of *Helicteres angustifolia* L. as a novel source for production of natural bioactive compounds. Acta Physiol Plant 41:171. https://doi.org/10.1007/s11738-019-2964-0

Yesil-Celiktas O, Nartop P, Gurel A, Bedir E, Vardar-Sukan F (2007) Determination of phenolic content and antioxidant activity of extracts obtained from *Rosmarinus officinalis*' calli. J Plant Physiol 164(11):1536–1542

Yildirim AB, Turker AU (2014) Effects of regeneration enhancers on micropropagation of *Fragaria vesca* L. and phenolic content comparison of field-grown and in vitro-grown plant materials by liquid chromatographyelectrospray tandem mass spectrometry (LC–ESI-MS/MS). Sci Hortic 169:169–178

Zielińska S, Dryś A, Piątczak E, Kolniak-Ostek J et al (2019) Effect of LED illumination and amino acid supplementation on phenolic compounds profile in *Agastache rugosa in vitro* cultures. Phytochem Lett 31:12–19. https://doi.org/10.1016/j.phytol.2019.02.029

In Vitro Production of Alkaloids

6

Supriya Meena, Bhanupriya Kanthaliya, Abhishek Joshi, Farhana Khan, Seema Choudhary, and Jaya Arora

Abstract

Plants are considered as a potent source of a wide variety of bioactive molecules that can be used for the development of the various pharmaceutical drugs. Alkaloids are the important class of secondary metabolites, known to exhibit therapeutic properties including anti-tumor, anti-viral, anti-inflammatory, and anti-malarial activities. Alkaloids are able to prevent various degenerative diseases by binding with the oxidative reaction catalyst or free radicals. The commercial extraction of alkaloids is reported from some major families like Apocynaceae, Papaveraceae, Rubiaceae, and Solanaceae. By this system, the yield of alkaloids is inconsistent due to genetic and geographical variations. Chemical synthesis is still not feasible system due to complex molecular structure of various metabolites. Therefore, in vitro system for production of alkaloids has become a promising biotechnological approach from a range of medicinal plants. Some of the medicinal plants such as *Nicotiana tobaccum* (nicotine), *Erythroxylum coca* (cocaine), *Cinchona officinalis* (quinine and quinidine), *Rauwolfia serpentina* (reserpine), and *Pilocarpine microphyllus* (pilocarpine) have been explored for in vitro production of their respective alkaloids. The present chapter provides brief information on various in vitro production systems and scale-up techniques used for alkaloid production.

Keywords

Alkaloids · Biosynthesis · Extraction · Biological activities · Bioreactor · Plant cell culture

S. Meena · B. Kanthaliya · A. Joshi · F. Khan · S. Choudhary · J. Arora (✉)
Laboratory of Biomolecular Technology, Department of Botany, Mohanlal Sukhadia University, Udaipur, Rajasthan, India

T. Belwal et al. (eds.), *Nutraceuticals Production from Plant Cell Factory*,
https://doi.org/10.1007/978-981-16-8858-4_6

6.1 Introduction

Plants have always been the major source for the traditional medicine systems, and they have provided various remedies in therapeutic application for thousands of years (Ramawat et al. 2009). Generally, plants are under selection pressure to protect themselves from pathogenic microbes and insects, whereas these pathogens struggle for their survival from plant defense to obtain food and reproduction site. Therefore, both plants and pathogens need to develop tactics to adapt or adjust with each other and the changing environment. Then the co-evolution of secondary metabolites is a consequence of the biological processes in plants that regulate defense mechanism (Ramawat and Goyal 2019). Alkaloids are one of the major secondary metabolites obtained from plants. Ubiquitous distribution of alkaloids is found in the plant kingdom mainly in higher plants, such as those belonging to Ranunculaceae, Leguminosae, Papaveraceae, Menispermaceae, and Loganiaceae (Ramawat and Mérillon 2013; Kandar 2021).

These compounds have been classified into various categories which include indole, piperidine, tropane, purine, pyrrolizidine, imidazole, quinolizidine, isoquinoline, and pyrrolidine alkaloids on the basis of their biosynthetic precursor and heterocyclic ring system (Waller 2012). In vitro cell, tissue, or organ culture has been employed as a probabilistic alternative to produce such industrial compounds. Tissue culture techniques can be used for the large-scale culture that provide continuous, reliable, and renewable source of valuable plant pharmaceuticals to extend commercial importance of plants to emerge or identify their biological activities which includes anti-tumor, anti-viral, anti-inflammatory, anti-malarial activities (Debnath et al. 2018). There are up to 80% of people in developing countries who are totally dependent on herbal drugs for their primary healthcare, and over 25% of prescribed medicines in developed countries are derived from wild plant species (Hamilton 2004). As there is an increasing demand of medicinal plants for herbal drugs, natural health products, and secondary metabolites, the use of medicinal plants is growing rapidly throughout the world. Consequently, some of them are increasingly being threatened even in their natural habitats, and some also face natural extinction (Chen et al. 2016). Therefore, in the production of desirable medicinal compounds from plants in search for alternatives, cell and tissue culture technologies were emerged as a possible tool for studying, and producing plant secondary metabolites as in vitro regeneration holds enormous potential for the synthesis of high-quality plant-based medicines (Rohini 2020). Techniques like somaclonal variations and genetic manipulations may be utilized to improve the production of alkaloids. The in vitro cell culture system is more beneficial than the conventional in vivo cultivation of whole plants in context of production of desirable compounds under controlled conditions independent to climatic factors or soil compositions and reduces labor costs and improves productivity as automated control of cell growth and rational regulation of metabolite processes. These cultured cells would be free of microbes and insects. Another benefit of cell culture is that the cells of any plants, tropical or alpine, could be easily multiplied to yield their specific metabolites in any artificial conditions. In vitro production of alkaloids is an impactful technique to develop the large-scale scenario in pharmaceutical industries. Callus culture assisted the optimization of alkaloid production. Media composition is significant for the callus induction to enhance the alkaloid content and conservation of threatened genotype (Hussain et al. 2012).

6.2 Biosynthetic Pathway

According to Ramawat and Mérillon (2013), alkaloids are classified into three major categories on the basis of their origin and structure:

1. **True-alkaloids**: These are derived from amino acids and containing nitrogen moiety in their heterocyclic ring and found to be basic in nature. These alkaloids are highly reactive molecules with biological activity even in low doses.
   Examples: Nicotine, morphine, ergotamine, quinine, atropine, etc.
2. **Proto-alkaloids: These are also derived from amino acids**, but nitrogen is absent in their heterocyclic ring.
   Examples: Ephedrine, mescaline, etc.
3. **Pseudo-alkaloids**: Alkaloid-like compounds that do not originate from amino acids. It includes mainly terpenoid, steroid, and purine-like alkaloids. So they are also called as steroidal alkaloids.
   Examples: caffeine, pinidine, coniceine, etc.

The precursors of alkaloids biosynthesis are mainly amino acids (Fig. 6.1). Furthermore, the diversity of alkaloids depends on their precursor molecule and their structure. Alkaloid biosynthesis is a sequenced process in context of plant development, controlling the expression of genes in pathways, inside specific cells or organelles. Their biosynthesis and accumulation depends on developmental stage and environmental conditions and also have cell or tissue specific regulations. The degree of expression of the genes involved in the biosynthetic pathway of a particular alkaloid affects the accumulation of that metabolite (Ziegler and Facchini 2008). Alkaloid biosynthesis and accumulation is increasing with the number of genes involved in it. Generally, biosynthesis starts from the fixation of atmospheric CO_2 in primary carbon metabolism. The erythrose-4-phosphate (PPP pathway intermediate) and phosphoenolpyruvate (glycolysis intermediate) go through the shikimic acid pathway (Fig. 6.2) to form aromatic amino acids and pyruvate (end product of glycolysis) followed by acetyl coA and then go through TCA cycle to form aliphatic amino acids. These two types of amino acid constitute for synthesis of a range of N-containing compounds (alkaloids). This is followed by a series of reactions such as bond formations, breakages, rearrangements, addition, and modification of functional groups, yielding a vast range of various alkaloids. The pathway begins with two substrates, phosphoenolpyruvate and erythrose-4-phosphate, and ends with chorismate (substrate for the three aromatic amino acids – Tyr, Phe, Trp). This pathway includes seven steps regulated by seven enzymes: DAHP (3-deoxy-D-arabino-heptulosonic acid-7-phosphate) synthase, 3-dehydroquinate synthase,

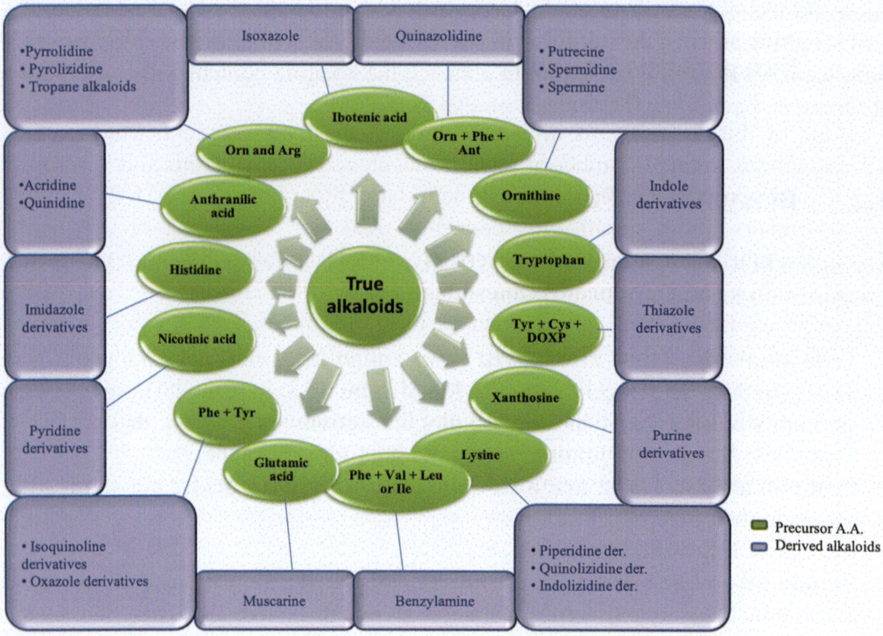

Fig. 6.1 Precursors of some true alkaloids for biosynthesis

3-hydratase, shikimate dehydrogenase, shikimate kinase, EPSP (3-enolpyruvylshikimate-5-phosphate) synthase, and chorismate synthase (Pathak et al. 2019). After that, a particular group of alkaloid followed a different pathway to complete its biosynthesis. Here are some examples of complete biosynthesis of alkaloid category.

1. **Amaryllidaceae** alkaloids use L-tyrosine and L-phenylalanine as precursors. From phenylalanine, the phenylpropanoid pathway leads the formation of 3, 4-dihydroxybenzaldehyde (3, 4-DHBA) to synthesize aldehyde (-CHO) moiety of Amaryllidaceae alkaloid. The other pathway to synthesize tyramine (the amine $-NH_2$ moiety of Amaryllidaceae alkaloid) is from tyrosine and its concretion with 3, 4-DHBA to form the central precursor norbelladine as well as its pursuant *O*-methylation occurs. The phenol coupling of the 4-*O*-methylnorbelladine in intermediate pathway ensued by a reduction step will progress in a series of unstable intermediates. The pathway(s) for the different Amaryllidaceae alkaloid's biosynthesis found in different plant species and their pathways remain still uncharacterized (Desgagné-Penix 2020).
2. The biosynthesis of **quinolizidine** alkaloids (QAs) starts with the formation of cadaverine as intermediate by the decarboxylation of L-lysine. The cadaverine then undergoes oxidative deamination, by a copper amine oxidase enzyme to yield 5-aminopentanal which is then spontaneously cyclized to 11-piperidine Schiff base. In addition to these reactions, a chain of reactions including Schiff

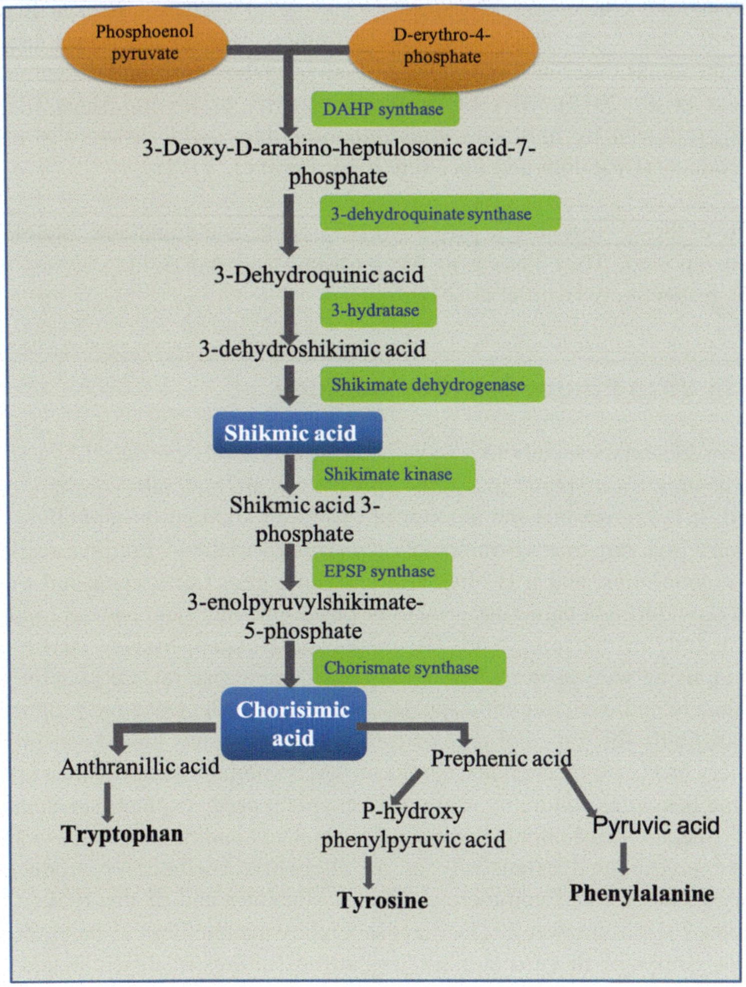

Fig. 6.2 Shikimic acid pathway

base formations, aldol-type reactions, hydrolysis, oxidative deamination, and coupling generates the major structural QAs. The alkaloidal group can be further modified by dehydrogenation, oxygenation, hydroxylation, glycosylation, or esterification to form a wide range of structurally related QAs (Frick et al. 2017).

3. The biosynthesis of **benzylisoquinoline alkaloids** (BIAs) initiates the formation of dopamine and 4-hydroxyphenylacetaldehyde by using L-lysine as precursor amino acid, which are then condensed to (S)-norcoclaurine by (S)-norcoclaurine synthase (NCS). One cytochrome P450 [(S)-*N*-methylcoclaurine 30-hydroxylase] and three methyltransferases [(S)-norcoclaurine/norlaudanosoline 6-Omethyltransferase, (S)-coclaurine-*N*-methyltransferase, and (S)-30-hydroxy-*N*-methylcoclaurine-40-omethyltransferase] are involved in

catalyzing the conversion of (S)-norcoclaurine to a central intermediate, (S)-reticuline, for the production of different BIAs, i.e., protoberberine, benzophenanthridine, and morphinan alkaloids (Beaudoin and Facchini 2014; Yamada et al. 2017). BIA biosynthesis consists of several steps (describes above), followed by multistep transformations that yield structurally different end products (Beaudoin and Facchini 2014; He et al. 2018).

In spite of these, biosynthetic pathway(s) of tropane and granatane alkaloids have also been reported. They belong to the pyrroline and piperidine classes of plant alkaloids, respectively (Kim et al. 2016).

6.3 In Vitro Production Methodology

Many plant secondary metabolites have been commercially produced by extraction and purification from plant materials either they are naturally present or field cultivated. It is known that the amount of produced alkaloid in naturally growing plant is very low, due to environmental or seasonal variations, nutrient availability, and stress conditions, and it is often restricted to a genus or species and might be activated only during a particular growth or developmental stage, which hinders the biological study of the compounds (Bagnères and Hossaert-McKey 2016). On the other hand, using wild plant materials has considerable risk related to extinction of many valuable and even endemic species. Hereupon, to increase the production of various alkaloids for remedial applications, two approaches have been proposed. The primary one is a total chemical synthesis that is complex and not much effective. The second is in vitro culture by using different plant parts (explants) to enhance the secondary metabolites content (Fig. 6.3). So, plant cell and tissue culture techniques have been investigated extensively as an alternative method for production of secondary metabolites of commercial interest since the end of the 1950s (Davies and Deroles 2014; Ramawat 2021). Plant secondary metabolites can be produced by two major groups of in vitro cultures: organized cultures of differentiated tissues (i.e., organ cultures as root, shoot, and embryo cultures) and unorganized cultures of undifferentiated cells (i.e., cell suspension and callus cultures) (Fig. 6.4). Investigations have showed that differentiated plant tissues produce the identical product as the plant produce itself, and they were relatively more stable in the production of secondary metabolites than the undifferentiated cells (Nielsen et al. 2019). Shoot cultures are used for many medicinal plants, which accumulate secondary metabolites much greater than that of natural plants. Besides, many of the valuable secondary metabolites like tropane alkaloids, hyoscyamine, and scopolamine are produced quite well in the root cultures (Filova 2014). However, plant roots in cultures generally grow slower than undifferentiated plant cells, and their harvesting is difficult. Therefore, plant hairy root cultures have been applied as an alternative method for the production of compounds synthesized in the plant roots. Hairy roots obtained by *Agrobacterium rhizogenes*-mediated transformation exhibits higher growth rates than cell suspension cultures and produce secondary

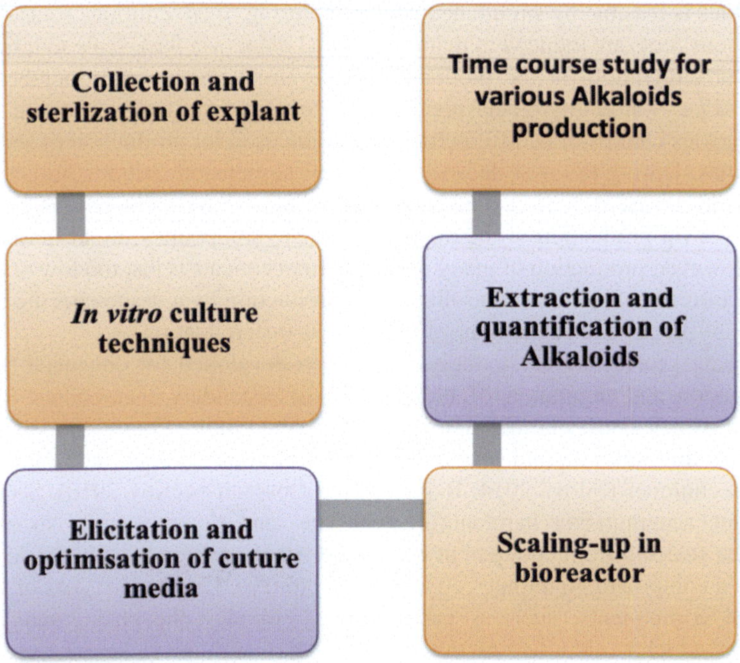

Fig. 6.3 Schematic presentation of in vitro alkaloids production

Fig. 6.4 Types of in vitro cultures used in alkaloids production from certain medicinal plants

metabolites over successive generations without losing genetic or biosynthesis stability (Mehrotra et al. 2015a, b). Huang et al. (2018) reported a sanguinarine alkaloid from *Macleaya cordata* hairy root cultures by co-cultivating leaf and stem explants with *A. rhizogenes*. Furthermore, production of two different secondary metabolites is feasible by simultaneously co-culturing of adventitious roots. Natural adventitious roots are induced in many medicinal plants via flask scale to large-scale bioreactor cultivation for the production of several bioactive compounds (Baque et al. 2012). Non-embryogenic plant callus cultures, consisting of more or less homogeneous clumps of dedifferentiated cells, are used for production of secondary metabolites. During the past decades, plant cell suspension cultures initiated from callus cultures have been extensively studied and have emerged as attractive alternatives for production of the range of valuable compounds found in the whole plant. However, production of many of the pharmaceuticals is just too low or may be zero in cultured cells due to controlling their production by a tissue-specific manner and loss of production capacity resulting from dedifferentiation.

In recent years, various strategies have been developed for enhanced biomass accumulation and regulation of biosynthesis of secondary metabolites, such as selection of cell lines, optimization of medium and culture conditions, elicitation, immobilization, nutrient and precursor feeding, permeabilization, and biotransformation techniques (Filova 2014; Bagnères and Hossaert-McKey 2016). Secondary metabolite accumulation in plants is genotype- and tissue-specific, so explants should be selected from elite parent plants and tissues, which have higher contents of desired compound, to initiate cell and organ cultures. Levels and types of various chemical components, such as carbohydrates, nitrate, phosphate, and growth regulators which could affect biomass accumulation and biosynthesis of secondary metabolites in plant cell and organ cultures, have been taken into consideration to optimize the medium. Agitation and aeration are also important factors that ought to be controlled in flask-scale to large-scale bioreactor cultures for optimization of biomass growth and secondary metabolite production (Murty et al. 2014). Secondary metabolites are being synthesized in plant cells to retort various abiotic (e.g., temperature, salinity, water, heavy metal, etc.) and biotic (e.g., pathogen or insects) stresses. Therefore, these stress factors have been designated as "elicitors" to induce biosynthesis of secondary metabolites. The number of parameters, such as concentration of elicitors, types, exposure duration, cell line, nutrient composition, and culture age or stage, is also crucial factors influencing the successful production of biomass and alkaloid accumulation (Naik and Al-Khayri 2016). Unfortunately, elicitation does not always result in product of interest because it activates certain plant species in the specific pathway. By utilizing preexisting enzyme systems, many plant cell cultures have also been used to convert precursors into products. Biotransformation is another technique which can be utilized for the high production of selected metabolites using plant cell and organ cultures (Bhatia and Bera 2015). In spite of this, the yield of secondary metabolites continues to be economically insufficient and expensive in many cases. Therefore, metabolic and genetic engineering techniques have also been incorporated into plant cell cultures to boost the production of secondary metabolites via regulation of their biosynthesis (Wilson and Roberts 2012). Some alkaloidal plant sources that have been used for in vitro cultures are mentioned in Table 6.1.

Table 6.1 Types of in vitro cultures of alkaloid producing various medicinal plants

Plant name	Active alkaloids	Culture medium	Culture type	Results	Reference
Atropa belladonna	Tropane alkaloids	Seeds + MS medium with 3% sucrose, 0.1 mg/l indole acetic acid (IAA) and 1 mg/l benzyladenine (BA)	Hairy root culture	The highest amount of atropine was observed; however, there were no differences in the amount of scopolamine, where the scopolamine content was significantly decreased	Chashmi et al. (2010)
Alstonia scholaris (Apocynaceae)	Indole alkaloids (echitamine, acetylechitamine, tubotaiwine, and picrinine)	Leaf + MS medium, 0.3 mg/l 2,4-D, 0.5 mg/l FAP and 3% sucrose. Elicited by MJ, PEG, and CHI	Callus culture	Enrichments of acetylechitamine (6.3780 mg/g DW, i.e., ~ 15-fold) and echitamine (1.6513 mg/g DW, i.e., ~ 12-fold) were found with 4.5 g/L KCl in 10 days incubation period, followed by tubotaiwine (0.0952 mg/g DW, i.e., ~ fourfold) with 3.0 g/L KCl in 10 days and picrinine (0.3784 mg/g DW, i.e., ~ fourfold) with 4.5 g/L KCl	Jeet et al. (2020)
Nicotiana rustica (Solanaceae)	Nicotine	Gamborg's B5 medium with 20 g/l sucrose and 3 g/l phytagel	Hairy root culture	Alteration of aeration results decreases nicotine accumulation	Zhao et al. (2013)
Papaver orientale (Papaveraceae)	Morphine, thebaine, codeine	Seed + Gamborg's B5 liquid medium, 3% sucrose, 300 mg/l cefotaxime, 1.0 g/l PVP, 15 mg/l ascorbic acid. Elicited by methyl jasmonate	Hairy root culture	Enhanced thebaine, codeine, and morphine by 2.63-fold (3.08 mg g-1), 3.67-fold (2.57 mg g-1), and 6.18-fold (5.38 mg g-1), respectively	Hashemi and Naghavi (2016)
Catharanthus roseus (Apocynaceae)	Vinca alkaloids	Leaf + MS medium supplemented with 1.5 mg/L BAP and 1.5 mg/L 2,4D	Callus culture	Vinblastine showed a 3.39-fold increase compared to the wild plant	Mekky et al. (2018)

(continued)

Table 6.1 (continued)

Plant name	Active alkaloids	Culture medium	Culture type	Results	Reference
Hyoscyamus niger (Solanaceae)	Tropane alkaloids	Leaf explants + MS medium, 3% sucrose supplemented with antibiotics (cefotaxime and amoxicillin, 500 mg/l)	Hairy root culture	The amount of cuscohygrine was (7.079 mg/g dry wt) more than 20-fold higher than the concentrations of anisodamine, therefore in order to determine the anisodamine content	Jaremicz et al. (2014)
Securinega suffruticosa (Phyllanthaceae)	Indolizine alkaloids	Callus + SH medium, 3% sucrose, 5.0 mg/l 2,4-D, 5.0 mg/l kinetin	Callus culture	The highest concentrations of securinine (1.73 mg g-1 DW) and allosecurinine (3.11 mg g-1 DW) were observed	Raj et al. (2015a, b)
Macleaya cordata (Papaveraceae)	Benzylisoquinoline alkaloids (protopine, sanguinarine, dihydrosanguinarine)	Leaf and stem + MS solid medium, 30 g/L sucrose, 8 g/L agar	Hairy root culture	The contents of 3 alkaloids (PROT, DHSAN, SAN) were significantly higher in hairy root cultures than in wild plant	Huang et al. (2018)
Hyoscyamus reticulatus (Solanaceae)	Hyoscyamine and scopolamine	Seeds + MS medium, 3% sucrose, 7.2 g/l agar, and 0.1 g/l myo-inositol and 200 mg/l cefotaxime elicited by iron oxide nanoparticles (FeNPs) at different concentrations (0, 450, 900, 1800, and 3600 mg L^{-1})	Hairy root culture	Highest hyoscyamine and scopolamine production (about fivefold increase over the control) was achieved with 900 and 450 mg L^{-1} FeNPs	Moharrami et al. (2017)
Hyoscyamus muticus (Solanaceae)	Hyoscyamine	Shoot tip + MS media +0.5 mg/l BAP, 0.5, 1 and 2 mg/l NAA, pH (5.7–5.8)	Callus culture	Total alkaloids increased by twofold at 10 dS/m compared to control or wild leaves	Abdelrazik et al. (2019)
Pancratium maritimum (Amaryllidaceae)	Amaryllidaceae alkaloids	Fruit slice + MS medium, 3% sucrose, 1.15 mg/L NAA and 2.0 mg/L BAP	Shoot culture	Twenty-two compounds of different structural types of the Amaryllidaceae alkaloids (tyramine, narciclasine, galanthamine, haemanthamine, lycorine, pancracine, tazettine, and homolycorine types) were detected in the studied samples	Georgiev et al. (2011)

6.4 Scale-Up Techniques and Bioreactors

The extraction method of alkaloids from the plant sources merely depends upon the objective and scale of the operation (pilot-scale or laboratory scale). It is also based on the quantum and bulk of stuff to be employed in the operation. For using commercially, it is required to develop the sufficient amount of alkaloids. A scale-up technique must be needed to obtain the plant's by-products, which is accomplished with no reduction in alkaloid productivity and bioactivity. A bioreactor (Figs. 6.5 and 6.6) is a device that supports a biologically active environment

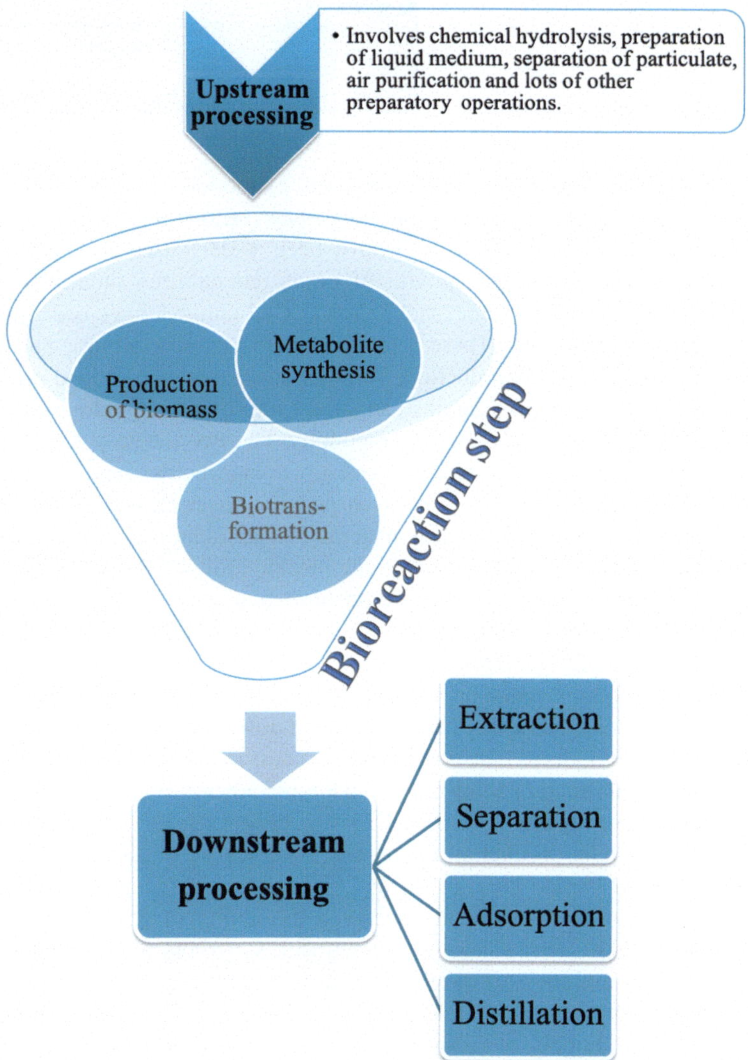

Fig. 6.5 Bioreactor processing

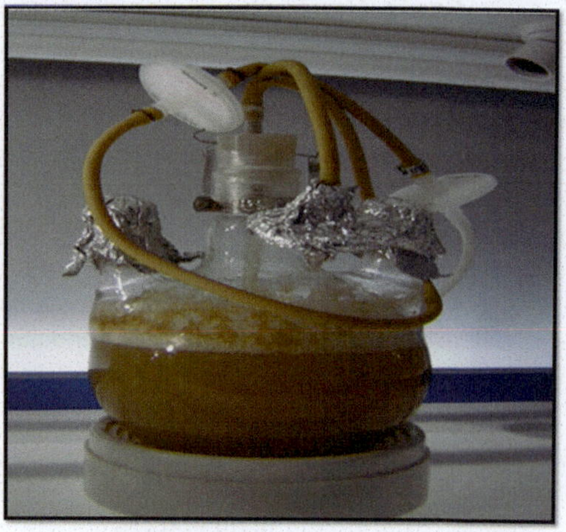

Fig. 6.6 Scale-up techniques for enhanced production of secondary metabolites

(aerobic or anaerobic) and allowing continuous extraction of alkaloids by using tissue culture techniques, i.e., hairy roots, suspension culture, etc. Various capacity and designs of bioreactors (Table 6.2) have been widely used for growing cell cultures of different plants, but growth of organized cultures should be started from a smaller capacity bioreactor (shake flasks). Cell cultures have been grown in both static and liquid cultures. These cell cultures are exposed to biotic and abiotic elicitors to increase alkaloid production. The biotransformation of added precursors and exploitation of variant cell strains can surely improve the employment of cell cultures for the production of desired compounds. The two facets of the examined bioreactor are upstream (elicitation, scale-up experiments) and downstream processing which involves permeabilization and in situ extraction (Ruffoni et al. 2010). In upstream processing, the raw material of alkaloidal source becomes more suitable for the processing which involves chemical hydrolysis, preparation of liquid medium, particulate separation, air purification, and lots of other preparatory operations. After that, the resulting feed is transferred to multiple bioreaction stages (Rosser and Thomas 2018). Three operations, production of biomass, metabolite biosynthesis and biotransformation, are included in the bioreaction step. Finally, the produced material must be further processed in the downstream section to transform it into a more beneficial form. The downstream process mainly comprises of physical separation operations such as solid-liquid separation, adsorption, distillation, liquid-liquid extraction, drying, etc. (Hatti-Kaul 2010).

A research study on production of tropane alkaloids by transformed hairy root cultures of *Atropa belladonna* in stirred bioreactors is reported. In this, the transformed roots of *A. belladonna* conserved the ability of growth and tropane alkaloid biosynthesis after a random cut treatment. Cut roots were inoculated and immobilized on a stainless-steel mesh, which resulted in the good distribution in the modified stirred bioreactor for a scale-up culture. This sort of bioreactor would help provide a sufficient supply of oxygen and nutrition for root growth and alkaloid production (Lee et al. 1999). Hairy root cultures of *Hyoscyamus niger* (black

Table 6.2 Different types of bioreactor used for enhancement of the alkaloid content

Bioreactor type		Plant source	Bioreactor conditions	Enhanced alkaloid	Reference
Liquid phase Submerged connective flow bioreactor	Stirred tank bioreactor	*Catharanthus roseus*	Air flow rate 4 vvm and stirring speed 100–120 rpm	Ajmalicine, catharanthine, serpentine	Verma et al. (2012)
	Bubble column bioreactor	*Datura stramonium*	Aeration rate 15.0 vvm	Tropane alkaloids	Marchev et al. (2012); Pavlov (2012)
		Papaver somniferum	Air flow rate 2 vvm and rotation speed 70–100 rpm	Sanguinarine	Verma et al. (2014)
		Uncaria tomentosa	Impeller tip speed 95 cm/s and agitation speed 400 rpm	Monoterpenoid oxindole alkaloid	Trejo-Tapia et al. (2005, 2007)
		Brugmansia candida	Air flow rate 0.5 vvm and agitation speed 50 rpm	Scopolamine, anisodamine, and hyoscyamine	Cardillo et al. (2010)
	Bubble column bioreactor	*Stephania glabra*	Air flow rate 0.1–1.0 vvm and agitation speed 30–65 rpm	Stepharine alkaloid	Titova et al. (2012)
		Securinega suffruticosa	Aeration rate 800 ml/min	Indolizidine alkaloids	Raj et al. (2015a, b)
		Catharanthus roseus	Aeration rate 0.3 vvm	Ajmalicine	Thakore et al. (2017); Fulzele and Namdeo (2018)
		Tripterygium wilfordii	Air flow rate 5 L/min, pressure 0.05 MPa	Wilforgine and wilforine (sesquiterpene)	Miao et al. (2013)
		Leucojum aestivum	Shaking at 50 rpm, immersion and gassing (continuous and discontinuous)	Galanthamine	Georgiev et al. (2012); Schumann et al. (2012); Ptak et al. (2013)
	Bubble column and spray bioreactor	*Hyoscyamus niger*	Aeration rate 0.8 vvm	Tropane alkaloids (scopolamine, cuscohygrine, anisodamine)	Jaremicz et al. (2014)

(continued)

Table 6.2 (continued)

Bioreactor type		Plant source	Bioreactor conditions	Enhanced alkaloid	Reference
	Air sparged and mechanically agitated bioreactor	*Rauwolfia serpentine*	Aeration rate 0.2–1.2 vvm and agitation rate 50–200 rpm	Indole alkaloids	Mehrotra et al. (2015a, b)
Gas phase	Balloon-type airlift bioreactor	*Dendrobium candidum*	Aeration rate 0.1 vvm, temp. 25 ± 2 C with 70% relative humidity	Alkaloids	Yang et al. (2015)
	Liquid-liquid impelled loop bioreactor	*Atropa belladonna*	Agitation (40, 70, 110 rpm) and aeration (0.75, 1.25, 1.75 vvm)	Scopolamine	Habibi et al. (2015)
	Siphon-mist bioreactor	*Pseudostellaria heterophylla*	Air flow rate 0.1 to 0.7 vvm by adjusting the gas pump	Alkaloids	Wang and Qi (2010)

henbane) were cultivated in shake flasks, a bubble-column bioreactor, and a hybrid bubble-column/spray bioreactor for anisodamine, scopolamine, hyoscyamine, and cuscohygrine alkaloids production (Jaremicz et al. 2014). *Brugmansia candida* produces tropane alkaloids (hyoscyamine, 6β-hydroxyhyoscyamine (anisodamine), and scopolamine) that have been widely applied in medicines (Cardillo et al. 2016). The chemical synthesis of alkaloids is complex and expensive; thereby the in vitro production of alkaloids by hairy roots cultures in bioreactor presents certain advantages over the natural source and chemical synthesis. Besides, the scaling-up of hairy root cultures makes this technology an attractive tool for industrial or commercial scale. The production of alkaloids in bioreactor guarantees that the process has been done under defined and controlled conditions, thus preventing or reducing the variations in the quality and yield of alkaloid compounds.

6.5 Extraction and Detection Techniques

Due to the high value of alkaloids, the worldwide researchers have tried to search out new and reliable methods for the extraction and detection of those compounds. Special methods have been developed for isolating commercially useful alkaloids.

In most cases, plant tissue is processed to get aqueous solutions of the alkaloids. The alkaloids are then recovered from the solution by a process called extraction, which involves dissolving some components of the mixture with compatible or suitable solvents/reagents that may be polar or nonpolar. This process requires either an acidic or alkaline/basic environment. Extraction techniques (Fig. 6.7), such as solid-liquid extraction (SLE), supercritical fluid extraction (SFE), microwave-assisted extraction (MAE), pressurized liquid extraction (PLE), solid-phase microextraction (SPME), supercritical carbon dioxide extraction method, and ultrasound-assisted method, have been used. Then, different alkaloids can be separated and purified from the mixture. A range of chromatographic techniques may be used for the efficient quantitative and qualitative analysis of alkaloids. Alkaloids in crystalline form are also obtained using certain solvents (Gupta et al. 2012; Zhu et al. 2018). Extraction of pure alkaloids from crude extract needs to be performed with multi-step chromatographic techniques.

It can be started with paper chromatography that is the easier way for the quantification of alkaloids. This method is rapid and cheaper. Further thin-layer chromatography is used. It is a reproducible method and has a low detection limit as compared to paper chromatography. After that, highly efficient chromatographic techniques can be employed, i.e., gas chromatography (GC), high-performance liquid chromatography (HPLC), capillary electrophoresis (CE), etc. (Maciel et al. 2019). These techniques are chosen accordingly to the nature of the alkaloidal sources.

Detection/analysis of the particular alkaloid with some specifications, mass spectrometry techniques can be used. MS (mass spectrometry) technique now plays a valuable role in the analysis of biomolecules, i.e., alkaloids, flavonoids, terpenes, etc. This revolution is realized by ESI-MS (electrospray ionization mass

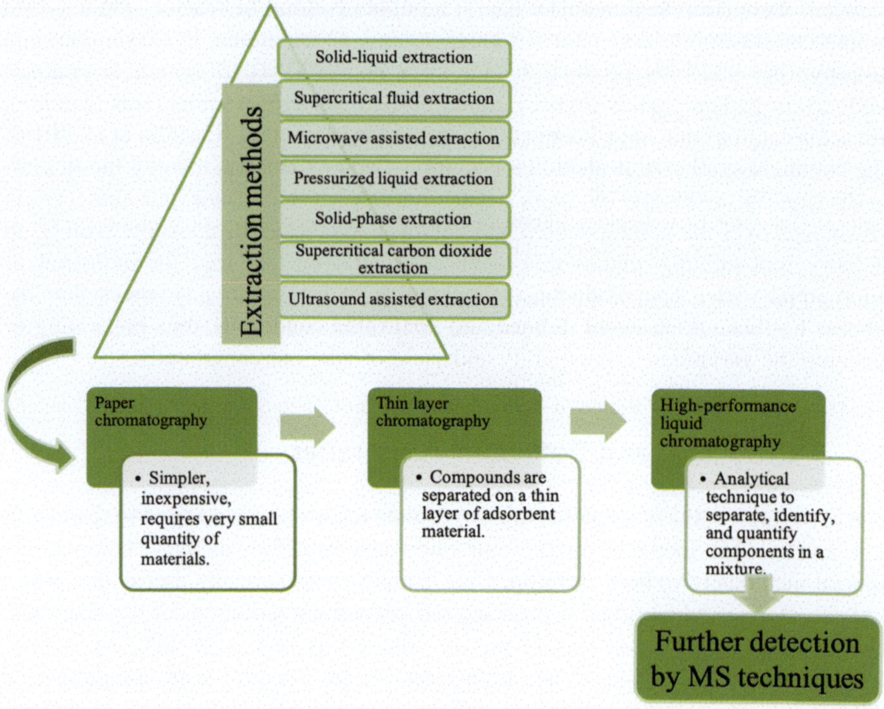

Fig. 6.7 Methods of extraction, purification, and detection of alkaloids

spectrometry) and MALDI-MS (matrix-assisted laser desorption ionization mass spectrometry) in the analysis of bio-polymeric products (Sasidharan et al. 2011). A completely unique method was developed for extraction and enrichment of the four alkaloids (nuciferine, *O*-nornuciferine, armepavine, and *N*-nornuciferine) from lotus leaf by coupling microwave-assisted extraction (MAE) with solid-phase microextraction (SPME) before ultra-high-performance liquid chromatography (UHPLC) analysis (Zou et al. 2020). In this recent report, the newly MAE-SPME is concluded as an efficient method for the extraction and enrichment for alkaloids from herbs (Zou et al. 2020). A two-dimensional analysis method endorsed high-performance liquid chromatography (HPLC) separation and electrospray ionization-ion mobility spectrometry (ESI-IMS) detection was developed for the evaluation of alkaloid compounds from *Peganum harmala* L. seeds. Their results reveal that this method is recognized to be advantageous over traditional absorbance detection methods for resolving complex mixtures due to complementary separation steps, elevated peak capacity, and better sensitivity (Wang et al. 2018). A simple, cost-effective salting-out assisted liquid-liquid extraction-based method for HPLC–DAD determination of khat (*Catha edulis*) alkaloids has been found to endow cleaner chromatogram with good selectivity and reproducibility. The salting-out assisted liquid-liquid extraction (SALLE)-based protocol provided good results as the conventional extraction method (ultrasonic-assisted extraction followed by solid-phase extraction, UAE–SPE), and hence the method can be applied in forensic and biomedical sectors (Atlabachew et al. 2017).

6.6 Biological Activities

The medicinal properties of alkaloids are quite diverse. Alkaloids generally exert biological activities particularly in humans (Koleva et al. 2012). Even today, many of the used drugs are natural alkaloids or made by them, and new alkaloidal drugs are still being developed for clinical uses. The activity of alkaloids against herbivores, cytotoxic activity, the molecular targets of alkaloids, mutagenic or carcinogenic activity, antibacterial, antifungal, and antiviral properties and their possible roles as phytoalexins have been evaluated (Debnath et al. 2018). Some alkaloids, i.e., morphine, codeine, nicotine, cocaine, etc., can be extremely harmful to animals/humans to cause death due to their dose-dependent toxicity if taken orally (Matsuura and Fett-Neto 2015).

6.6.1 Biological Activities of Pyridine Alkaloids Group

Nicotine obtained from the tobacco plant (*Nicotiana tabacum*) is the principal alkaloid and main ingredient of the tobacco smoked in cigarettes, cigars, and pipes. Nicotine binds to nicotinic cholinergic receptors. It facilitates neurotransmitter release and is liable for behavior modifying effects in individual. Stimulation of central nAChRs (nicotinic acetylcholine receptor) by nicotine leads to the release of a range of neurotransmitters in the brain. Nicotine-containing product is obtainable in the market to interrupt the habit of smoking. Nasal mucosa irritation, arthralgia, nausea, vomiting, and mild headache are most common adverse effect of nicotine (Benowitz 2009; Pang et al. 2016). Cytisine could be a selective nicotinic cholinergic agonist obtained from the seeds of *Laburnum anagyroides* of Leguminosae family. It also acts as nicotine for smoking cessation (Perez et al. 2012).

6.6.2 Biological Activities of Tropane Alkaloids Group

Many tropane alkaloids possess local anesthetic properties. **Atropine** (obtained from *Atropa belladonna*) is anticholinergic (Çaksen et al. 2003). It reduces the secretion such as sweat, saliva, and gastric juice. It competitively inhibits muscarinic acetylcholine receptor. The most prominent effect of atropine is tachycardia due to blockade of the M2 receptor present on SA node through which vagal tone decreases (Tripathi 2013). **Scopolamine** is available in the leaves of plant *Hyoscyamus niger* (Solanaceae). It is also known as hyoscine. It competitively inhibits muscarinic receptors and acts as a nonselective muscarinic antagonist. It produces both peripheral anti-muscarinic properties and also sedative, antiemetic, and amnestic effects

(Ullrich et al. 2017). **Cocaine** is isolated from the dried leaves of *Erythroxylum coca* and *Erythroxylon truxillense*, belonging to the family Erythroxylaceae, a very potent local anesthetic. The central action of cocaine is sympathetic and works as a CNS stimulant agent. Loss of sense in taste and smell (after given in the nose or mouth) are the most common side effects of cocaine (Manna et al. 2020). **Catuabine** is a tropane alkaloid obtained from the bark of *Trichilia catigua* belonging to the family Meliaceae. A pure catuabine found antidepressant-like effects on forced swim model of depression in mice and rats (Campos et al. 2005).

6.6.3 Biological Activities of Quinoline Alkaloids Group

Quinine and **quinidine** are obtained from the bark of *Cinchona officinalis* belonging to the family Rubiaceae. Quinine is used to treat malaria. It was the first anti-malarial drug used in the early 1600s (Achan et al. 2011). It has rapid schizonticidal action against intra-erythrocytic malaria parasites. Quinidine is the dextro isomer of the quinine alkaloid. It blocks myocardial Naþ channels and acts as antiarrhythmic drug to treat irregular rhythms of the heartbeat. It is effective antimalarial drug against *Plasmodium falciparum*. Reported adverse reactions of quinidine are diarrhea, nausea, and vomiting (Diaz et al. 2015). **Dihydroquinine** is a natural impure compound found in commercial pharmaceutical formulations of quinine. **Dihydroquinidine** also have similar bioactivity (antimalarial). Both alkaloids inhibit the actions of parasympathetic nervous system. Therefore, biological source of dihydroquinine and dihydroquinidine are same with quinine as these are obtained from the bark of *Cinchona officinalis* (Mehrotra et al. 2018).

6.6.4 Biological Activities of Isoquinoline Alkaloids Group

Papaverine is a benzylisoquinoline alkaloid that occurs in the plant *Papaver somniferum* belonging to family Papaveraceae. It acts on smooth muscle throughout the body and causes vasodilation and relaxation of smooth muscle tone (Shimizu et al. 2000). **Berberine** occurs in roots and stem bark of different species of *Berberis* belonging to the family Berberidaceae. *Berberis aristata, B. lyceum, B. petiolaris,* and *B. tinctoria* are the main sources of berberine (Srivastava et al. 2015). The most important biological activity of berberine is its anti-diabetic effect. It activates AMPK and improves insulin sensitivity in rodent models of insulin resistance (Turner et al. 2008). Berberine-induced apoptosis is associated with upregulated expressions of p53, and decreased vimentin expression. These results suggest that berberine can suppress cell growth (Han and Qi 2012). Other important pharmacological activities are anti-hypertensive, anti-inflammatory, antioxidant, antidepressant, and hepatoprotective activities (Amritpal et al. 2010).

6.6.5 Biological Activities of Phenanthrene Alkaloids Group

Codeine and morphine are present in dried latex of unripe capsules of *Papaver somniferum*. These are used as opioid analgesic. Morphine (10%) and codeine (0.5%) are present in opium. Morphine and codeine depress respiratory center in a dose-dependent manner. Morphine is an egregious narcotic used for the pain relief, though its addictive properties limit its usefulness. Codeine is a wonderful analgesic that is relatively nonaddictive. Death may occur due to respiratory failure at its high doses (Dehghan et al. 2010).

6.6.6 Biological Activities of Phenylethylamine Alkaloids Group

Ephedrine is obtained naturally from the plants *Ephedra vulgaris*, *E. sinica*, *E. major*, *E. gerardiana*, etc. of genus ephedra (Family: Ephedraceae). It is a sympathetic stimulant that directly acts on α- and β-receptor. It can be used to prevent low blood pressure during spinal anesthesia. It is also used as bronchodilator in asthmatic condition. Allergic condition like hay fever can be treated with ephedrine (Ma et al. 2007). **Hordenine** is a natural phenethylamine compound that occurs in barley grass (*Hordeum vulgare*), a cereal crop belonging to the family Poaceae. It is a nootropic (non-pharmaceutical cognitive enhancers) compound that enhances cognitive ability. It is an effective MAO-B inhibitor. Since it helps to increase the level of norepinephrine, it is considered as norepinephrine and noradrenaline uptake inhibitor (Debnath et al. 2018).

6.6.7 Biological Activities of Indole Alkaloids Group

Reserpine is isolated mostly from the root of *Rauwolfia serpentina* and *Rauvolfia vomitoria*. It is known as antipsychotic and antihypertensive (Bunkar 2017). **Ergotamine** and **ergometrine** are obtained from the rye fungus *Claviceps purpurea*. It can be used for uterine contraction, uterine bleeding, and postpartum hemorrhage after delivery, incomplete recovery of uterus, retrogression, etc. It causes constriction of peripheral and cranial blood vessels to control extra blood flow and produces depression of central vasomotor centers (Ma et al. 2018). **Yohimbine** is isolated from the bark of *Pausinystalia yohimbe* belonging to the family Rubiaceae. It is chemically identical to reserpine. It increases parasympathetic (cholinergic) activity and decreases sympathetic (adrenergic) activity by acting on peripheral autonomic nervous system. It has a mild anti-diuretic action and has effect on blood pressure. Headache and excessive sweating are common side effects of yohimbine (Cohen et al. 2016). **Vinblastine** and **vincristine** are extracted from the pink periwinkle plant, *Catharanthus roseus*, belonging to the family Apocynaceae (Das and Sharangi 2017). Vinblastine is an antineoplastic agent, and it inhibits mitosis at metaphase by interacting with tubulin (Alam et al. 2017). It also has immunosuppressant effect. Major side effects of vinblastine are cough, fever, and painful urination. Vincristine is employed for the treatment of some types of cancer like breast cancer, Hodgkin's disease, Kaposi's sarcoma, and testicular cancer. The antitumor activity of vincristine is same to vinblastine. Most common side effects of vincristine are blurred or double vision, constipation, difficulty in walking, drooping eyelids, headache, jaw pain, joint pain, lower back or side pain, and stomach cramps (Alam et al. 2017). **Ergine** is a D-lysergic acid amide (LSA) found in various species of vines belonging to the family Convolvulaceae and *Argyreia nervosa*. It is also isolated from rye fungus *Claviceps purpurea*. Ergine has psychedelic effects (Paulke et al. 2013).

6.6.8 Biological Activities of Purine Alkaloids Group

Caffeine is a purine alkaloid. It is found naturally in the seeds and leaves of the plants *Theobroma cacao* (Malvaceae) and *Thea sinensis* (Theaceae), respectively (Rusconi and Conti 2010). Caffeine is the most widely consumed stimulant drug in the world. It is also consumed in cold medications, analgesics, and anorectants and in CNS stimulant. CNS stimulation is the main pharmacological action of caffeine because it can also act on the peripheral adenosine receptor (A1) on adipocyte that suppresses lipolysis by inhibition of adenylate cyclase activity (Cappelletti et al. 2015).

6.6.9 Biological Activities of Imidazole Alkaloids Group

Pilocarpine is the main alkaloid of imidazole group, and L-histidine is the biosynthetic precursor of the imidazole moiety. Pilocarpine is isolated from the leaves of *Pilocarpus microphyllus* that belongs to the family Rutaceae. It has cholinergic properties to stimulate the parasympathetic system (bladder, tear ducts, sudoriferous, and salivary glands). This alkaloid is an elected drug for glaucoma treatment. It has been exploited to treat the xerostomy (dry mouth) of throat cancer caused by the chemotherapy. Small doses of pilocarpine generally cause fall in blood pressure but in higher doses elicit rise in blood pressure (Santos and Moreno 2004).

6.6.10 Biological Activities of Terpenoid Alkaloids Group

Capsaicin is a unique alkaloid found primarily in the fruit of the *Capsicum* genus like *Capsicum annum* and *Capsicum frutescens* belonging to the family Solanaceae. Capsaicin can be bonded to TRPV1, which is mainly expressed in the sensory neurons. It also acts in the gastrointestinal tract, for weight loss and as an analgesic. The common side effects of capsaicin are burning, itching, dryness, pain, redness, swelling, or soreness (Reyes-Escogido et al. 2011). **Choline** is found in diverse plant foods in small amounts. It is a constituent of cell and mitochondrial membranes and of the synaptical neurotransmitter acetylcholine. Hence, this supplement impacts different cycles, for example, lipid metabolism, signaling through secondary messengers, and methylation-dependent biosynthesis of molecules. Major side effects of choline are constipation, diarrhea, dizziness, drowsiness, and migraine (Corbin and Zeisel 2012).

6.7 Commercial Utilization and Prospects

Plant biotechnology techniques provide valuable tools to synthesize a wide range of alkaloids as obtained from plants, as well as novel compounds are also synthesized via biotransformation and genetic engineering tools. These alkaloids have been used in the various commercial products. The in vitro cultures (shoot, callus, suspension, and hairy root cultures) are found to be used as sustainable system for the production of various secondary metabolites. Over the past twenty years, the concept of plant-based production of high-quality pharmaceutical alkaloids has increased the research interest and offered critical advantages over traditional extraction systems. In this chapter, some approaches (techniques) discussed have proven that medicinal plants can be used efficiently to produce various pharmaceutical alkaloids for remedial applications.

6.8 Conclusions and Recommendations

This chapter gives an insight to the different aspects of tissue culture for the production of alkaloids under in vitro conditions and their biosynthesis scenario. The extended use of plant cell culture systems in recent years is probably due to a benignant understanding of the alkaloid pathway in economically important plants. Advancement in plant cell culture system could provide the cost-effective, commercial production of rare, endangered, or even exotic plants, their cells, and the bioactive molecules that they will produce. These discussed alkaloids are found beneficial for certain life-threatening disease and will serve to extend and enhance the continued usefulness of higher plants as renewable sources of chemicals, especially alkaloids.

References

Abdelrazik E, El-hamahmy M, Eltamany EE, Abuseidah A (2019) Enhancement of growth and alkaloids accumulation in *Hyoscyamus muticus* L. Callus cultures by high salt concentration. Rec Pharmaceut Biomed Sci 3(1):26–37

Achan AO, Talisuna AE, Yeka A, Tibenderana JK, Baliraine FN, Rosenthal PJ, D'Alessandro U (2011) Quinine, an old anti-malarial drug in a modern world: role in the treatment of malaria. Malar J 10:144

Alam MM, Naeem M, Khan MMA, Uddin M (2017) Vincristine and vinblastine anticancer catharanthus alkaloids: pharmacological applications and strategies for yield improvement. In: Naeem M, Aftab T, Khan M (eds) *Catharanthus roseus*. Springer, Cham, pp 277–307

Amritpal S, Sanjiv D, Navpreet K, Jaswinder S (2010) Berberine: alkaloid with wide spectrum of pharmacological activities. J Nat Prod (India) 3:64–75

Atlabachew M, Chandravanshi BS, Redi-Abshiro M (2017) Preparative HPLC for large scale isolation, and salting-out assisted liquid–liquid extraction based method for HPLC–DAD determination of khat (*Catha edulis* Forsk) alkaloids. Chem Cent J 11(1):1–10

Bagnères AG, Hossaert-McKey M (eds) (2016) Chemical ecology. ISTE Limited

Baque MA, Moh SH, Lee EJ, Zhong JJ, Paek KY (2012) Production of biomass and useful compounds from adventitious roots of high-value added medicinal plants in bioreactor. Biotechnol Adv 30:1255–1267

Beaudoin GA, Facchini PJ (2014) Benzylisoquinoline alkaloid biosynthesis in opium poppy. Planta 240:19–32

Benowitz NL (2009) Pharmacology of nicotine: addiction, smoking-induced disease, and therapeutics. Annu Rev Pharmacol Toxicol 49:57–71

Bhatia S, Bera T (2015) Classical and nonclassical techniques for secondary metabolite production in plant cell culture. In: Bhatia S, Bera T, Dahiya R, Bera T (eds) Modern applications of plant biotechnology in pharmaceutical sciences. Springer-Verlag, pp 231–291

Bunkar AR (2017) Therapeutic uses of *Rauwolfia serpentina*. Int J Adv Sci Res 2:23–26

Çaksen H, Odabas D, Akbayram S, Cesur Y, Arslan S, Üner A, Oner AF (2003) Deadly nightshade (*Atropa belladonna*) intoxication: an analysis of 49 children. Hum Exp Toxicol 22:665–668

Campos MM, Fernandes ES, Ferreira J, Santos AR, Calixto JB (2005) Antidepressant-like effects of *Trichilia catigua* (Catuaba) extract: evidence for dopaminergic-mediated mechanisms. Psychopharmacology (Berl) 182:45–53

Cappelletti S, Daria P, Sani G, Aromatario M (2015) Caffeine: cognitive and physical performance enhancer or psychoactive drug? Curr Neuropharmacol 13:71–88

Cardillo AB, Otálvaro AÁM, Busto VD, Talou JR, Velásquez LME, Giulietti AM (2010) Scopolamine, anisodamine and hyoscyamine production by *Brugmansia candida* hairy root cultures in bioreactors. Process Biochem 45(9):1577–1581

Cardillo AB, Talou JR, Giulietti AM (2016) Establishment, culture, and scale-up of *Brugmansia candida* hairy roots for the production of tropane alkaloids. In: Jain S (ed) Protocols for *in vitro* cultures and secondary metabolite analysis of aromatic and medicinal plants, 2nd edn. Humana Press, New York, pp 173–186

Chashmi NA, Sharifi M, Karimi F, Rahnama H (2010) Differential production of tropane alkaloids in hairy roots and *in vitro* cultured two accessions of *Atropa belladonna* L. under nitrate treatments. Z Naturforsch C 65(5–6):373–379

Chen SL, Yu H, Luo HM, Wu Q, Li CF, Steinmetz A (2016) Conservation and sustainable use of medicinal plants: problems, progress, and prospects. Chin Med 11(1):1–10

Cohen PA, Wang YH, Maller G, DeSouza R, Khan IA (2016) Pharmaceutical quantities of yohimbine found in dietary supplements in the USA. Drug Test Anal 8(3–4):357–369

Corbin KD, Zeisel SH (2012) Choline metabolism provides novel insights into nonalcoholic fatty liver disease and its progression. Curr Opin Gastroenterol 28:159

Das S, Sharangi AB (2017) Madagascar periwinkle (*Catharanthus roseus* L.): diverse medicinal and therapeutic benefits to humankind. J Pharmacogn Phytochem 6(5):1695–1701

Davies KM, Deroles SC (2014) Prospects for the use of plant cell cultures in food biotechnology. Curr Opin Biotechnol 26:133–140

Debnath B, Singh WS, Das M, Goswami S, Singh MK, Maiti D, Manna K (2018) Role of plant alkaloids on human health: a review of biological activities. Mater Today Chem 9:56–72

Dehghan E, Hosseini B, Naghdi BH, Shahriari AF (2010) Application of conventional and new biotechnological approaches for improving of morphinane alkaloids production. J Med Plants 9(35):33–50

Desgagné-Penix I (2020) Biosynthesis of alkaloids in Amaryllidaceae plants: a review. Phytochem Rev:1–23

Diaz G, Miranda IL, Diaz MAN (2015) Quinolines, isoquinolines, angustureine, and congeneric alkaloids—occurrence, chemistry, and biological activity. In: Phytochemicals—isolation, characterisation and role in human health. Intech, pp 141–162

Filova A (2014) Production of secondary metabolities in plant tissue cultures. Res J Agric Sci 46(1): 236–245

Frick KM, Kamphuis LG, Siddique KH, Singh KB, Foley RC (2017) Quinolizidine alkaloid biosynthesis in lupins and prospects for grain quality improvement. Front Plant Sci 8:87

Fulzele DP, Namdeo AG (2018) Cost effective pilot-scale ajmalicine production by *Catharanthus roseus* cell suspension cultures in a 100 lit bioreactor. Biotechnol J Int:1–15

Georgiev V, Ivanov I, Berkov S, Pavlov A (2011) Alkaloids biosynthesis by *Pancratium maritimum* L. shoots in liquid culture. Acta Physiol Plant 33(3):927–933

Georgiev V, Ivanov I, Berkov S, Ilieva M, Georgiev M, Gocheva T, Pavlov A (2012) Galanthamine production by *Leucojum aestivum* L. shoot culture in a modified bubble column bioreactor with internal sections. Eng Life Sci 12(5):534–543

Gupta A, Naraniwal M, Kothari V (2012) Modern extraction methods for preparation of bioactive plant extracts. Int J Appl Nat Sci 1(1):8–26

Habibi P, Piri K, Deljo A, Moghadam YA, Ghiasvand T (2015) Increasing scopolamine content in hairy roots of *Atropa belladonna* using bioreactor. Braz Arch Biol Technol 58(2):166–174

Hamilton AC (2004) Medicinal plants, conservation and livelihoods. Biodivers Conserv 13(8): 1477–1517

Han X, Qi Y (2012) Mechanisms involved in the cytotoxic effects of berberine on human colon cancer HCT-8 cells. Biocell 36:113e12

Hashemi SM, Naghavi MR (2016) Production and gene expression of morphinan alkaloids in hairy root culture of *Papaver orientale* L. using abiotic elicitors. Plant Cell Tiss Org Cult 125(1): 31–41

Hatti-Kaul R (2010) Downstream processing in industrial biotechnology. In: Vandamme EJ, Soetaert W (eds) Industrial biotechnology. Wiley, pp 279–322

He SM, Liang YL, Cong K, Chen G, Zhao X, Zhao QM, Zhang JJ, Wang X, Dong Y, Yang JL, Zhang GH (2018) Identification and characterization of genes involved in benzylisoquinoline alkaloid biosynthesis in *Coptis* species. Front Plant Sci 9:731

Huang P, Xia L, Liu W et al (2018) Hairy root induction and benzylisoquinoline alkaloid production in *Macleaya cordata*. Sci Rep 8(1):1–9

Hussain A, Qarshi IA, Nazir H, Ullah I (2012) Plant tissue culture: current status and opportunities. In: Leva A, Rinaldi L (eds) Recent advances in plant *in vitro* culture, pp 1–28

Jaremicz Z, Luczkiewicz M, Kokotkiewicz A, Krolicka A, Sowinski P (2014) Production of tropane alkaloids in *Hyoscyamus niger* (black henbane) hairy roots grown in bubble-column and spray bioreactors. Biotechnol Lett 36(4):843–853

Jeet A, Singh Y, Singh P, Nimoriya R, Bilung CJ, Kanojiya S, Mishra DK (2020) Strategies for indole alkaloids enrichment through callus culture from *Alstonia scholaris* (L.) R. Br. Plant Growth Regul 90(2):383–392

Kandar CC (2021) Secondary metabolites from plant sources. In: Pal D, Nayak AK (eds) Bioactive natural products for pharmaceutical applications, Advanced structured materials, vol vol 140. Springer, Cham, pp 329–377. https://doi.org/10.1007/978-3-030-54027-2_10

Kim N, Estrada O, Chavez B, Stewart C, D'Auria JC (2016) Tropane and granatane alkaloid biosynthesis: a systematic analysis. Molecules 21(11):1510

Koleva II, van Beek TA, Soffers AE, Dusemund B, Rietjens IM (2012) Alkaloids in the human food chain–natural occurrence and possible adverse effects. Mol Nutr Food Res 56(1):30–52

Lee KT, Suzuki T, Yamakawa T, Kodama T, Igarashi Y, Shimomura K (1999) Production of tropane alkaloids by transformed root cultures of *Atropa belladonna* in stirred bioreactors with a stainless steel net. Plant Cell Rep 18(7–8):567–571

Ma G, Bavadekar SA, Davis YM, Lalchandani SG, Nagmani R, Schaneberg BT, Khan IA, Feller DR (2007) Pharmacological effects of ephedrine alkaloids on human a1-and a2-adrenergic receptor subtypes. J Pharmacol Exp Ther 322:214–221

Ma YZ, Qiang GF, Du GH (2018) Ergometrine and ergotamine. In: Du GH (ed) Natural small molecule drugs from plants. Springer, Singapore, pp 237–242

Maciel EVS, de Toffoli AL, Lanças FM (2019) Current status and future trends on automated multidimensional separation techniques employing sorbent-based extraction columns. J Sep Sci 42(1):258–272

Manna K, Debnath B, Singh WS (2020) Major metabolites of certain marketed plant alkaloids. Front Nat Prod Chem 6(6):124–150

Marchev A, Georgiev V, Ivanov I, Pavlov A (2012) Cultivation of diploid and tetraploid hairy roots of *Datura stramonium* L in stirred tank bioreactor for tropane alkaloids production. J BioSci Biotechnol 1(3)

Matsuura HN, Fett-Neto AG (2015) Plant alkaloids: main features, toxicity, and mechanisms of action. Plant Toxins 2(7):1–15

Mehrotra S, Goel MK, Srivastava V, Rahman LU (2015a) Hairy root biotechnology of *Rauwolfia serpentina*: a potent approach for the production of pharmaceutically important terpenoid indole alkaloids. Biotechnol Lett 37(2):253–263

Mehrotra S, Srivastava V, Rahman LU, Kukreja AK (2015b) Hairy root biotechnology—indicative timeline to understand missing links and future outlook. Protoplasma 252(5):1189–1201

Mehrotra S, Mishra S, Srivastava V (2018) Hairy root cultures for monoterpene indole alkaloid pathway: investigation and biotechnological production. In: Mehrotra S, Mishra S, Srivastava V (eds) Hairy roots. Springer, Singapore, pp 95–121

Mekky H, Al-Sabahi J, Abdel-Kreem MFM (2018) Potentiating biosynthesis of the anticancer alkaloids vincristine and vinblastine in callus cultures of *Catharanthus roseus*. S Afr J Bot 114: 29–31

Miao GP, Zhu CS, Feng JT, Han J, Song XW, Zhang X (2013) Aggregate cell suspension cultures of *Tripterygium wilfordii* Hook. f. for triptolide, wilforgine, and wilforine production. Plant Cell Tiss Org Cult 112(1):109–116

Moharrami F, Hosseini B, Sharafi A, Farjaminezhad M (2017) Enhanced production of hyoscyamine and scopolamine from genetically transformed root culture of *Hyoscyamus reticulatus* L. elicited by iron oxide nanoparticles. In Vitro Cell Dev Biol Plant 53(2):104–111

Murty HN, Lee EJ, Paek KY (2014) Production of secondary metabolites from cell and organ cultures: strategies and approaches for biomass improvement and metabolite accumulation. Plant Cell Tiss Org Cult 118:1–16

Naik PM, Al-Khayri JM (2016) Abiotic and biotic elicitors–role in secondary metabolites production through *in vitro* culture of medicinal plants. In: Shanker AK, Shanker C (eds) Abiotic and biotic stress in plants-recent advances and future perspectives. IntechOpen, pp 247–277

Nielsen E, Temporiti MEE, Cella R (2019) Improvement of phytochemical production by plant cells and organ culture and by genetic engineering. Plant Cell Rep 38(10):1199–1215

Pang X, Liu L, Ngolab J, Zhao-Shea R, McIntosh JM, Gardner PD, Tapper AR (2016) Habenula cholinergic neurons regulate anxiety during nicotine withdrawal via nicotinic acetylcholine receptors. Neuropharmacology 107:294–304

Pathak S, Agarwal AV, Agarwal P, Trivedi PK (2019) Secondary metabolite pathways in medicinal plants: approaches in reconstruction and analysis. In: Singh S, Upadhyay S, Pandey A, Kumar S (eds) Molecular approaches in plant biology and environmental challenges. Energy, environment, and sustainability. Springer, Singapore, pp 339–364. https://doi.org/10.1007/978-981-15-0690-1_16

Paulke A, Kremer C, Wunder C, Achenbach J, Djahanschiri B, Elias A, Schwed JS, Hübner H, Gmeiner P, Proschak E, Toennes SW (2013) *Argyreia nervosa* (Burm. f.): receptor profiling of lysergic acid amide and other potential psychedelic LSD-like compounds by computational and binding assay approaches. J Ethnopharmacol 148(2):492–e497

Pavlov A (2012) Cultivation of diploid and tetraploid hairy roots of *Datura stramonium* L. in stirred tank bioreactor for tropane alkaloids production. J BioSci Biotechnol 1(3):211–216

Perez EG, Endez-Galvez CM, Cassels BK (2012) Cytisine: a natural product leads for the development of drugs acting at nicotinic acetylcholine receptors. Nat Prod Rep 29:555e567

Ptak A, Simlat M, Kwiecień M, Laurain-Mattar D (2013) *Leucojum aestivum* plants propagated in in vitro bioreactor culture and on solid media containing cytokinins. Eng Life Sci 13(3):261–270

Raj D, Kokotkiewicz A, Drys A, Luczkiewicz M (2015a) Effect of plant growth regulators on the accumulation of indolizidine alkaloids in *Securinega suffruticosa* callus cultures. Plant Cell Tiss Org Cult 123(1):39–45

Raj D, Kokotkiewicz A, Luczkiewicz M (2015b) Production of therapeutically relevant indolizidine alkaloids in *Securinega suffruticosa* in vitro shoots maintained in liquid culture systems. Appl Biochem Biotechnol 175(3):1576–1587

Ramawat KG (2021) An introduction to the process of cell, tissue, and organ differentiation, and production of secondary metabolites. In: Ramawat KG, Ekiert HM, Goyal S (eds) Plant cell and tissue differentiation and secondary metabolites, Reference series in phytochemistry. Springer, Cham, pp 1–22. https://doi.org/10.1007/978-3-030-30185-9_35

Ramawat KG, Goyal S (2019) Co-evolution of secondary metabolites during biological competition for survival and advantage: an overview. In: Mérillon JM, Ramawat KG (eds) Co-evolution of secondary metabolites, Reference series in phytochemistry. Springer, Berlin/Heidelberg, Germany, pp 3–17. https://doi.org/10.1007/978-3-319-76887-8_45-1

Ramawat KG, Mérillon JM (eds) (2013) Natural products: phytochemistry, botany and metabolism of alkaloids, phenolics and terpenes. Springer, Heidelberg, pp 1541–2662

Ramawat KG, Dass S, Mathur M (2009) The chemical diversity of bioactive molecules and therapeutic potential of medicinal plants. In: Ramawat K (ed) Herbal drugs: ethnomedicine to modern medicine. Springer, Berlin, pp 7–32. https://doi.org/10.1007/978-3-540-79116-4_2

Reyes-Escogido MD, Gonzalez-Mondragon EG, Vazquez-Tzompantzi E (2011) Chemical and pharmacological aspects of capsaicin. Molecules 16:1253–e1270

Rohini MR (2020) Biotechnological interventions for conservation and multiplication of threatened medicinal plants. In: Rajasekharan P, Wani S (eds) Conservation and utilization of threatened medicinal plants. Springer, Cham, pp 135–158

Rosser J, Thomas DJ (2018) Bioreactor processes for maturation of 3D bioprinted tissue. In: Thomas DJ, Jessop ZM, Whitaker IS (eds) 3D bioprinting for reconstructive surgery. Woodhead publishing, pp 191–215

Ruffoni B, Pistelli L, Bertoli A, Pistelli L (2010) Plant cell cultures: bioreactors for industrial production. In: Giardi MT, Rea G, Berra B (eds) Bio-Farms for Nutraceuticals, Advances in experimental medicine and biology, vol 698. Springer, Boston, pp 203–221

Rusconi M, Conti A (2010) *Theobroma cacao* L., the food of the gods: a scientific approach beyond myths and claims. Pharmacol Res 61(1):5–13

Santos AP, Moreno PR (2004) *Pilocarpus* spp.: a survey of its chemical constituents and biological activities. Rev Bras Ciencias Farm 40(2004):116–137

Sasidharan S, Chen Y, Saravanan D, Sundram KM, Latha LY (2011) Extraction, isolation and characterization of bioactive compounds from plants' extracts. Afr J Tradit Complement Altern Med 8(1)

Schumann A, Berkov S, Claus D, Gerth A, Bastida J, Codina C (2012) Production of galanthamine by *Leucojum aestivum* shoots grown in different bioreactor systems. Appl Biochem Biotechnol 167(7):1907–1920

Shimizu K, Yoshihara E, Takahashi M, Gotoh K, Orita S, Urakawa N, Nakajyo S (2000) Mechanism of relaxant response to papaverine on the smooth muscle of non-pregnant rat uterus. J Smooth Muscle Res 36:83–91

Srivastava S, Srivastava M, Misra A, Pandey G, Rawat AK (2015) A review on biological and chemical diversity in Berberis (Berberidaceae). EXCLI J 14:247–267

Thakore D, Srivastava AK, Sinha AK (2017) Mass production of Ajmalicine by bioreactor cultivation of hairy roots of *Catharanthus roseus*. Biochem Eng J 119:84–91

Titova MV, Reshetnyak OV, Osipova EA, Osip'yants AI, Shumilo NA, Oreshnikov AV, Nosov AM (2012) Submerged cultivation of *Stephania glabra* (Roxb.) Miers cells in different systems: specific features of growth and accumulation of alkaloid stepharine. Appl Biochem Microbiol 48(7):645–649

Trejo-Tapia G, Cerda-García-Rojas CM, Rodríguez-Monroy M, Ramos-Valdivia AC (2005) Monoterpenoid oxindole alkaloid production by *Uncaria tomentosa* (Willd) DC cell suspension cultures in a stirred tank bioreactor. Biotechnol Prog 21(3):786–792

Trejo-Tapia G, Sepúlveda-Jiménez G, Trejo-Espino JL, Cerda-García-Rojas CM, De La Torre M, Rodríguez-Monroy M, Ramos-Valdivia AC (2007) Hydrodynamic stress induces monoterpenoid oxindole alkaloid accumulation by *Uncaria tomentosa* (Willd) DC cell suspension cultures via oxidative burst. Biotechnol Bioeng 98(1):230–238

Tripathi KD (ed) (2013) Essentials of medical pharmacology, 7th edn. JP Medical Ltd, New Delhi

Turner LJY, Gosby A, To SW, Cheng Z, Miyoshi H, Taketo MM, Cooney GJ, Kraegen EW, James DE, Hu LH (2008) Berberine and its more biologically available derivative, dihydroberberine, inhibit mitochondrial respiratory complex I: a mechanism for the action of berberine to activate AMPactivated protein kinase and improve insulin action. Diabetes 57:1414–1418

Ullrich SF, Hagels H, Kayser O (2017) Scopolamine: a journey from the field to clinics. Phytochem Rev 16(2):333–353

Verma P, Mathur AK, Shanker K (2012) Growth, alkaloid production, rol genes integration, bioreactor up-scaling and plant regeneration studies in hairy root lines of *Catharanthus roseus*. Plant Biosyst 146(sup1):27–40

Verma P, Khan SA, Mathur AK, Ghosh S, Shanker K, Kalra A (2014) Improved sanguinarine production via biotic and abiotic elicitations and precursor feeding in cell suspensions of latex-less variety of *Papaver somniferum* with their gene expression studies and upscaling in bioreactor. Protoplasma 251(6):1359–1371

Waller GR (ed) (2012) Alkaloid biology and metabolism in plants. Springer

Wang GR, Qi NM (2010) Influence of mist intervals and aeration rate on growth and second metabolite production of *Pseudostellaria heterophylla* adventitious roots in a siphon-mist bioreactor. Biotechnol Bioprocess Eng 15(6):1059–1064

Wang Z, Kang D, Jia X, Zhang H, Guo J, Liu C, Liu W (2018) Analysis of alkaloids from *Peganum harmala* L. sequential extracts by liquid chromatography coupled to ion mobility spectrometry. J Chromatogr B 1096:73–79

Wilson SA, Roberts SC (2012) Recent advances towards development and commercialization of plant cell culture processes for the synthesis of biomolecules. Plant Biotechnol J 10(3):249–268

Yamada Y, Shimada T, Motomura Y, Sato F (2017) Modulation of benzylisoquinoline alkaloid biosynthesis by heterologous expression of CjWRKY1 in *Eschscholzia californica* cells. PLoS One 12(10):e0186953

Yang F, Wei NN, Gao R, Piao XC, Lian ML (2015) Effect of several medium factors on polysaccharide and alkaloid accumulation in protocorm-like bodies of *Dendrobium candidum* during bioreactor culture. Acta Physiol Plant 37(5):94

Zhao B, Agblevor FA, Ritesh KC, Jelesko JG (2013) Enhanced production of the alkaloid nicotine in hairy root cultures of *Nicotiana tabacum* L. Plant Cell Tiss Org Cult 113(1):121–129

Zhu MZ, Chen GL, Wu JL, Li N, Liu ZH, Guo MQ (2018) Recent development in mass spectrometry and its hyphenated techniques for the analysis of medicinal plants. Phytochem Anal 29(4):365–374

Ziegler J, Facchini PJ (2008) Alkaloid biosynthesis: metabolism and trafficking. Annu Rev Plant Biol 59:735–769

Zou S, Wang C, Li J, He J, Gao X, Chang YX (2020) Microwave assisted solid phase microextraction for extraction and selective enrichment of four alkaloids in lotus leaf. Sustain Chem Pharm 18:100345

In Vitro Production of Coumarins

7

Muneera Q. Al-Mssallem and Fatima Mohamed Alissa

Abstract

Coumarin is a multitudinous nutraceutical compound naturally presented in various part of the plant and can be also produced by microorganisms. Additionally, coumarin can be produced at large-scale through in vitro plant cell culture systems. Coumarins can be found in simple, linear, and angular form, and the biosynthesis pathway of coumarins involves hydroxylation, glycolysis, and cyclization. The extraction process of coumarins requires using polar and nonpolar solvents, and the identification and quantification of coumarins can be performed by chromatographic techniques. Coumarins have attracted considerable attention in recent years due to their potential biological and pharmaceutical properties such as anticancer, anti-inflammatory, antimicrobial, antioxidant, and anticonvulsant properties. These medicinal properties of coumarin and its derivatives demonstrate their significance as promising nutraceutical for multifunctional applications. This chapter mainly discusses biosynthesis, in vitro production, different extractions, and detection methods of coumarins. In addition, the biological activities and the commercial applications of coumarins are also discussed.

M. Q. Al-Mssallem (✉) · F. M. Alissa
Department of Food and Nutrition Sciences, College of Agriculture and Food Sciences, King Faisal University, Al-Ahsa, Saudi Arabia
e-mail: mmssallem@kfu.edu.sa; falissa@kfu.edu.sa

T. Belwal et al. (eds.), *Nutraceuticals Production from Plant Cell Factory*,
https://doi.org/10.1007/978-981-16-8858-4_7

Keywords

Anticancer · Anticonvulsant · Antioxidant · Biosynthesis · Chromatographic method · Coumarin · Extraction · Pharmaceutical

Abbreviations

bw	Body weight
DAD	Diode array detector
FTIR	Fourier transform mid-infrared
GC-MS	Chromatography-mass spectrometry
HPLC	High-performance liquid chromatography
HPLC-MS/MS	High-performance liquid chromatography-mass spectrometry
UHPLC	Ultra-high-performance liquid chromatography
UHPLC-DAD	Ultra-high-performance liquid chromatography-diode array detector
UV	Ultraviolet
V	Volt
W	Watt

7.1 Introduction

Coumarin is an essentially neutral molecule and characterized with its pleasant sweet-bitter taste (Venugopala et al. 2013; Arn and Acree 1998). It is an organic compound from a family of benzopyrones known as 1,2-benzopyrone (2H-1-benzopyran-2-one). Coumarin was first isolated from the tonka bean (*Dipteryx odorata*) in which the name coumarin comes from (Borges et al. 2005; Arn and Acree 1998).

Coumarins are formed of merged benzene and α-pyrone rings. Therefore, the prototypical compound of coumarin is known as 1,2-benzopyrone (Venugopala et al. 2013).

Coumarins naturally occur in a variety of higher plants such as cinnamon, *Angelica archangelica*, *Cinnamomum cassia* Blume, *Mikania laevigata*, *Lavandula angustifolia*, and ***Melilotus officinalis*** (Passari et al. 2014; Bourgaud et al. 2006; Miller et al. 1996). Coumarins are distributed throughout all parts of the plant including the roots, leaves, seeds, and fruits. The highest levels of coumarins have been isolated from fruits followed by seeds (Venugopala et al. 2013). It is evident that the accumulation of coumarins in plants can be triggered by various abiotic and biotic stresses (Bourgaud et al. 2006). Research on the effect of coumarin in plants have demonstrated that coumarin may possess a hormonal property (Bourgaud et al. 2006). Additionally, coumarins have been identified from fungi and bacteria (Venugopala et al. 2013; Arn and Acree 1998).

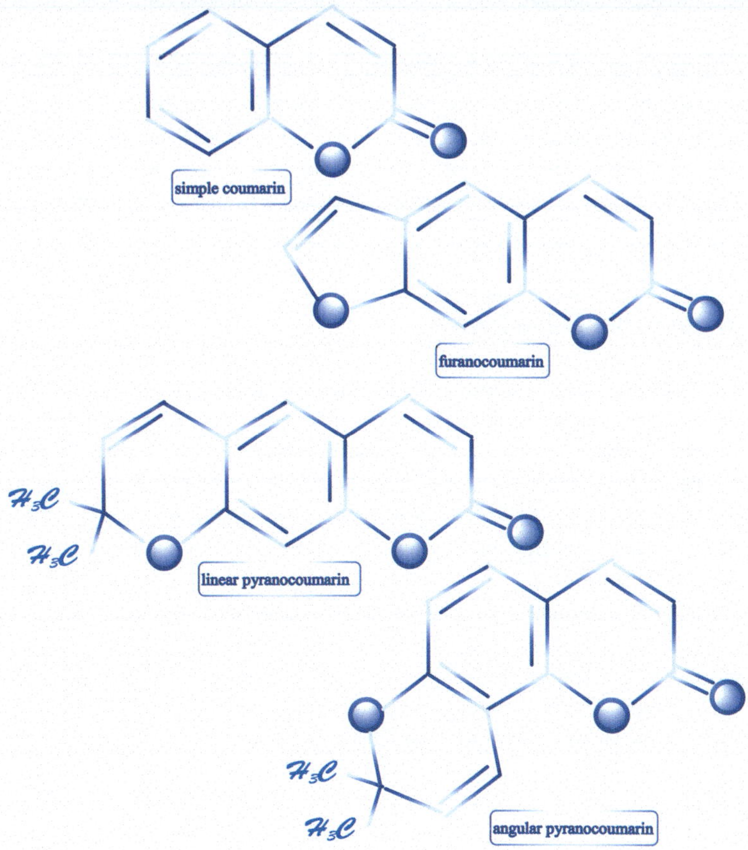

Fig. 7.1 Most common coumarin derivatives

Coumarins can be subcategorized into simple coumarins, pyranocoumarins, phenylcoumarins, and 7-oxygenated coumarin (Bourgaud et al. 2006). The most common hydroxylated coumarins are umbelliferone, herniarin and scoparone, isofraxidin, esculetin, fraxetin, isoscopoletin, daphnetin, and their corresponding glucosides (Bourgaud et al. 2006). However, the most common derivatives of coumarins are simple coumarins, lineal furocoumarins and pyranocoumarins, angular furocoumarins, and pyranocoumarins (Borges et al. 2005, 2009; Bourgaud et al. 2006; Dewick 2002). Simple coumarins are regarded as the most widespread in plants (Fig. 7.1).

Coumarins can bind to glycone by a glycosidic bond to form coumarin derivative called coumarin glycosides. This occurs when anthocyanin is converted to carbinol then into chalcone and finally cleaved into a coumarin derivative. Moreover, coumarin 3,5-diglycoside is also a common derivative that is produced from the degradation of anthocyanins (Rodriguez-Amaya 2019).

The dominant coumarin derivatives are isoimperatorin, oxypeucedanin, imperatorin, ostruthol, angelicin, bergapten, scopoletin, isopimpinellin, and xanthotoxin (Harmala and Vuorela 1990; Kumar et al. 2013).

In terms of health benefits of coumarins, they have exhibited multitudinous pharmacological activities such as anticancer, anti-inflammatory, antimicrobial, antihypertensive, anticoagulant, antioxidant, and anticonvulsant properties. However, isolated coumarin from cassia leaf exhibited cytotoxic effect (Venugopala et al. 2013).

7.2 Biosynthesis Pathway

Coumarins have been synthesized by a number of reactions such as Perkin, Knoevenagel, Reformatsky, and Pechmann reactions. The most popular example of Perkin reaction is that between acetic anhydride and salicylaldehyde (Venugopala et al. 2013; Dippy and Evans 1950). However, Knoevenagel reaction includes synthesis of coumarins by a solvent-free reaction under microwave irradiation and in the presence of piperidine through the condensation of salicylaldehyde or its derivatives with different derivatives of ethyl acetate (Bogdal 1998). The synthesis of coumarins by Pechmann reaction involves a condensation of a phenol and a carboxylic acid or ester group under acidic condition (Daru and Stirling 2011). It has been reported that the required conditions for these reactions include high temperature and strong basic or acidic reaction (Venugopala et al. 2013).

On the other hand, the biosynthesis pathway of coumarins consists of hydroxylation, glycolysis, and cyclization (Bourgaud et al. 2006; Dewick 2002). Hydroxylated coumarins are commonly stress induced in higher plant species (Bourgaud et al. 2006). In fact, coumarins biosynthesis is mostly triggered by various abiotic and biotic stresses in many plants. Additionally, coumarins and their derivative biosynthesis require specific enzymes and genes to be implicated in the biosynthesis process. It has been revealed that lactonization requires ortho-hydroxylation which is preceded by the para-hydroxylation (Bourgaud et al. 2006). The most common hydroxylated coumarins are umbelliferone, herniarin and scoparone, isofraxidin, esculetin, fraxetin, isoscopoletin, daphnetin, and their corresponding glucosides. Umbelliferone is obtained from cis-p-coumaric acid, while coumarin is derived from cis-coumaric acid (Bourgaud et al. 2006). In addition, complex coumarins could be also biosynthesized from prenylated simple coumarin (Bourgaud et al. 2006; Dewick 2002).

The key step of coumarin biosynthesis is ortho-hydroxylation. Nevertheless, coumarins, furocoumarins, and pyranocoumarins are genetically biosynthesized from cinnamic acid. Furanocoumarins can be found as linear furocoumarins and angular furocoumarins. Similarly, linear pyranocoumarins and angular pyranocoumarins have been distinguished as well. The classification is based on the location where the isopentenyl pyrophosphate is condensed to form the heterocycle. In linear type, the dihydrofuran ring is attached at C6 and C7, while it is linked at C7 and C8 in angular furocoumarins (Venugopala et al. 2013).

7.3 In Vitro Production Methodology

Several techniques have been applied for in vitro production of coumarins. In vitro plant cell culture techniques are commonly employed for production of coumarins. Conventional micropropagation approach is applied in vitro plant culture protocols. This approach involves a series of phases (Fig. 7.2). First phase is the selection of an appropriate plant when it is preferable to select insect- and disease-free plant. This first step is called pre-propagation. The second step is known as initiation of the in vitro plant culture. Explants or seeds are selected and their surface sterilization with chemicals. An appropriate culture media was used for the explants which were shortly incubated and any contaminated ones are disposed of, while the rest was used for the next step. The next step is the propagation phase when the shoot, root, or callus is proliferated. Then, the bioreactors are used for scaling up root and callus cultures. The micropropagation phase is used when the propagated shoots are transferred to culture media. Final step is called hardening when the

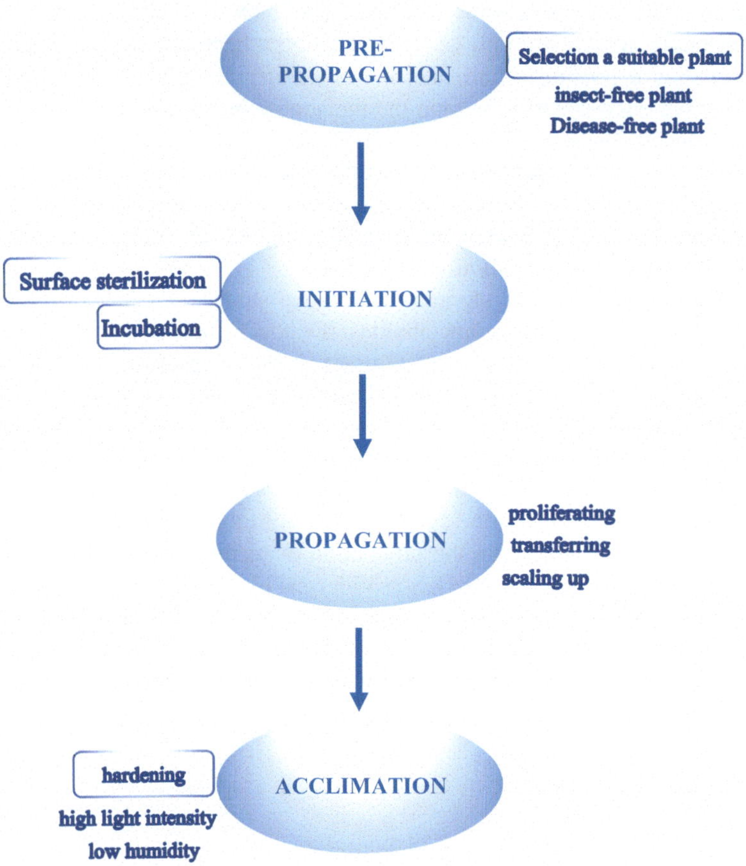

Fig. 7.2 Phases of in vitro production of coumarins

micropropagated plants are gradually hardened to allow the plants to acclimate to ex vitro conditions. In this last phase, the plants are moved from low to high light intensity and from high to low humidity (Espinosa-Leal et al. 2018).

In the unconventional micropropagation techniques, autotrophy of the explants is permitted to be developed in the last phase. An improvement in the quality of the multiplied shoots can be achieved through this autotrophic growth which in turn facilitates the acclimatization of the plantlets (Lucchesini and Mensuali-Sodi 2010).

The best in vitro plant cell culture technique for in vitro production of secondary metabolites including coumarins is cell suspension cultures which offer the most reliable and productive technique to generate phytochemical compounds (Ochoa-Villarreal et al. 2016). In this technique, the calli are first enhanced in a solid medium, and then all cells are transferred to a liquid medium allowing them for growing in shaking flasks and later transferred to a large-scale liquid-phase bioreactor (Furusaki and Takeda 2017). The in vitro production of coumarins can be enhanced by applying elicitors. Abiotic and biotic elicitors are available. Biotic elicitors can be extracted from plant cell wall molecules, plant immune signaling constituents, and microorganisms (Namdeo 2007; Ochoa-Villarreal et al. 2016). Abiotic elicitors include high pressure, high salinity, heavy metals, inorganic salts, and UV irradiation (Ochoa-Villarreal et al. 2016; Luo and He 2004). The selection of the appropriate elicitor will rely upon the metabolite being produced and the plant culture employed (Espinosa-Leal et al. 2018).

There are some factors that play an essential role on the in vitro plants production of coumarins such as the availability of nutrients, type and concentration of the growth regulators, and the type of culture media applied (Espinosa-Leal et al. 2018; Murthy et al. 2014; Fargoso Monfort et al. 2018).

Hairy root cultures are an important alternative strategy for in vitro production with high value of secondary metabolites (Ochoa-Villarreal et al. 2016). This technique is generated by transferring *Agrobacterium rhizogenes* T-DNA into the infected plant genome. To some extent, hairy root systems are genetically stable and easy to manage. A successful evidence in the production of a wide range of secondary metabolites has been achieved through hairy root cultures (Guillon et al. 2006; Ochoa-Villarreal et al. 2016). However, hairy root technique requires that the target molecule has to be synthesized within the given source plant roots. This is an essential limitation in the production of secondary metabolites by hairy root cultures (Guillon et al. 2006; Ochoa-Villarreal et al. 2016).

There is also a major limitation facing the selection of a suitable cell line because cell line is usually loose in its ability to produce the desired secondary metabolites due to genetic instability resulting from somaclonal variation (Ochoa-Villarreal et al. 2016).

7.4 Scale-Up Techniques and Bioreactors

In vitro plant production systems through applying biotechnologies are considered as an attractive approach for scaling up the production of coumarins. During the in vitro productions of coumarins, the propagated cells are scaled up using bioreactors (Ahloowalia et al., 2003; Furusaki and Takeda 2017). Scaling up plant cell cultures from laboratories to industrial large scales is an essential step for commercialization of coumarins production. Indeed, scaling up from laboratories to bioreactors is generally not straightforward as a result of modifications in the cell growth conditions in terms of rheological properties and hydrodynamic shear stresses (Sajc et al. 2000; Ochoa-Villarreal et al. 2016).

Several bioreactors have been successfully adapted for growing plant cells in vitro such as stirred tank bioreactor which is considered as the most widely exploited bioreactor. Recently, there are more improved bioreactor designs such as membrane bioreactors, wave bioreactors, and rotating drum bioreactors (Furusaki and Takeda 2017; Huang and McDonald 2012; Ochoa-Villarreal et al. 2016).

7.5 Extraction and Detection Techniques

Beside their existence in roots, leaves, seeds, and fruits of the plants, coumarins are also found in fungi and bacteria (Venugopala et al. 2013; Arn and Acree 1998). Therefore, extraction of coumarins from their natural sources requires different techniques including ultrasound-facilitated extraction, kinetic maceration, and microwaves-facilitated extraction. Three styles for these techniques are applied including serial ascendingly and descendingly ordered in polarity and non-serial. The solvents used in the extraction include n-hexane, chloroform, methanol, and water (Khalil and Mustafa 2020).

Coumarins are purified and isolated from the selected extract, and the isolated coumarins are chemically characterized to determine their biological activities (Khalil and Mustafa 2020; Mohammed and Mustafa 2020).

When serial style is applied, the coumarins are extracted from the powder of selected plant with water, methanol, chloroform, and n-hexane solvents in order. The residues are extracted by the next solvent in order, and the same methods are applied with the other solvents. Then, the three techniques of extractions were applied. Firstly, kinetic maceration extraction technique is conducted using a shaker water bath of the extracted mixture at 30 °C for 72 h. Secondly, ultrasound extraction technique is applied using ultrasonic water bath to sonicate the extracted mixture for 30 min at 30 °C (Fiorito et al. 2019; Khalil and Mustafa 2020; Mohammed and Mustafa 2020). Thirdly, microwaves-facilitated extraction technique is performed using a domestic microwave oven to irradiate the extracted mixture at 100 W for 5 min (Fiorito et al. 2019; Khalil and Mustafa 2020; Mohammed and Mustafa 2020). Finally, all the extracts from the three techniques are filtered through filter paper, concentrated by a rotary vacuum evaporator at 40 °C, and kept at 4 °C until further analysis. It has been reported that the use of the reduced pressure vacuum generated about 2.53%, 2.81%, 6.03%, 3.89%, and 8.27% of n-hexane, chloroform, acetone, ethyl acetate, and methanol extracts, respectively (Alagesan et al. 2019). It is important to note that polar solvents are more efficient than nonpolar solvents. Methanol possesses efficient extraction yield among other solvents and low toxicity; therefore, it can be used in food and pharmaceutical industry (Fiorito et al. 2019).

Chromatographic methods are recently considered as the most effective and rapid techniques used to detect and quantify coumarins. For example, gas chromatography-mass spectrometry (GC-MS) system was used to analyze coumarins (Alagesan et al. 2019). In addition, high-performance liquid chromatography (HPLC) technique was applied to quantify coumarins in *Cinnamomum cassia* Blume (Solaiman and Al-Zehouri 2017). It is important to note that simple coumarins have similar chemical structures and polarity, and it is hard to be separated precisely. Therefore, ultra-HPLC (UHPLC) method has been developed, and it applies a less organic solvent, runs rapidly, and gives a good peak separation. Furthermore, UHPLC-DAD method can be used for coumarin detection rapidly and routinely (Lončar et al. 2020). Moreover, HPLC-MS/MS has been applied for characterization and identification of coumarins (Li et al. 2019). The HPLC conditions with C 18 column include mobile phase: 0.1% aqueous formic acid (A) and 0.1% formic acid in acetonitrile (B). The MS triple quad was operated at 3500 V positive mode. MS has a selective detector to resolve the bioactive components and shorten employed gradient. However, the MS's purchase and maintenance are very expensive and unavailable for daily use (Li et al. 2019; Lončar et al. 2020). Since some of coumarins' structures include fluorophore, UV, fluorescence, and DAD detectors can be combined with HPLC for coumarins' detections. These kinds of detectors are widely available, can be devoted routinely, and are inexpensive. Proper monitoring of wavelengths is required to achieve high sensitivity and selectivity of the detection. Suitable wavelength is 320 nm for coumarins such as esculin, daphnetin, fraxetin, umbelliferone, 4-methylumbelliferone, and herniarin, while 280 nm was proven to be proper for coumarins such as 4-hydroxycoumarin, coumarin, and scoparone (Lončar et al. 2020; Li et al. 2019).

In addition, coumarins can be analyzed using Fourier transform mid-infrared spectroscopy (MID-FITR) combined with chemometric analysis (Moreno-Ley et al. 2019; Lončar et al. 2020). This method is simple and fast and had no need for reagents or pretreatments; however, it is more expensive than HPLC and unaffordable for commercial treatments. Therefore, analyzing coumarins by combining MID-FTIR spectroscopy and the HPLC-DAD is affordable for chimerical samples (Moreno-Ley et al. 2019). Scientists have been attentive in developing new, fast, non-expensive, and less use of organic solvents technique.

All mentioned above methods have been applied for their efficacy to detect and quantify coumarins, demonstrating their benefits and boundaries including cost, time of preparation and detection, availability, and use of organic solvents. The suitable method for quantification and detection must be proper for various sample of coumarins since they have diverse structures and polarity. In addition, highly sensitive technique is required to detect very low concentrations, selective, reproducible, accurate, non-expensive, and able to devote routinely.

7.6 Biological Activities

From chemical standpoint, coumarins are chemically represented by benzo-α-pyrone (2H-1-benzopiran-2-one), and they are organic heterocycle. Based on the chemical diversity and complexity of coumarins, they subdivided to multiple classes including simple coumarins, isocoumarins, furanocoumarins and pyranocoumarins, biscoumarins, and phenyl coumarins (Annunziata et al. 2020). Coumarins have been attracting scientists due to their simple structure, low molecular weight, and bioavailability. Moreover, they are highly soluble in polar and nonpolar organic solvents such as ethanol, methanol, water, chloroform, and n-hexane with low toxicity (Annunziata et al. 2020; Khalil and Mustafa 2020). Along with their numerous biological activities, they exhibit pharmacological and industrial potential including anticancer, antimicrobial, antioxidants, antidiabetic, anticonvulsant, anticoagulant, anti-inflammatory, neuroprotective, antiproliferative, anti-algal activity, luminescent properties, and fungicide properties (Pereira et al. 2018; Stefanachi et al. 2018; Zhu and Jiang 2018; Mark et al. 2019; Santra and Banerjee 2020; Annunziata et al. 2020).

It is well documented that human body cells are exposed to oxidative stresses through the accumulation of free radicals and highly reactive species including ions, oxygen, carbon, nitrogen, and sulfur species, which can lead to various pathologies such as inflammation, cardiovascular diseases, cancer, diabetes, and neurodegenerative disorders. It is important to maintain the radical's concentration, reduce the production of free radicals, and potentially inhibit the oxidative stress. Therefore, the natural exogenous antioxidants such as synthetic coumarins (polyhydroxy or phenolic coumarins) have been demonstrated as a potential antioxidant (Galano et al. 2016; Wahy et al. 2017; Annunziata et al. 2020). New series of coumarinyl pyrazolinyl thioamide derivatives have been designed to act as free radical scavengers and inhibitors of jack beam urease (Singh et al. 2019).

The applications of coumarin and its derivatives as anticancer agent are summarized underneath. It is well known that cancer is a term which includes various types of diseases, which can be caused by a multistep process involving different factors resulting in increasing genetic mutation and accumulation chance. Annually, it causes about nine million mortalities, and it is considered the second cause of death around the world (Zhang and Xu 2019). The effect of synthesized O-prenylated coumarin derivatives on HeLa cervical cancer has been studied, and the results showed that the substitutes 6-geranyloxycoumarin and 8-geranyloxycoumarin were the most effective anticancer activities (Maleki et al. 2020). Coumarin derivatives are highly promising anticancer drug due to their antiproliferative mechanisms of action, biodiversity, and versatility. Irosustat is a promising anticancer drug under clinical trials. Moreover, the synergic effect of combination coumarin motive with other available anticancer drugs can be a significant approach to decrease the side effect of drugs and the emergence of drug-resistant (Zhang and Xu 2019). A series of novel isatin-coumarin hybrid molecules are synthesized and designed through triazole ring. All isatin-coumarin hybrid exhibited anti-cancer activities against various cell cancer including leukemia cancer cells and colon cancer cells (COLO-205 and HCT-116). The reason behind this significant effect might be because of the structural feature of the isatin motility substitution and the linker between isatin and triazole length (Singh et al. 2017; Singh et al. 2019).

Coumarin and its derivatives can be also applied as a drug for preventing or alleviating degenerative diseases such as Alzheimer. The main causes of Alzheimer's disease can be β-amyloids aggregation, tau proteins formation, brain's neurotransmitter (Acetylcholine) degradation, and free radicals leading to oxidative stress (Singh et al. 2019). A novel series of coumarin-N-benzyl pyridinium hybrids were designed, and it demonstrated potential effect against Alzheimer's disease by inhibiting the acetylcholinesterase (AChE) enzyme, which is responsible for acetylcholine breaking down in the brain and MAO-B enzyme (Lan et al. 2017). It has been reported that there was an effect of synthesized and organized acrine-coumarin hybrids linked to 1,2,3-triazole on Alzheimer's disease (Najafi et al. 2019).

Moreover, coumarin and its derivatives exhibit antimicrobial activity. In fact, the emergence of microbial resistant has led the scientists to develop and investigate a novel of new antimicrobial drug to protect the future of human life and health (Singh et al. 2019). For example, a series of coumarins (dithioacetals derivatives) are characterized, synthesized, and investigated (Zhao et al. 2019). These compounds have a potential role against anti-tobacco mosaic virus biological activities. Also, coumarin-pargyline hybrids and multi-target tacrine-coumarin hybrids inhibited AChE, MAO-B enzymes, and aggregation of β-amyloid (Yang et al. 2017; Xie et al. 2015). Additionally, coumarin-based antimicrobial hybrid molecules exhibited antimicrobial effect against gram-positive *S. aureus* (Chavan et al. 2018). Coumarin-theophylline hybrid series exhibited antimicrobial effect against gram-positive bacteria (*S. aureus*) and gram-negative bacteria (*E. coli*, *S. Typhi*) and fungi (*C. albicans*) (Mangasuli et al. 2018). Scaffolds containing conjugates of 4-hydroxy coumarin and phenyl glyoxal series were generated and reported for their potential role as antimicrobial agents (Gupta et al. 2016). In addition, coumarins can be a great alternative among available conventional antibiotic drugs (Reen et al. 2018) because it has been proven to block the cell-to-cell communication, which is known as quorum sensing (QS), and inhibited biofilm formation. Biofilm formation presents the most antibiotic resistance, and it is hard to control. Coumarins have been reported to have potential effect not only on human infections but also can govern plant's pathogens, aquaculture infections, preventing food spoilage, and biofilm formation (Reen et al. 2018). Coumarin derivative 7-hydroxy-6-nitro-2H-1-benzopyran-2-one have been investigated for its ability as antifungal agents of various taxonomies of fungi such as Phycomycetes, Ascomycetes, Basidiomycetes, mycelial growth and development, and conidia formation in Aspergillus spp. (Dietrich and Valio 1973; Guerra et al. 2015; Prusty and Kumar 2019).

Coumarin and its derivatives also play an important role in some chronic diseases such as diabetes mellitus. It is well known that diabetes mellitus is known as a metabolic disorder, which presents due to the resistance or deficiency of insulin, and this can cause severe damages to body' organs. There are some enzymes that play a significant role in hyperglycemia such as α-glucosidase, α-amylase, and aldose reductase (Singh et al. 2019). It has been identified a novel hybrid molecule of natural flavonoids and coumarin for their antidiabetic effect including hybrids of apigenin, chrysin, quercetin, and luteolin (Pan et al. 2016). Also, a new series of 3-thiazolylcoumarin hybrids were designed by incorporating three pharmacophores (coumarin, thiazole, hydrazide) into a single entity due to their potential role in reducing α-glucosidase activity (Salar et al. 2016).

Beside their mentioned activities, coumarin and its derivatives exhibit anti-inflammatory property. Steroidal anti-inflammatory drugs (NSAIDs) are the major anti-inflammatory agent due to their effect in preventing prostaglandins release. The existent anti-inflammatory agents are primly targeting cyclooxygenase (COX) enzyme, which presents into forms of isoforms (Singh et al. 2019). The active pharmacophores of COX-2 inhibitors (celecoxib and valdecoxib) and 5-LOX enzyme inhibitor coumarin derivatives were combined and incorporated by hybridization method into single molecule, and it demonstrated a great effect as anti- inflammatory agents by inhibiting the COX and 5-LOX enzymes (Shen et al. 2010).

Coumarin hybridization with different active pharmacophores has demonstrated a potential role in enhancing potency of the novel molecules and general pharmacological profile.

These hybrid molecules have multi-target effects, which can increase the development of small therapeutics molecule, and can act as anticancer, antioxidant, antidiabetic, antimicrobial, and anti-inflammatory effects.

7.7 Commercial Utilization and Prospects

Coumarins and its derivatives are significantly used in diverse marketing sectors including food and pharmaceutical industries due to their multifunctional pharmaceutical and biological activities including antimicrobial, antiviral, antidiabetic, anticoagulant, estrogenic, dermal photosensitizing, vasodilator, molluscicidal, antithelmintic, sedative and hypnotic, analgesic, hypothermic, anticancer, antioxidant, antiparasitic, antihelmintic, antiproliferative, anticonvulsant, anti-inflammatory, and antihypertensive activities (Hussain et al. 2019; Sharma and Katiyar 2019). New products of coumarin and its derivatives with therapeutic efficacy against metabolic disorders and various kinds of human infections have been introduced (Nadeem and Ahmad 2019; Hussain et al. 2019), and there is a huge demand on accepted extraction methods to elicit most of coumarin and its derivatives (Nadeem and Ahmad 2019). They can be extracted from their natural sources such as different parts of plants, fungi, and microorganisms in addition to the in vitro production of coumarins (Hussain et al. 2019). Recently, coumarins hybrid molecules were newly designed and synthesized with different available pharmacophores. This strategy enhanced the biological activities because they are synergically acting through multiple modes of actions, which is more beneficial than single molecule (Peters et al. 2009; Solomon et al. 2009; Hussain et al. 2019). For example, the combination of stilbene and coumarin compound excreted significant anti-cancer properties through different mechanism of actions such as telomerase enzyme, protein kinase inhibition activities, and oncogene expression downregulating (Wu et al. 2014; Bronikowska et al. 2012; Hussain et al. 2019).

There is a potential effort to efficiently incorporate coumarins and its derivatives into dietary human intake as natural flavoring agent such as *Cinnamomum cassia*, *Anthoxanthum odoratum*, and *Dipteryx odorata* (Hussain et al. 2019; Sharma and Katiyar 2019). However, coumarins can be added only to food as naturally extracted flavoring agent from raw materials such as cinnamon, or it is naturally found in the food based on European regulations (Lončar et al. 2020). Therefore, the most exposure of coumarins in human diet comes from cinnamon since it's highly used in dessert such as cakes, cookies, and beverages. Cinnamon has been used widely in food, beverages, and pharmaceutical industry because of its culinary and medicinal characteristics. Because a great quantity of coumarins might be hepatotoxic, it is prohibited to supplement food with a pure coumarin. Due to the fact that great quantity of coumarins might be hepatotoxic, supplement of pure coumarin is prohibited. Therefore, the daily intake of coumarins from diet should not be exceeded 0.07 mg/kg bw day (Lončar et al. 2020).

7.8 Conclusions and Recommendations

Coumarins are polar heterocyclic secondary metabolites extracted from their natural sources of plants and microorganism. Additionally, coumarins can be produced in vitro through several approaches including in vitro cell suspension and hairy root cultures. However, scaling up plant cell cultures from laboratories to bioreactors is generally not straightforward due to modifications in the cell growth conditions. Several bioreactors have been successfully adapted for growing plant cells in vitro including membrane bioreactors and wave bioreactors. The extraction of coumarins from their natural sources requires different techniques including kinetic maceration and ultrasound-facilitated extraction. The most rapid and effective techniques for characterization and identification of coumarins are chromatographic approaches such as GC-MS, ultra-HPLC, HPLC-MS/MS, and MID-FITR technologies. Besides their benefit to plants as antipathogenic compounds, coumarin and its derivatives also play a crucial role as antioxidant, anticancer, anti-inflammatory, antimicrobial, antidiabetic, and anti-Alzheimer. The multifunctional bioactive compounds of coumarins and its derivatives are promising applications in food and pharmaceutical industries.

References

Ahloowalia BS, Prakash J, Savangikar VA, Savangikar C (2003) Plant tissue culture. Low cost options for tissue culture technology in developing countries. In: Proceedings of a technical meeting organized by the Joint FAO/IAEA Division of Nuclear Techniques in Food and Agriculture and held in Vienna, 26–30 August 2002. International Atomic Energy Agency, Vienna, pp 3–10. ISBN 92–0–115903–X

Alagesan V, Ramalingam S, Kim M, Venugopal S (2019) Antioxidant activity guided isolation of a coumarin compound from Ipomoea pescaprea (Convolvulaceae) leaves acetone extract and its biological and molecular docking studies. Eur J Integr Med 32:100984. https://doi.org/10.1016/j.eujim.2019.100984

Annunziata F, Pinna C, Dallavalle S, Tamborini L, Pinto A (2020) An overview of coumarin as a versatile and readily accessible scaffold with broad-ranging biological activities. Int J Mol Sci 21(13):4618. https://doi.org/10.3390/ijms21134618

Arn H, Acree TE (1998) Flavornet: a database of aroma compounds based on odor potency in natural products. Dev Food Sci 40:27. https://doi.org/10.1016/S0167-4501(98)80029-0

Bogdal D (1998) Coumarins: fast synthesis by Knoevenagel condensation under microwave irradiation. J Chem Res Synop (8):468–469. https://doi.org/10.1039/A801724G

Borges F, Roleira F, Milhazes N, Santana L, Uriarte E (2005) Simple coumarins and analogues in medicinal chemistry: occurrence, synthesis and biological activity. Curr Med Chem 12(8): 887–916. https://doi.org/10.2174/0929867053507315

Borges F, Roleira F, Milhazes N, Uriarte E, Santana L (2009) Simple coumarins: privileged scaffolds in medicinal chemistry. Front Med Chem Biol Inter 4:23–85. https://doi.org/10.2174/978160805207310904010023

Bourgaud F, Hehn A, Larbat R, Doerper S, Gonier E, Kellner S, Matern U (2006) Biosynthesis of coumarins in plants: a major pathway still to be unravelled for cytochrome P450 enzymes. Phytochem Rev 5:293–308. https://doi.org/10.1007/s11101-006-9040-2

Bronikowska J, Szliszka E, Jaworska D, Czuba ZP, Krol W (2012) The coumarin psoralidin enhances anticancer effect of tumor necrosis factor-related apoptosis-inducing ligand (TRAIL). Molecules 17(6):6449–6464. https://doi.org/10.3390/molecules17066449

Chavan RR, Hosamani KM, Kulkarni BD, Joshi SD (2018) Molecular docking studies and facile synthesis of most potent biologically active N-tert-butyl-4-(4-substituted phenyl)-2-((substituted-2-oxo-2H-chromen-4-yl) methylthio)-6-oxo-1, 6-dihydropyrimidine-5-carboxamide hybrids: an approach for microwave-assisted syntheses and biological evaluation. Bioorg Chem 78:185–194. https://doi.org/10.1016/j.bioorg.2018.03.007

Daru J, Stirling A (2011) Mechanism of the Pechmann reaction: a theoretical study. J Org Chem 76(21):8749–8755. https://doi.org/10.1021/jo201439u

Dewick PM (2002) Medicinal natural products: a biosynthetic approach, 2nd edn. Wiley, West Sussex. https://doi.org/10.1002/0470846275

Dietrich SM, Valio IFM (1973) Effect of coumarin and its derivatives on the growth of Pythium and other fungi. Trans Br Mycol Soc 61(3):461–469. https://doi.org/10.1016/S0007-1536(73)80116-0

Dippy JFJ, Evans RM (1950) The nature of the catalyst in the Perkin condensation. J Org Chem 15(3):451–456. https://doi.org/10.1021/jo01149a001

Espinosa-Leal CA, Puente-Garza CA, García-Lara S (2018) In vitro plant tissue culture: means for production of biological active compounds. Planta 248(1):1–18. https://doi.org/10.1007/s00425-018-2910-1

Fargoso Monfort LE, Vilela Bertolucci SK, Fabri Lima A, Alves de Carvalho A, Mohammed A, Fitzgerard Blank A et al (2018) Effects of plant growth regulators, different culture media and strength MS on production of volatile fraction com- position in shoot cultures of *Ocimum basilicum*. Ind Crop Prod 116:231–239. https://doi.org/10.1016/j.indcrop.2018.02.075

Fiorito S, Ianni F, Preziuso F, Epifano F, Scotti L, Bucciarelli T et al (2019) UHPLC-UV/Vis quantitative analysis of hydroxylated and O-prenylated coumarins in pomegranate seed extracts. Molecules 24(10):1963. https://doi.org/10.3390/molecules24101963

Furusaki S, Takeda T (2017) Bioreactors for plant cell culture. In: Reference module in life sciences. Elsevier, Dordrecht. https://doi.org/10.1016/b978-0-12-809633-8.09076-2

Galano A, Mazzone G, Alvarez-Diduk R, Marino T, Alvarez-Idaboy JR, Russo N (2016) Food antioxidants: chemical insights at the molecular level. Annu Rev Food Sci Technol 7:335–352. https://doi.org/10.1146/annurev-food.041715-033206

Guerra FQ, Araújo RS, Sousa JP, Pereira FD, Mendonça-Junior FJ, Barbosa-Filho JM et al (2015) Evaluation of antifungal activity and mode of action of new coumarin derivative, 7-hydroxy-6-nitro-2h-1-benzopyran-2-one, against Aspergillus spp. Evid Based Complement Alternat Med 2015. https://doi.org/10.1155/2015/925096

Guillon S, Trémouillaux-Guiller J, Pati PK, Rideau M, Gantet P (2006) Hairy root research: recent scenario and exciting prospects. Curr Opin Plant Biol 9:341–346. https://doi.org/10.1016/j.pbi.2006.03.008

Gupta S, Kushwaha B, Srivastava A, Maikhuri JP, Sankhwar SN, Gupta G et al (2016) Design and synthesis of coumarin–glyoxal hybrids for spermicidal and antimicrobial actions: a dual approach to contraception. RSC Adv 6(80):76288–76297. https://doi.org/10.1039/C6RA12156J

Harmala P, Vuorela H (1990) Optimization of the high-performance liquid chromatography of coumarins in Angelica archangelica with reference to molecular structure. J Chromatogr 507:367–380

Huang TK, McDonald KA (2012) Bioreactor systems for in vitro production of foreign proteins using plant cell cultures. Biotechnol Adv 30(2):398–409. https://doi.org/10.1016/j.biotechadv.2011.07.016

Hussain MI, Syed QA, Khattak MN, Hafez B, Reigosa MJ, El-Keblawy A (2019) Natural product coumarins: biological and pharmacological perspectives. Biologia 74:863–888. https://doi.org/10.2478/s11756-019-00242-x

Khalil RR, Mustafa YF (2020) Phytochemical, antioxidant and antitumor studies of coumarins extracted from granny smith apple seeds by different methods. Syst Rev Pharm 11(2):57–63. https://doi.org/10.5530/srp.2020.2.10

Kumar D, Bhat ZA, Kumar V, Shah MY (2013) Coumarins from Angelica archangelica Linn. and their effects on anxiety-like behavior. Prog Neuropsychopharmacol Biol Psychiatry 40:180–186. https://doi.org/10.1016/j.pnpbp.2012.08.004

Lan JS, Ding Y, Liu Y, Kang P, Hou JW, Zhang XY et al (2017) Design, synthesis and biological evaluation of novel coumarin-N-benzyl pyridinium hybrids as multi-target agents for the treatment of Alzheimer's disease. Eur J Med Chem 139:48–59. https://doi.org/10.1016/j.ejmech.2017.07.055

Li GJ, Wu H, Wang Y, Hung WL, Rouseff RL (2019) Determination of citrus juice coumarins, furanocoumarins and methoxylated flavones using solid phase extraction and HPLC with photodiode array and fluorescence detection. Food Chem 271:29–38. https://doi.org/10.1016/j.foodchem.2018.07.130

Lončar M, Jakovljević M, Šubarić D, Pavlić M, Buzjak Služek V, Cindrić I et al (2020) Coumarins in food and methods of their determination. Foods 9(5):645. https://doi.org/10.3390/foods9050645

Lucchesini M, Mensuali-Sodi A (2010) Plant tissue culture—an opportunity for the production of nutraceuticals. In: Giardi MT, Rea G, Berra B (eds) Bio-farms for nutraceuticals, Advances in experimental medicine and biology, vol 698. Springer, Boston, pp 185–202. https://doi.org/10.1007/978-1-4419-7347-4_14

Luo J, He GY (2004) Optimization of elicitors and precursors for paclitaxel production in cell suspension culture of Taxus chinensis in the presence of nutrient feeding. Process Biochem 39:1073–1079

Maleki EH, Bahrami AR, Sadeghian H, Matin MM (2020) Discovering the structure–activity relationships of different O-prenylated coumarin derivatives as effective anticancer agents in human cervical cancer cells. Toxicol In Vitro 63:104745. https://doi.org/10.1016/j.tiv.2019.104745

Mangasuli SN, Hosamani KM, Devarajegowda HC, Kurjogi MM, Joshi SD (2018) Synthesis of coumarin-theophylline hybrids as a new class of anti-tubercular and anti-microbial agents. Eur J Med Chem 146:747–756. https://doi.org/10.1016/j.ejmech.2018.01.025

Mark R, Lyu X, Lee JJ, Parra-Saldívar R, Chen WN (2019) Sustainable production of natural phenolics for functional food applications. J Funct Foods 57:233–254. https://doi.org/10.1016/j.jff.2019.04.008

Miller KG, Poole CF, Pawlowskí TMP (1996) Classification of the botanical origin of cinnamon by solid-phase microextraction and gas chromatography. Chromatographia 42:639–646. https://doi.org/10.1007/BF02267695

Mohammed ET, Mustafa YF (2020) Coumarins from red delicious apple seeds: extraction, phytochemical analysis, and evaluation as antimicrobial agents. Sys Rev Pharm 11(2):64–70. https://doi.org/10.5530/srp.2020.2.11

Moreno-Ley CM, Hernández-Martínez DM, Osorio-Revilla G, Tapia-Ochoategui AP, Dávila-Ortiz G, Gallardo-Velázquez T (2019) Prediction of coumarin and ethyl vanillin in pure vanilla extracts using MID-FTIR spectroscopy and chemometrics. Talanta 197:264–269. https://doi.org/10.1016/j.talanta.2019.01.033

Murthy HN, Lee EJ, Paek KY (2014) Production of secondary metabolites from cell and organ cultures: strategies and approaches for biomass improvement and metabolite accumulation. Plant Cell Tiss Org Cult 118:1–16. https://doi.org/10.1007/s11240-014-0467-7

Nadeem H, Ahmad F (2019) Prospects for the use of plant cell culture as alternatives to produce secondary metabolites. In: Akhtar M, Swamy M (eds) Natural bioactive compounds. Springer, Singapore, pp 153–182. https://doi.org/10.1007/978-981-13-7438-8_6

Najafi Z, Mahdavi M, Saeedi M, Karimpour-Razkenari E, Edraki N, Sharifzadeh M et al (2019) Novel tacrine-coumarin hybrids linked to 1,2,3-triazole as anti-Alzheimer's compounds: in vitro and in vivo biological evaluation and docking study. Bioorg Chem 83:303–316. https://doi.org/10.1016/j.bioorg.2018.10.056

Namdeo AG (2007) Plant cell elicitation for production of secondary metabolites: a review. Pharmacogn Rev 1:69–79

Ochoa-Villarreal M, Howat S, Hong S, Jang MO, Jin YW, Lee EK, Loake GJ (2016) Plant cell culture strategies for the production of natural products. BMB Rep 49:149–158. https://doi.org/10.5483/BMBRep.2016.49.3.264

Pan G, Zhao L, Xiao N, Yang K, Ma Y, Zhao X et al (2016) Total synthesis of 8-(6″-umbelliferyl)-apigenin and its analogs as anti-diabetic reagents. Eur J Med Chem 122:674–683. https://doi.org/10.1016/j.ejmech.2016.07.015

Passari LM, Scarminio IS, Bruns RE (2014) Experimental designs characterizing seasonal variations and solvent effects on the quantities of coumarin and related metabolites from Mikania laevigata. Anal Chim Acta 821:89–96. https://doi.org/10.1016/j.aca.2014.03.003

Pereira TM, Franco DP, Vitorio F, Kummerle AE (2018) Coumarin compounds in medicinal chemistry: some important examples from the last years. Curr Top Med Chem 18(2):124–148. https://doi.org/10.2174/1568026618666180329115523

Peters GJ, Rajendra Prasad VV, Lemos C, Sathish NK (2009) Design of new drug molecules to be used in reversing multidrug resistance in cancer cells. Curr Cancer Drug Targets 9(3):298–306. https://doi.org/10.2174/156800909788166619

Prusty JS, Kumar A (2019) Coumarins: antifungal effectiveness and future therapeutic scope. Mol Divers 24:1367–1383. https://doi.org/10.1007/s11030-019-09992-x

Reen FJ, Gutiérrez-Barranquero JA, Parages ML (2018) Coumarin: a novel player in microbial quorum sensing and biofilm formation inhibition. Appl Microbiol Biotechnol 102(5):2063–2073. https://doi.org/10.1007/s00253-018-8787-x

Rodriguez-Amaya DB (2019) Natural food pigments and colorants. In: Mérillon J-M, Ramawat KG (eds) Bioactive molecules in food, Reference series in phytochemistry. https://doi.org/10.1007/978-3-319-78030-6_12

Sajc L, Grubisic D, Vunjak-Novakovic G (2000) Bioreactors for plant engineering: an outlook for further research. Biochem Eng J 4:89–99

Salar U, Taha M, Khan KM, Ismail NH, Imran S, Perveen S et al (2016) Syntheses of new 3-thiazolyl coumarin derivatives, in vitro α-glucosidase inhibitory activity, and molecular modeling studies. Eur J Med Chem 122:196–204. https://doi.org/10.1016/j.ejmech.2016.06.037

Santra HK, Banerjee D (2020) Natural products as fungicide and their role in crop protection. In: Singh J, Yadav A (eds) Natural bioactive products in sustainable agriculture. Springer, Singapore, pp 131–219. https://doi.org/10.1007/978-981-15-3024-1_9

Sharma RK, Katiyar D (2019) Recent advances in the development of coumarin derivatives as antifungal agents. In: Singh K, Srivastava N (eds) Recent trends in human and animal mycology. Springer, Singapore, pp 235–263. https://doi.org/10.1007/978-981-13-9435-5_10

Shen Q, Shao J, Peng Q, Zhang W, Ma L, Chan AS et al (2010) Hydroxycoumarin derivatives: novel and potent α-glucosidase inhibitors. J Med Chem 53(23):8252–8259. https://doi.org/10.1021/jm100757r

Singh H, Singh JV, Bhagat K, Gulati HK, Sanduja M, Kumar N et al (2019) Rational approaches, design strategies, structure activity relationship and mechanistic insights for therapeutic coumarin hybrids. Bioorg Med Chem 27(16):3477–3510. https://doi.org/10.1016/j.bmc.2019.06.033

Singh H, Singh JV, Gupta MK, Saxena AK, Sharma S, Nepali K et al (2017) Triazole tethered isatin-coumarin based molecular hybrids as novel antitubulin agents: design, synthesis, biological investigation and docking studies. Bioorg Med Chem Lett 27(17):3974–3979. https://doi.org/10.1016/j.bmcl.2017.07.069

Solaiman R, Al-Zehouri J (2017) Determination of coumarin in methanol extract of cinnamon (Cinnamomum cassia Blume) using reversed-phase high performance liquid chromatography. J Pharmacogn Phytochem 6:726–729

Solomon VR, Hu C, Lee H (2009) Hybrid pharmacophore design and synthesis of isatin–benzothiazole analogs for their anti-breast cancer activity. Bioorg Med Chem 17(21): 7585–7592. https://doi.org/10.1016/j.bmc.2009.08.068

Stefanachi A, Leonetti F, Pisani L, Catto M, Carotti A (2018) Coumarin: a natural, privileged and versatile scaffold for bioactive compounds. Molecules 23(2):250. https://doi.org/10.3390/molecules23020250

Venugopala KN, Rashmi V, Odhav B (2013) Review on natural coumarin lead compounds for their pharmacological activity. Biomed Res Int 2013:963248. https://doi.org/10.1155/2013/963248

Wahy AH, Ismail AR, Abou Kana MT, Negm NA (2017) Synthesis and characterization of novel bis-(4-methylcoumarin) derivatives as photosensitizers in antimicrobial photodynamic therapy. J Taiwan Inst Chem Eng 77:83–91. https://doi.org/10.1016/j.jtice.2017.05.006

Wu XQ, Huang C, Jia YM, Song BA, Li J, Liu XH (2014) Novel coumarin-dihydropyrazole thioethanone derivatives: design, synthesis and anticancer activity. Eur J Med Chem 74:717–725. https://doi.org/10.1016/j.ejmech.2013.06.014

Xie SS, Wang X, Jiang N, Yu W, Wang KD, Lan JS et al (2015) Multi-target tacrine-coumarin hybrids: cholinesterase and monoamine oxidase B inhibition properties against Alzheimer's disease. Eur J Med Chem 95:153–165. https://doi.org/10.1016/j.ejmech.2015.03.040

Yang HL, Cai P, Liu QH, Yang XL, Li F, Wang J et al (2017) Design, synthesis and evaluation of coumarin-pargyline hybrids as novel dual inhibitors of monoamine oxidases and amyloid-β aggregation for the treatment of Alzheimer's disease. Eur J Med Chem 138:715–728. https://doi.org/10.1016/j.ejmech.2017.07.008

Zhang L, Xu Z (2019) Coumarin-containing hybrids and their anticancer activities. Eur J Med Chem 181:111587. https://doi.org/10.1016/j.ejmech.2019.111587

Zhao L, Zhang J, Liu T, Mou H, Wei C, Hu D et al (2019) Design, synthesis, and antiviral activities of coumarin derivatives containing dithioacetal structures. J Agric Food Chem 68(4):975–981. https://doi.org/10.1021/acs.jafc.9b06861

Zhu JJ, Jiang JG (2018) Pharmacological and nutritional effects of natural coumarins and their structure–activity relationships. Mol Nutr Food Res 62(14):1701073. https://doi.org/10.1002/mnfr.201701073

In Vitro Production of Terpenoids

8

Sandeep Ramchandra Pai

Abstract

Plants produce diverse groups of secondary metabolites (SMs); terpenoids are one such large group of SMs. Terpenoids are found commonly in related and unrelated plant taxa; however, some specific terpenoids are also reported in lower and higher taxa. The pharmacological importance and commercial utilization of terpenoids are on the top among the plant SMs. A number of reviews on biosynthesis of terpenoids have been published suggesting 2-*C*-methyl-D-erythritol-4-phosphate (MEP) and the mevalonate (MVA) as common pathways. Today, researchers are working on target-specific production of terpenoids by altering metabolic pathways and expressing genes in microsystems. Recent high-throughput analytical techniques coupled to functional genomics approach has geared up biosynthesis and overproduction of terpenoids. In this chapter, terpenoids have been reviewed in detail for their sources, biosynthetic pathways, in vitro production technologies, scale-up techniques, and biological activities. The ecological and environmental perspectives for function of terpenoids have also been discussed. Considering commercial implications of terpenoids in therapeutic, perfumery, food, flavor, and fuel industries, a comprehensive account on their prospective future has been concentrated upon. Extraction and detection methodologies for terpenoids have been focused. Attention has also been drawn toward the need for designing possible roadmap for its sustainable utilization.

Keywords

Terpenoids · Medicinal plants · Extraction · Scale-up · Activities · Utilization

S. R. Pai (✉)
Department of Botany, Rayat Shikshan Shanstha's Dada Patil Mahavidyalaya, Karjat, Maharashtra, India

T. Belwal et al. (eds.), *Nutraceuticals Production from Plant Cell Factory*,
https://doi.org/10.1007/978-981-16-8858-4_8

8.1 Introduction

Interest in medicinal plants has increased in past few decades due to their low cost and safety. Herbal drug is a quick growing sector worldwide; it is estimated that the world trade in medicinal plants will reach US$ 5 trillion by 2050 (Anonymous 2000). Terpenoids are naturally occurring hydrocarbons of plant origin with general formula $(C_5H_8)_n$. The C_5H_8 unit is called isoprene unit (2-methyl-1,3-butadiene). Terpenoids constitute about 30,000 identified compounds (Dzubak et al. 2006; Mufflera et al. 2011). It is reported that terpenoids are probably the largest group of phytochemicals. They include essential oils phytohormones (cytokinin and gibberellins), resins, steroids, carotenoids, and others (Lohr et al. 2012). Terpenoids or isoprenoids are a group of structurally diverse phytochemicals known for their wide range of pharmacological activities. The term terpene is used for compounds with C_5 (isoprene) units. Terpenoids are classified on the basis of number of isoprene units (Table 8.1).

The isoprene units have a branched and an unbranched ends, earlier termed as head and the later as tail. Although head-tail arrangements of the isoprene units are the commonest type of linking, e.g., in mono-, sesqui-, di-, and sesterterpenoids, tail-tail arrangements are at center of tri- and tetraterpene molecules. Polyisoprenoids like rubber and gutta are two of the many other naturally occurring compounds in which non-terpenoid structural parts are linked to terpenoids. Structure and examples of terpenoids are depicted in Table 8.2.

Apart from this, lower plants like bryophytes are also reviewed for the presence of terpenoids. It is estimated that in the past 40 years, around 1600 terpenoids have been reported from this plant group (Asakawa 1982, 1995; Asakawa et al. 2013). Ludwiczuk and Asakawa (2019), in their review on bryophytes, have enlisted over 128 studied pharmacologically active terpenoids from liverworts and mosses. Many of them are species specific like dumortane type (*Dumortiera hirsute)*, hodgsonoxanes (*Lepidolaena hodgsoniae*), bergamotanes, and clavigerins (*L. clavigera*) (Chen et al. 2018). A mini review by Pai and Joshi (2014) have enlisted about 56 plant species producing triterpenoid betulinic acid. Very recently, 7 undescribed terpenoids, alongside 26 known compounds with isolated from aerial parts of *Elsholtzia rugulosa* (Yang et al. 2021). In a separate study, Yu et al. (2020) reported six new terpenoids from *Eclipta prostrata*, thus indicating continuation of

Table 8.1 Classification of terpenoids

Classification	Isoprene unit/s	Carbon atoms
Hemiterpenoids	1	C_5
Monoterpenoids	2	C_{10}
Sesquiterpenoids	3	C_{15}
Diterpenoids	4	C_{20}
Sesterterpenoids	5	C_{25}
Triterpenoids	6	C_{30}
Tetraterpenoids	8	C_{40}
Polyterpenoids	>8	$(C5)_n$

Table 8.2 Few examples and structures of terpenoids with molecular weight and compound identity (CID) from PubChem (National Center for Biotechnology Information 2021a, b, c, d, e, f, g, h, i, j, k, l, m, n, o, p)

Group	Monoterpenoids			
Structure	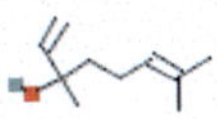		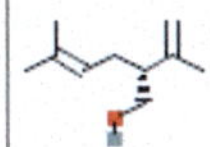	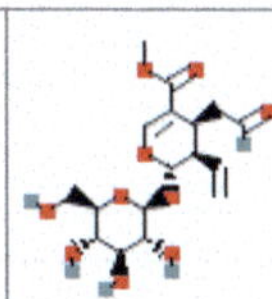
Example	Linalool	Menthol	Lavandulol	Secologanin
MW	154.25	156.26	154.25	388.4
CID	6549	1254	5,464,156	161,276
Group	Sesquiterpenoids		Diterpenoids	
Structure	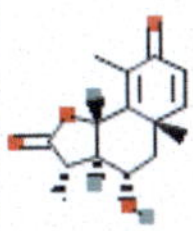		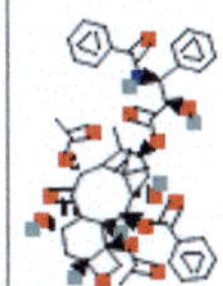	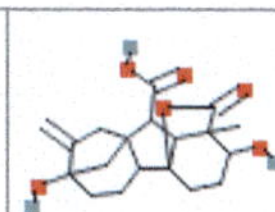
Example	Artemisinin	Zingiberene	Taxol	Gibberellin 1
MW	262.3	204.35	853.9	348.4
CID	65,030	92,776	36,314	3,509,874
Group	Triterpenoids			
Structure	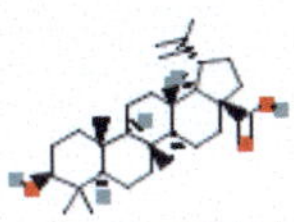	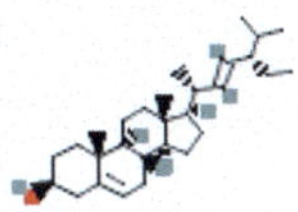	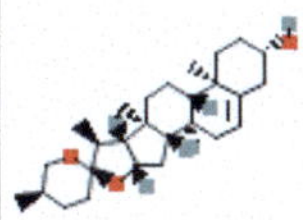	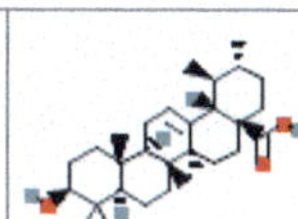
Example	Betulinic acid	Stigmasterol	Diosgenin	Ursolic acid
MW	456.7	412.7	414.6	456.7
CID	64,971	5,280,794	99,474	64,945
Group	Tetraterpenoids		Polyprenols	Dolichols
Structure	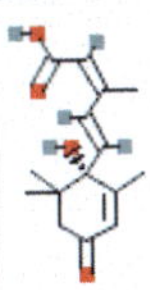	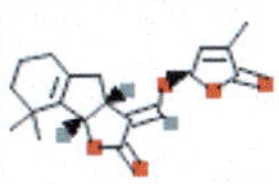	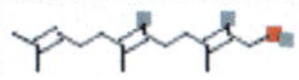	
Example	Abscisic acid	Deoxystrigol	Farnesol	Dolichol-20
MW	264.32	330.4	222.37	1382.4
CID	5,280,896	15,102,684	445,070	6,433,320

adding new terpenoids to the family. Below is the list of some plants, parts, and terpenoids reported (Table 8.3).

Isoprenoids have reported to be having functionalities in plant cells. They serve as starting material in the formation of various plant hormones. Differential expression of pathway isozymes; metabolic regulation and network, alongside regulation to light, external stimuli; and metabolic exchanges across the subcellular compartments have been well documented by Tholl (2015). Terpenoids like sterols function in

Table 8.3 Various terpenoids of plant origin

Plant name	Part	Terpenoids	Reference
Origanum majorana, Rosmarinus officinalis, Ocimum basilicum, Juniperus communis, Piper nigrum, Laurus nobilis, Lavandula latifolia, Mentha spicata, Zingiber officinale	WP	a-Terpinene	Ercioglua et al. (2018)
		g-Terpinene	
		Limonene	
		r-cymene	
		Myrcene	
		!-Pinene	
		b-Pinene	
		Sabinene	
		Carene	
		Camphene	
		Terpineol	
		Linalool	
		Linalyl acetate	
		Borneol	
		Methyl eugenol	
		Carvone	
		Camphor	
		Ar-Curcumene	
		Eucalyptol	
		Thymol	
		Carvacrol	
Perovskia atriplicifolia	WP	Biperovskatone B	Liu et al. (2018)
		1α- hydroxyl demethylsalvicanol quinine	
Heteroscyphus coalitus	WP	Heteroscyphsic acid A - I	Wang et al. (2020)
		Heteroscyphin A - E	
Lactuca orientalis	S	Leucodin	Stojakowska et al. (2018)
Achyranthes aspera	L	Betulinic acid,	Pai et al. (2014)
		Oleanolic acid	
Achyranthes coynei	L, St, I	Betulinic acid	Upadhya et al. (2014)
		Oleanolic acid	
		Ursolic acid	
Swertia minor, S. densifolia, S. lawii, S. corymbosa, S. angustifolia var. pulchella	WP	Betulinic acid	Kshirsagar et al. (2015)
		Oleanolic acid	
		Ursolic acid	
Ocimum basilicum, O. gratissimum, O. kilimandscharicum, O. tenuiflorum	L, St, I	Betulinic acid	Pai and Joshi (2016)
		Oleanolic acid	
		Ursolic acid	
Mentha longifolia	L, St	1,8-cineole	Bertoli et al. (2011)
	L	4-Terpineol	

(continued)

Table 8.3 (continued)

Plant name	Part	Terpenoids	Reference
	L, St	Menthol	
	AP	Longifene	
	L	Borneol	Mkaddem et al. (2009)
	AP	Dihydrocarvone	Motamed and Naghibi (2010)
Withania coagulans	AP, R	Withanolides	Gupta et al. (2021)

L leaves, *WP* whole plant, *S* seeds, *St* stem, *I* inflorescence, *AP* aerial parts

fluidity of plasma membrane; polyprenyl dolichol is and sugar carrier lipid used in protein glycosylation; phytol is a part of chlorophyll, tocopherols, phylloquinones, carotenoids thus functioning in photosynthesis, and polyprenyl plastoquinone in electron transport chain (Lohr et al. 2012).

A number of genes coding enzymes and regulators in biosynthesis of terpene were studied with respect to their location, genome, and expression (Chen et al. 2011). Though much is understood about the terpene metabolism, recent findings are revealing new substrates and enzymatic reactions in its biosynthesis (Zhou and Pichersky 2020). On other hand, alternative strategies like chemical synthesis are impractical due to the complex nature of these compounds (Misawa 2011; Oksman-Caldentey and Inzé 2004). With limited commercial success, biotechnological methods give an attractive alternative to obtain these drugs. Comprehensive reviews have already been published on microbial transformation and its production using tissue culture methods (Parra et al. 2009; Malinowska et al. 2013). This chapter concentrates on in vitro production of terpenoids.

8.2 Biosynthesis Pathway

It is one of the large groups of phytochemicals in plant kingdom. They include representations in large number of plant molecules such as vitamins and hormones. Isopentenyl pyrophosphate (IPP) and dimethylallyl pyrophosphate (DMAPP) are the two precursors which significantly contribute into the biosynthesis of terpenoids. Classical studies by a number of workers have elucidated biosynthesis of terpenoids via acetyl-CoA and mevalonate (Qureshi and Porter 1981; Bloch 1992; Bach 1995; Bochar et al. 1999). Thus, mevalonate is the best studied pathway for terpenoid biosynthesis which includes mevalonic acid as an intermediate. Recent studies have revealed variations in biosynthesis of terpenoids production from the common 2-*C*-methyl-D-erythritol 4-phosphate (MEP) and the mevalonate (MVA) pathways. The earlier (MEP) being working in plastids and later one (MVA) in cytoplasm, endoplasmic reticulum and peroxisomes (Vranová et al. 2013).

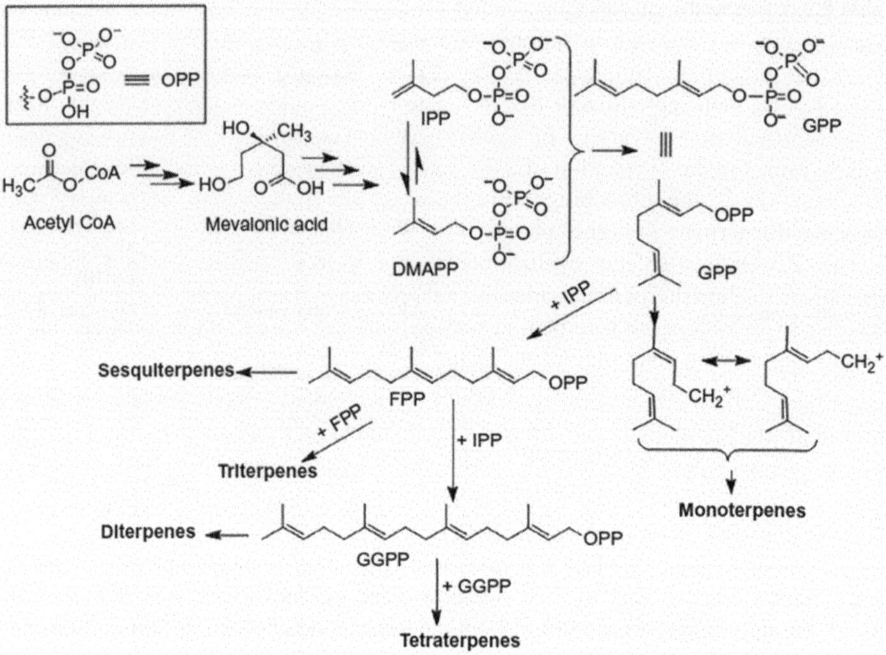

Fig. 8.1 An overview of terpenoid biosynthesis (source: Habtemariam 2019)

As reported by Habtemariam (2019), acetyl-CoA a primary metabolite undergoes a series of reaction to produce a six carbon molecule mevalonic acid (Fig. 8.1). This further produces DMAPP and IPP an interchangeable 5 carbon (C) building blocks. Isopentenyl diphosphate (IPP) and dimethylallyl pyrophosphate (DMAPP) condense to produce geranyl diphosphate (GPP), which is considered as the intermediate precursor for all monoterpenes. The GPP serves precursor to monoterpene production, and further addition of a 5 carbon unit (IPP/DMAPP) produces farnesyl pyrophosphate (FPP) a precursor of sesquiterpenes (15 C) and its dimer to give triterpenes (30 carbon). The addition of another 5 carbon unit to FPP gives geranylgeranyl diphosphate (GGPP) as an originator of diterpenes (20 C). Dimerization of GGPP gives tetraterpenes (40 C). Mostly, head-to-tail arrangements by addition of 5 carbon units are observed, exceptionally head to head in tri- and tetraterpenes. Though we consider 5 carbon unit (IPP and DMAPP) for biosynthesis terpenoids, monoterpenes (10 C) are considered to be the smallest unit.

Studies also reveal that enzymes such as Nudix hydrolases and isopentenyl phosphate kinase (IPK) affect biosynthesis of terpene in cytosol as well as plastids by regulating IPP and DMAPP ratios (Zhou and Pichersky 2020).

8.3 In Vitro Production Methodology

Plants are looked as inexhaustible sources of new drugs. In vitro methods as alternative strategies are applied in two ways for secondary metabolite production: (1) biomass and (2) synthesis. Biomass aggregation would rely on in vitro culturing of shoot, root, callus, etc. (Guerriero et al. 2018; Hussain et al. 2012). Terpenoids are no different, and they play vital role in the organism. Apart from pharmacological properties of terpenoids, their cosmeceuticals, antioxidant, and nutraceutical properties of terpenoids are studied vastly (Bonfill et al. 2013). In vitro culture techniques coupled with bioengineering has demonstrated great applications. Artemisinin a sesquiterpene lactone drug from *Artemisia annua* is known for its antimalarial property. Rationally, overexpression of farnesyl diphosphate synthase (FDS) enzyme which condenses IPP with DMAPP to form (FPP) farnesyl diphosphate yields higher artemisinin. The same was achieved by placing a 35S CaMV promoter with FPS gene using *Agrobacterium* mediated transformation. This study showed about fourfold increase in the content (Exposito et al. 2010). Farzaei et al. (2017) included 55 mono- and sesquiterpenoids alongside other compounds in their review on pharmacological and phytochemical properties of *Mentha longifolia*. They also tabulated plant part used and the extraction method. Few of the plants alongside their in vitro methods studied and terpenoids identified are enlisted below (Table 8.4).

8.4 Scale-up Techniques and Bioreactors

Commercially important secondary metabolites (SMs) need scale-up using bioreactors, as the traditional way of production may not suffice the need. It takes help of modern methods like genomics, proteomics, metabolomics etc. to modify the metabolic routes in an organism for scaled-up production of SMs (DellaPenna 2001). Shikimate, polyketide, and terpenoids pathways are the main routes for production of secondary metabolites. Terpenoids comprise around one-third of the total secondary metabolites (Verpoorte 2000). Scaling-up may include metabolic engineering, up- and downregulating pathways, redirecting common precursors, and targeting metabolites to specific cell compartments (Chandran et al. 2020).

Overexpressing of FaNES1 (*Fragaria ananassa* nerolidol synthase 1) a protein identified as sensor for production of both mono and sesquiterpene precursors. High levels of monoterpene linalool was achieved by CaMV 35S-driven promoter and engineered with wild strawberry FvNES1 (*Fragaria vesca* nerolidol synthase 1) plastid targeting region (Aharoni et al. 2006). It was also reported that, compared to the monoterpenes, sesquiterpene engineering in plants is difficult as precursor pool required is less. In another study, Rodríguez et al. (2014) observed that downregulating production of terpene D-limonene is inversely correlated by lowering fungal infection in transgenic oranges. Redirecting common precursor, viz., prenyl pyrophosphate (C5), to provide an adequate flux of IPP and DMAPP can be achieved by manipulation of genes regulating MVA pathway or by inserting MEP

Table 8.4 Terpenoids with method of in vitro production

Plant name	In vitro method	Terpenoid	Reference
Lavandula angustifolia	CSC	Humulene	Banthorpe et al. (1995)
		Caryophyllene	
Ocimum basilicum	SC	Linalool	Monfort et al. (2018)
Achyranthes aspera	CC	Betulinic acid	Pai et al. (2018)
		Oleanolic acid	
		Ursolic acid	
Mentha longifolia	IVP, CC	1,8-cineole	Bertoli et al. (2011)
		β-Pinene	
		γ-Terpinene	
		Limonene	
		Linalool	
		Myrcene	
		Piperitenone oxide	
		α-Pinene	
	IVP	Piperitone	
	CC	α-Thujene	
	CC	Bicyclogermacrene	
	CC	Germacrene D	
	CC	α-Humulene	
	CC	β-Caryophyllene	
Withania coagulans	RC	Withaferin A	Gupta et al. (2021)
		Withanolides	

CSC cell suspension culture, *SC* shoot cultures, *CC* callus cultures, *IVP* in vitro plantlets, *RC* root cultures

pathway in organisms having only MVA pathway (Vavitsas et al. 2018; Yang et al. 2016). *Escherichia coli* and *Saccharomyces cerevisiae* are termed as industrial workhorses and are used to produce high-valued terpenoids in bioreactors. Zhang and Hong (2020) in their mini review have well documented a list of strategies for production of terpenoids by targeting it to specific cell compartments.

Gupta et al. (2021), in an overview on pharmaceutical properties and biotechnological advancements of *Withania coagulans*, have reported that the plant is on the verge of extinction. Withanolides are a group of steroidal lactone triterpenoids which can also be synthesized by terpenoids forming mevalonate pathway (Kreis and Muller-Uri 2010). Seven different methods have been highlighted for increased production of bioactive compounds from the plant; they include *Agrobacterium*-mediated transformation, metabolic engineering, plant tissue culture, germination by seeding, abiotic/biotic elicitors, gene transfer, and nanoparticle synthesis. To infer the work of Gupta et al. (2021), in vitro system is suggested to be the attractive option to achieve the goal.

Apart from the above, Zhang et al. (2019) used *Agrobacterium*-mediated transformation studies to functionally expedite terpenoids biosynthesis in *Tripterygium wilfordii*. In a separate study, pharmaceutically important metabolites were enhanced by adding silver nitrate (linalool, estragole) and yeast extract (rutin, isoquercetin) in *Ocimum basilicum* suspension cultures (Açıkgöz 2020). Polzin and Rorrer (2018) reported selective production of β-myrcene (monoterpene) from *Octodes secundiramea* on nutrient perfusion cultivation with bromide-free medium. High levels of amorphadiene a precursor of artemisinin were produced by using *Bacillus subtilis* as terpenoids cell factory (Pramastya et al. 2021). Finding of new genes and computational biology has boosted production of new terpenoids using microbial methods. Expression of plant enzymes in microorganisms has helped biosynthesis of diverse class of complex terpenoid derivatives (Belcher et al. 2020).

8.5 Extraction and Detection Techniques

A number of extraction and detection methods are employed for determining the contents of phytochemicals from plants (Fig. 8.2). Range of methods from conventional, simple maceration to recent ultra-sonication and microwave-assisted extractions are deployed for extracting terpenoids. In a study, phenolic and terpenoids were identified using reversed-phase high-performance liquid chromatography (RP-HPLC) and head space solid phase microextraction gas

Fig. 8.2 Various extraction methods reported

Table 8.5 Method of extraction and detection used in various plant terpenoids

Plant	Method of Extraction	Detection	Terpenoid	Reference
Ancistrocladus heyneanus	SE, CSE, UE, MAE	RP-HPLC	Betulinic acid	Pai et al. (2011)
Pine	M	GC	α- and β-pinene, camphene, and δ-carene	Harman-Ware et al. (2016)
Achyranthes coynei	RE	RP-UFLC	Betulinic acid, Oleanolic acid, Ursolic acid	Upadhya et al. (2014)
Swertia minor *S. densifolia* *S. lawii* *S. corymbosa* *S. angustifolia var. pulchella*	M	RP-HPLC	Betulinic acid, oleanolic acid, ursolic acid	Kshirsagar et al. (2015)
Ocimum basilicum *O. gratissimum* *O. kilimandscharicum* *O. tenuiflorum*	ME	RP-HPLC, HPTLC	Betulinic acid, oleanolic acid, ursolic acid	Pai and Joshi (2016)
Achyranthes aspera	CSE, MAE, UE	RP-UFLC-DAD	Betulinic acid, Oleanolic acid, Ursolic acid	Pai et al. (2016)
Vitex negundo	M	HPTLC	Betulinic acid, Oleanolic acid	Pai and Joshi (2018)
Sparkling wines	HSME	GC-MS	α-Terpineol, (−)-*β*-citronellol, *β*-cyclocitral	Muñoz-Redondo et al. (2020)

SE Soxhlet extraction, *CSE* continuous extraction, *UE* ultrasonic extraction, *MAE* microwave extraction; *RE* reflux extraction, *M* maceration, *ME* microextraction, *HSME* head-space microextraction

chromatography coupled with mass spectrometry (HS-SPME-GC/MS), respectively (Açıkgöz 2020; Muñoz-Redondo et al. 2020)

More recently, Vaníčková et al. (2020) have identified species-specific terpenoid markers to chemo-taxonomically distinguish four species and one subspecies of *Dracaena* using solid-phase microextraction – coupled to a gas chromatography – ion trap tandem mass spectrometry (SPME-GC × GC-MS). Percolation and maceration extraction methods have been reported for extraction of terpenoids from stem and root material of *Eurycoma longifolia*. Supercritical fluid extraction and microwave-assisted extraction are some of the non-conventional methods utilized and reported for extraction in many plants including *Andrographis paniculata* for terpenoids andrographolide, deoxyandrographolide, and its variants (Aziz et al. 2021). Plants reported with different extraction and detection methods alongside terpenoids identified is tabulated in Table 8.5.

8.6 Biological Activities

The significance of terpenoids is because of their anticancer, anti-inflammatory, antiplatelet, antibacterial, hypocholesterolemic, immune adjuvant, anti-viral, antibacterial, fungicidal, and antileishmanial agents like pharmacological properties (Abdelrahman and Jogaiah 2020). Anticancer and antimicrobial properties are some of the largely studied activities of terpenoids. Apart from cancer treatment and life style disorders, withanolides as bioactive terpenoid derivative from *Withania* has proved antianxiety activity in albino mice model (Gupta et al. 2021). On other hand, a range of biological activities may be attributed to terpenoids from cucurbit family Cucurbitacin, and its derivatives from *Cucurbitaceae* have been reviewed against inflammation, cancer, and many other properties (Montesano et al. 2018). The terpenoids are not only studied from plant system but are reported from marine bacteria and fungi with similar properties. Altemicidin, marinocyanins A-F, azamerone, and napyradiomycins were some of the potent monoterpenoids identified from bacteria (Gozari et al. 2021). Chen et al. (2012) reviewed *Sinularia* soft coral as a source of terpenoids for potential bioactivities such as antimicrobial, anti-inflammatory, and cytotoxic activities. Terpenoids like andrographolide, glycyrrhetinic acid, ursolic acid, costunolide, β-elemene, glaucocalyxin A, and cucurbitacin B have been reviewed and reported to show anti-liver fibrosis by Ma et al. (2020).

Plant essential oils are a media of communication within and between the plants. These are chemical terpenes containing repeated units of isoprenes. Depending on the occurrence in plant parts, they may be for attracting pollinators or to repel predators. They are generally present on leaves in secretary glands (e.g., mint, sage, basil, pine), in flowers (e.g., orange, chamomile, clove), in wood (e.g., sandalwood, balsam, camphor), in fruits (e.g., star anise, fennel, apple), in rhizomes (ginger, turmeric), and in seeds (cardamom, nutmeg, pepper). Geraniol (geranium), linalool (mint), myrcene (thyme), and ß-ocimene (basil) are some of the important essential oils (and plant name) having terpenes (Böttger et al. 2018). They are termed biopesticides due to their repellent and insecticidal properties.

The sequence of outbreaks of viral infection particularly corona, starting from the year 2003, SARS-CoV; 2012, MERS-CoV; and 2019, SARS-CoV-2, raised global health concerns. Conventional drug limitations forced researchers to look into plant-based bioactive compounds. Bhattacharya et al. (2021), in their review, updated on antiviral drugs of plant origin which are potential players in counteracting these viruses. Rajan et al. (2021) in a recent commentary on promising antivirals from ayurvedic herbs against COVID-19 have mentioned 26 natural compounds along with terpenoids as a potential candidate based on in silico approach. Table 8.6 enlists some important bioactivities of terpenoids or terpenoid-derived compounds majorly reviewed by Martin-Smith and Sneader (1969).

Table 8.6 Important bioactivities of terpenoids

Bioactivity	Terpenoid
Antibiotic	Fusidic acid
	Polyporenic acid A, C
	Pristimerin
Antibacterial	Eunicin
Antituberculosis	Pristimerin
Vincristine	Anticancer
Vinblastine	
Reserpine	Cytostatic activity
β-Thujaplicin	
Glycyrrhizin	Anti-inflammatory
Arjunolic acid	Anticold stress
β-Glycyrrhetinic acid	Antitussive agent
Alloferin	Muscle relaxant
Chamazulene	In periodontal infections
Methysergide	Migraine
Toxaphene	Pesticides
Ryanodine,	Insecticide
Trichothecin	Antifungal for plants
Nepetalactone	Insect repellent
(+)- cis-Verbenol	Insect attractant

8.7 Commercial Utilization and Prospects

Commercial utilization and prospects should include lucrative applications of terpenes in various industries. For example, rubber, a polyterpene with repeating units of isoprene, is extensively studied, known and one of the largest industries. Still, it is important to find and concentrate on such targets which are of real concern in the coming years. Industries like therapeutics, food, flavor, perfumery, or even fuel are highly commercially utilized, in relation to environment. However, without better environment and sustainable resources, all other industries will have poor prospects. Thus, environment and ecology being one of the three such points will be explored in this section.

8.7.1 Environment and Ecology

Terpenes are the most abundantly found and studied natural product. Out of the three main categories of natural products, it is estimated that terpenoid accounts for 55%, alkaloid 27%, and phenolic 18% (Croteau et al. 2000), thus overstating their importance in nature and for mankind. Cheng et al. (2007) have discussed the role of terpene in ecological point of view. Terpenes are assumed to have a role in earth's atmosphere by interacting with solar radiations. The 500 teragrams of isoprene emission annually is being estimated which influence formation of ozone (Zwenger and Basu 2008). Beyond plant-insect interaction, terpenes also act as messengers in plant defense mechanism, which can be further utilized commercially for their insect repellent or even attractants.

8.7.2 Therapeutic, Perfumery, Food, Flavor

Advent of bioinformatics and progressions in molecular databases have contributed in understanding the synthesis and pathways much better. Commercially, they have contributed largely to therapeutics, perfumery, food, and flavor industries. Also their role in essential oils and cosmetics can't be underestimated. Developing agronomic traits in medicinal plants is an important area of research explored in the past few decades. A combination of engineering to use "omics" technologies and computational and systems biology, along with synthetic chemistry, shows an imperative prospect in commercialization. Artemisinin, squalane, sandalwood, and patchouli oils are few of the examples discussed by Leavell et al. (2016) as potential terpenes in pharma and fine chemicals. The volatility of terpenoid supplies and prices has been attributed to smaller market size and the dependency on the plant source for its production. Thus, alternative strategies for stable production using fermentation are an attractive option. Amyris Inc. is a synthetic biotechnology company making its presence in all the four sectors for terpenoids, artemisinin (pharma), squalane, numerous fragrance oils (fine), isoprene, farnesene for lubricants, polymers (as commodity chemicals), and farnesene for diesel and jet fuel (fuels). Antimicrobial-resistant (AMR) bacteria or even superbug issues in present-day list of drugs can be updated by addition of phytocompounds chiefly terpenes (Mahizan et al. 2019; Zwenger and Basu 2008). Terpenes may reduce the rate of antibiotic resistance via livestock feed by substituting orthodox antibiotics.

8.7.3 Fuel

"Biofuel" a renewable source particularly from plant origin is another prospective application of terpenes in commercialization. Countries such as the USA, Brazil, and European Union are expressing interest and supporting research based on such alternative sources. Diesel obtained from *Copaifera langsdorffii* a natural biofuel largely contains terpenes. In a study, Mewalal et al. (2017) have summarized specific terpenes as appropriate standby or composite for present-day fuels. They have also made a comparison of such biofuels with much known fuels. It has been reported that terpenes could assist as standalone or blended up to 65% in diesel engines. There are more than a few challenges to leverage plant terpenes as a commercially viable source for replacing diesel or gasoline. A roadmap has been proposed by Mewalal et al. (2017) for commercial recovery of terpenes synthesized. Its tremendous potential as sustainable biofuel has been swotted and described.

The above-discussed points will raise concerns in the coming years. Thus, it was important to highlight the area of future research on terpenoids.

8.8 Conclusions

Terpenes are still being fully realized for their applications. Discovery of more and more number of terpenes will surely add to the ever-growing list of its utilization. Like most other bioactives, terpenoids from nature are supply limited. However, production at commercial level is driven by recent annotated technologies. Few important such terpenes are reviewed and discussed above. Differences in the challenges to produce terpenoids vary from one to another, thus making it non-comprehensive. It means that a method developed for one terpene cannot be applied to another. On the other hand, predictive engineering along with in vitro culture techniques can lead to selective production of terpenoids. With this, it is also projected that wide understanding of structure and activity liaisons of terpenes will open avenue for medicinally important phytocompounds. Further study on their activity as well as mechanism will be of great importance. Furthermore, work related to cost-effective and more contemporary way for production of biodiesels rather than the now conventional ethanol conversion from sugarcane or corn would be essential. Conclusively, owing to the ever-increasing number of terpenoids, a lot of study still can be undertaken for in vitro sustainable production, fundamental understanding of their role in biological processes, and commercial utilization.

Acknowledgments Author is indebted to Rayat Shikshan Sanstha, Satara, (MS) India, for providing support.

Conflict of Interest None.

References

Abdelrahman M, Jogaiah S (2020) Production of plant bioactive triterpenoid and steroidal Saponins. In: Bioactive molecules in plant defence. Springer, Cham, pp 5–13

Açıkgöz MA (2020) Establishment of cell suspension cultures of *Ocimum basilicum* L. and enhanced production of pharmaceutical active ingredients. Ind Crop Prod 148:112278

Aharoni A, Jongsma MA, Kim TY et al (2006) Metabolic engineering of terpenoid biosynthesis in plants. Phytochem Rev 5:49–58

Anonymous, Government of India, Medicinal plants introduction, Indian System of Medicine and Homoeopathy (ISMH), Department of ISMH, Ministry of Health and Family Welfare, Governmet of India (2000). http://indianmedicine.nic.in/html/plants/mimain.htm

Asakawa Y (1982) Chemical constituents of the hepaticae. In: Herz W, Grisebach H, Kirby GW (eds) Progress in the chemistry of organic natural products, vol vol. 42. Springer, Vienna, pp 1–285

Asakawa Y (1995) Chemical constituents of the bryophytes. In: Herz W, Kirby GW, Moore RE, Steglich W, Tamm C (eds) Progress in the chemistry of organic natural products, vol vol. 65. Springer, Vienna, pp 1–618

Asakawa Y, Ludwiczuk A, Nagashima F (2013) Chemical constituents of bryophytes: bio- and chemical diversity, biological activity, and chemosystematics. In: Kinghorn AD, Falk H, Kobayashi J (eds) Progress in the chemistry of organic natural products, vol vol. 95. Springer-Verlag, Vienna, pp 1–796

Aziz NAA, Hasham R, Sarmidi MR et al (2021) A review on extraction techniques and therapeutic value of polar bioactives from Asian medicinal herbs: case study on *Orthosiphon aristatus*, *Eurycoma longifolia* and *Andrographis paniculata*. Saudi Pharm J 29(2):143–165

Bach TJ (1995) Some aspects of isoprenoid biosynthesis in plants. A review. Lipids 30:191–202

Banthorpe DV, Bates MJ, Ireland MJ (1995) Stimulation of accumulation of terpenoids by cell suspensions of *Lavandula angustifolia* following pre-treatment of parent callus. Phytochemistry 40(1):83–87

Belcher MS, Mahinthakumar J, Keasling JD (2020) New frontiers: harnessing pivotal advances in microbial engineering for the biosynthesis of plant-derived terpenoids. Curr Opin Biotechnol 65:88–93

Bertoli A, Leonardi M, Krzyzanowska J et al (2011) Mentha longifolia in vitro cultures as safe source of flavouring: ingredients. Acta Biochim Pol 58(4):581–587

Bhattacharya R, Dev K, Sourirajan A (2021) Antiviral activity of bioactive phytocompounds against coronavirus: an update. J Virological Met 290:114070

Bloch K (1992) Sterol molecule: structure, biosynthesis and function. Steroids 57:378–382

Bochar DA, Friesen JA, Stauffacher CV et al (1999) Biosynthesis of mevalonic acid from acetyl-CoA. In: Cane D (ed) Comprehensive natural product chemistry, vol Vol. 2. Pergamon, Oxford, pp 15–44

Bonfill M, Malik S, Mirjalili MH et al (2013) Production and genetic engineering of Terpenoids production in plant cell and organ cultures. In: Ramawat K, Mérillon JM (eds) Natural products. Springer, Berlin, Heidelberg, pp 2761–2796

Böttger A, Vothknecht U, Bolle C et al (2018) Terpenes and terpenoids. In: Lessons on caffeine, cannabis & co. learning materials in biosciences. Springer, Cham. https://doi.org/10.1007/978-3-319-99546-5_10

Chandran H, Meena M, Barupal T et al (2020) Plant tissue culture as a perpetual source for production of industrially important bioactive compounds. Biotech Rep 26:e00450

Chen F, Tholl D, Bohlmann J et al (2011) The family of terpene synthases in plants: a mid-size family of genes for specialized metabolism that is highly diversified throughout the kingdom. Plant J 66:212–229

Chen W, Li Y, Guo Y (2012) Terpenoids of *Sinularia* soft corals: chemistry and bioactivity. Acta Pharm Sin B 2(3):227–237

Chen F, Ludwiczuk A, Wei G et al (2018) Terpenoid secondary metabolites in bryophytes: chemical diversity, biosynthesis and biological functions. Crit Rev Plant Sci 37:210–231

Cheng A, Lou Y, Mao Y et al (2007) Plant terpenoids: biosynthesis and ecological functions. J Integr Plant Biol 49:179–186

Croteau R, Kutchan TM, Lewis NG (2000) Natural products (secondary metabolites). In: Buchanan B, Gruissem W, Jones R (eds) Biochemistry and molecular biology of plants. American Society of Plant Physiologists, Rockville, pp 1250–1318

DellaPenna D (2001) Plant metabolic engineering. Plant Physiol 125(1):160–163

Dzubak P, Hajduch M, Vydra D et al (2006) Pharmacological activities of natural triterpenoids and their therapeutic implications. Nat Prod Rep 23:394–341

Ercioglua E, Velioglub HM, Boyaci IH (2018) Determination of terpenoid contents of aromatic plants using NIRS. Talanta 178:716–721

Exposito O, Syklowska-Baranek K, Moyano E et al (2010) Metabolic responses of *Taxus* media transformed cell cultures to the addition of methyl jasmonate. Biotechnol Prog 26:1145–1153

Farzaei MH, Bahramsoltani R, Ghobadi A et al (2017) Pharmacological activity of *Mentha longifolia* and its phytoconstituents. J Tradit Chin Med 37(5):710–720

Gozari M, Alborz M, El-Seedi HR (2021) Chemistry, biosynthesis and biological activity of terpenoids and meroterpenoids in bacteria and fungi isolated from different marine habitats. Eur J Med Chem 210:112957

Guerriero G, Berni R, Muñoz-Sanchez JA et al (2018) Production of plant secondary metabolites: examples, tips and suggestions for biotechnologists. Genes (Basel) 9(6):E309

Gupta R, Sonawane T, Pai SR (2021) An overview on pharmaceutical properties and biotechnological advancement of *Withania coagulans*. Adv Tradit Med. https://doi.org/10.1007/s13596-021-00558-7

Habtemariam S (2019) Introduction to plant secondary metabolites - from biosynthesis to chemistry and antidiabetic action. In: Medicinal foods as potential therapies for Type-2 diabetes and associated diseases, the chemical and pharmacological basis of their action, pp 109–132

Harman-Ware AE, Sykes R, Peter GF et al (2016) Determination of terpenoid content in pine by organic solvent extraction and fast-GC analysis. Front Energy Res 4:2. https://doi.org/10.3389/fenrg.2016.00002

Hussain MS, Fareed S, Ansari MS et al (2012) Current approaches toward production of secondary plant metabolites. J Pharm Bioallied Sci 4(1):10–20

Kreis W, Muller-Uri F (2010) Biochemistry of sterols, cardiac glycosides, brassinosteroids, phytoecdysteroids and steroid saponins. In: Michael W (ed) Annual plant reviews. Wiley, Singapore, pp 304–363

Kshirsagar PR, Pai SR, Nimbalkar MS et al (2015) Quantitative determination of three pentacyclic triterpenes from five *Swertia* L. species endemic to Western Ghats, India, using RP-HPLC analysis. Nat Prod Res 29(19):1783–1788

Leavell MD, McPhee DJ, Paddon CJ (2016) Developing fermentative terpenoid production for commercial usage. Curr Opin Biotechnol 37:114–119

Liu WX, Zhao JW, Zuo AX (2018) Two novel terpenoids from the cultured *Perovskia atriplicifolia*. Fitoterapia 130:152–155

Lohr M, Schwender J, Polle JEW (2012) Isoprenoid biosynthesis in eukaryotic phototrophs: a spotlight on algae. Plant Sci 185–186:9–22

Ludwiczuk A, Asakawa Y (2019) Bryophytes as a source of bioactive volatile terpenoids – a review. Food Chem Toxicol 132:110649

Ma X, Jiang Y, Wen J et al (2020) A comprehensive review of natural products to fight liver fibrosis: alkaloids, terpenoids, glycosides, coumarins and other compounds. Eur J Pharmacol 888:173578

Mahizan NA, Yang SK, Moo CL et al (2019) Terpene derivatives as a potential agent against antimicrobial resistance (AMR) pathogens. Molecules 24(14):2631

Malinowska M, Sikora E, Ogonowski J (2013) Production of triterpenoids with cell and tissue cultures. Acta Biochim Pol 60(4):731–735

Martin-Smith M, Sneader WE (1969) Biological activity of the terpenoids and their derivatives--recent advances. Prog Drug Res 13:11–100

Mewalal R, Rai DK, Kainer DA et al (2017) Plant-derived terpenes: a feed stock for specialty biofuels. Trends Biotechnol 35(3):227–240

Misawa N (2011) Pathway engineering for functional isoprenoid. Curr Opin Biotechnol 22:627–633

Mkaddem M, Bouajila J, Ennajar et al (2009) Chemical composition and antimicrobial and antioxidant activities of *Mentha* (*longifolia* L. and *viridis*) essential oils. J Food Sci 74(7): M358–M363

Monfort LEF, Bertolucci SKV, Lima AF, de Carvalho AA, Mohammed A, Blank AF, Pinto JEBP (2018) Effects of plant growth regulators, different culture media and strength MS on production of volatile fraction composition in shoot cultures of Ocimum basilicum. Ind Crop Prod 116:231–239

Montesano D, Rocchetti G, Putnik P et al (2018) Bioactive profile of pumpkin: an overview on terpenoids and their health-promoting properties. Curr Opin Food Sci 22:81–87

Motamed SM, Naghibi F (2010) Antioxidant activity of some edible plants of the Turkmen Sahra region in northern Iran. Food Chem 119(4):1637–1642

Mufflera K, Leipolda D, Schellera MC et al (2011) Biotransformation of triterpenes. Process Biochem 46:1–15

Muñoz-Redondo JM, Ruiz-Moreno MJ, Puertas B et al (2020) Multivariate optimization of headspace solid-phase microextraction coupled to gas chromatography-mass spectrometry for the analysis of terpenoids in sparkling wines. Talanta 208:120483

National Center for Biotechnology Information (2021a) PubChem compound summary for CID 3509874, Gibberellin 1. Retrieved February 6, 2021 from https://pubchem.ncbi.nlm.nih.gov/compound/Gibberellin-1

National Center for Biotechnology Information (2021b) PubChem compound summary for CID 6549, Linalool. Retrieved February 6, 2021 from https://pubchem.ncbi.nlm.nih.gov/compound/Linalool

National Center for Biotechnology Information (2021c) PubChem compound summary for CID 1254, Menthol. Retrieved February 6, 2021 from https://pubchem.ncbi.nlm.nih.gov/compound/Menthol

National Center for Biotechnology Information (2021d) PubChem compound summary for CID 5464156, (R)-Lavandulol. Retrieved February 6, 2021 from https://pubchem.ncbi.nlm.nih.gov/compound/R_-Lavandulol

National Center for Biotechnology Information (2021e) PubChem compound summary for CID 161276, Secologanin. Retrieved February 6, 2021 from https://pubchem.ncbi.nlm.nih.gov/compound/Secologanin

National Center for Biotechnology Information (2021f) PubChem compound summary for CID 92776, Zingiberene. Retrieved February 6, 2021 from https://pubchem.ncbi.nlm.nih.gov/compound/Zingiberene

National Center for Biotechnology Information (2021g) PubChem compound summary for CID 36314, Paclitaxel. Retrieved February 6, 2021 from https://pubchem.ncbi.nlm.nih.gov/compound/taxol

National Center for Biotechnology Information (2021h) PubChem compound summary for CID 64971, Betulinic acid. Retrieved February 6, 2021 from https://pubchem.ncbi.nlm.nih.gov/compound/Betulinic-acid

National Center for Biotechnology Information (2021i) PubChem compound summary for CID 5280794, Stigmasterol. Retrieved February 6, 2021 from https://pubchem.ncbi.nlm.nih.gov/compound/Stigmasterol

National Center for Biotechnology Information (2021j) PubChem compound summary for CID 99474, Diosgenin. Retrieved February 6, 2021 from https://pubchem.ncbi.nlm.nih.gov/compound/Diosgenin

National Center for Biotechnology Information (2021k) PubChem compound summary for CID 64945, Ursolic acid. Retrieved February 6, 2021 from https://pubchem.ncbi.nlm.nih.gov/compound/Ursolic-acid

National Center for Biotechnology Information (2021l) PubChem compound summary for CID 5280896, Abscisic acid. Retrieved February 6, 2021 from https://pubchem.ncbi.nlm.nih.gov/compound/S_-_-Abscisic-acid

National Center for Biotechnology Information (2021m) PubChem compound summary for CID 15102684, 5-Deoxystrigol. Retrieved February 6, 2021 from https://pubchem.ncbi.nlm.nih.gov/compound/5-Deoxystrigol

National Center for Biotechnology Information (2021n) PubChem compound summary for CID 445070, Farnesol. Retrieved February 6, 2021 from https://pubchem.ncbi.nlm.nih.gov/compound/Farnesol

National Center for Biotechnology Information (2021o) PubChem compound summary for CID 6433320, Dolichol-20. Retrieved February 6, 2021 from https://pubchem.ncbi.nlm.nih.gov/compound/Dolichol-20

National Center for Biotechnology Information (2021p) PubChem compound summary for CID 65030, Artemisin. Retrieved February 6, 2021 from https://pubchem.ncbi.nlm.nih.gov/compound/Artemisin

Oksman-Caldentey KM, Inzé D (2004) Plant cell factories in the post-genomic era: new ways to produce designer secondary metabolites. Trends Plant Sci 9:433–440

Pai SR, Joshi RK (2014) Distribution of betulinic acid in plant kingdom. Plant Sci Today 1(3): 103–107

Pai SR, Joshi RK (2016) Variations in pentacyclic triterpenoids in different parts of four *Ocimum* species using reverse phase-high performance liquid chromatography. Proc Natl Acad Sci India Sect B 87:1153–1158

Pai SR, Joshi RK (2018) Optimized densitometric analysis for determination of triterpenoid isomers in *Vitex negundo* L. leaf. Natl Acad Sci Lett 41(5):323–327

Pai SR, Nimbalkar MS, Pawar NV et al (2011) Optimization of extraction techniques and quantification of Betulinic acid (BA) by RP-HPLC method from *Ancistrocladus heyneanus* wall. Ex Grah. Ind Crop Prod 34:1458–1464

Pai SR, Upadhya V, Hegde HV (2014) New report of triterpenoid betulinic acid along with Oleanolic acid from *Achyranthes aspera* by reversed- phase–ultra flow liquid chromatographic analysis and confirmation using high-performance thin-layer chromatographic and Fourier transform–infrared. J Planar Chromatogr 27:38–41

Pai SR, Upadhya V, Hegde HV et al (2016) Determination of betulinic acid, oleanolic acid and ursolic acid from *Achyranthes aspera* L. using RP-UFLC-DAD analysis and evaluation of various parameters for their optimum yield. Indian J Exp Biol 54:196–202

Pai SR, Upadhya V, Hegade HV et al (2018) *In vitro* rapid multiplication and determination of terpenoids in callus cultures of *Achyranthes aspera* Linn. Indian J Biotechnol 17:151–159

Parra A, Rivas F, Garcia-Granados A et al (2009) Microbial transformation of triterpenoids. Mini-Rev Org Chem 6:307–320

Polzin J, Rorrer GL (2018) Selective production of the acyclic monoterpene β-myrcene by microplantlet suspension cultures of the macrophytic marine red alga *Ochtodes secundiramea* under nutrient perfusion cultivation with bromide-free medium. Algal Res 36:159–166

Pramastya H, Xuea D, Abdallah II et al (2021) High level production of amorphadiene using *Bacillus subtilis* as an optimized terpenoid cell factory. New Biotechnol 60:159–167

Qureshi N, Porter JW (1981) Conversion of acetyl-coenzyme a to isopentenyl pyrophosphate. In: Porter JW, Spurgeon SL (eds) Biosynthesis of isoprenoid compounds, vol vol. 1. Wiley, New York, pp 47–94

Rajan M, Gupta P, Kumar A (2021) Promising antiviral molecules from Ayurvedic herbs and spices against COVID-19. Chin J Integr Med. https://doi.org/10.1007/s11655-021-3331-8

Rodríguez A, Shimada T, Cervera M et al (2014) Terpene Down-regulation triggers Defence responses in transgenic Orange leading to resistance against fungal Pathogens[1[W]]. Plant Physiol 164:321–339

Stojakowska A, Michalska K, Kłeczek N et al (2018) Phenolics and terpenoids from a wild edible plant *Lactuca orientalis* (Boiss.) Boiss.: a preliminary study. J Food Comp Anal 69:20–24

Tholl D (2015) Biosynthesis and biological functions of terpenoids in plants. Adv Biochem Eng Biotechnol 148:63–106

Upadhya V, Ankad GM, Pai SR et al (2014) Accumulation and trends in distribution of three triterpenoids in various parts of *Achyranthes coynei* determined using RP-UFLC analysis. Phcog Mag 10:398–401

Vaníčková L, Pompeiano A, Mader P et al (2020) Terpenoid profiles of resin in the genus *Dracaena* are species specific. Phytochemistry 170:112197

Vavitsas K, Fabris M, Vickers CE (2018) Terpenoid metabolic engineering in photosynthetic microorganisms. Genes 9:520

Verpoorte R (2000) Plant secondary metabolites. In: Verpoorte R, Alfermann AW (eds) Metabolic engineering of plant secondary metabolism. Kluwer Academic Publisher, Dodrecht, Boston, London, pp 1–30

Vranová E, Coman D, Gruissem W (2013) Network analysis of the MVA and MEP pathways for isoprenoid synthesis. Annu Rev Plant Biol 64:665–700

Wang X, Jin XY, Zhou JC et al (2020) Terpenoids from the Chinese liverwort *Heteroscyphus coalitus* and their antivirulence activity against *Candida albicans*. Phytochemistry 174:112324

Yang C, Gao X, Jiang Y et al (2016) Synergy between methylerythritol phosphate pathway and mevalonate pathway for isoprene production in *Escherichia coli*. Metab Eng 37:79–91

Yang F, Pu HY, Yaseen A et al (2021) Terpenoid and phenolic derivatives from the aerial parts of *Elsholtzia rugulosa* and their anti-inflammatory activity. Phytochemistry 181:112543

Yu SJ, Yu JH, Yu ZP et al (2020) Bioactive terpenoid constituents from *Eclipta prostrate*. Phytochemistry 170:112192

Zhang C, Hong K (2020) Production of Terpenoids by synthetic biology approaches. Front Bioeng Biotechnol 8:347. https://doi.org/10.3389/fbioe.2020.00347

Zhang B, Huo Y, Zhang J, Zhang X et al (2019) *Agrobacterium rhizogenes*-mediated RNAi of *Tripterygium wilfordii* and application for functional study of terpenoid biosynthesis pathway genes. Ind Crops Prod 139:111509

Zhou F, Pichersky E (2020) More is better: the diversity of terpene metabolism in plants. Curr Opin Plant Biol 55:1–10

Zwenger S, Basu C (2008) Plant terpenoids: applications and future potentials. Biotech Mol Biol Rev 3(1):1–7

In Vitro Production of Anthocyanins and Carotenoids

9

Randah M. Al-Qurashi and Muneera Q. Al-Mssallem

Abstract

There is increasing attention on anthocyanins (ACs) and carotenoids as significant sources of antioxidants with biological activity regarding human health. However, the roles of these compounds in human health depend on their availability in food, extraction methods, and stability. The results of this chapter indicate the structures, extraction and analytical methods, stability, bioavailability, biological activity, and finally commercial utilization of ACs and carotenoids. The effect of these compounds in the food industry is significantly attributed. Finally, with their high-level content of antioxidants, these bioactive compounds can be very attractive in biotechnology, used as fortified and natural additives to foods delivering special benefits in both health and the wider economy.

Keywords

Anthocyanin · Carotenoids · Fruits · Bioavailability · Biological · Production

Abbreviations

ACs	Anthocyanins
ANS	Anthocyanidin synthase
ASE	Accelerated solvent extraction
CHI	Chalcone isomerase
CHS	Chalcone synthase

R. M. Al-Qurashi (✉) · M. Q. Al-Mssallem
Department of Food and Nutrition, College of Agriculture and Food Sciences, King Faisal University, Al-Ahsa, Saudi Arabia
e-mail: ralqurashi@kfu.edu.sa; mmssallem@kfu.edu.sa

T. Belwal et al. (eds.), *Nutraceuticals Production from Plant Cell Factory*,
https://doi.org/10.1007/978-981-16-8858-4_9

CoA	Coenzyme A
CVD	Cardiovascular disease
DFR	Dihydroflavonol 4-reductase
DMAPP	Dimethylallyl pyrophosphate
DNA	Deoxyribonucleic acid
F3′5′H	Flavonoid 3′,5′-hydroxylase
F3′H	Flavonoid 3′-hydroxylase
FPP	Farnesyl pyrophosphate
GGPP	Geranylgeranyl pyrophosphate
GPP	Geranyl pyrophosphate
HMG-CoA	3-Hydroxy-3-methylglutaryl-CoA
IPP	Isopentenyl pyrophosphate
LC	Liquid chromatography
LDL	Low-density lipoproteins
MAE	Microwave-assisted extractions
MeOH	Methanol
MS	Mass spectrometry
PLE	Pressurized liquid extraction
RNS	Reactive nitrogen species
ROS	Reactive oxygen species
SFE	Supercritical fluid extraction
SLE	Solid-liquid extraction
UAE	Ultrasound-assisted extraction
UFGT	Flavonoid 3-O-glucosyltransferase
UV	Ultraviolet

9.1 Introduction

Anthocyanins (ACs) are one of the six subclasses of flavonoids and a water-soluble plant pigment responsible for the blue, purple, and red colors in many fruits and vegetables, generally found in the form of glycosides (Karakaya 2004; Pandey and Rizvi 2009). ACs are an important class of flavonoids that represent a large group of plant secondary metabolites (Liu et al. 2018a). The backbone of anthocyanins is anthocyanidin, and the basic core of anthocyanidin is called flavylium which is composed of two aromatic benzene rings (Fig. 9.1). The most common anthocyanidins in nature are pelargonidin, cyanidin, delphinidin, peonidin, petunidin, and malvidin (Liu et al. 2018a; Zhao et al. 2014).

It is evident that the ACs are known to vary in their number and position of hydroxyl and methoxy1 groups regarding their basic antioxidant skeleton, the identified number and position at which sugars are attached (Wu et al. 2006). Some common foods that are high in ACs include cherries, strawberries, raspberries, blackberries, blackcurrants, blueberries, and grapes, and the concentration of ACs

Anthocyanidins	R_1	R_2
Pelargonidin	H	H
Delphinidin	OH	OH
Cyanidin	OH	H
Peonidin	OCH_3	H
Petunidin	OH	OCH_3
Malvidin	OCH_3	OCH_3

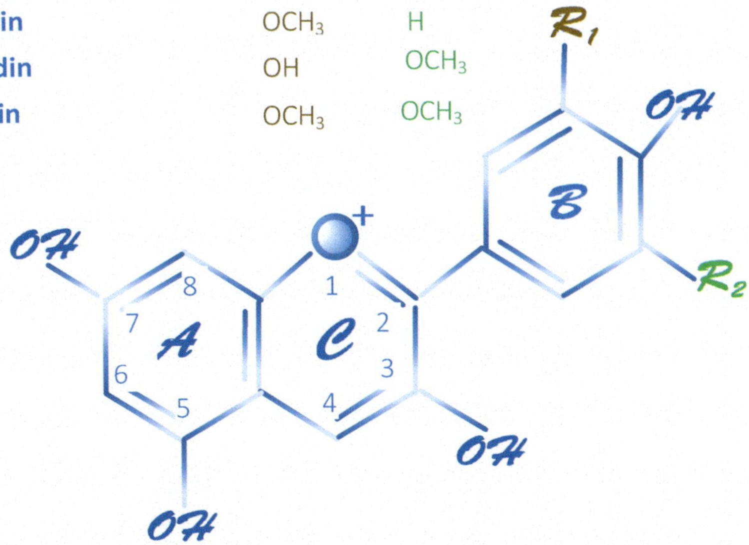

Fig. 9.1 Structure of anthocyanidin backbone and the most common of anthocyanidins. Source: Liu et al. (2018a) and Zhao et al. (2014)

varies widely between fruit. AC intake has a wide range of health-related benefits including cardiovascular disease risk reduction (Pandey and Rizvi 2009; Schewe et al. 2008).

It has been reported there are more than 600 different compounds characterized in plants. One of these are carotenoids, which are natural oil pigments that can be found in many fruits and vegetables and have an essential physiological function as a vitamin A precursor (Ngamwonglumlert and Devahastin 2018). Carotenoids are responsible for many red, orange, and yellow colors found in plant leaves, fruits, vegetables, and flowers (Amorim-Carrilho et al. 2014). The chemical structure of carotenoids is unique due to their 40-carbon basal structure which includes a system of bonds (Yabuzaki 2017). Carotenoids are divided into two classes. First, there are oxygen-containing xanthophylls, where the oxygen atom can be present in the form of alcohol, such as lutein and zeaxanthin (Dutta et al. 2005). Second, there are non-oxygenated carotenes, which include alpha-carotene, beta-carotene, and lycopene (see Fig. 9.2) (Ngamwonglumlert and Devahastin 2018). Fruits and vegetables are important sources of carotenoids such as apricots, cantaloupe, carrots, pumpkin, and sweet potato; all contain alpha-carotene and beta-carotene. However, grapefruit,

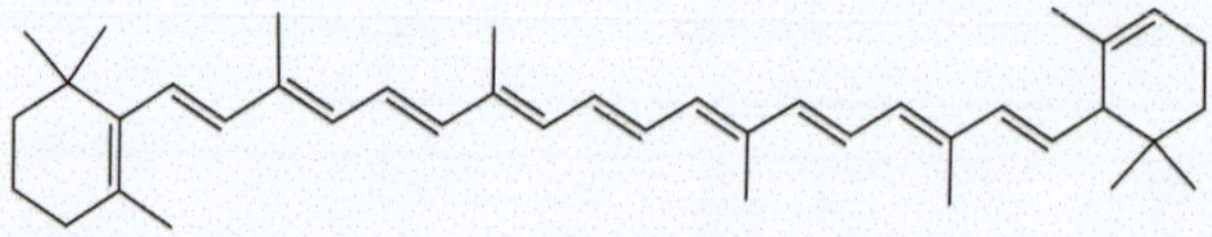

α -carotene

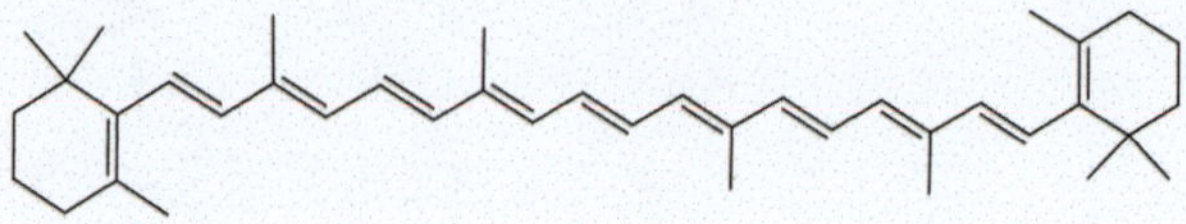

β -carotene

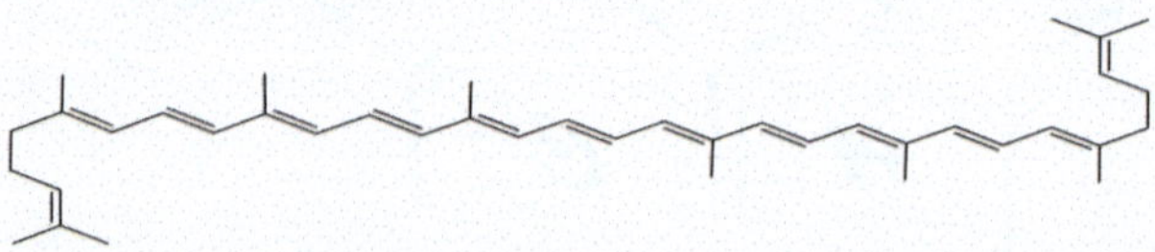

Lycopene

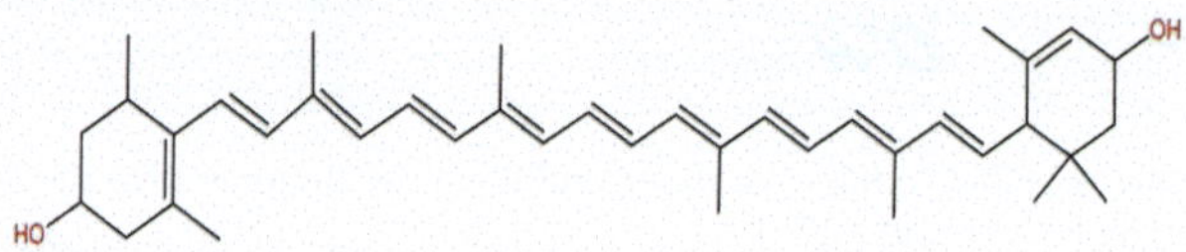

Lutein

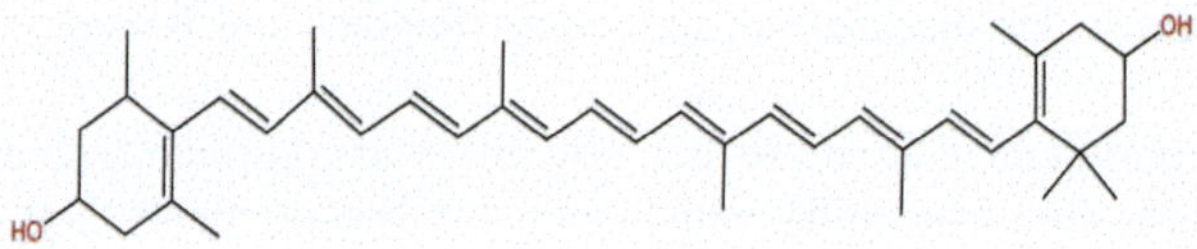

Zeaxanthin

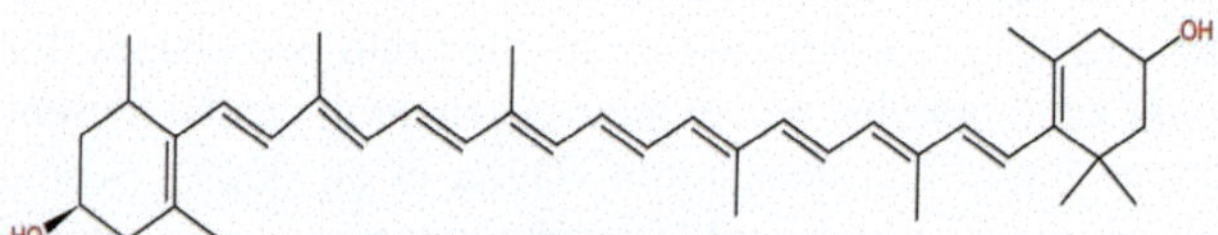

β-cryptoxanthin

Fig. 9.2 Chemical structure of carotenoids

tomatoes, and watermelon are sources of lycopene, zeta-carotene, beta-carotene, phytofluene, and phytoene. Additionally, green vegetables and fruits including broccoli, cabbage, spinach, lettuce, kale, and kiwi all contain lutein, zeaxanthin, alpha-carotene, and beta-carotene (Zakynthinos and Varzakas 2016). The concentration of carotenoids varies between different types of fruits and vegetables and are also influenced by their time of harvest and storage conditions (Ngamwonglumlert and Devahastin 2018).

Carotenoids play an important roles in human health due to their antioxidant properties (Eggersdorfer and Wyss 2018). A number of epidemiological studies have reported that a diet high in carotenoids correlates with a decreased risk of many diseases including cancer and cardiovascular disease (Rao and Rao 2007; Wang 2012). Their mechanism of effect is associated with their antioxidant activity in protecting cells and tissues from damage (Lucini et al. 2012). This chapter provides an overview of the in vitro production of ACs and carotenoids with a particular focus on the stability, extraction, bioavailability, biological activity, and commercial utilization of these compounds.

9.2 Biosynthesis Pathway

Anthocyanins biosynthesis is an essential part of the general flavonoid pathway. The flavan skeleton consists of two aromatic rings A and B which are obtained from acetate and phenylpropanoid precursors, respectively. It is well documented that the regulation of anthocyanins biosynthesis is achieved on the levels of enzymes and genes (Zhang and Furusaki 1999).

Several enzymes are involved in the biosynthesis pathway of anthocyanins including chalcone synthase (CHS), chalcone isomerase (CHI), flavonoid 3-′-hydroxylase (F3′H), flavonoid 3′,5′-hydroxylase (F3′5′H), dihydroflavonol 4-reductase (DFR), anthocyanidin synthase (ANS), and flavonoid 3-O-glucosyltransferase (UFGT). CHS enzyme is regarded as the main enzyme of flavonoid biosynthesis; F3′H and F3′5′H enzymes are the primary enzymes responsible for the diversification of anthocyanins by determining their B-ring hydroxylation and consequently their color. DFR enzymes play a role in influence anthocyanin composition and pigmentation (Liu et al. 2018a; Tanaka and Brugliera 2013).

The biosynthesis of anthocyanins involves three condensation steps, hydroxylation, glycosylation, and acylation. Initially, there are two active precursors involved in the biosynthesis of anthocyanins (4-coumaryl and malonyl-CoA). A condensation occurred between three molecules of malonyl-CoA and one molecule of 4-coumaryl to synthesize naringenin chalcone by the action of CHS enzyme. This step is considered as the central reaction of anthocyanins biosynthesis. Next, naringenin is formed through isomerizing naringenin chalcone by CHI enzyme. Then, naringenin is converted into dihydrokaempferol. Further hydroxylation occurred on F3H to create dihydroflavonols, dihydroquercetin, or dihydromyricetin by F3′H or F3′5′H enzyme, respectively. Colorless leucoanthocyanidins are formed from three molecules of dihydroflavonols by the enzyme DFR. Then, colored anthocyanidins are formed by ANS. The final step includes a glycosylation of anthocyanidins which is considered as an essential step of anthocyanins biosynthesis. In this step, sugar molecules are joined to anthocyanidins by the action of UFGT enzyme (Liu et al. 2018a; Tanaka and Brugliera 2013; Zhang and Furusaki 1999; Zhao et al. 2014).

There are environmental and genetic factors that influence the biosynthesis of anthocyanins such as temperature, light intensity, and UV light. It has been found that production of anthocyanins can be increased at low temperature and high light intensity (Wang et al. 2016). Genetic modification also plays a role in increasing the production anthocyanins and reducing their degradation (Liu et al. 2018a).

In terms of carotenoids biosynthesis, they are synthesized through isoprenoid pathway. Firstly, acetyl-coenzyme A (CoA) is converted to isopentenyl pyrophosphate (IPP). This IPP is considered as a precursor for generation of geranyl pyrophosphate (GPP), farnesyl pyrophosphate (FPP), and geranylgeranyl pyrophosphate (GGPP) using various enzymes of prenyl transferases. These intermediates may self-condense, and cyclization is performed to create the basic skeletons of the terpenoids. Initially, three molecules of acetyl-CoA are fused to produce 3-hydroxy-3-methylglutaryl-CoA (HMG-CoA) by using acetyl-CoA acetyltransferase and HMG-CoA synthase enzymes in both animals and yeast. In plants, however, these reactions are activated in the presence of Fe^{2+} and quinone cofactors by using a single enzyme (McGarvey and Croteau 1995; Weber and Bach 1994).

The simplest terpenoids is isoprene which is synthesized directly from dimethylallyl pyrophosphate (DMAPP). This reaction requires isoprene synthase enzyme and the presence of Mn^{2+} or Mg^{2+} (McGarvey and Croteau 1995). However, for the production of higher terpenoids, multistep reactions are performed starting with DMAPP and IPP using GPP synthase enzyme to form the C10 intermediate (GPP). Then, the C15 intermediate (FPP) is formed via FPP synthase in two separate steps; GPP is firstly formed from DMAPP and IPP; then another IPP is combined to produce FPP. Three condensation steps are carried out to form the C20 intermediate by using GCPP synthase enzyme and activated with either Mn^{2+} or MG^{2+} (Delgado-Vargas et al. 2000; McGarvey and Croteau 1995).

Modern biotechnological approaches have been applied to increase anthocyanin and carotenoids synthesis and the pigmentation of tissues that are not normally pigmented. These great efforts have contributed very well to meet industrial and academic interests (Zhang and Furusaki 1999).

9.3 In Vitro Production Methodology

In vitro plant cell and tissue cultures are regarded as an efficient approach for production of valuable natural products of both anthocyanins and carotenoids (Fischer et al. 2015). Most of the in vitro production techniques of anthocyanins and carotenoids are based on callogenesis and organogenesis approaches including callus and cell suspension cultures. Cell suspension culture is recognized as a potential source for production of anthocyanins and carotenoids. Different media conditions are required in most in vitro plant cultures to induce an alteration from the growth situation to the metabolite production condition. Therefore, biotechnological methods have been applied to increase the productivity of these secondary metabolites (Meyer and Van Staden 1995; Simões et al. 2012).

In organogenesis approach, callus cultures are initiated from an appropriate stem or leaf of chosen plant after sterilization, and then cell lines are generated under in vitro conditions (Lazăr et al. 2010; Meyer and Van Staden 1995; Sakamoto et al. 1994). In vitro conditions include physical and chemical factors. The explants are incubated and then transferred to culture media for propagation. Finally, plants are gradually hardened under ex vitro circumstances (Debnath 2009; Espinosa-Leal et al. 2018; Simões et al. 2012).

The productivity can be changed over time due to inherent genetic and epigenetic instability of plant cell cultures (Phillips et al. 1994; Yuan et al. 2009). There are several factors that modulate the in vitro production such as physical and chemical factors. Physical factors include light, temperature, and medium pH. Light can enhance the production of anthocyanins and carotenoids through increasing the accumulation due to the effect of light in activating different biosynthesis enzymes. Additionally, another important environmental factor is temperature which plays a role in increasing the pigment production. However, high temperature can cause changing in the color of the callus and inhibition of pigment biosynthesis (Lai et al. 2011; Nakatsuka et al. 2009; Wang et al. 2016). There are also chemical factors such as nutrients composition, and elicitors including biotic and abiotic stresses (Simões et al. 2012).

In addition, an improved yield of these products has been achieved through applying new biotechnologies such as gene cloning (Zhang and Furusaki 1999).

Undoubtedly, in vitro plant cell and tissue cultures are an effective system for in vitro productions of both anthocyanins and carotenoids. However, new approach based on biotechnological techniques is effectively applied for obtaining improved yields for these natural colorants in order to be used instead of artificial food colors.

9.4 Scale-Up Techniques and Bioreactors

In vitro plant cell and tissue cultures strategies, for the production of carotenoids and anthocyanins at large scale for commercial purposes, remain as one of the considerable challenges for scientists (Simões et al. 2012). It is well known that scaling up the plant cell and tissue cultures is usually accomplished through using bioreactors. Therefore, well-designed bioreactor is required for the maintenance of a constant and high production of these valuable pigments. This procedure can optimize the in vitro conditions for effective cell growth and reduce variations in product quality and yield (Eibl and Eibl 2008).

Several bioreactors are being used successfully for in vitro plant cell and tissue culture systems. They are bubble column reactors, stainless steel stirred bioreactors, drum reactors, and airlift reactors. Practically, the bioreactors are applied for scaling up in vitro production of secondary metabolites including anthocyanins and carotenoids (Furusaki and Takeda 2017; Zhong et al. 1993). However, in vitro production in bioreactor at commercial level encountered some technological limitations such as the heterogeneity of media of plant cell suspensions which produce large aggregates causing oxygen diffusion and reduced growth rates (Delgado-Vargas et al. 2000). The exploitation of modern biotechnological tools to enhance the production of secondary metabolites may contribute to the scale-up and commercialization of plant cell cultures. Further scientific efforts are successfully carried out for developing improved cell lines not just for sustaining and enhancing high productivity of the secondary metabolite but also for economic purposes such as reducing process costs (Eibl and Eibl 2008). It is evident that the utilization of automation and large-scale liquid cultures significantly reduced the production costs (Simões et al. 2012).

9.5 Stability of Anthocyanins and Carotenoids

Many factors play a role in the stability of anthocyanin in foods, including the type of anthocyanin pigment present; its exposure to light, heat, different pH levels, metal ions, enzymes, and oxygen; the B-ring in the AC's structure; and its hydroxyl or methomyl groups. Furthermore, depending on their structure, some types of ACs are inherently more stable than others. For example, malvidin glycosides, found in high levels in grapefruit, are more stable due to its dimethyl oxylation structure (Laleh et al. 2006; Morais et al. 2002). Temperature is one important factor affecting AC's stability in foods; high temperatures can increase the degradation of ACs which can lead to the hydrolyzation of 3-glycoside (Laleh et al. 2006). One previous study assessed the impact of heat on foods containing high levels of anthocyanin, such as blueberries, and reported that the total anthocyanin level decreased rapidly at 60 °C (Liu et al. 2018b). However, research has addressed the optimum temperature and time condition to preserve anthocyanin stability during hibiscus extraction; the study, completed by Maciel et al. (Maciel et al. 2018), indicated that the anthocyanin present in *Hibiscus sabdariffa* was highest at a temperature of 60 °C for an extraction time of 20 min.

A number of techniques have been used to increase the stability of anthocyanin when processing it as a natural pigment to be added to food. Some of these techniques include encapsulation, spray drying, and freeze-drying enzymatic methods (de Moura et al. 2019; De Souza et al. 2015). Encapsulation is used to decrease the degradation of compounds by reducing their reactivity with environmental influences such as oxygen, light, and temperature (da Fonseca Machado et al. 2018). The advantage of encapsulation when using natural compounds such as ACs is in maintaining their stability for longer periods (Santos et al. 2013). Some of the materials used to encapsulate ACs include carbohydrates (starch, maltodextrins, and cyclodextrins), cellulose (methyl cellulose, carboxymethyl cellulose), gum (gum acacia, gum arabic), lipids (wax, paraffin), and finally proteins (gluten, casein)

(Rai et al. 2009; Robert and Fredes 2015). The most effective methods used to encapsulate ACs are spray drying (SD) and freeze-drying (FD). A study carried out by de Fonseca Machado et al. on the stability of ACs and their antioxidant activity as a result of encapsulation, using blackberries, compared the content of ACs across three methods: spray drying, freeze-drying, and supercritical antisolvent (da Fonseca Machado et al. 2018).

9.6 Extraction and Detection Techniques

The extraction of ACs and carotenoids should garner attention from food and nutrition sciences. There are several methods to extract ACs and carotenoids from plants, including ultrasound-assisted extraction, microwave-assisted extraction, enzyme-assisted extraction, pressurized liquid extraction, supercritical fluid extraction, and using high hydrostatic pressure (Altemimi et al. 2017; Azmir et al. 2013).

9.6.1 Extraction of Anthocyanin

Many factors play a role in extracting samples of AC, including the type of solvent used, temperature, extraction time, and pH level (Chandrasekhar et al. 2012). Among these factors, the type of solvent used is one of the most important affecting extractions, and selecting the type of solvent to be used depends on the chemical structure and polarity of the compounds to be extracted. A numbers of solvents can be used to extract anthocyanin compounds such as water, ethanol, methanol, and acetone (Khalifa et al. 2018; Silva et al. 2017). However, the most effective technique to extract anthocyanin pigments from plants is using acidified water at 0.01% HCL with a pH ~2.3 and using high temperatures between 120 to 160 °C under a constant pressure of 40 bars (Khoo et al. 2017). Furthermore, AC pigments can also be extracted by adding sulfur dioxide to water which can help stabilize the AC's structure and enhance the diffusion coefficient of AC molecules throughout their solid state, which can in turn increase the stability of the AC pigments (Maciel et al. 2018). Other authors have suggested that to increase AC yield, several extraction steps should be performed to make sure maximum extraction is completed (Javier David Vega et al. 2017).

9.6.2 Extraction of Carotenoids

Carotenoids are lipid-soluble compounds that can be found in a wide variety of foods and plants (Eggersdorfer and Wyss 2018). There are no specific or standard methods for carotenoid extraction (Saini and Keum 2018). However, the most common methods use an organic solvent which can allow for the extraction of the compounds from the matrix. The most common solvents used to extract carotenoids from plants use a mixture of hexane with ethanol, methanol, and acetone or a mixture of ethyl acetate with ethanol, methanol, and acetone (Saini and Keum 2018). Hexane, a nonpolar solvent, is usually used with nonpolar carotenes or esterified carotenoids, whereas a polar solvent such as ethanol and acetone can be used for carotenes and xanthophylls. After extracting the carotenoids using the solution, extra care should be taken because the extraction becomes very sensitive to heat, oxygen, acids, and light (Butnariu 2016). Extraction of carotenoids can be done using a mixture of extraction solutions such as methanol (MeOH) and diethyl ether (1:1 v/v), MeOH and chloroform, MeOH and acetone, MeOH and acetone-hexane, or MeOH and hexane-acetone-toluene. These solvents can also be used in the extraction of carotenoids from human samples (Amorim-Carrilho et al. 2014). Furthermore, one recent study found that hexane is more effective in the extraction of carotenoid samples when mixed with other solvents such as methanol and ethanol (Mäki-Arvela et al. 2014).

Similarly, Ranveer et al. (Ranveer et al. 2013) reported that extracting lycopene from tomato waste using a combination of hexane, acetone, and ethanol (50:25:25, v/v/v) solution produced better results than other solvents used individually. Another study done by Kamffer et al. (Kamffer et al. 2010) attempted to develop a method to extract carotenoids and chlorophylls from grape berries, reporting diethyl ether and hexane as a more stable alternative to using acetone. A further study published by Lucini et al. (Lucini et al. 2012) examined the most effective extraction method for carotenoids using tomatoes and observed that the best solvent for extracting b-carotene was ethyl acetate, whereas dichloromethane was preferred for other types of carotenoids. Moreover, extraction from tomato waste using a combination of ethyl acetate and hexane (55:45, v/v) yielded high concentrations of carotenoids (Strati and Oreopoulou 2011). Finally, there are a number of other carotenoid extraction techniques that can be applied in both human and food samples; of these, liquid-liquid and solid-liquid extraction is the most frequently used technique.

Further extraction methods exist for carotenoids and other bioactive compounds, including solid-liquid extraction (SLE), supercritical fluid extraction (SFE), ultrasound-assisted extraction (UAE), pressurized liquid extraction (PLE), and microwave-assisted extraction (MAE) (Jaime et al. 2010; Mustafa et al. 2012; Stout et al. 2018; Sun et al. 2012). SLE is used to separate compounds from food mixtures, and it is widely used in the extraction of polyphenols from foods such as carbohydrates, protein, and aromas (Garcia-Salas et al. 2010). This method is very expensive and requires a long time to process and extract samples; however, the high temperatures used in SLE may destroy some bioactive compounds including ACs (Naviglio et al. 2019). SFE is an alternative method for extracting from food samples which can greatly reduce the amount of solvent used as well as the processing time. Furthermore, this method uses carbon dioxide which may increase the yield of phenolic compounds such as ACs. This method is used to extract compounds from anthocyanin-rich foods such as blueberries (Paes et al. 2014). UAE uses sound waves beyond human hearing and is an effective technique in obtaining the highest extraction yields of bioactive compounds from plants (Chemat et al. 2017). Bubble formation is based on the principle of this technique, using temperatures of 5000 K, pressures of 1000 atm, and heating and cooling rates above 10^{10} K/s (Khan et al.

2010). Studies have shown that UAE is effective in extracting isoflavones, anthocyanins, and other bioactive compounds. Advantages of this methodology include reduced extraction time and the use of a solvent, which helps to reduce the degradation of different bioactive compounds and their antioxidant activity (Saini and Keum 2018). Factors such as moisture content, solvent type, particle size, temperature, pressure used, and time taken are all very important in achieving a high extraction yield with this method (Chemat et al. 2016). PLE, also known as accelerated solvent extraction (ASE), allows for the rapid extraction of the desired compounds and uses only a small amount of solvent. This technique is combined with temperature and pressure to increase the solubility of the sample and therefore improve extraction efficiency. This method can be used to extract bioactive compounds by using high pressures to ensure the solvents remain as liquids at high temperatures, which improves the solubility of the samples and in turn increases their concentration. MAE uses microwave technology as a rapid extraction method which can be effective in the extraction of active ingredients from plants. The main advantages of using this method are the reduction in extraction time and the ability to use a less organic solvent (Liazid et al. 2011). Duan et al. (2015) reported that MAE was a highly efficient extraction technique in extracting anthocyanins from Chinese bayberry. The factors influencing MAE performance include solvent type, extraction time, microwave power, temperature, sample characteristics, and food maturity (Chan et al. 2011).

9.7 Analytical Methods

9.7.1 Liquid Chromatography (LC)

LC is the most commonly use method for the separation of bioactive compounds, including anthocyanins and carotenoids. This technique uses liquid as mobile phases to separate different compounds. The type of mobile phases used typically consists of water, methanol, and acetonitrile and are activated with acidic compounds such as formic acid or hydrochloric acid (HCl). Using different types of columns depends on the compounds identified in foods. The most commonly used columns to identify anthocyanins and carotenoids are C_{18} and XBridge Amide. The basic principle of LC is that samples are injected into the LC and carried through a column packed with a tiny particle 3 to 5 μm in diameter called the stationary phase (see Fig. 9.3.) (McCreedy 2003). Individual components of the sample move down the column carried by a liquid interaction with the stationary phase that creates different constituents within the mixture and different retention times. These separated chemical compounds are detected at the exit of the column by a flow-through fluorescence detector, where fluorescent stimuli are translated into a visual chromatogram for interpretation (Meyer 2006; Talcott 2013). There are several factors that affect the retention times and peaks including temperature and concentration of the acid. Studies have reported that increasing the concentration of formic acid used to identify anthocyanins results in the peak area of cyanidin-3-O-diglucoside and the

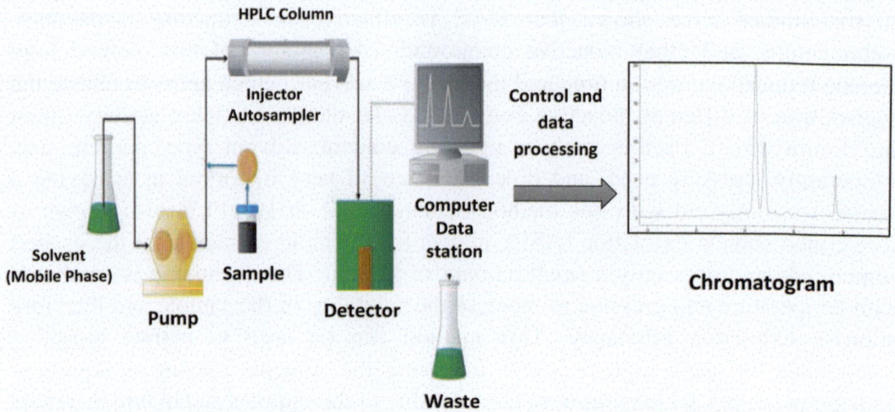

Fig. 9.3 The fundamentals of high-performance liquid chromatography (HPLC)

retention times being increased; this may be due to the increased polarity of the mobile phases (Ongkowijoyo et al. 2018; Sang et al. 2017). High-performance liquid chromatography (HPLC) is the most popular separation technique using liquids as mobile phases to analyze compounds in plant foods such as anthocyanins and carotenoids (Anderson 1999). HPLC is also used to analyze the metabolites of anthocyanins in blood samples in legal situations such as urine drug analysis and for both quantitative and qualitative analysis of a compound. HPLC can be often be identified on the basis of retention time and either UV-visible or mass spectra (Stauffer 2013). The most widely used tool in the identification and quantification of anthocyanin profiles is reversed phase HPLC coupled with photodiode array detection. Analyzing different types of anthocyanin depends on their polarity which causes elution at different times. Provisos studies have reported that the measured anthocyanin content in food matrixes can be influenced by the method used, which can also affect the quantity of the anthocyanins (Lee et al. 2008; McCreedy 2003) (Kamffer et al. 2010; Kong 2003; Lee et al. 2008).

9.7.2 Mass Spectrometry (MS)

MS measurements are carried out in the gas phase on ionized analytes. The MS method is high resolution and sensitive and can be used to analyze the separation of multi-food mixtures such as essential oils and bioactive compounds. Moreover, MS can be used to identify and qualify unknown compounds which are determent by molecular weight. Gas chromatography-mass spectrometry (GC-MS) is widely used to separate compounds such as anthocyanins and carotenoids which can be analyzed in food mixtures (Technologies 2010). Flame ionization detects any electron capture in GC, which make the technique highly sensitive (Al-Rubaye et al. 2017). This method can detect low concentrations of compounds in foods or plants. GC can also be used to identify and quantitatively analyze food mixtures and the purity of compounds and is able to determine the thermochemical constants of solutions such as heat and vapor pressure (Eiceman 2006). Recently, GC-MS studies have been increasingly used in the analysis of plants and bioactive compounds, as well as nonpolar compounds such as lipids and alkaloids (Al-Rubaye et al. 2017; Stashenko and Martínez 2012).

9.8 Bioavailability and Biological Activities

9.8.1 Anthocyanins

The beneficial health effects of polyphenols are dependent upon both intake and bioavailability. The bioavailability and bioaccessibility of polyphenols such as anthocyanins are largely influenced by their chemical structure, and they may be absorbed to varying degrees throughout the gastrointestinal tract (Haminiuk et al. 2012; Pandey and Rizvi 2009). Aglycones can be absorbed in the small intestine; however, most polyphenols are present in food in the form of esters, glycosides, or polymers that cannot be absorbed in their native form (Mertens-Talcott et al. 2008). These substances must be hydrolyzed by intestinal enzymes or by colonic microflora before they can be absorbed (Alqurashi et al. 2017; de Pascual-Teresa et al. 2010; Johnson et al. 2013; Spencer 2003). Following absorption, polyphenols are conjugated, either in the small intestinal epithelial cells or later in the liver (Kay 2006). This process mainly includes methylation, sulfation, and glucuronidation (Dall'Asta et al. 2012). The conjugation process is highly efficient, and aglycones are generally either absent in blood or present in low concentrations after consumption (Kay et al. 2004). Generally, non-anthocyanin flavonoids are better absorbed than ACs and proanthocyanidins, although absorption may be influenced by various factors such as the food matrix in which these compounds are distributed and consumed; the heterogeneity of sugars or other functional groups on the flavan nucleus; gender and genetic differences; dietary habits; dosage; and intestinal microflora (González-Barrio et al. 2010). Previous research into bioactive compounds has reported that anthocyanins appear in the blood in micromolar concentrations (Spencer 2003). ACs are recovered in urine as glycosides; however, they are very unstable in the pH range of the intestinal tract and may be mostly degraded before absorption can take place (Bermúdez-Soto et al. 2007; Castañeda-Ovando et al. 2009). One study in rats reported that ACs were rapidly absorbed in the stomach and small intestine (He et al. 2009). In humans, the level of ACs in plasma after cranberry consumption is very low, when measured between 1 to 3 h following consumption (Milbury et al. 2010). Elsewhere, after volunteers consumed 300 g of raspberries, ACs were excreted in urine after 4 h in quantities accounting for $<0.1\%$ of the ACs present in the consumed raspberries (González-Barrio et al. 2010). ACs are detected in plasma at concentrations between 1 and 120 nM after consumption (Kay et al. 2009; Nielsen et al. 2003). One study of pharmacokinetics in humans, whereby participants consumed ACs from acai pulp and juice, estimated the C_{max} of ACs in plasma to be between 2321 and 1138 ng/L between 2 and 2.2 h (Mertens-Talcott et al. 2008). AC-rich foods are often presented as functional foods or superfoods due to associations between their consumption and the prevention of diseases. Fruits such as berries are associated with the reduced risk of cardiovascular disease (CVD), certain cancers, type II diabetes, and cerebrovascular diseases (Alqurashi et al. 2016; Mazza 2007; Schantz et al. 2010). The mechanisms of protection may be due to their antioxidant properties, anti-inflammatory effects, or uncertain effects on other pathways (Nijveldt et al. 2001). The biological system's reactive oxygen species (ROS) and reactive nitrogen species (RNS), such as nitric oxide radicals and superoxide, can damage the deoxyribonucleic acid (DNA) and lead to oxidation of both lipid and proteins in human cells. Antioxidant activity can prevent the human body from free radical scavengers by helping to maintain low levels of oxidation and antioxidation (Luis-Villaroya et al. 2015).

9.8.2 Carotenoids

Carotenoids from fruits, vegetables, and animal products are fat-soluble, which are associated with lipid in human tissues, cells, and membranes (Yabuzaki 2017). Carotenoids may be attached or esterified to proteins and during the digestion process released from the proteins and aggregate with lipid in the human body. The bioavailability of carotenoids in the human body has been investigated in many studies, which have reported that 5% of carotenoids are absorbed by the intestine, whereas 50% or more carotenoids are absorbed from micellar solutions (Saini et al. 2015; Zakynthinos and Varzakas 2016). Other factors affecting the absorption of carotenoids in humans include dietary intake of fat and protein; dietary fat intake is important for the effective absorption of carotenoids (Saini et al. 2015). The type of carotenoids in the diet can also affect human absorption; for example, lutein and other xanthophylls are known for their improved absorption and bioavailability over beta-carotene or lycopene (Desmarchelier and Borel 2017). Food mixture interactions, vitamin E intake, fiber intake, and nutrient status can also affect carotenoid bioavailability in the human body (Van Het Hof et al. 2000).

The biological activity of carotenoid intake in the human body has been studied. Intake of carotenoids have been found to reduce free radical scavengers, reduce lipid oxidation, prevent CVD, reduce the risk of certain cancers, and protect against low-density lipoproteins (LDL) from oxidation; this may be due to the antioxidant properties contained in carotenoids (Rao and Rao 2007; Stahl and Sies 2003). Moreover, the mechanisms governing these compounds that are specifically responsible for these health benefits are not understood but could be due to antioxidant activities contained within the compounds (Dutta et al. 2005). One recent study reported that red cherries contain high levels of carotenoids that can deliver health-promoting and protective antioxidant and antiproliferative effects (Campestrini et al. 2019).

9.9 Commercial Utilization and Prospects

The number of products including natural food pigments continues to grow worldwide. Global food and beverage markets have dramatically increased in recent years including the trend of using food colors as natural food pigments in many foods and beverages (Rodriguez-Amaya 2019). This may be due to increased public awareness of using artificial colorants in foods which can have potential side effects on human health. There are a number of natural food pigments very commonly used in the food industry which include anthocyanins, carotenoids, betalains, and chlorophylls, among many others (Delgado-Vargas et al. 2000).

9.9.1 Anthocyanins

Anthocyanin extracted from foods and plants is used in the food industry in many ways such as food additives and natural food colorants. Anthocyanins are a group of naturally phenolic compounds responsible for the red, blue, and purple colors in many fruits and vegetables. One of the most frequently used food additives from anthocyanins is E163 which is derived from fruits such as grapes and the skin of berries. E163 is used in the production of the purple color in beverages and other foods (Cortez et al. 2017; Sampaio et al. 2021). Several studies have reported that artificial food colorings have created public concern regarding human safety, as it has been suggested that artificial colorants may significantly increase hyperactivity in children (McCann et al. 2007; Pollock and Warner 1990). The same results can be achieved through natural food colorants from plants such as anthocyanin. Furthermore, autolysins extracted from food and plants have been used in food and drinks such as yogurt, milk, fruit juice, and jam (Galaffu et al. 2015; Solymosi et al. 2015). However, there are a number of factors influencing anthocyanin pigment stability including pH level, temperature, level of ascorbic acid, sugar content, and metal ion content (Laleh et al. 2006). Therefore, the biggest challenge in using anthocyanins as natural pigments is their stability.

9.9.2 Carotenoids

Carotenoids are beneficial to human health since they provide antioxidant activity to the human body. Carotenoids can be used as ingredients in functional foods and also as natural pigments. As previously mentioned in this chapter, carotenoids can prevent diseases, therefore making them useful bioactive compounds in functional foods (Eggersdorfer and Wyss 2018). There is strong evidence that carotenoids are more bioavailable compared to other bioactive compounds (Schweiggert and Carle 2017). Generally, carotenoids in foods are stable and can be used as food additives. However, during storage or during the food process, carotenoids can be degraded due to light and oxygen exposure (Van Het Hof et al. 2000). For these reasons, encapsulation of carotenoids can be done to reduce loss of these compounds.

Encapsulation of the beta-carotene form of carotenoids was developed for supplementation and food fortification (Gul et al. 2015). Encapsulation techniques offer the potential to improve the bioavailability and suitability of carotenoid compounds that are used as functional ingredients in the food industry (Coronel-Aguilera and San Martín-González 2015). In the food industry, different types of carotenoids such as α-carotene, β-carotene, lutein, and zeaxanthin are used in nutraceuticals, pharmaceuticals, animal feed additives, and food pigmentation (Zakynthinos and Varzakas 2016).

9.10 Conclusions and Recommendations

This chapter has highlighted anthocyanins and carotenoids as bioactive compounds. They are the phytochemicals that impart the distinctive orange, red, and purple colors in many fruits and vegetables. Anthocyanins and carotenoids are reported to contain high sources of active antioxidants that have a positive biological effect on human health. The consensus among scientific opinions is that a high fruit consumption of plant foods rich in polyphenols is beneficial to human health and that these foods protect against CVD, cancer, and other chronic diseases. There is a pressing need to further understand the mechanism effects and roles of these compounds as beneficial sources of antioxidants. However, the biological benefits are intimately related to their structures, extraction methods, factors that affect their stability in food processing, and finally the bioavailability of these compounds in the human body. Foods with a high level of bioactive compounds including anthocyanins and carotenoids are increasingly popular as superfoods found in food markets and marketed as having high antioxidant levels with generally unsubstantiated messages regarding their health benefits. Moreover, anthocyanins and carotenoids as natural pigments are used as food additives and colorants in cosmetics and special foods. This chapter is therefore timely in its contribution to addressing the knowledge gap around the bioavailability and biological effect of anthocyanins and carotenoids in human health. Further research is clearly needed to fully justify the stability, bioavailability, and biological effect of anthocyanins and carotenoids in order to achieve a wider consensus in using these compounds in food markets and the human diet.

References

Alqurashi RM, Galante LA, Rowland IR, Spencer JPE, Commane DM (2016) Consumption of a flavonoid-rich açai meal is associated with acute improvements in vascular function and a reduction in total oxidative status in healthy overweight men. Am J Clin Nutr 104(5): 1227–1235. https://doi.org/10.3945/ajcn.115.128728

Alqurashi RM, Alarifi SN, Walton GE, Costabile AF, Rowland IR, Commane DM (2017) In vitro approaches to assess the effects of acai (Euterpe oleracea) digestion on polyphenol availability and the subsequent impact on the faecal microbiota. Food Chem 234(1):190–198. https://doi.org/10.1016/j.foodchem.2017.04.164

Al-Rubaye AF, Hameed IH, Kadhim MJ (2017) A review: uses of gas chromatography-mass spectrometry (GC-MS) technique for analysis of bioactive natural compounds of some plants. Int J Toxicol Pharmacol Res 9(1):81–85. https://doi.org/10.25258/ijtpr.v9i01.9042

Altemimi A, Lakhssassi N, Baharlouei A, Watson DG, Lightfoot DA (2017) Phytochemicals: extraction, isolation, and identification of bioactive compounds from plant extracts. Plan Theory 6(4):42. https://doi.org/10.3390/plants6040042

Amorim-Carrilho KT, Cepeda A, Fente C, Regal P (2014) Review of methods for analysis of carotenoids. Trends Anal Chem 56:49–73. https://doi.org/10.1016/j.trac.2013.12.011

Anderson DJ (1999) High-performance liquid chromatography in clinical analysis. Anal Chem 71(12):314–327. https://doi.org/10.1021/a1999904m

Azmir J, Zaidul ISM, Rahman MM, Sharif KM, Mohamed A, Sahena F, Jahurul MHA, Ghafoor K, Norulaini NAN, Omar AKM (2013) Techniques for extraction of bioactive compounds from plant materials: a review. J Food Eng 117(4):426–436. https://doi.org/10.1016/j.jfoodeng.2013.01.014

Bermúdez-Soto MJ, Tomás-Barberán FA, García-Conesa MT (2007) Stability of polyphenols in chokeberry (Aronia melanocarpa) subjected to in vitro gastric and pancreatic digestion. Food Chem 102(3):865–874. https://doi.org/10.1016/j.foodchem.2006.06.025

Butnariu M (2016) Methods of analysis (extraction, separation, identification and quantification) of carotenoids from natural products. J Ecosystem Ecography 6(2):1000193. https://doi.org/10.4172/2157-7625.1000193

Campestrini LH, Melo PS, Peres LEP, Calhelha RC, Ferreira ICFR, Alencar SM (2019) A new variety of purple tomato as a rich source of bioactive carotenoids and its potential health benefits. Heliyon 5(11):e02831. https://doi.org/10.1016/j.heliyon.2019.e02831

Castañeda-Ovando A, de Pacheco-Hernández ML, Páez-Hernández ME, Rodríguez JA, Galán-Vidal CA (2009) Chemical studies of anthocyanins: a review. Food Chem 113(4):859–871. https://doi.org/10.1016/j.foodchem.2008.09.001

Chan CH, Yusoff R, Ngoh GC, Kung FWL (2011) Microwave-assisted extractions of active ingredients from plants. J Chromatogr A 1218(37):6213–6225. https://doi.org/10.1016/j.chroma.2011.07.040

Chandrasekhar J, Madhusudhan MC, Raghavarao KSMS (2012) Extraction of anthocyanins from red cabbage and purification using adsorption. Food Bioprod Process 90(4):615–623. https://doi.org/10.1016/j.fbp.2012.07.004

Chemat F, Tomao V, Virot M (2016) Ultrasound-assisted extraction in food analysis. In: Handbook of food analysis instruments. https://doi.org/10.1201/9781420045673.ch5

Chemat F, Rombaut N, Sicaire AG, Meullemiestre A, Fabiano-Tixier AS, Abert-Vian M (2017) Ultrasound assisted extraction of food and natural products. Mechanisms, techniques, combinations, protocols and applications. a review. Ultrason Sonochem 34:540–560. https://doi.org/10.1016/j.ultsonch.2016.06.035

Coronel-Aguilera CP, San Martín-González MF (2015) Encapsulation of spray dried β-carotene emulsion by fluidized bed coating technology. LWT Food Sci Technol 62(1):187–193. https://doi.org/10.1016/j.lwt.2014.12.036

Cortez R, Luna-Vital DA, Margulis D, Gonzalez de Mejia E (2017) Natural pigments: stabilization methods of anthocyanins for food applications. Compr Rev Food Saf 16(1):180–198. https://doi.org/10.1111/1541-4337.12244

da Fonseca Machado AP, Alves Rezende C, Alexandre Rodrigues R, Fernández Barbero G, de Tarso Vieirae Rosa P, Martínez J (2018) Encapsulation of anthocyanin-rich extract from blackberry residues by spray-drying, freeze-drying and supercritical antisolvent. Powder Technol 340:553–562. https://doi.org/10.1016/j.powtec.2018.09.063

Dall'Asta M, Calani L, Tedeschi M, Jechiu L, Brighenti F, Del Rio D (2012) Identification of microbial metabolites derived from in vitro fecal fermentation of different polyphenolic food sources. Nutrition 28(2):197–203. https://doi.org/10.1016/j.nut.2011.06.005

de Moura SCSR, Schettini GN, Garcia AO, Gallina DA, Alvim ID, Hubinger MD (2019) Stability of hibiscus extract encapsulated by ionic gelation incorporated in yogurt. Food Bioprocess Technol 12:1500–1515. https://doi.org/10.1007/s11947-019-02308-9

de Pascual-Teresa S, Moreno DA, García-Viguera C (2010) Flavanols and anthocyanins in cardiovascular health: a review of current evidence. Int J Mol Sci 11(4):1679–1703. https://doi.org/10.3390/ijms11041679

De Souza VB, Thomazini M, Balieiro JCDC, Fávaro-Trindade CS (2015) Effect of spray drying on the physicochemical properties and color stability of the powdered pigment obtained from vinification byproducts of the Bordo grape (Vitis labrusca). Food Bioprod Process 93:39–50. https://doi.org/10.1016/j.fbp.2013.11.001

Debnath SC (2009) Characteristics of strawberry plants propagated by *in vitro* bioreactor culture and *ex vitro* propagation method. Eng Life Sci 9(3):239–246. https://doi.org/10.1002/elsc.200800095

Delgado-Vargas F, Jiménez AR, Paredes-López O, Francis FJ (2000) Natural pigments: carotenoids, anthocyanins, and betalains—characteristics, biosynthesis, processing, and stability. Crit Rev Food Sci Nutr 40(3):173–289. https://doi.org/10.1080/10408690091189257

Desmarchelier C, Borel P (2017) Overview of carotenoid bioavailability determinants: from dietary factors to host genetic variations. Trends Food Sci Technol 69:270–280. https://doi.org/10.1016/j.tifs.2017.03.002

Duan W, Jin S, Zhao G, Sun P (2015) Microwave-assisted extraction of anthocyanin from Chinese bayberry and its effects on anthocyanin stability. Food Sci Technol 35:524–530

Dutta D, Chaudhuri UR, Chakraborty R (2005) Structure, health benefits, antioxidant property and processing and storage of carotenoids. Afr J Biotechnol 4(12):1510–1520. https://doi.org/10.4314/ajfand.v4i13.71773

Eggersdorfer M, Wyss A (2018) Carotenoids in human nutrition and health. Arch Biochem Biophys 652(15):18–26. https://doi.org/10.1016/j.abb.2018.06.001

Eibl R, Eibl D (2008) Design of bioreactors suitable for plant cell and tissue cultures. Phytochem Rev 7:593–598. https://doi.org/10.1007/s11101-007-9083-z

Eiceman GA (2006) Instrumentation of gas chromatography. In: Encyclopedia of analytical chemistry. https://doi.org/10.1002/9780470027318.a5505

Espinosa-Leal CA, Puente-Garza CA, García-Lara S (2018) In vitro plant tissue culture: means for production of biological active compounds. Planta 248(1):1–18. https://doi.org/10.1007/s00425-018-2910-1

Fischer R, Vasilev N, Twyman RM, Schillberg S (2015) High-value products from plants: the challenges of process optimization. Curr Opin Biotechnol 32:156–162. https://doi.org/10.1016/j.copbio.2014.12.018

Furusaki S, Takeda T (2017) Bioreactors for plant cell culture, reference module in life sciences. Elsevier, Dordrecht. https://doi.org/10.1016/b978-0-12-809633-8.09076-2

Galaffu N, Bortlik K, Michel M (2015) An industry perspective on natural food colour stability. In: Colour additives for foods and beverages, pp 91–130. https://doi.org/10.1016/B978-1-78242-011-8.00005-2

Garcia-Salas P, Morales-Soto A, Segura-Carretero A, Fernández-Gutiérrez A (2010) Phenolic-compound-extraction systems for fruit and vegetable samples. Molecules 15(12):8812–8826. https://doi.org/10.3390/molecules15128813

González-Barrio R, Borges G, Mullen W, Crozier A (2010) Bioavailability of anthocyanins and ellagitannins following consumption of raspberries by healthy humans and subjects with an ileostomy. J Agric Food Chem 58(7):3933–3939. https://doi.org/10.1021/jf100315d

Gul K, Tak A, Singh AK, Singh P, Yousuf B, Wani AA (2015) Chemistry, encapsulation, and health benefits of β-carotene—a review. Cogent Food Agric 1. https://doi.org/10.1080/23311932.2015.1018696

Haminiuk CWI, Maciel GM, Plata-Oviedo MSV, Peralta RM (2012) Phenolic compounds in fruits—an overview. Int J Food Sci Technol 47:2023–2044. https://doi.org/10.1111/j.1365-2621.2012.03067.x

He J, Wallace TC, Keatley KE, Failla ML, Giusti MM (2009) Stability of black raspberry anthocyanins in the digestive tract lumen and transport efficiency into gastric and small intestinal tissues in the rat. J Agric Food Chem 57(8):3141–3148. https://doi.org/10.1021/jf900567t

Jaime L, Rodríguez-Meizoso I, Cifuentes A, Santoyo S, Suarez S, Ibáñez E, Señorans FJ (2010) Pressurized liquids as an alternative process to antioxidant carotenoids' extraction from Haematococcus pluvialis microalgae. LWT Food Sci Technol 43(1):105–112. https://doi.org/10.1016/j.lwt.2009.06.023

Javier David Vega A, Hector R-E, Juan Jose L-G, Maria LL-G, Paola H-C, Raúl Á-S, Carlos Enrique O-V (2017) Effect of solvents and extraction methods on total anthocyanins, phenolic compounds and antioxidant capacity of Renealmia alpinia (Rottb.) Maas peel. Czech J Food Sci 35(5):456–465. https://doi.org/10.17221/316/2016-cjfs

Johnson MH, De Mejia EG, Fan J, Lila MA, Yousef GG (2013) Anthocyanins and proanthocyanidins from blueberry-blackberry fermented beverages inhibit markers of inflammation in macrophages and carbohydrate-utilizing enzymes in vitro. Mol Nutr Food Res 57(7):1182–1197. https://doi.org/10.1002/mnfr.201200678

Kamffer Z, Bindon KA, Oberholster A (2010) Optimization of a method for the extraction and quantification of carotenoids and chlorophylls during ripening in grape berries (Vitis vinifera cv. Merlot). J Agric Food Chem 58(11):6578–6586. https://doi.org/10.1021/jf1004308

Karakaya S (2004) Bioavailability of phenolic compounds. Crit Rev Food Sci Nutr 44:453–464. https://doi.org/10.1080/10408690490886683

Kay CD (2006) Aspects of anthocyanin absorption, metabolism and pharmacokinetics in humans. Nutr Res Rev 19(1):137–146. https://doi.org/10.1079/NRR2005116

Kay CD, Mazza G, Holub BJ, Wang J (2004) Anthocyanin metabolites in human urine and serum. Br J Nutr 91(6):933–942. https://doi.org/10.1079/BJN20041126

Kay CD, Kroon PA, Cassidy A (2009) The bioactivity of dietary anthocyanins is likely to be mediated by their degradation products. Mol Nutr Food Res 53(1):92–101. https://doi.org/10.1002/mnfr.200800461

Khalifa I, Zhu W, Li K-k, Li C-m (2018) Polyphenols of mulberry fruits as multifaceted compounds: compositions, metabolism, health benefits, and stability—a structural review. J Funct Foods 40:28–43. https://doi.org/10.1016/j.jff.2017.10.041

Khan MK, Abert-Vian M, Fabiano-Tixier AS, Dangles O, Chemat F (2010) Ultrasound-assisted extraction of polyphenols (flavanone glycosides) from orange (Citrus sinensis L.) peel. Food Chem 119(2):851–585. https://doi.org/10.1016/j.foodchem.2009.08.046

Khoo HE, Azlan A, Tang ST, Lim SM (2017) Anthocyanidins and anthocyanins: colored pigments as food, pharmaceutical ingredients, and the potential health benefits. Food Nutr. Res 61(1). https://doi.org/10.1080/16546628.2017.1361779

Kong J (2003) Analysis and biological activities of anthocyanins. Phytochemistry 64(5):923–933. https://doi.org/10.1016/S0031-9422(03)00438-2

Lai Y-S, Yamagishi M, Suzuki T (2011) Elevated temperature inhibits anthocyanin biosynthesis in the tepals of an oriental hybrid lily via the suppression of LhMYB12 transcription. Sci Hortic 132:59–65. https://doi.org/10.1016/j.scienta.2011.09.030

Laleh GH, Frydoonfar H, Heidary R, Jameei R, Zare S (2006) The effect of light, temperature, pH and species on stability of anthocyanin pigments in four Berberis species. Pak J Nutr 5(1):90–92. https://doi.org/10.3923/pjn.2006.90.92

Lazăr A, Petolescu C, Popescu S (2010) Experimental results concerning the effect of photoperiod and callus culture duration on anthocyanin synthesis. J Hortic For Biotechnol 14:100–104

Lee J, Rennaker C, Wrolstad RE (2008) Correlation of two anthocyanin quantification methods: HPLC and spectrophotometric methods. Food Chem 110(3):782–786. https://doi.org/10.1016/j.foodchem.2008.03.010

Liazid A, Guerrero RF, Cantos E, Palma M, Barroso CG (2011) Microwave assisted extraction of anthocyanins from grape skins. Food Chem 124(3):1238–1243. https://doi.org/10.1016/j.foodchem.2010.07.053

Liu Y, Tikunov Y, Schouten RE, Marcelis LFM, Visser RGF, Bovy A (2018a) Anthocyanin biosynthesis and degradation mechanisms in Solanaceous vegetables: a review. Front Chem 6:52. https://doi.org/10.3389/fchem.2018.00052
Liu Y, Liu Y, Tao C, Liu M, Pan Y, Lv Z (2018b) Effect of temperature and pH on stability of anthocyanin obtained from blueberry. J Food Meas Charact 12:1744–1753. https://doi.org/10.1007/s11694-018-9789-1
Lucini L, Pellizzoni M, Baffi C, Molinari GP (2012) Rapid determination of lycopene and β-carotene in tomato by liquid chromatography/electrospray tandem mass spectrometry. J Sci Food Agric 92:1297–1303. https://doi.org/10.1002/jsfa.4698
Luis-Villaroya A, Espina L, García-Gonzalo D, Bayarri S, Pérez C, Pagán R (2015) Bioactive properties of a propolis-based dietary supplement and its use in combination with mild heat for apple juice preservation. Int J Food Microbiol 205:90–97. https://doi.org/10.1016/j.ijfoodmicro.2015.03.020
Maciel LG, do Carmo MAV, Azevedo L, Daguer H, Molognoni L, de Almeida MM, Granato D, Rosso ND (2018) Hibiscus sabdariffa anthocyanins-rich extract: chemical stability, in vitro antioxidant and antiproliferative activities. Food Chem Toxicol 113:187–197. https://doi.org/10.1016/j.fct.2018.01.053
Mäki-Arvela P, Hachemi I, Murzin DY (2014) Comparative study of the extraction methods for recovery of carotenoids from algae: extraction kinetics and effect of different extraction parameters. J Chem Technol Biotechnol 89:1607–1626. https://doi.org/10.1002/jctb.4461
Mazza G (2007) Anthocyanins and heart health. Ann Ist Super Sanita 43(4):369–374
McCann D, Barrett A, Cooper A, Crumpler D, Dalen L, Grimshaw K, Kitchin E, Lok K, Porteous L, Prince E, Sonuga-Barke E, Warner JO, Stevenson J (2007) Food additives and hyperactive behaviour in 3-year-old and 8/9-year-old children in the community: a randomised, double-blinded, placebo-controlled trial. Lancet 370(9598):1560–1567. https://doi.org/10.1016/S0140-6736(07)61306-3
McCreedy T (2003) High performance liquid chromatography fundamental principles and practice. Anal Chim Acta. https://doi.org/10.1016/s0003-2670(97)81192-0
McGarvey DJ, Croteau R (1995) Terpenoid metabolism. Plant Cell 7:1015–1026. https://doi.org/10.1105/tpc.7.7.1015
Mertens-Talcott SU, Rios J, Jilma-Stohlawetz P, Pacheco-Palencia LA, Meibohm B, Talcott ST, Derendorf H (2008) Pharmacokinetics of anthocyanins and antioxidant effects after the consumption of anthocyanin-rich acai juice and pulp (Euterpe oleracea Mart.) in human healthy volunteers. J Agric Food Chem 56(17):7796–7802. http://www.ncbi.nlm.nih.gov/pubmed/18693743
Meyer VR (2006) Practical high-performance liquid chromatography, 4th edn. https://doi.org/10.1002/0470032677
Meyer HJ, Van Staden J (1995) The in vitro production of an anthocyanin from callus cultures of *Oxalis linearis*. Plant Cell Tissue Organ Cult 40:55–58. https://doi.org/10.1007/BF00041119
Milbury PE, Vita JA, Blumberg JB (2010) Anthocyanins are bioavailable in humans following an acute dose of cranberry juice. J Nutr 140(6):1099–1104. https://doi.org/10.3945/jn.109.117168
Morais H, Ramos C, Forgasc E, Cserháti T, Matos N, Almeida V, Oliveira J (2002) Stability of anthocyanins extracted from grape skins. Chromatographia 56:S-173–S-175. https://doi.org/10.1007/bf02494133
Mustafa A, Trevino LM, Turner C (2012) Pressurized hot ethanol extraction of carotenoids from carrot by-products. Molecules 17(2):1809–1818. https://doi.org/10.3390/molecules17021809
Nakatsuka A, Yamagishi M, Nakano M, Tasaki K, Kobayashi N (2009) Light-induced expression of basic helix-loop-helix genes involved in anthocyanin biosynthesis in flowers and leaves of Asiatic hybrid lily. Sci Hortic 121(1):84–91. https://doi.org/10.1016/j.scienta.2009.01.008
Naviglio D, Scarano P, Ciaravolo M, Gallo M (2019) Rapid solid-liquid dynamic extraction (RSLDE): a powerful and greener alternative to the latest solid-liquid extraction techniques. Foods 8(7):245. https://doi.org/10.3390/foods8070245

Ngamwonglumlert L, Devahastin S (2018) Carotenoids. In: Encyclopedia of food chemistry. https://doi.org/10.1016/B978-0-08-100596-5.21608-9

Nielsen ILF, Dragsted LO, Ravn-Haren G, Freese R, Rasmussen SE (2003) Absorption and excretion of black currant anthocyanins in humans and {Watanabe} heritable hyperlipidemic rabbits. J Agric Food Chem 51(9):2813–2820. https://doi.org/10.1021/jf025947u

Nijveldt RJ, Van Nood E, Van Hoorn DE, Boelens PG, Van Norren K, Van Leeuwen PA (2001) Flavonoids: a review of probable mechanisms of action and potential applications. Am J Clin Nutr 74(4):418–425. http://www.ncbi.nlm.nih.gov/pubmed/11566638

Ongkowijoyo P, Luna-Vital DA, Gonzalez de Mejia E (2018) Extraction techniques and analysis of anthocyanins from food sources by mass spectrometry: an update. Food Chem 250(1):113–126. https://doi.org/10.1016/j.foodchem.2018.01.055

Paes J, Dotta R, Barbero GF, Martínez J (2014) Extraction of phenolic compounds and anthocyanins from blueberry (Vaccinium myrtillus L.) residues using supercritical CO2 and pressurized liquids. J Supercrit Fluids 95:8–16. https://doi.org/10.1016/j.supflu.2014.07.025

Pandey KB, Rizvi SI (2009) Plant polyphenols as dietary antioxidants in human health and disease. Oxidative Med Cell Longev 2:897484. https://doi.org/10.1302/0301-620X.98B8.37623

Phillips RL, Kaepplert SM, Olhoft P (1994) Genetic in stability of plant tissue cultures: break down of normal controls. Proc Natl Acad Sci 91:5222–5226

Pollock I, Warner JO (1990) Effect of artificial food colours on childhood behaviour. Arch Dis Child 65(1). https://doi.org/10.1136/adc.65.1.74

Rai MK, Asthana P, Singh SK, Jaiswal VS, Jaiswal U (2009) The encapsulation technology in fruit plants—a review. Biotechnol Adv 27(6):671–679. https://doi.org/10.1016/j.biotechadv.2009.04.025

Ranveer RC, Patil SN, Sahoo AK (2013) Effect of different parameters on enzyme-assisted extraction of lycopene from tomato processing waste. Food Bioprod Process 91(4):370–375. https://doi.org/10.1016/j.fbp.2013.01.006

Rao AV, Rao LG (2007) Carotenoids and human health. Pharmacol Res 55(3):207–216. https://doi.org/10.1016/j.phrs.2007.01.012

Robert P, Fredes C (2015) The encapsulation of anthocyanins from berry-type fruits. Trends in foods. Molecules 20(4):5875–5888. https://doi.org/10.3390/molecules20045875

Rodriguez-Amaya DB (2019) Update on natural food pigments—a mini-review on carotenoids, anthocyanins, and betalains. Food Res Int 124:200–205. https://doi.org/10.1016/j.foodres.2018.05.028

Saini RK, Keum YS (2018) Carotenoid extraction methods: a review of recent developments. Food Chem 240(1):90–103. https://doi.org/10.1016/j.foodchem.2017.07.099

Saini RK, Nile SH, Park SW (2015) Carotenoids from fruits and vegetables: chemistry, analysis, occurrence, bioavailability and biological activities. Food Res Int 76(3):735–750. https://doi.org/10.1016/j.foodres.2015.07.047

Sakamoto K, Kumiko I, Sawamura K, Hajiro K, Asada Y, Yoshikawa T et al (1994) Anthocyanin production in cultured cells of *Aralia cordata* Thunb. Plant Cell Tissue Organ Cult 36:21–26. https://doi.org/10.1007/BF00048311

Sampaio SL, Lonchamp J, Dias MI, Liddle C, Petropoulos SA, Glamočlija J, Alexopoulos A, Santos-Buelga C, Ferreira ICFR, Barros L (2021) Anthocyanin-rich extracts from purple and red potatoes as natural colourants: bioactive properties, application in a soft drink formulation and sensory analysis. Food Chem 342(16). https://doi.org/10.1016/j.foodchem.2020.128526

Sang J, Sang J, Ma Q, Hou X-F, Li C-Q (2017) Extraction optimization and identification of anthocyanins from Nitraria tangutorun Bobr. seed meal and establishment of a green analytical method of anthocyanins. Food Chem 218(1):386–395. https://doi.org/10.1016/j.foodchem.2016.09.093

Santos DT, Albarelli JQ, Beppu MM, Meireles MAA (2013) Stabilization of anthocyanin extract from jabuticaba skins by encapsulation using supercritical CO2 as solvent. Food Res Int 50(2):617–624. https://doi.org/10.1016/j.foodres.2011.04.019

Schantz M, Mohn C, Baum M, Richling E (2010) Antioxidative efficiency of an anthocyanin rich bilberry extract in the human colon tumor cell lines Caco-2 and HT-29. J Berry Res 1(1):25–33. https://doi.org/10.3233/BR-2010-003

Schewe T, Steffen Y, Sies H (2008) How do dietary flavanols improve vascular function? A position paper. Arch Biochem Biophys 476(2):102–106. https://doi.org/10.1016/j.abb.2008.03.004

Schweiggert RM, Carle R (2017) Carotenoid deposition in plant and animal foods and its impact on bioavailability. Crit Rev Food Sci Nutr 13(57):1807–1830. https://doi.org/10.1080/10408398.2015.1012756

Silva S, Costa EM, Calhau C, Morais RM, Pintado ME (2017) Anthocyanin extraction from plant tissues: a review. Crit Rev Food Sci Nutr 57(14):3072–3083. https://doi.org/10.1080/10408398.2015.1087963

Simões C, Albarello N, de Castro TC, Elisabeth M (2012) Production of anthocyanins by plant cell and tissue culture strategies. In: Orhan IE (ed) Biotechnological production of plant secondary metabolites. Bentham Science, Sharjah, pp 67–86. https://doi.org/10.2174/978160805114411201010067

Solymosi K, Latruffe N, Morant-Manceau A, Schoefs B (2015) Food colour additives of natural origin. In: Colour additives for foods and beverages. https://doi.org/10.1016/B978-1-78242-011-8.00001-5

Spencer JPE (2003) Metabolism of tea flavonoids in the gastrointestinal tract. J Nutr 133(10):3255–3261. https://doi.org/10.1093/jn/133.10.3255s

Stahl W, Sies H (2003) Antioxidant activity of carotenoids. Mol Asp Med 24(6):345–351. https://doi.org/10.1016/S0098-2997(03)00030-X

Stashenko EE, Martínez JR (2012) GC-MS analysis of volatile plant secondary metabolites. In: Gas chromatography in plant science, wine technology, toxicology and some specific applications. https://doi.org/10.5772/32246

Stauffer E (2013) Gas chromatography-mass spectrometry. In: Encyclopedia of forensic sciences, 2nd edn. https://doi.org/10.1016/B978-0-12-382165-2.00249-X

Stout MA, Benoist DM, Drake MA (2018) Technical note: simultaneous carotenoid and vitamin analysis of milk from total mixed ration-fed cows optimized for xanthophyll detection. J Daily Sci 101(6):4906–4913. https://doi.org/10.3168/jds.2017-13092

Strati IF, Oreopoulou V (2011) Process optimisation for recovery of carotenoids from tomato waste. Food Chem 129(3):747–752. https://doi.org/10.1016/j.foodchem.2011.05.015

Sun H, Ge X, Lv Y, Wang A (2012) Application of accelerated solvent extraction in the analysis of organic contaminants, bioactive and nutritional compounds in food and feed. J Chromatogr A 1237(11):1–23. https://doi.org/10.1016/j.chroma.2012.03.003

Talcott S (2013) High performance liquid chromatography. https://doi.org/10.1007/978-1-4757-5250-2_17

Tanaka Y, Brugliera F (2013) Flower colour and cytochromes P450. Philos Trans R Soc Lond Ser B Biol Sci 368(1612):20120432. https://doi.org/10.1098/rstb.2012.0432

Technologies A (2010) Agilent 7890A gas chromatograph. Technology.

Van Het Hof KH, West CE, Weststrate JA, Hautvast JGAJ (2000) Dietary factors that affect the bioavailability of carotenoids. J Nutr 130(3):503–506. https://doi.org/10.1093/jn/130.3.503

Wang XD (2012) Carotenoids. In: Modern nutrition in health and disease, 11th edn. https://doi.org/10.5650/jos1956.40.893

Wang N, Zhang Z, Jiang S, Xu H, Wang Y, Feng S et al (2016) Synergistic effects of light and temperature on anthocyanin biosynthesis in callus cultures of red-fleshed apple (*Malus sieversii* f. *niedzwetzkyana*). Plant Cell Tissue Organ Cult 127:217–227. https://doi.org/10.1007/s11240-016-1044-z

Weber T, Bach TJ (1994) Conversion of acetyl coenzyme A into 3-hydroxy-3-methyl glutaryl coenzyme A in radish seedlings: evidence of a single monomeric protein catalyzing a Fe^{II}/quinone- stimulated double condensation reaction. Biochim Biophys Acta 1211:85–96. https://doi.org/10.1016/0005-2760(94)90142-2

Wu X, Beecher GR, Holden JM, Haytowitz DB, Gebhardt SE, Prior RL (2006) Concentrations of anthocyanins in common foods in the United States and estimation of normal consumption. J Agric Food Chem 54(11):4069–4075. https://doi.org/10.1021/jf0603001

Yabuzaki J (2017) Carotenoids database: structures, chemical fingerprints and distribution among organisms. Database 2017:1–11. https://doi.org/10.1093/database/bax004

Yuan XF, Dai ZH, Wang XD, Zhao B (2009) Assessment of genetic stability in tissue-cultured products and seedlings of Saussurea involucrata by RAPD and ISSR markers. Biotechnol Lett 31(8):1279–1287. https://doi.org/10.1007/s10529-009-9984-6

Zakynthinos G, Varzakas T (2016) Carotenoids: from plants to food industry. Curr Res Nutr Food Sci 4(1):38–51. https://doi.org/10.12944/CRNFSJ.4.Special-Issue1.04

Zhang W, Furusaki S (1999) Production of anthocyanins by plant cell cultures. Biotechnol Bioprocess Eng 4:231–252. https://doi.org/10.1007/BF02933747

Zhao CL, Chen ZJ, Bai XS, Ding C, Long TJ, Wei FG et al (2014) Structure-activity relationships of anthocyanidin glycosylation. Mol Divers 18(3):687–700. https://doi.org/10.1007/s11030-014-9520-z

Zhong J-J, Yoshida M, Fujiyama K, Seki T, Yoshida T (1993) Enhancement of anthocyanin production by *Perilla frutescens* cells a stirred bioreactor with internal light irradiation. J Ferment Bioeng 75(4):299–303. https://doi.org/10.1016/0922-338X(93)90155-2

In Vitro Production of Saponins

10

Poornananda M. Naik, W. N. Sudheer, Sakshi Dubey, Rutwick Surya Ulhas, and N. Praveen

Abstract

Plants have been utilized as food, feed, and fodder since the dawn of civilization. Plants are also thought to be a rich source of bioactive compounds with a variety of pharmacological actions. Saponins are one such group of molecules which are present in various plant species. As triterpenoid glycosides, they have a 30C oxidosqualene precursor aglycone moiety (sapogenin), which is then linked with glycosyl residues to form saponin. These saponins have a unique platform in the field of pharmaceutical and nutraceutical industries. Saponins are used for the treatment of various diseases which include cancer, diabetic, cardiac, hepatic, and nervous disorders. The production of saponins through conventional approaches is time-consuming and hard to extract pure compounds, and thus to achieve this, in vitro methods have been developed and enhanced the production and extraction of the metabolites. The present chapter focuses on the in vitro production of saponins through various tissue culture techniques such as shoot, callus, cell suspension, adventitious root, hairy root culture, and applications of bioreactors at commercial level. The chapter also focuses on biosynthetic pathway, extraction methods, and biological activities of saponins.

Keywords

Saponins · Plant secondary metabolite · In vitro production · Biosynthesis · Pharmacology

P. M. Naik
Department of Botany, Karnatak University, Dharwad, Karnataka, India

W. N. Sudheer · S. Dubey · R. Surya Ulhas · N. Praveen (✉)
Department of Life Sciences, CHRIST (Deemed to be University), Bengaluru, Karnataka, India
e-mail: wudali.sudheer@res.christuniversity.in; sakshi.dubey@science.christuniversity.in; rutwick.u@science.christuniversity.in; praveen.n@christuniversity.in

T. Belwal et al. (eds.), *Nutraceuticals Production from Plant Cell Factory*,
https://doi.org/10.1007/978-981-16-8858-4_10

10.1 Introduction

Saponins are plant secondary metabolites derived from mevalonic acid pathway which are amphipathic glycosides of triterpenes and steroids, often also called as steroidal glycoalkaloids. 'Sapo' in Latin refers to soap as these compounds produce foams when shaken with aqueous solutions (Mugford and Osbourn 2012). Saponins are large molecules containing a hydrophilic sugar moiety at one end separated from hydrophobic (lipophilic) non-sugar triterpene or steroid moiety. This property enables them to form a micelle and act as a detergent (Mishra et al. 2017). The non-sugar component is called aglycone (sapogenin) and composed of triterpenoid or steroidal backbone and the sugar component is called glycone and composed of molecules like arabinose, xylose, glucose, galactose, fructose, rhamnose, and glucuronic acid (Moghimipour and Handali 2015) Fig. 10.1. Based on their molecular and chemical nature, they are divided into triterpenoid (30 carbon atoms) and steroidal saponins (27 carbon atoms with 6-ringed spirostane or 5-ringed furostane skeleton). Dammaranes, tirucallanes, lupanes, hopanes, oleananes, taraxasteranes, ursanes, cycloartanes, lanostanes, cucurbitanes, and steroids are among the 11 saponin classes. In the plant kingdom, oleananes are found more often (Kregiel et al. 2017) Fig. 10.2.

Triterpenoid saponins play a vital role in plant physiology during the external stress conditions. Apart from these due to their pharmacological properties, they are extensively used in various fields of medicine (Yao et al. 2020). The antiviral properties of saponins inhibit the replication of Herpes simplex virus type-1 and

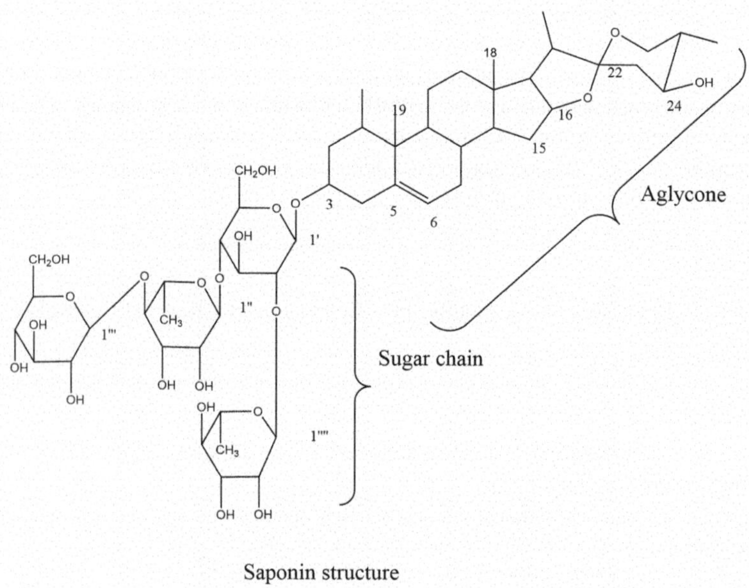

Fig. 10.1 Structure of Saponin. (Source: Moghimipour and Handali 2015)

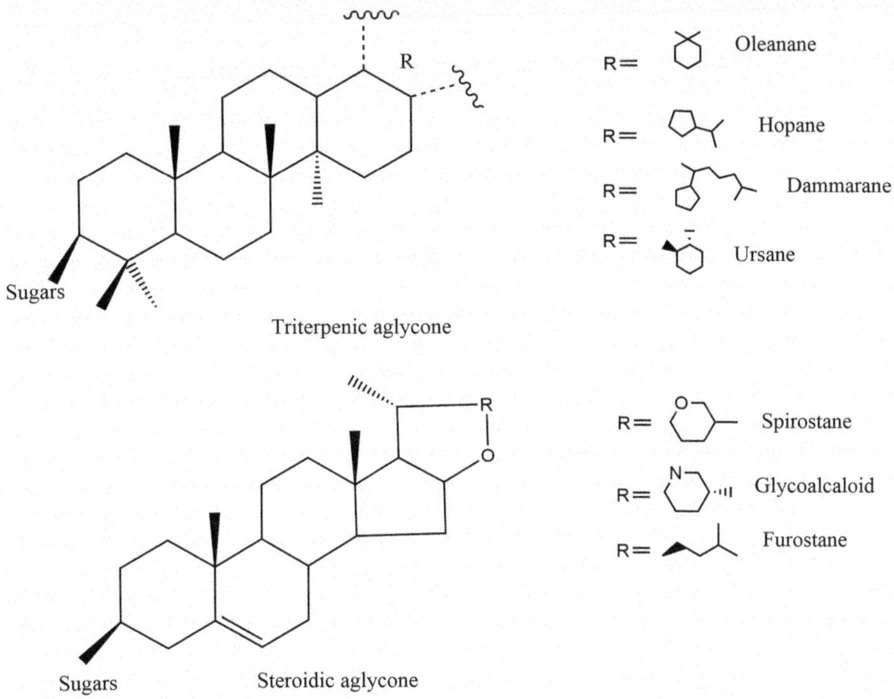

Fig. 10.2 Structure of triterpenoid and steroidal saponins. (Source: Moghimipour and Handali 2015)

Polio virus type-2 (Amoros et al. 1987). Plants extracts with saponins are used as animal feed for dairy and beef cattle as they inhibit rumen ciliate protozoans (Holtshausen et al. 2009), also suppress methane emission and change the fermentation patterns (Hu et al. 2005). Saponins in plants act as 'phytoprotectants' or 'phytoanticipins' as they act as a defence molecule against microbes and pest attack. They also play an important role in cell membrane permeability. Apart from these, they possess anti-inflammatory, anti-microbial, hyperlipidemic, and hypoglycemic activity (Desai et al. 2009). The triterpenoids play a major role in preventive and curative healthcare, perfume industries, cosmetics, flavouring, food, and beverage industries. (Biswas and Dwivedi 2019).

Depending on its growth phase and development, as well as seasonal variations, various plant species synthesize and accumulate varying quantities of saponins in different plant regions (Table 10.1). Saponin production may also be stimulated in response to external biotic stress, such as herbivores and disease attacks. Abiotic stresses such as light, temperature, and nutritional deficiency can all have an impact on both the quality and amount of saponin content. It has been revealed that at the molecular level, their synthesis is assisted by the transcriptional activation of their respective biosynthetic genes via a complicated signalling cascade including the

Table 10.1 Saponin content of some selected plant materials

Sl. No.	Name of the plant	Source	Saponin content	Reference
1	*Chenopodium quinoa* Willd	Seed	7.51–12.12 mg OAE/g	Han et al. (2019)
2	*Sapindus mukorossi* Gaertn.	Pericarp	280.55 ± 6.81 mg/g	Deng et al. (2019)
3	*Aloe vera* (L.) Burm. f.	Leaves	65.89 mg OAE/g	Akbari et al. (2021)
4	*Aesculus hippocastanum* L.	Seed	3–6%	Güçlü-Ustündağ and Mazza 2007)
5	*Primula grandis* L.	Roots	15–20%	Włodarczyk et al. (2020)
6	*Glycyrrhiza glabra* L.	Roots	3.6 g/200 g	Hajimohammadi et al. (2017)
7	*Yucca schidigera* Roezl.	Truck	10	Oleszek et al. (2001)
8	*Trigonella foenum-graecum* L.	Seed	0.98%	Chaudhary et al. (2018)
9	*Panax notoginseng* (Burkill) F.H.Chen	Root	9.26–46.52 mg/g	Cui et al. (2019)
10	*Glycine max* (L.) Merr.	Seed	1173.5 to 3582.3 mg/100 g	Lee et al. (2020)
11	*Avena sativa* L.	Seed bran	4.6%	Ralla et al. (2018)
12	*Medicago sativa* L.	Aerial parts	6.5–9.5 mg/g DW	Zhang et al. (2021)
13	*Bacopa monnieri* (L.) Pennell	Leaves	13–38.12 mg/g DW	Bhardwaj et al. (2019)
14	*Centella asiatica* (L.) Urban	Leaves	1.2–2 mg/g DW	Mangas et al. (2008)
15	*Gymnema sylvestre* R. Br.	Leaves	397.9 mg/g DW	Sheoran et al. (2015)
16	*Ziziphus joazeiro*	Bark	2–10%	Ribeiro et al. (2014)

jasmonate and salicylate hormones. Exogenous effect of stress-causative factors (elicitors) on secondary metabolite synthesis is frequently used to upregulate production of these essential bioactive metabolites (Lambert et al. 2011; Biswas and Dwivedi 2019).

However, the synthesis of this therapeutically significant class of bioactive compounds in the plant system in vivo is exceedingly low, leading to massive overuse of wild plant populations for their procurement by the pharmaceutical companies. Furthermore, with little or no planned re-cultivation, these techniques frequently endanger the plant populations. Plant tissue culture techniques like cell and organ cultures serve as an alternative for sustained and quality synthesis of these metabolites. The possibility of scaling up these cultures to the commercial level contributes to the industrial potential of metabolite synthesis using tissue cultures (Namdeo 2007; Biswas and Dwivedi 2019).

The present chapter deals with the production of saponins from cell and organ cultures and engineering strategies for enhanced metabolite content. Also, emphasis has been given for the biosynthesis of saponins and the key genes involved in the biosynthetic pathway have been mentioned. Furthermore, the different extraction and quantification methods employed for saponins and their potential biological activities have been discussed.

10.2 Biosynthesis of Saponins

Saponins are broadly classified as steroidal saponins or triterpenoid saponins based on the type of aglycone backbone from which it is derived. While triterpenoid saponins are synthesized majorly by dicotyledonous plants and are the most abundant class of saponins, steroidal saponins are mainly synthesized by monocotyledonous plants. Both triterpenoid and steroidal saponins are synthesized from the same oxidosqualene precursor which is 30 carbon long and linear. While the triterpenoid aglycone backbone contains all 30 carbons of oxidosqualene, the steroidal aglycone backbone retains only 27 carbons and loses 3 methyl groups. If additional nitrogen is incorporated to the steroidal aglycone backbone, it functions as a precursor for the biosynthesis of steroidal glycoalkaloids which may also be sometimes considered as saponins (Augustin et al. 2011; Friedman 2006; Ginzberg et al. 2009; Itkin et al. 2013).

The steroidal aglycone and triterpenoid aglycone being isoprenoids are synthesized using Isopentenyl pyrophosphate (IPP) as precursors that are converted from Acetyl CoA through the Mevalonate (MVA) pathway which is also known as 3-Hydroxy-3-methylglutaryl-CoA-reductase (HMGR) pathway. Further, Isopentenyl diphosphate isomerase (IDI) isomerizes IPP to allylic form dimethylallyl pyrophosphate (DMAPP). One molecule of the 5 carbon DMAPP then condenses with two molecules of 5 carbon IPP to give a 15-carbon immediate prenylated precursor of saponins, called Farnesyl pyrophosphate (FPP). Squalene synthase is a key enzyme in saponin biosynthesis. It catalyses the formation of a 30-carbon precursor, squalene, by the condensation of two molecules of FPP. Squalene is then epoxidized by the action of squalene epoxidase to form 2,3-oxidosqualene. A variety of cyclizing enzymes of the class oxidosqualene cyclase carry out the cyclization of 2,3-Oxidosqualene to form polycyclic structures. This reaction is the branching point between the metabolism of primary and specialized triterpene in higher plants.

The 2,3-Oxidosqualene cyclization results in one of the earliest inherent diversities to the triterpenoid saponins aglycones, accounting for its tendency to give rise to a vast array of triterpenoid scaffolds arising from a single substrate due to several carbocation rearrangements in the cyclization process. Majorly, there are 9 classes of triterpene backbones in plants synthesized by either specific or multifunctional Oxidosqualene cyclase to produce either single or multiple products through a single cyclization reaction of 2,3-Oxidosqualene (Vincken et al. 2007). One such cyclized structure, namely Cycloartenol, is a tetracyclic precursor of primary terpene formed by 2,3-Oxidosqualine cyclization catalysed by Cycloartenol synthase. Cycloartenol is a precursor for numerous phytosterols in angiosperms including carbon sitosterol, while all other cyclization products get involved as precursors for specialized triterpenes synthesis. The cholesterol backbone of these Cycloartenol derivatives undergoes several glycosylations and oxygenations to give rise to Spirostanol or Furostanol derivatives containing an oxygen heteroatom in their aglycone structure that further forms steroidal saponins (Thakur et al. 2011). Aglycones such as Solanidine, Solasodine, Tomatidine, and Demissidine are formed by cholesterol precursors that are utilized by steroidal glycoalkaloids, where an amine group is incorporated as a heteroatom instead of oxygen, through a series of side-chain modifications (Itkin et al. 2013; Ginzberg et al. 2009).

These triterpene aglycones are, although, majorly oxidized by multiple Cytochrome P450-dependent monooxygenases (P450s), several other modifications are also carried out that contribute to the extended diversity in the structure of the aglycone backbone by adding a second level of complexity. Various transferases including UDP-dependent glycosyltransferases (UGTs) and acyltransferases catalyses the modification of reactive functional groups and normalize the polarity of scaffolds that were introduced as a consequence of repeated oxidations on the triterpene backbone in order to enhance its structural diversity.

Throughout this biosynthesis process of triterpenoid, steroidal saponins, and steroidal glycoalkaloids, the key classes of enzymes are oxidosqualenecyclase, P450-dependent monooxygenase, and UDP-dependent glycosyltransferases. Additionally, numerous transferases and other tailoring enzymes also play a significant role. The saponins' biosynthesis pathway is illustrated in Fig. 10.3.

10.3 In Vitro Production of Saponins

In vitro technique is unique approach for the production of plant secondary metabolites. There are various reports regarding the production of secondary metabolites from plants through in vitro culture of plant tissues/explants (Verpoorte et al. 2002; Murthy et al. 2014a). Saponins are one of the major pharmaceutically important compounds found in various parts of the plant spp. and various in vitro cultures such as shoot, callus, cell, and root cultures reported the presence of saponins (Murthy et al. 2014a; Biswas and Dwivedi 2019). This section is concentrating on the production of saponins via shoot culture, callus culture, cell suspension culture, adventitious root culture, and hairy root culture (Table 10.2).

10.3.1 Shoot Culture

In vitro shoot culture is one of the best-known tissue culture methods to isolate saponins. Some important plant spp. producing saponins via shoot cultures have been presented. Praveen et al. (2009) established *Bacopa monnieri* shoot culture in both semisolid and liquid medium and evaluated the bacoside A content and found

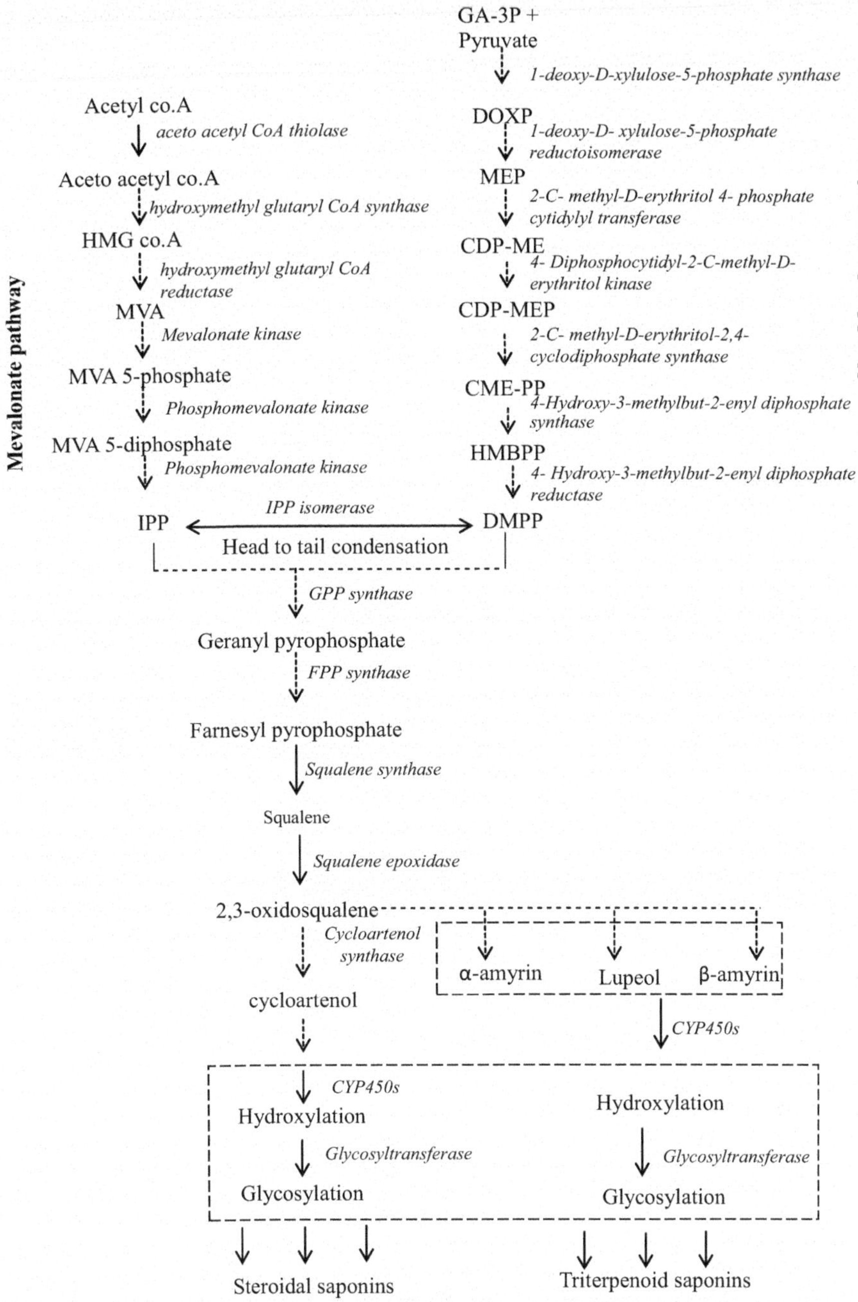

Fig. 10.3 Biosynthetic pathway of Saponins. (Source: Reproduced from Kumar et al. 2016)

Table 10.2 List of various explants and culture system used for the production of saponins

Sl. No.	Plant source	Explants	Culture system	Saponins	Reference
1	*Agave salmiana* Otto ex Salm-Dyck	Axillary shoot	In vitro plants	Tigogenin glycoside	Puente-Garza et al. (2017)
2	*Astragalus glycyphyllos* L.	Shoot	Callus, shoot, and suspension cultures	Cycloartane saponin	Shkondrov et al. (2019)
3	*Bacopa monnieri* (L.) Pennell	Leaf and stem	Shoot culture	Bacoside A	Sharma et al. (2013)
					Sharma et al. (2015)
		Aerial part	Shoot culture	Triterpenoid saponin glycosides	Watcharatanon et al. (2019)
		Leaf	Cell suspension culture	Bacosides	Koul and Mallubhotla (2020)
4	*Calendula officinalis* L.	Seedlings and Young leaf	Cell suspension culture	Oleanolic acid	Wiktorowska et al. (2010)
		Mature embryo	Hairy root culture		Alsoufi et al. (2019a)
					Alsoufi et al. (2019b)
5	*Centella asiatica* (L.) Urban	Node	Shoot culture	Asiaticoside	Prasad et al. (2013)
		Leaf	Callus culture	Centellosides	Mangas et al. (2008)
		Leaf	Cell suspension culture		Bonfill et al. (2011)
		Leaf	Hairy root culture		Kim et al. (2010)
6	*Chlorophytum borivilianum* Santapau and R. R.Fern	Leaf sheath	Callus culture	Stigmasterol and Hecogenin	Bathoju and Giri (2012)
7	*Codonopsis pilosula* Franch.	Seedling	Hairy root culture	Total saponins	Yang et al. (2020)
8	*Eryngium campestre* L.	Epicotyl	Shoot culture	Triterpenoid saponins	Kikowska et al. (2016)
9	*Eryngium maritimum* L.	Apical and axillary buds	Adventitious root cultures	Triterpenoid saponins	Kikowska et al. (2014)
10	*Eryngium planum* L.	Axillary buds	Shoot/callus/cell suspension	Triterpenoid saponins	Kikowska et al. (2019)

(continued)

Table 10.2 (continued)

Sl. No.	Plant source	Explants	Culture system	Saponins	Reference
11	*Gymnema sylvestre* R. Br.	Leaves and stalks	Cell suspension culture	Gymnemic acids	Chodisetti et al. 2015, Chodisetti et al. (2013)
		Cotyledons and young leaves	Hairy root culture		Nagella et al. (2013), Praveen et al. (2014)
12	*Helicteres angustifolia*	Young leaves	Callus suspension cultures	Total saponins	Yang et al. (2019)
13	*Panax ginseng* C.A. Meyer	Stem	Cell suspension culture	Ginsenoside	Huang et al. (2013)
					Huang and Zhong (2013)
		Root	Adventitious root cultures		Wang et al. (2013)
			Hairy root cultures		Liang et al. (2009)
14	*Panax quinquefolium* L.	Seedlings	Hairy root cultures	Ginsenoside	Kochan et al. (2018)
15	*Panax vietnamensis* Ha & Grushv.	Leaf segments	Callus culture/ in vitro plants	Ginsenoside	Nhut et al. (2015)
		Shoot	Hairy root culture	Majonoside R2, dammarane	Ha et al. (2016)
16	*Ruscus aculeatus* L.	Phylloclades, rhizomes, and seeds	Root-rhizome culture	Ruscogenin	Khojasteh et al. (2019)
17	*Silene vulgaris* (Moench) Garcke	Leaf	Hairy root culture	Segetalic acid and gypsogenic acid	Kim et al. (2015)
18	*Zingiber montanum* (J. König) Link ex A. Dietr.	Rhizome buds	Callus/cell suspension	Total saponins	Rajkumari and Sanatombi (2020)

maximum in shoots cultured in liquid medium (2.2-fold) when compared to shoots grown on semisolid medium. Effect of various concentrations of sucrose and pH on in vitro shoot culture and bacoside A production in *Bacopa monnieri* was examined by Naik et al. (2010) and reported lower concentrations of sucrose and that pH treatments enhance the production of bacoside A. Continuation of earlier work, Naik et al. (2011) studied the effect of macroelements and nitrogen source, and an increased amount of bacoside A content was reported in the shoot cultures grown in the higher concentration of nitrogen. Heavy metals such as manganese, zinc, and copper also affect the production of bacoside A in the shoot culture of *Bacopa monnieri* (Naik et al. 2015). Various media, medium strength, and carbon source alter the accumulation of bacoside A content in the shoot culture of *Bacopa monnieri* and found maximum in the treatments with full-strength MS medium, 2% sucrose and carbon source in combination with glucose and fructose (Naik et al. 2017). Prasad et al. (2013) worked on the accumulation of biomass and asiaticoside in *Centella asiatica* multiple shoot culture using fungal elicitors at different doses and culture age. Treatment with 3% v/v culture filtrates of *Trichoderma harzianum* in a culture medium on the tenth day influenced the biomass and asiaticoside accumulation by 2.53 and 2.35-fold when compared to untreated shoots in the culture cycle of 35 days. Interestingly, Nhut et al. (2015) showed that light-emitting diodes influence the accumulation of ginsenosides in the in vitro plant culture of *Panax vietnamensis.* Shkondrov et al. (2019) found in vitro shoot culture of *Astragalus glycyphyllos* yields double the amount of cycloartane saponins when compared to wild grown plants.

10.3.2 Callus Culture

In tissue culture, callus is an undifferentiated mass of cells that serves as the basic structure and the most important stage for development of embryos, shoots/roots, and friable cells by modifying the cultural conditions. The callus is also a source of secondary metabolites of particular plant spp. from which it has originated. Researchers have induced the callus culture of *Centella asiatica* for the production of centellosides and also studied the genes responsible for the biosynthesis of centellosides and found expression of the 5.8S rRNA gene (Mangas et al. 2008). In a callus culture of *Eryngium planum,* the application of methyl jasmonate elicitor in Murashige and Skoog (MS) medium fortified with 3% sucrose accumulated 1.2-fold triterpenoid saponins when compared to untreated callus culture (Kikowska et al. 2019). Yang et al. (2019) investigated the *Helicteres angustifolia* callus suspension culture for the estimation of phytochemical contents and found the potential source of total saponins in the culture.

10.3.3 Cell Suspension Culture

Cell suspension culture creates the avenue to obtain plant-based metabolites. It has the advantage over other culture method as it gets even/sufficient quantities of nutrients and cultural conditions, which induces the cells to grow faster and to maintain stability. *Bacopa monnieri* cell suspension cultures were tested with various elicitors (salicylic acid and jasmonic acid) and precursors (sodium nitroprusside, calcium pantothenate, and cholesterol) at different concentrations for the induction of bacoside content, and in elicitor treated cultures, elevated biomass and bacoside content was observed on 6th–9th day (Koul and Mallubhotla 2020) and salicylic acid found to be the best suitable for the induction of bacoside content among the tested elicitors and precursors. Shkondrov et al. (2019) determined the cycloartane saponins' accumulation in suspension culture of *Astragalus glycyphyllos*. Wiktorowska et al. (2010) studied the effect of various elicitors (Jasmonic acid, chitosan, yeast extract, pectin, and fungal strain *Trichoderma viride*) on the production of oleanolic acid using cell suspension cultures of *Calendula officinalis*. Cell suspension cultures treated with jasmonic acid after 72 h found to be the most efficient elicitors used and accumulated 9.4-fold oleanolic acid when compared to untreated culture. In another study, biotic elicitors (extracts of *Agrobacterium rhizogenes, Aspergillus niger, Bacillus subtilis, Escherichia coli,* and *Saccharomyces cerevisiae*) were used to induce gymnemic acids from the cell suspension culture of *Gymnema sylvestre* and all the elicitors treated cultures showed positive response on the accumulation of gymnemic acids (Chodisetti et al. 2013). In continuation of earlier study, researchers applied methyl jasmonate and salicylic acid in cell suspension culture and yielded optimum gymnemic acid content at 72 h after methyl jasmonate treatment (Chodisetti et al. 2015).

10.3.4 Adventitious Root Culture

Adventitious root culture as a differentiated organ culture serves as an excellent system for the production of secondary metabolites as it grows relatively fast and stably without any harmful molecules (Murthy et al. 2014b). Kikowska et al. (2014) performed an experiment with adventitious root cultures of *Eryngium maritimum* to test the nutritional factor and plant growth regulators on the production of saponins. Results found that elevated accumulation of triterpenoid saponins of about 3.2-fold in in vitro-derived roots was observed compared to field grown plant roots. Adventitious root cultures grow continuously in the liquid media and presence or absence of exogenous auxins does not affect. Wang et al. (2013) studied the effect of methyl jasmonate in adventitious root cultures of *Panax ginseng* for the production of ginsenoside content, and 10 mg/L methyl jasmonate for 24 h treatment increased the accumulation of ginsenoside with 4.76-fold higher than control.

10.3.5 Hairy Root Culture

Transformed hairy roots are induced by infecting the explants with *Agrobacterium rhizogenes*, which has the higher capacity biosynthesis of secondary metabolite compared to non-transformed roots (Chandra and Chandra 2011). Kim et al. (2015) induced hairy root culture using leaf explants of *Silene vulgaris*, and analysed the triterpenoid saponins like segetalic acid and gypsogenic acid accumulation was found to be 5 and 2-fold higher after the treatment of methyl jasmonate respectively when compared to the control roots. *Panax vietnamensis* is an important plant considering its peculiar pharmacological active saponins, namely majonoside R2 and dammarane. Hairy root culture of *P. vietnamensis* contains ginsenosides and majonoside R2, and dammarane saponins (Ha et al. 2016). Ginsenosides were induced in hairy root cultures *Panax quinquefolium* using trans-anethole elicitor, and trans-anethole activates the synthesis of saponins irrespective of exposure time (Kochan et al. 2018). Hairy root culture of *Calendula officinalis* is subjected to elicitor treatment (jasmonic acid and chitosan) for the stimulation of triterpenoid biosynthesis (Alsoufi et al. 2019a). Elicitor such as jasmonic acid, was very effective with respect to the accumulation of 20-fold oleanolic acid saponins in the hairy root tissue and 113-fold in the medium. Alsoufi et al. (2019b) selected abiotic elicitors (cadmium and silver ions, UV-C irradiation, and ultrasound) to induce triterpenoid biosynthesis from hairy root culture of *Calendula officinalis.* Heavy metals, UV-C irradiation, and ultrasound stimulated 12-fold, 8.5-fold, and 11-fold of triterpenoid biosynthesis, respectively. Recently, Yang et al. (2020) established the hairy root culture of *Codonopsis pilosula* and determined the total saponins' content from the grown hairy roots.

10.4 Bioreactors: Scale-up Techniques

The secondary metabolites production requires the optimum cell growth in the cell/organ culture. Adequate nutrients, balanced mixing, and oxygen supply minimize the plant cell damage which also depends on the species and cell lines selected (Georgiev et al. 2013). Other parameters such as temperature, pH, oxygen concentration, carbon dioxide, and substrate concentrations also control the production of secondary metabolites. Bioreactors in general possess the basic function which provides low shear stress, sufficient oxygen supply, and better mixing system of cells to maintain their optimum physiological conditions and regulate the metabolism of different environmental factors (Murthy et al. 2014b). Bioreactors are the important tools in the field of bioprocessing industry, which allow the optimum rate of multiplication of quality grade metabolites in a short duration of time with lesser cost. Various kinds of bioreactors are utilized for the production of secondary metabolites/saponins among stirred tank bioreactor (STB) which is most common. The advantage of these bioreactors is that they provide enough space for the cells to accumulate in different stages and also have the capacity to scale up nutrients because of their huge sizes (Gantait et al. 2020). For large-scale production of secondary metabolites/saponins, airlift bioreactors (ALB's) are also selected as one of most favourite bioreactors by researchers and industrialists as they provide maximum oxygen transfer and reduced cell shearing, which in turn give maximum yield (Gantait et al. 2020). For the reference purpose, some of the bioreactor types, culture system, and saponins are listed in the Table 10.3.

Applications of bioreactor enhance the multiplication of *Bacopa monnieri* in in vitro shoot culture rate and which in turn lead to propagate at commercial level (Saha et al. 2020). Jain et al. (2012) investigated shoot culture of *Bacopa monnieri* using nodal explants and found increased biomass accumulation when 10% aeration

Table 10.3 List of various types of bioreactors and culture system used for the production of saponins

Sl. No	Plant source	Culture system	Bioreactor type	Saponins	Reference
1	*Astralagus membranaceous* (Fisch.) Bunge	Hairy root culture	Airlift bioreactors	Astragalosides	Ionkova et al. (2010)
2	*Bacopa monnieri* (L.) Pennell	Shoot culture	Airlift bioreactors	Bacosides	Sharma et al. (2015)
					Saha et al. (2020)
3	*Centella asiatica* (L.) Urban	Cell suspension cultures	Bioreactor	Centellosides	Loc and Nhat (2013)
4	*Glycyrrhiza glabra* L.	Hairy root culture	Stirred tank bioreactor	Glyrhizzin	Mehrotra et al. (2008)
5	*Panax ginseng* C.A.Mey	Adventitious root cultures	Bioreactor (with sparging air)	Ginsenosides	Jeong et al. (2009)
		Cell suspension and adventitious root cultures	Stirred tank bioreactor, airlift bioreactors, bubble bioreactor		Murthy et al. (2014a, b)
					Murthy et al. (2017), Adil and Jeong (2018)
		Hairy root cultures	Airlift bioreactors, bubble bioreactor		Gantait et al. (2020)
6	*Panax quinquefolium* L.	Cell suspension cultures	Stirred tank bioreactor	Ginsenoside	Wang et al. (2012)
7	*Solanum chrysotrichum* (Schldl.)	Cell suspension cultures	Airlift bioreactors	Antifungal saponins	Salazar-Magallón and de la Peña (2020)

was applied in Growtek® bioreactor. In another study, two different bioreactors, Growtek® bioreactor and ALB, were used for bacosides production from *Bacopa monnieri* in in vitro shoot cultures, and optimum bacoside content was obtained in the biomass cultured in ALB system when compared to Growtek® bioreactor culture. As ALB provides enough aeration, it supplies the maximum oxygen for the synthesis of bacosides (Sharma et al. 2015).

Embryogenic tissues of *Panax ginseng* were cultured in two different types of bioreactors, STB and ALB, using flat-blade turbine and a paddle impeller; interestingly, higher biomass and optimum ginsenoside saponins production was observed in the ALB compared to STB (Asaka et al. 1993). The use of STB in a two-stage culture mode of cell suspension culture of *Panax quinquefolium* yields maximum ginsenosides (Wang et al. 2012). Using 5-L bioreactor, Loc and Nhat (2013) standardized the protocol for the production of asiaticoside in cell suspension culture of *Centella asiatica*, the parameters included rate of aeration, size of inoculum, and speed of agitation. Recently, Salazar-Magallón and de la Peña (2020) carried an experiment for the production of antifungal saponins with a transformed cell line from cell suspension culture of *Solanum chrysotrichum* using ALB and found in vitro and in vivo antifungal activity of saponins against fungal pathogens.

Kim et al. (2004) studied the adventitious root cultures of *Panax ginseng* using various vessels (cone, bulb, balloon, and cylinder type) in ALB's and proved the balloon type of ALB is most appropriate for the accumulation of biomass and ginsenosides. Kim et al. (2005) worked on the aeration rate and sparger pore size and diameter and found great influence on the ginsenoside accumulation. Paek et al. (2009) found that ALB's are most suitable bioreactors for the production of biomass and ginsenosides from the adventitious root cultures of *Panax ginseng*. Different bioreactors, STB's and bubble column bioreactors (BCB's) with various capacities, were applied for the production of ginseng hairy root culture by Jeong et al. (2003) and maximum yield of hairy roots in 5-L and 19-L BCB's with 38 and 55-folds increment of biomass was obtained on 40 and 39 days of culture, respectively. Palazón et al. (2003) carried an experiment for the production of ginsenosides from ginseng hairy roots; it includes wave bioreactors or spray bioreactors, culture period, and medium exchange. In this study, wave bioreactor emerged as a promising system to grow hairy roots and found 28-fold biomass accumulation and enhanced ginsenoside content at the end of 56 days of culture when medium exchange was carried for every 14 days. Yu et al. (2003) found ALB's are the most promising bioreactors for the production of ginsenosides from ginseng hairy root culture.

10.5 Extraction and Detection Techniques of Saponins

Saponins are being isolated from different plant sources for their diversified economic and pharmacological activities. The saponins are distributed all over the plant body right from aerial parts and root regions. Presence of different functional groups and sugar units bound to the aglycone component of the saponins makes it very difficult in extraction technique. There are chances in which saponins may get hydrolysed and esterified in extractions. So it is bit challenging in the extraction of saponins (Runner 2006). Quantification of saponins is done similar to that of other metabolites using spectroscopic and chromatographic methods. For isolating the pure saponins from the plant source, preparative liquid chromatography is employed (Kim and Park 2001). But, for the estimation, HPLC (High-performance liquid chromatography) is the most accepted technique.

10.5.1 Extraction Techniques of Saponins

Right from initial times, researchers were using conventional methods to isolate the saponins from different plant sources. Different plant sources show different efficiencies with respect to the various extraction methods.

10.5.1.1 Conventional Methods

Soxhlet extraction was carried out using various types of solvents suitable for particular plant sources. There are many reported research articles which used soxhlet extraction for the isolation of saponins (Bajad et al. 2019). Different parameters influence the extraction procedure like time taken to extract the metabolites from the plant sample, nature of solvent, and its boiling point. In this extraction, the sample will be placed in the porous chamber made out of cheese cloth and vapours of the desired solvents are sent through it which ultimately takes out the target metabolites from the tissue. Polar solvents like water, ethanol, and methanol are suitable for the extraction of saponins through soxhlet extraction (Cheok et al. 2014). Reflux extraction is another method of extraction of metabolites where the plant sample is always kept in contact with desired polar solvents and subjected to continuous boiling, and once desired change and time are reached, it can be filtered and the extracts are used for the quantification of the desired metabolites with specific techniques (Tao et al. 2013). Maceration is also employed for extracting saponins and the principle behind it is solid-liquid extraction (Takeuchi et al. 2009). In this technique, the sample is soaked along with the desired solvents and chemicals for the required time, temperature, and stirring to aid the extraction of the metabolites (Verza et al. 2012). Solvent partitioning (liquid-liquid isolation) is also employed to effectively isolate the saponins (Kim and Park 2001; Cheok et al. 2014). Along with the above-mentioned conventional methods, there are many advanced techniques which can help in extraction of desired saponins. These techniques are environment-friendly and help us to conserve solvents. The main advantage is that the time taken for the isolation of the metabolites is comparatively less when compared to conventional methods.

10.5.1.2 Advanced Techniques

In ultrasound-assisted extraction, specific frequencies of sound waves are given to the tissue which helps us to extract the desired metabolites. Conceptually, it is similar to that of sonication. The time given for the extraction, solvent, and the frequency are the important parameters. The time taken for the extraction of saponins and related compounds is relatively less when compared to conventional methods like soxhlet extraction (Jadhav et al. 2009). Microwave-assisted extraction is most accepted technique for the isolation of metabolites in the recent times (Deore et al. 2015). The time required for the extraction of desired metabolite is comparatively less when compared to other techniques, usually the time taken for extraction will be around few minutes (6–8 min). Microwaves are given for the sample which is already associated with desired polar solvents like ethanol and methanol where extraction becomes very easy (Kerem et al. 2005). Accelerated solvent extraction is another technique employed to isolate saponins and other metabolites (Zhang et al. 2013). The conventional and advanced techniques employed for extraction studies have been mentioned in Table 10.4.

10.5.2 Techniques for Detection of Saponins

Detection of saponins from the plant sample is done both qualitatively and quantitatively. Qualitatively to detect the saponins from the plant sample, foam test is conducted. If the foam persists for more than 15 min it confirms, the presence of saponins (Tadhani and Subhash 2006). Quantitatively, it is detected using chromatographic and spectrophotometric techniques.

Chromatographic techniques are employed for the quantification of plant secondary metabolites. In the same way, saponins are also quantified using high-performance liquid chromatography (HPLC) and are most widely accepted. Type of column used, flow rate, wavelength, and solvents used in the mobile phase are the important parameters considered for the quantification of saponins. Chromatogram and the peaks graduated help us to understand the presence of desired metabolites. Furthermore, these chromatographic techniques are modified by adding different components like mass spectroscopy, diode array detection, and evaporative light scattering detector, which help us to study the isolated molecules at atomic level (Guajardo-flores et al. 2012). Acetonitrile and water are most widely used solvents for eluting saponins from the samples.

In the spectrometric method, measuring the colour developed upon chemical reaction between desired metabolite and chemicals is the basic principle. Chemical standard used and wavelength at which colour measured are the important parameters in spectrometric methods. Vanillin (8%) along with sulphuric acid (72%) develops red-purple colour with plant sample which is later measured, which proves the presence of saponins (Le et al. 2018). The details of some of the detection methods for saponins are mentioned in Table 10.5.

Since saponins have such diversified economic and pharmacological uses, scientific community should design advanced techniques for the extraction of metabolites at large scale with minimum expenses. In the same way, the quantifying techniques can also be revised which can give more information of the desired molecule at less expenses.

10.6 Biological Activities of Saponins

Among various phytochemicals of the plant kingdom, saponins constitute an important class of bioactive molecules; this is attributed to the numerous biological activities exhibited by different aglycones of saponins, majorly the steroidal and the triterpenoid saponins (Bruneton 1999). These activities range from antitumor, gastroprotective, and antihyperlipidemic potential to their ability to regulate blood glucose and promote bone marrow haematopoiesis.

Table 10.4 Extraction techniques for saponins

Sl. No	Plant source	Part used	Saponin type	Extraction method	Reagents	Reference
1	*Bupleurum falcatum* L.	Dried roots	Saikosaponin c, a, and d	Solvent partitioning	70% ethanol- for extraction of the sample, diethyl ether, n-butanol, acetone and methanol- for solvent partitioning of the sample and LC	Kim and Park (2001)
2	*Panax ginseng* L.	Roots	Ginsenosides	Ultrasound-assisted extraction for 2 h using ultrasound bath (38.5 kHz) or ultrasonic probe (20 kHz)	Methanol, aqueous n-butanol, and 10% methanol- for extraction, acetonitrile- for TLC and HPLC	Wu et al. (2001)
3	*Cicer arietinum* L.	Seed powder	Soyasaponin B	Microwave-assisted extraction (2450 Mhz) for 20 min	Methanol, ethanol, butanol or aqueous ethanol, and butanol- for extraction of the sample, water and methanol- for HPLC	Kerem et al. (2005)
4	*Vigna radiata* L.	Whole plant	Soyasaponins	Soxhlet extraction for 24 h	Chloroform, 80% ethanol, butanol- for extraction of the sample	Waller et al. (1999)
5	*Ziziphus mauritiana* L.	Leaves	Saponin	Reflux extraction for 48 h	70% methanol and distilled water- for extraction of the sample	Dubey et al. (2019)
6	*Glycyrrhiza glabra* L.	Root powder	Triterpene saponins	Reflux extraction	50% aqueous methanol- for extraction of the sample	Tao et al. (2013)
	Glycyrrhiza inflate Bat.					
	Glycyrrhiza uralensis Fisch.					
7	*Momordica charantia* L.	Vine and leaves	Triterpene saponins	Reflux extraction	Ethanol- for extraction of the sample	Chen et al. (2009)
8	*Momordica charantia* L.	Root powder	Triterpene saponins	Maceration for 6 h at 60 °C	Methanol- for the first digestion, ethanol, n-butanol- for the subsequent maceration	Chen et al. (2008)

(continued)

Table 10.4 (continued)

Sl. No	Plant source	Part used	Saponin type	Extraction method	Reagents	Reference
9	*Ipomoea batatas* L.	Tuber flour	Triterpene saponins	Maceration for 4 h at room temperature	80%methanol- for the maceration	Dini et al. (2009)
10	*Tribulus terrestris* L.	Whole plant powder	Steroid saponins	Maceration for 1 h followed by reflux extraction for 3 times around 1 h	Chloroform- for maceration, 70% ethanol- for reflux extraction	Dinchev et al. (2008)
11	*Bacopa monnieri* (L.) Pennell	Plant dry powder	Triterpenoid saponins	Ultrasound sonication for 10 min	Methanol- for the extraction of the sample	Ganzera et al. (2004)
12	*Ziziphus jujuba* Mill.	Leaves dry powder	Zizyphussaponins I, II	Ultrasound sonication at 40 kHz for 30 min at room temperature	80% ethanol- for the extraction of sample	Guo et al. (2011)
	Ziziphus jujuba var. spinosa Bunge.					
13	*Panax notoginseng* (Burkill) F.H. Chen	Dried powder	Notoginseng saponins	Microwave-assisted extraction (2450 MHz) for 6 min	Water saturated n-butanol- for the extraction of the sample	Vongsangnak et al. (2004)
14	*Litchi chinensis* Sonn.	Seed powder	Litchinoside and Sapondoside	Liquid-phase pulsed discharge and ultrasonic extraction (276 W at 47 °C), 3 mm hollow electrode, 123 ml/min flow velocity.	30% ethanol- for the extraction of the sample	Fan et al. (2020)
15	*Gymnema sylvestre* R. Br.	Dry leaf powder	Oleanolic acid	Microwave-assisted extraction (2450 MHz) for 8 min	Ethanol- for the extraction of the sample	Mandal and Mandal (2010)
16	*Glycyrrhiza glabra* L.	Dry powder	Triterpenoid saponins	Ionic liquid-based ultrasonic-assisted extraction along with in situ		Ji et al. (2020)

				alkaline aqueous biphasic system for 20 min.	NaOH and $[C_4MIM]BF_4$ (imidazolium type)-for alkaline aqueous biphasic system	
17	*Trigonella foenum-graecum* L.	Seed dry powder	Common saponins	Microwave-assisted extracted	40–80% ethanol- for the extraction	Akbari et al. (2020)

Table 10.5 Estimation methods for saponins

S. No	Plant source	Part used	Saponin type	Estimation method	Reagents	Yield	Reference
1	*Bupleurum falcatum* L.	Dried roots	Saikosaponin c, a, and d	HPLC C18 column flow rate- 1 mL/min wavelength- 203 nm	70% ethanol- for extraction of the sample, acetonitrile: Water- for eluting the samples	144.25 mg/100 g and 247.25 mg/100 g of saikosaponin c and d (145 days)	Park et al. (2000)
2	*Ilex paraguariensis* A.St.-Hil.	Aerial parts	Matesaponin 1, 2, 3, 4, 5	HPLC C18 column flow rate- 1 mL/min wavelength- 203 nm	Water- for extraction of the sample, acetonitrile: Water(70:30)- for eluting the samples	352 μg/ mL in saponin fraction	Gnoatto et al. (2005)
3	*Bryonia laciniosa* L.	Seed powder	Saponin	Spectrophotometric method at 538 nm	Ethanol- for extraction of the sample, vanillin (8%) + sulfuric acid (72%)- for colour development	15 μg/mg of oleanolic acid equivalent	Patel et al. (2012)
4	*Phaseolus vulgaris* L.	Dry powder	Soya saponins	HPLC-DAD-ELSD and TOF (diode array detection- evaporative light scattering detector and time of flight) flow- 0.5 ml/ minwavelength- 295 nm	80% methanol- for extraction of the sample, Trifluoroacetic acid and HPLC-grade acetonitrile- for eluting the samples	1063.62 mg/100 g in (soaked) and 455.95 mg/100 g in (raw)	Guajardo-Flores et al. (2012)
5	*Bacopa monnieri* (L.) Pennell	Dry powder	Bacoside and Bacopasides	HPLC column- C-8(2) flow rate-0.5 ml/min, wavelength- 205 nm.	Methanol- for the sample extraction water and methanol- for the eluting the sample	1.1 to 13% of dry weight	Ganzera et al. (2004)
6	*Asparagus species*						

		Freeze dried lyophilized powders of spears	Protodioscin, Saponins	HPLC-DAD–MS, particle size-5 μm flow- 0.3 mL/min	80% ethanol- for the sample extraction 1% formic acid and acetonitrile- for eluting the samples	1.09 to 2.73 mg/ 100 g of fresh weight	Vázquez-Castilla et al. (2013)
7	*Tribulus terrestris* L., *Achyranthus aspara* L., *Albizia lebbeck* (L.) Benth.	Seed dry powder	Quillaja saponin	Spectrophotometric method at 544 nm	Petroleum ether and methanol- for the extraction of the sample, vanillin (8%) + sulfuric acid (72%)- for colour development	25.65, 45.75, and 48.26% (w/w)	Goel et al. (2012)
8	*Solanum nigrum* L.	Dry leaf powder	General sapogenins	Ultra-high-performance liquid chromatography-MS flow rate- 0.4 mL/min	Water and acetonitrile, 0.1% formic acid- for eluting the sample	0.4 to 0.8% of dry mass	Yuan et al. (2019)

Saponins are amphiphilic molecules, consisting of both hydrophilic and hydrophobic moiety. The structural complexity, particularly the molecular structure, enables it to interact with the lipid components of the plasma membrane including sterols, proteins, and phospholipids and these interactions result in the formation of aggregates; this property majorly accounts for the wide range of biological activities that saponins have (Lorent et al. 2014) (Table 10.6). In addition to this, the ability of saponins to undergo chemical changes during processing or storage and their glycosylation pattern also determine the biological activities (Augustin et al. 2011; Güçlü-Ustündağ and Mazza 2007). Antifungal, anti-parasitic, antibacterial, and antioxidant are few of the minor activities of saponins, in addition to the activities mentioned below:

10.6.1 Anti-tumor Activity

One of the most important biological activities of saponins include antitumor activity. This property is exhibited by major saponins like dammarane or oleanane as well as their derivatives such as theasaponin. The activity is operated by different mechanisms by different classes of saponins, some of the major ones include reducing the synthesis of DNA, causing damage to DNA, inhibiting tumour angiogenesis, altering the host susceptibility to mutations, and by increasing immunosurveillance and apoptosis (Shibata 2001). Saponins are found to act on some of the major types of cancers such as breast cancer, lung cancer, liver cancer, colon cancer, and gastric cancer. The impact of aglycones on the antitumor activities has been studied and it is concluded that factors such as number of hydroxyl groups (Wang et al. 2007), site of hydroxyl group, lipophilicity of sugars (Mimaki et al. 2001), and sequence of sugars present (Bang et al. 2005) affect the activity of saponins. According to a study, spirostanes such as polyphyllin D show strong anticancer activity by inducing endoplasmic reticulum stress-mediated apoptotic pathway by the accumulation of unfolded or misfolded proteins, followed by mitochondria-mediated pathways by the downregulation of anti-apoptotic and upregulation of pro-apoptotic factors, which eventually leads to the apoptosis of tumour cells (Siu et al. 2008; Cheung et al. 2005); dioscin shows similar activity in addition to the antiproliferative activity against cancer cells (Wang et al. 2006). In contrast to this, dammarane saponins like OSW 1 damage the mitochondria and its cristae and triggers the calcium-dependent apoptotic pathway (Zhou et al. 2005). The cytotoxic effect of avicin D has been demonstrated by Haridas et al. (2009) where it was shown that saponin downregulates some of the major factors involved in apoptosis such as cyclin D1, c-myc, VEGF, and Bcl-2; this is done by decreasing the level of IL-6 and dephosphorylation of Stat-3. Platycodon D is another antineoplastic agent that operates by generating reactive oxygen species and activating CASPASE 3, thereby inducing apoptosis (Shin et al. 2009). Few other mechanisms underlying antitumour activity include inhibition of COX-2/PGE-2 pathway (Han et al. 2013), upregulation of proapoptotic proteins like Bcl-2 and Bax, generation of ROS, inhibition of wnt/beta catenin signalling pathway, downregulation of few other

Table 10.6 Saponins and their pharmacological activities

Sl. No.	Saponins	Activity	Mechanism	Reference
1	Notoginsenoside R1	Anti-atherosclerosis	Inhibition of plasminogen activator inhibitor 1	Zhang & Wang (2006)
2	Ginsenosides	Anticoagulant activity	Increasing the synthesis of plasminogen activators and antagonistic activity of platelet activating factors	Jung et al. (1998) Zhang et al. (1997)
3	Sea cucumber saponins	Anti-hyperuricemic activity	Inhibition of enzymes xanthine oxidase and adenosine deaminase	Xu et al. (2011)
4	Platycosides	Anti-inflammatory activity	Inhibition of NF-κB activation and MAPK signalling pathways	Jang et al. (2013)
5	Maesa Saponins, oleanolic saponin	Haemolytic activity	Bursting of erythrocyte membrane due to the interaction between sterols of membrane and saponins	Voutquenne et al. (2003) Baumann et al. (2000) Sindambiwe et al. (1998)
6	Cucumarioside A2-2, and Frondoside A	Immunomodulatory activity	Stimulation of cytosolic calcium concentration, lysosomal activity, ROS formation, and natural cellular defence barrier	Aminin et al. (2009)
7	Oleanolic acid	Anti-HIV activity	Inhibition of in vitro HIV-1 protease activity	Mengoni et al. (2002)
8	Ziyu glycoside	Promotion of bone marrow hematopoiesis	Reduction of suppressive cytokines and activation of FAK and Erk1/2 pathways	Chen et al. (2017)
9	Ginsenoside Rg 1	Neuroprotective activity	Inhibition of mitochondrial apoptotic pathway and increase in activity of choline acetyltransferase	Leung et al. (2007) Yamaguchi et al. (1997)
10	Oleanolic acid	Molluscicidal activity	Formation of pores resulting in leakage of liquids due to the interaction between saponins and cholesterol of the membrane	de Paula Barbosa (2014)

proteins like cyclin D, cdk-4, and MDM2, and upregulation of caspase 3/9, p21, p53 expression which eventually leads to cell cycle arrest or apoptosis (Wang et al. 2018; Cui et al. 2018). Furthermore, targeting the three main family members of MAPKs, namely, p38, JNK, and ERK, and inducing the formation of massive vacuoles containing lysosomes and autolysosomes that is characteristic of autophagy is another way of killing the cancer cells which is mainly performed by the saponin jujuboside B (Xu et al. 2014).

10.6.2 Anti-hyperglycemic Activity

Triterpene saponins such as saponins of *Panax ginseng*, Charantin, and saponins from *Asparagus officinalis* L., have shown the ability to improve the uptake of glucose and insulin sensitivity in the liver cells (Hu et al. 2014; Kim et al. 2009; Zhu et al. 2020). Ginsenoside Rb1 acts as a hypoglycemic agent by regulating glycolipid metabolism and increasing the insulin sensitivity, which are achieved by activation of the Peroxisome proliferator-activated receptor gamma (PPAR-γ), thereby improving glucose homeostasis (Kwon et al. 2012). In vitro studies have revealed that ginsenoside activates insulin signalling pathways such as phosphorylation of insulin receptor substrate-1 and protein kinase B (PKB) and the activity of phosphatidylinositol 3-kinase (PI3K); this in turn leads to the increased translocation of GLUT receptors in the adipose tissue (Shang et al. 2008). In addition to this, saponins target glucose metabolizing enzymes in order to normalize blood glucose level. Diosgenin is a steroid saponin which is known to promote glycogenolysis by increasing the levels of enzymes such as Phosphofructokinase and pyruvate kinase (Raju and Chinthapally Rao 2012. Lee et al. (2011) have reported that ginsenoside Rb2 inhibits gluconeogenesis by upregulating SHP (Short heterodimer partner), which inhibits the mRNA expression of gluconeogenic enzymes such as Glucose-6-phosphatases and phosphoenolpyruvate carboxykinase. The inhibitory effect of saponins against carbohydrate hydrolysing enzymes including pancreatic alpha amylase and alpha glucosidase and their potential to reverse the atrophic pancreatic beta cells as shown by Oleifera Saponin A1 (Di et al. 2017) further establishes saponins as an effective antidiabetic agent. Another hypoglycemic mechanism of saponins involves regulation of AMPK/NF-κB signal pathway, improvement of lipid metabolism in diabetic subjects, and inhibition of reactive oxygen species formation resulting in reduced oxidative stress and normal functioning of the kidneys (El Barky et al. 2016; Wang et al. 2019).

10.6.3 Anti-hyperlipidemic Activity

One of the common lipoprotein abnormalities involves decreased levels of HDL cholesterol and increased levels of LDL cholesterol and triglycerides (Gupta et al. 1994); these conditions are closely associated with diseases like coronary heart disease, atherosclerosis, and diabetes. Thus, modulating lipid metabolism or levels of TC, TG, HDL, and LDL is one way of functioning as an antihyperlipidemic agent and this is efficiently carried out by saponins such as soyasaponins, ginsenosides, and trigonelline. One way of doing this is by influencing the lipid biosynthetic pathway; this mechanism was demonstrated by Hu et al. (2010), where the dietary saponins could inhibit the mRNA expression of SREBP-1c; inhibition of this transcription factor leads to a reduced expression of lipogenic genes like fatty acid synthase (FAS) and glycerol-3-phosphate acyltransferase (GPAT) (Horton et al. 2002), thereby inhibiting enzymatic activity and eventually decreasing lipid synthesis. Another mechanism is by increasing the activity of carnitine palmitoyl transferase (CPT), an important enzyme in beta oxidation of fatty acids; increased activity of CPT results in reduced triglyceride synthesis due to reduced flux of fatty acids. Hypocholesterolemia is another activity of saponins which is operated either by inhibiting the enzymes of cholesterol biosynthetic pathway such as 3-hydroxy-3-methyl-glutaryl-CoA reductase (Elekofehinti et al. 2012) or by increasing the dual transporters in the liver, namely ABCG55 and ABCG8, which transport free cholesterol into bile thereby reducing their level in the liver (Ji and Gong 2007). Further, activation of AMPK/ACC signalling pathway by total saponins (Xu et al. 2018) increased expression of lipoprotein lipase (Eu et al. 2010) and reduced expression of fatty acid binding protein 4 (FABP4) (Bhavsar et al. 2009), which are few other mechanisms involved in hypolipidemic effects of saponins.

10.7 Commercial Utilization and Prospects

Biotechnology makes possible to use the plant-based compounds and their derivatives in the pharmaceutical/nutraceutical industries. It is necessary to follow the suitable method of cultivation to achieve enhanced production of secondary metabolites from plants. Commercial scale bioreactors were manufactured by South Korean company Kihyung Plant Co., Ltd. The CBN Biotech Company, South Korea, used a commercial scale bioreactor to produce ginseng adventitious root. The company produces nearly 35 tons of adventitious roots of ginseng per year, which are used in industries like food, cosmetics, and pharmaceutical (Murthy et al. 2014b). This is one of the fine examples of application of biotechnological tool to meet the commercial need. By looking at the biosafety and toxicological evaluation of ginseng adventitious roots for the human consumption, the United States Food and Drug Administration (USFDA) and Korean Food and Drug Administration (KFDA), ISO (9001/2000), have approved (2,030,950, dated: 06/07/2002) products of ginseng adventitious roots and their commercial production. The ginsenosides are one of the most important saponins produced from ginseng roots which have the important medicinal value and high global market. These metabolites achieve the total revenue of 2 billion American dollars. United States of America, Canada, China, South Korea, Japan, and European countries are the major producers and commercial users of ginsenosides (Kim et al. 2013; Gantait et al. 2020). In Japan, company named NITTO DENKO CO. produces ginseng cell culture using large scale bioreactors (20,000 and 25,000 L) and markets the food products from ginseng which are very healthy and nutritious since 1988 (Adil and Jeong 2018). Hairy roots are one of important sources of the saponins and other metabolites; the company named ROOTec Bioactives AG in Witterswil, Switzerland, is one such company which produces hairy roots at industrial level (Talano et al. 2012). The company produces high quality compounds using applied biotechnological tool to meet the consumers' need and world market, especially pharma and cosmetics. The compounds are more efficient and production is rapid and is of reasonable cost compared to chemical synthesis and conventional production. For the production of optimum/increased level of hairy roots biomass, ROOTec came with a new bioreactor named "ROOTec Mist Bioreactor" (Talano et al. 2012). Some of the saponin compounds which are used in the pharmaceutical field are Madessol®, Centellase®, and Blastoestimulina®. These extracts are used in the form of tablets, drops, ointments, powder, and injections (Gallego et al. 2014).

10.8 Conclusion and Recommendation

Saponins are plant-derived triterpene glycosides that have commercial applications in the food, pharmaceutical, and cosmetic industry. Saponins being a chief metabolite present in many medicinally important plants, there is a need to prove their pharmacological ability which in turn helps in formulating the medicines in high market demand. There is a huge demand for these metabolites (saponins) because of their pharmacological importance and there is a need for their extensive mass production using biotechnological applications. There are ample of research reports on the production of saponins via in vitro method, and their applications at the industrial/commercial level. With the help of plant tissue culture approaches like callus and organ cultures, the secondary metabolite production can be enhanced. In the recent times, bioreactors serve as important biotechnological tools for the metabolite enhancement at the cellular level. Further, saponins production can be elevated by the identification and manipulation of genes which encode the key enzymes and alter the gene expressions, which in turn can be achieved through metabolomics, proteomics, and transcriptomics. Metabolic engineering strategy can be applied for the enhanced production of metabolites in less time and space.

References

Adil M, Jeong BR (2018) In vitro cultivation of *Panax ginseng* CA Meyer. Ind Crop Prod 122:239–251

Akbari S, Nour AH, Yunus RM (2020) Determination of phenolics and saponins in fenugreek seed extracted via microwave-assisted extraction method at the optimal condition. Energy Security and Chemical Engineering Congress. IOP Conf Ser Mater Sci Eng 736:022024. IOP Publishing

Akbari S, Abdurahman NH, Yunus RM et al (2021) LC-QTOF-MS analysis of phenolics and saponins extracted from *Aloe vera* leaves via microwave technology in optimal condition. S Afr J Bot 139:362–373

Alsoufi ASM, Pączkowski C, Długosz M, Szakiel A (2019a) Influence of selected abiotic factors on triterpenoid biosynthesis and saponin secretion in marigold (*Calendula officinalis* L.) in-vitro hairy root cultures. Molecules 24:2907

Alsoufi ASM, Pączkowski C, Szakiel A, Długosz M (2019b) Effect of jasmonic acid and chitosan on triterpenoid production in *Calendula officinalis* hairy root cultures. Phytochem Lett 31:5–11

Aminin DL, Koy C, Dmitrenok PS et al (2009) Immunomodulatory effects of holothurian triterpene glycosides on mammalian splenocytes determined by mass spectrometric proteome analysis. J Proteome 72:886–906

Amoros M, Fauconnier B, Girre RL (1987) In vitro antiviral activity of a saponin from *Anagallis arvensis*, Primulaceae, against herpes simplex virus and poliovirus. Antivir Res 8:13–25

Asaka I, Ii I, Hirotani M, Asada Y, Furuya T (1993) Production of ginsenoside saponins by culturing ginseng (*Panax ginseng*) embryogenic tissues in bioreactors. Biotechnol Lett 15 (12):1259–1264

Augustin JM, Kuzina V, Andersen SB, Bak S (2011) Molecular activities, biosynthesis and evolution of triterpenoid saponins. Phytochemistry 72:435–457

Bajad PN, Pardeshi AB, Pagore VP (2019) Extraction, isolation and quantification of saponin from *Dodonaea viscosa* JACQ. Pharm Innov J 8(5):41–44

Bang S-C, Lee J-H, Song G-Y et al (2005) Antitumor activity of *Pulsatilla koreana* saponins and their structure-activity relationship. Chem Pharm Bull 53:1451–1454

Bathoju G, Giri A (2012) Production of pharmaceutically important saponins from in vitro regenerated plants of *Chlorophytum borivilianum*. Curr Tren Biotechnol Pharm 6(4):458–463

Baumann E, Stoya G, Völkner A et al (2000) Hemolysis of human erythrocytes with saponin affects the membrane structure. Acta Histochem 102:21–35

Bhardwaj P, Jain CK, Mathur A (2019) Comparative analysis of saponins, flavonoids, phenolics and antioxidant activities of field acclimatized and in vitro propagated *Bacopa monnieri* (L.) Pennell from different locations in India. Indian J Exp Biol 57:259–268

Bhavsar SK, Singh S, Giri S, Jain MR, Santani DD (2009) Effect of saponins from *Helicteres isora* on lipid and glucose metabolism regulating genes expression. J Ethnopharmacol 124(3):426–433

Biswas T, Dwivedi UN (2019) Plant triterpenoid saponins: biosynthesis, in vitro production, and pharmacological relevance. Protoplasma 256:1463–1486

Bonfill M, Mangas S, Moyano E et al (2011) Production of centellosides and phytosterols in cell suspension cultures of *Centella asiatica*. Plant Cell Tissue Organ Cult 104:61–67

Bruneton J (1999) Pharmacognosy, phytochemistry, medicinal plants. Intercept Limited

Chandra S, Chandra R (2011) Engineering secondary metabolite production in hairy roots. Phytochem Rev 10:371–395

Chaudhary SA, Chaudhary PS, Syed BA et al (2018) Validation of a method for diosgenin extraction from fenugreek (*Trigonella foenum-graecum* L.). Acta Sci Pol Technol Aliment 17 (4):377–385

Chen J, Tian R, Qiu M et al (2008) Trinorcucurbitane and cucurbitane triterpenoids from the roots of *Momordica charantia*. Phytochemistry 69:1043–1048

Chen JC, Liu WQ, Lu L, Qiu MH, Zheng YT, Yang LM, Zhang XM, Zhou L, Li ZR (2009) Kuguacins F–S, cucurbitane triterpenoids from *Momordica charantia*. Phytochemistry 70 (1):133–140

Chen X, Li B, Gao Y et al (2017) Saponins from improve hematopoiesis by promoting survival through FAK and Erk1/2 activation and modulating cytokine production in bone marrow. Front Pharmacol 8:130

Cheok CY, Salman HAK, Sulaiman R (2014) Extraction and quantification of saponins: a review. Food Res Int 59:16–40

Cheung JY, Ong RC, Suen YK, Ooi V, Wong HN, Mak TC, Fung KP, Yu B, Kong SK (2005) Polyphyllin D is a potent apoptosis inducer in drug-resistant HepG2 cells. Cancer Lett 217 (2):203–211

Chodisetti B, Rao K, Gandi S, Giri A (2013) Improved gymnemic acid production in the suspension cultures of *Gymnema sylvestre* through biotic elicitation. Plant Biotechnol Rep 7(4):519–525

Chodisetti B, Rao K, Gandi S, Giri A (2015) Gymnemic acid enhancement in the suspension cultures of *Gymnema sylvestre* by using the signaling molecules—methyl jasmonate and salicylic acid. In Vitro Cell Dev-Pl 51:88–92

Cui C, Zong J, Sun Y et al (2018) Triterpenoid saponins from the genus camellia: structures, biological activities, and molecular simulation for structure-activity relationship. Food Funct 9: 3069–3091

Cui Q, Liu J, Xu W et al (2019) Enhanced extraction and preconcentration of main target saponins from *Panax notoginseng* root using green and efficient formulated surfactant aqueous systems. J Clean Prod 210:1507–1516

Deng B, Liu Z, Zou Z (2019) Optimization of microwave-assisted extraction saponins from *Sapindus mukorossi* pericarps and an evaluation of their inhibitory activity on xanthine oxidase. J Chemother 2019(4):5204534

Deore SL, Baviskar BA, Rangari AS (2015) Rapid and high yield extraction method for saponins from Safed musli. Pharm J 7(4):210–214

Desai SD, Desai DG, Kaur H (2009) Saponins and their biological activities. Pharma Times 41:13–16

de Paula BA (2014) Saponins as immunoadjuvant agent: a review. Afr J Pharm Pharmacol 8 (41):1049–1057

Di T-M, Yang S-L, Du F-Y et al (2017) Cytotoxic and hypoglycemic activity of triterpenoid saponins from *Camellia oleifera* Abel seed pomace. Molecules 22:1562

Dinchev D, Janda B, Evstatieva L et al (2008) Distribution of steroidal saponins in *Tribulus terrestris* from different geographical regions. Phytochemistry 69:176–186

Dini I, Tenore GC, Dini A (2009) Saponins in *Ipomoea batatas* tubers: isolation, characterization, quantification and antioxidant properties. Food Chem 113(2):411–419

Dubey K, Dubey R, Gupta AR, Gupta AK (2019) Anti-diabetic and antioxidant potential of saponin extract of leaves of *Ziziphus mauritiana*. J Drug Deliv Ther 9:75–77

El Barky AR, Hussein SA, Alm-Eldeen AA et al (2016) Anti-diabetic activity of *Holothuria thomasi* saponin. Biomed Pharmacother 84:1472–1487

Elekofehinti OO, Adanlawo IG, Saliu JA, Sodehinde SA (2012) Saponins from *Solanum anguivi* fruits exhibit hypolipidemic potential in *Rattus norvegicus*. Pharm Lett 4(3):811–814

Eu CHA, Lim WYA, Ton SH, bin Abdul Kadir K (2010) Glycyrrhizic acid improved lipoprotein lipase expression, insulin sensitivity, serum lipid and lipid deposition in high-fat diet-induced obese rats. Lipids Health Dis 9:81

Fan Y, Li Z, Liu L, Xi J (2020) Combination of liquid-phase pulsed discharge and ultrasonic for saponins extraction from lychee seeds. Ultrason Sonochem 69:105264

Friedman M (2006) Potato glycoalkaloids and metabolites: roles in the plant and in the diet. J Agric Food Chem 54:8655–8681

Gallego A, Ramirez-Estrada K, Vidal-Limon HR et al (2014) Biotechnological production of centellosides in cell cultures of *Centella asiatica* (L) Urban. Eng Life Sci 14:633–642

Gantait S, Mitra M, Chen J-T (2020) Biotechnological interventions for ginsenosides production. Biomol Ther 10

Ganzera M, Gampenrieder J, Pawar RS, Khan IA, Stuppner H (2004) Separation of the major triterpenoid saponins in *Bacopa monnieri* by high-performance liquid chromatography. Anal Chim Acta 516(1–2):149–154

Georgiev MI, Eibl R, Zhong J-J (2013) Hosting the plant cells in-vitro: recent trends in bioreactors. Appl Microbiol Biotechnol 97:3787–3800

Ginzberg I, Tokuhisa JG, Veilleux RE (2009) Potato steroidal glycoalkaloids: biosynthesis and genetic manipulation. Potato Res 52:1–15. https://doi.org/10.1007/s11540-008-9103-4

Gnoatto SCB, Schenkel EP, Bassani VL (2005) HPLC method to assay total saponins in *Ilex paraguariensis* aqueous extract. J Braz Chem Soc 16(4):723–726

Goel N, Sirohi SK, Dwivedi J (2012) Estimation of total saponins and evaluate their effect on in-vitro methanogenesis and rumen fermentation pattern in wheat straw-based diet. J Adv Vet Res 2:120–126

Guajardo-Flores D, García-Patiño M, Serna-Guerrero D et al (2012) Characterization and quantification of saponins and flavonoids in sprouts, seed coats and cotyledons of germinated black beans. Food Chem 134:1312–1319

Güçlü-Ustündağ O, Mazza G (2007) Saponins: properties, applications and processing. Crit Rev Food Sci Nutr 47:231–258

Guo S, Duan J-A, Tang Y et al (2011) Simultaneous qualitative and quantitative analysis of triterpenic acids, saponins and flavonoids in the leaves of two Ziziphus species by HPLC-PDA-MS/ELSD. J Pharm Biomed Anal 56:264–270

Gupta R, Gupta HP, Kumar N et al (1994) Lipoprotein lipids and the prevalence of hyperlipidaemia in rural India. J Cardiovasc Risk 1:179–184

Ha LT, Pawlicki-Jullian N, Pillon-Lequart M et al (2016) Hairy root cultures of *Panax vietnamensis*, a promising approach for the production of ocotillol-type ginsenosides. Plant Cell Tissue Organ Cult 126:93–103

Hajimohammadi R, Hosseini M, Amani H et al (2017) Experimental design procedure for optimization of saponin extraction from *Glycyrrhiza glabra*: a biosurfactant for emulsification of heavy crude oil. Tenside Surfactant Deterg 54(4):308–314

Han L-T, Fang Y, Li M-M et al (2013) The antitumor effects of triterpenoid saponins from the *Anemone flaccida* and the underlying mechanism. Evid Based Complement Alternat Med 2013:517931

Han Y, Chi J, Zhang M et al (2019) Characterization of saponins and phenolic compounds: antioxidant activity and inhibitory effects on α-glucosidase in different varieties of colored quinoa (*Chenopodium quinoa* Willd). Biosci Biotechnol Biochem 83(11):2128–2139

Haridas V, Nishimura G, Xu Z-X et al (2009) Avicin D: a protein reactive plant isoprenoid dephosphorylates Stat 3 by regulating both kinase and phosphatase activities. PLoS One 4: e5578

Holtshausen L, Chaves AV, Beauchemin KA et al (2009) Feeding saponin-containing *Yucca schidigera* and *Quillaja saponaria* to decrease enteric methane production in dairy cows. J Dairy Sci 92:2809–2821

Horton JD, Goldstein JL, Brown MS (2002) SREBPs: activators of the complete program of cholesterol and fatty acid synthesis in the liver. J Clin Invest 109:1125–1131

Hu W-L, Wu Y-M, Liu J-X et al (2005) Tea saponins affect in vitro fermentation and methanogenesis in faunated and defaunated rumen fluid. J Zhejiang Univ Sci B 6:787–792

Hu X-Q, Wang Y-M, Wang J-F et al (2010) Dietary saponins of sea cucumber alleviate orotic acid-induced fatty liver in rats via PPARalpha and SREBP-1c signaling. Lipids Health Dis 9:25

Hu X, Wang S, Xu J et al (2014) Triterpenoid saponins from *Stauntonia chinensis* ameliorate insulin resistance via the AMPactivated protein kinase and IR/IRS-1/PI3K/Akt pathways in insulin-resistant HepG2 cells. Int J Mol Sci 15:10446–10458

Huang C, Zhong J-J (2013) Elicitation of ginsenoside biosynthesis in cell cultures of *Panax ginseng* by vanadate. Process Biochem 48:1227–1234

Huang C, Qian ZG, Zhong JJ (2013) Enhancement of ginsenoside biosynthesis in cell cultures of *Panax ginseng* by N, N′- dicyclohexylcarbodiimide elicitation. J Biotechnol 165(1):30–36

Ionkova I, Momekov G, Proksch P (2010) Effects of cycloartane saponins from hairy roots of *Astragalus membranaceus* Bge., on human tumor cell targets. Fitoterapia 81:447–451

Itkin M, Heinig U, Tzfadia O et al (2013) Biosynthesis of antinutritional alkaloids in solanaceous crops is mediated by clustered genes. Science 341:175–179

Jadhav D, Rekha BN, Gogate PR, Rathod VK (2009) Extraction of vanillin from vanilla pods: a comparison study of conventional soxhlet and ultrasound assisted extraction. J Food Eng 93(4): 421–426

Jain N, Sharma V, Ramawat KG (2012) Shoot culture of *Bacopa monnieri*: standardization of explant, vessels and bioreactor for growth and antioxidant capacity. Physiol Mol Biol Plants 18:185–190

Jang K-J, Kim HK, Han MH et al (2013) Anti-inflammatory effects of saponins derived from the roots of *Platycodon grandiflorus* in lipopolysaccharide-stimulated BV2 microglial cells. Int J Mol Med 31:1357–1366

Jeong G-T, Park D-H, Hwang B, Woo J-C (2003) Comparison of growth characteristics of *Panax ginseng* hairy roots in various bioreactors. Appl Biochem Biotechnol 105–108:493–503

Jeong CS, Murthy HN, Hahn EJ et al (2009) Inoculum size and auxin concentration influence the growth of adventitious roots and accumulation of ginsenosides in suspension cultures of ginseng (*Panax ginseng* C.A. Meyer). Acta Physiol Plant 31:219–222

Ji W, Gong BQ (2007) Hypolipidemic effects and mechanisms of *Panax notoginseng* on lipid profile in hyperlipidemic rats. J Ethnopharmacol 113:318–324

Ji S, Wang Y, Shao X, Zhu C, Lin Y, Gao S, Tang D (2020) Extraction and purification of triterpenoid saponins from licorice by ionic liquid-based extraction combined with in situ alkaline aqueous biphasic systems. Sep Purif Technol 247:116953

Jung KY, Kim DS, Oh SR et al (1998) Platelet activating factor antagonist activity of ginsenosides. Biol Pharm Bull 21:79–80

Kerem Z, German-Shashoua H, Yarden O (2005) Microwave-assisted extraction of bioactive saponins from chickpea (*Cicer arietinum* L). J Sci Food Agric 85(3):406–412

Khojasteh A, Sanchez-Muñoz R, Moyano E et al (2019) Biotechnological production of ruscogenins in plant cell and organ cultures of *Ruscus aculeatus*. Plant Physiol Biochem 141:133–141

Kikowska M, Thiem B, Sliwinska E et al (2014) The effect of nutritional factors and plant growth regulators on micropropagation and production of phenolic acids and saponins from plantlets and adventitious root cultures of *Eryngium maritimum* L. J Plant Growth Regul 33:809–819

Kikowska M, Thiem B, Sliwinska E et al (2016) Micropropagation of *Eryngium campestre* L. via shoot culture provides valuable uniform plant material with enhanced content of phenolic acids and antimicrobial activity. Acta Biol Cracov Ser Bot 58:43–56

Kikowska M, Kowalczyk M, Stochmal A, Thiem B (2019) Enhanced accumulation of triterpenoid saponins in in vitro plantlets and dedifferentiated cultures of *Eryngium planum* L.: a medicinal plant. Hortic Environ Biotechnol 60:147–154

Kim N, Park IS (2001) Purification of saponin compounds in *Bupleurum falcatum* by solvent partitioning and preparative LC. Biosci Biotechnol Biochem 65:1648–1651

Kim YS, Hahn EJ, Paek KY (2004) Effects of various bioreactors on growth and ginsenoside accumulation in ginseng adventitious root cultures (*Panax ginseng* C.A. Meyer). Korean J Plant Biotechnol 31:249–253

Kim YS, Hahn EJ, Shin CG, Paek KY (2005) Effects of aeration rate and sparger type on growth and ginsenoside accumulation in bioreactor cultures of ginseng adventitious root (*Panax ginseng* C.A. Meyer). Korean J Plant Biotechnol 32:111–116

Kim JJY, Xiao H, Tan Y et al (2009) The effects and mechanism of saponins of *Panax notoginseng* on glucose metabolism in 3T3-L1 cells. Am J Chin Med 37:1179–1189

Kim OT, Kim SH, Ohyama K et al (2010) Upregulation of phytosterol and triterpene biosynthesis in *Centella asiatica* hairy roots overexpressed ginseng farnesyl diphosphate synthase. Plant Cell Rep 29:403–411

Kim DS, Song M, Kim S-H et al (2013) The improvement of ginsenoside accumulation in *Panax ginseng* as a result of γ-irradiation. J Ginseng Res 37:332–340

Kim YB, Reed DW, Covello PS (2015) Production of triterpenoid sapogenins in hairy root cultures of *Silene vulgaris*. Nat Prod Commun 10:1919–1922

Kochan E, Szymczyk P, Kuźma Ł et al (2018) The increase of triterpene saponin production induced by trans-anethole in hairy root cultures of *Panax quinquefolium*. Molecules 23:2674

Koul A, Mallubhotla S (2020) Elicitation and enhancement of bacoside production using suspension cultures of *Bacopa monnieri* (L.) Wettst. 3 Biotech 10:256

Kregiel D, Berlowska J, Witonska I et al (2017) Saponin-based, biological-active surfactants from plants. Appl Charact Surfact 6(1):184–205

Kumar S, Kalra S, Singh B et al (2016) RNA-Seq mediated root transcriptome analysis of *Chlorophytum borivilianum* for identification of genes involved in saponin biosynthesis. Funct Integr Genomics 16:37–55. https://doi.org/10.1007/s10142-015-0465-9

Kwon DY, Kim YS, Ryu SY et al (2012) Platyconic acid, a saponin from *Platycodi radix*, improves glucose homeostasis by enhancing insulin sensitivity in vitro and in vivo. Eur J Nutr 51:529–540

Lambert E, Faizal A, Geelen D (2011) Modulation of triterpene saponin production: in vitro cultures, elicitation, and metabolic engineering. Appl Biochem Biotechnol 164:220–237

Le AV, Parks SE, Nguyen MH, Roach PD (2018) Improving the vanillin-sulphuric acid method for quantifying total saponins. Technologies 6(3):84

Lee K-T, Jung TW, Lee H-J et al (2011) The antidiabetic effect of ginsenoside Rb2 via activation of AMPK. Arch Pharm Res 34:1201–1208

Lee KS, Woo SY, Lee MJ et al (2020) Isoflavones and soyasaponins in the germ of Korean soybean [*Glycine max* (L.) Merr.] cultivars and their compound-enhanced BMP-2-induced bone formation. Appl Biol Chem 63(1):1–8

Leung KW, Yung KKL, Mak NK et al (2007) Neuroprotective effects of ginsenoside-Rg1 in primary nigral neurons against rotenone toxicity. Neuropharmacology 52:827–835

Liang Y, Zhao S, Zhang X (2009) Antisense suppression of cycloartenol synthase results in elevated ginsenoside levels in *Panax ginseng* hairy roots. Plant Mol Biol Report 27:298–304

Loc NH, Nhat NTD (2013) Production of asiaticoside from centella (*Centella asiatica* L. Urban) cells in bioreactor. Asian Pac J Trop Biomed 3:806–810

Lorent JH, Quetin-Leclercq J, Mingeot-Leclercq M-P (2014) The amphiphilic nature of saponins and their effects on artificial and biological membranes and potential consequences for red blood and cancer cells. Org Biomol Chem 12:8803–8822

Mandal V, Mandal SC (2010) Design and performance evaluation of a microwave based low carbon yielding extraction technique for naturally occurring bioactive triterpenoid: oleanolic acid. Biochem Eng J 50(1–2):63–70

Mangas S, Moyano E, Osuna L et al (2008) Triterpenoid saponin content and the expression level of some related genes in calli of *Centella asiatica*. Biotechnol Lett 30:1853–1859

Mehrotra S, Kukreja AK, Khanuja SPS, Mishra BN (2008) Genetic transformation studies and scale up of hairy root culture of *Glycyrrhiza glabra* in bioreactor. Electron J Biotechnol 11:0–0. https://doi.org/10.2225/vol11-issue2-fulltext-6

Mengoni F, Lichtner M, Battinelli L et al (2002) In vitro anti-HIV activity of oleanolic acid on infected human mononuclear cells. Planta Med 68:111–114

Mimaki Y, Yokosuka A, Kuroda M, Sashida Y (2001) Cytotoxic activities and structure-cytotoxic relationships of steroidal saponins. Biol Pharm Bull 24:1286–1289

Mishra N, Rajput PS, Mishra RC (2017) Phytomedicinal uses of saponin containing herbs. Int J Chem Stud 1:12–16

Moghimipour E, Handali S (2015) Saponin: properties, methods of evaluation and applications. Annu Res Rev Biol 5:207–220

Mugford ST, Osbourn A (2012) Saponin synthesis and function. In: Bach T, Rohmer M (eds) Isoprenoid synthesis in plants and microorganisms. Springer, New York, pp 405–424

Murthy HN, Georgiev MI, Kim Y-S et al (2014a) Ginsenosides: prospective for sustainable biotechnological production. Appl Microbiol Biotechnol 98:6243–6254

Murthy HN, Lee E-J, Paek K-Y (2014b) Production of secondary metabolites from cell and organ cultures: strategies and approaches for biomass improvement and metabolite accumulation. Plant Cell Tissue Organ Cult 118:1–16

Murthy HN, Park SY, Paek KY (2017) Production of ginsenosides by hairy root cultures of *Panax ginseng*. In: Production of plant derived natural compounds through hairy root culture. Springer, Cham, pp 203–216

Nagella P, Thiruvengadam M, Jung SJ et al (2013) Establishment of *Gymnema sylvestre* hairy root cultures for the production of gymnemic acid. Acta Physiol Plant 35:3067–3073

Naik PM, Manohar SH, Praveen N, Murthy HN (2010) Effects of sucrose and pH levels on in vitro shoot regeneration from leaf explants of *Bacopa monnieri* and accumulation of bacoside A in regenerated shoots. Plant Cell Tissue Organ Cult 100:235–239

Naik PM, Manohar SH, Murthy HN (2011) Effects of macro elements and nitrogen source on biomass accumulation and bacoside A production from adventitious shoot cultures of *Bacopa monnieri* (L.). Acta Physiol Plant 33:1553–1557

Naik PM, Godbole M, Praveen N, Murthy HN (2015) The effect of heavy metals on in-vitro adventitious shoot production and Bacoside A content in *Bacopa monnieri* (L). Mapana J Sci 14:1–10

Naik PM, Godbole M, Nagella P, Murthy HN (2017) Influence of different media, medium strength and carbon sources on adventitious shoot cultures and production of bacoside A in *Bacopa monnieri* (L.). Ceylon J Sci 46:97

Namdeo AG (2007) Plant cell elicitation for production of secondary metabolites: a review. Phcog Rev 1:69–79

Nhut DT, Huy NP, Tai NT et al (2015) Light-emitting diodes and their potential in callus growth, plantlet development and saponin accumulation during somatic embryogenesis of *Panax vietnamensis* Haet Grushv. Biotechnol Biotechnol Equip 29:299–308

Oleszek W, Sitek M, Stochmal A, Piacente S, Pizza C, Cheeke P (2001) Steroidal saponins of *Yucca schidigera* Roezl. J Agric Food Chem 49:4392–4396

Paek K-Y, Murthy HN, Hahn E-J, Zhong J-J (2009) Large scale culture of ginseng adventitious roots for production of ginsenosides. Adv Biochem Eng Biotechnol 113:151–176

Palazón J, Mallol A, Eibl R et al (2003) Growth and ginsenoside production in hairy root cultures of *Panax ginseng* using a novel bioreactor. Planta Med 69:344–349

Park IS, Kang EM, Kim N (2000) High-performance liquid chromatographic analysis of saponin compounds in *Bupleurum falcatum*. J Chromatogr Sci 38:229–233

Patel S, Santani D, Shah M, Patel V (2012) Anti-hyperglycemic and anti-hyperlipidemic effects of *Bryonia laciniosa* seed extract and its saponin fraction in streptozotocin-induced diabetes in rats. J Young Pharm 4:171–176

Prasad A, Mathur A, Kalra A et al (2013) Fungal elicitor-mediated enhancement in growth and asiaticoside content of *Centella asiatica* L. shoot cultures. Plant Growth Regul 69:265–273

Praveen N, Naik PM, Manohar SH et al (2009) In vitro regeneration of brahmi shoots using semisolid and liquid cultures and quantitative analysis of bacoside A. Acta Physiol Plant 31: 723–728

Praveen N, Thiruvengadam M, Yang YS, Kim SH, Murthy HN, Chung IM (2014) Production of gymnemic acid from hairy root cultures of *Gymnema sylvestre* R. Br. as influenced by polyunsaturated fatty acids (PUFAs) and their antioxidant activity. Ind Crop Prod 54:54–61

Puente-Garza CA, García-Lara S, Gutiérrez-Uribe JA (2017) Enhancement of saponins and flavonols by micropropagation of *Agave salmiana*. Ind Crop Prod 105:225–230

Rajkumari S, Sanatombi K (2020) Secondary metabolites content and essential oil composition of in vitro cultures of *Zingiber montanum* (Koenig) Link ex A. Dietr Biotechnol Lett 42(7):1237–1245

Raju J, Chinthapally Rao V (2012) Diosgenin, a steroid saponin constituent of yams and fenugreek: emerging evidence for applications in medicine. In: Rasooli I (ed) Bioactive compounds in phytomedicine, pp 125–142

Ralla T, Salminen H, Edelmann M et al (2018) Oat bran extract (*Avena sativa* L.) from food by-product streams as new natural emulsifier. Food Hydrocoll 81:253–262

Ribeiro BD, Barreto DW, Coelho MAZ (2014) Recovery of saponins from Jua (*Ziziphus joazeiro*) by micellar extraction and cloud point preconcentration. J Surfactant Deterg 17(3):553–561

Runner RTM (2006) Extraction and isolation of saponins. Methods Mol Biol 864:415–426

Saha PS, Sarkar S, Jeyasri R et al (2020) In vitro propagation, phytochemical and neuropharmacological profiles of *Bacopa monnieri* (L.) Wettst.: a review. Plan Theory 9(4):411

Salazar-Magallón JA, de la Peña AH (2020) Production of antifungal saponins in an airlift bioreactor with a cell line transformed from *Solanum chrysotrichum* and its activity against strawberry phytopathogens. Prep Biochem Biotechnol 50:204–214

Shang W, Yang Y, Zhou L et al (2008) Ginsenoside Rb1 stimulates glucose uptake through insulin-like signaling pathway in 3T3-L1 adipocytes. J Endocrinol 198:561–569

Sharma P, Yadav S, Srivastava A, Shrivastava N (2013) Methyl jasmonate mediates upregulation of bacoside A production in shoot cultures of *Bacopa monnieri*. Biotechnol Lett 35:1121–1125

Sharma M, Ahuja A, Gupta R, Mallubhotla S (2015) Enhanced bacoside production in shoot cultures of *Bacopa monnieri* under the influence of abiotic elicitors. Nat Prod Res 29:745–749

Sheoran S, Panda BP, Admane PS et al (2015) Ultrasound-assisted extraction of gymnemic acids from *Gymnema sylvestre* leaves and its effect on insulin-producing RINm-5 F β cell lines. Phytochem Anal 26(2):97–104

Shibata S (2001) Chemistry and cancer preventing activities of ginseng saponins and some related triterpenoid compounds. J Korean Med Sci 16(Suppl):S28–S37

Shin DY, Kim GY, Li W et al (2009) Implication of intracellular ROS formation, caspase-3 activation and Egr-1 induction in platycodon D-induced apoptosis of U937 human leukemia cells. Biomed Pharmacother 63:86–94

Shkondrov A, Krasteva I, Ionkova I et al (2019) Production of saponins from in vitro cultures of *Astragalus glycyphyllos* and their antineoplastic activity. Biotechnol Biotechnol Equip 33:1413–1418

Sindambiwe JB, Calomme M, Geerts S et al (1998) Evaluation of biological activities of triterpenoid saponins from *Maesa lanceolata*. J Nat Prod 61:585–590

Siu FM, Ma DL, Cheung YW, Lok CN, Yan K, Yang Z, Yang M, Xu S, Ko BC, He QY, Che CM (2008) Proteomic and transcriptomic study on the action of a cytotoxic saponin (Polyphyllin D): induction of endoplasmic reticulum stress and mitochondria-mediated apoptotic pathways. Proteomics 8(15):3105–3117. https://doi.org/10.1002/pmic.200700829

Tadhani M, Subhash R (2006) Preliminary studies in *Stevia rebaudiana* leaves: proximal composition, mineral analysis and phytochemical screening. J Med Sci 6:321–326

Takeuchi TM, Pereira CG, Braga ME, Maróstica MR, Leal PF, Meireles MAA (2009) Low pressure solvent extraction (solid-liquid extraction, microwave assisted, and ultrasound assisted) from condimentary plants. In: Extracting bioactive compounds for food products, pp 137–218

Talano MA, Oller ALW, González PS, Agostini E (2012) Hairy roots, their multiple applications and recent patents. Recent Pat Biotechnol 6:115–133

Tao W, Duan J, Zhao R et al (2013) Comparison of three official Chinese pharmacopoeia species of Glycyrrhiza based on separation and quantification of triterpene saponins and chemometrics analysis. Food Chem 141:1681–1689

Thakur M, Melzig MF, Fuchs H, Weng A (2011) Chemistry and pharmacology of saponins: special focus on cytotoxic properties. Bot Targets Ther 1:19–29

Vázquez-Castilla S, Jaramillo-Carmona S, Fuentes-Alventosa JM et al (2013) Optimization of a method for the profiling and quantification of saponins in different green Asparagus genotypes. J Agric Food Chem 61:6250–6258

Verpoorte R, Contin A, Memelink J (2002) Biotechnology for the production of plant secondary metabolites. Phytochem Rev 1(1):13–25

Verza SG, Silveira F, Cibulski S et al (2012) Immunoadjuvant activity, toxicity assays, and determination by UPLC/Q-TOF-MS of triterpenic saponins from *Chenopodium quinoa* seeds. J Agric Food Chem 60:3113–3118

Vincken J-P, Heng L, de Groot A, Gruppen H (2007) Saponins, classification and occurrence in the plant kingdom. Phytochemistry 68:275–297

Vongsangnak W, Gua J, Chauvatcharin S, Zhong JJ (2004) Towards efficient extraction of notoginseng saponins from cultured cells of *Panax notoginseng*. Biochem Eng J 18(2):115–120

Voutquenne L, Guinot P, Thoison O et al (2003) Oleanolic glycosides from *Pometia ridleyi*. Phytochemistry 64:781–789

Waller GR, Yang CF, Chen LF, Su CH, Liou RM, Wu SC, Young CC, Lee MR, Lee JS, Cheng CS, Chou CS, Kim D (1999) Saponins produced during the life cycle of mungbeans and their role as allelochemicals. Stud Plant Sci:105–130

Wang Y, Cheung YH, Yang Z et al (2006) Proteomic approach to study the cytotoxicity of dioscin (saponin). Proteomics 6:2422–2432

Wang W, Zhao Y, Rayburn ER et al (2007) In vitro anti-cancer activity and structure-activity relationships of natural products isolated from fruits of *Panax ginseng*. Cancer Chemother Pharmacol 59:589–601

Wang J, Gao W-Y, Zhang J et al (2012) Production of ginsenoside and polysaccharide by two-stage cultivation of *Panax quinquefolium* L. cells. In Vitro Cell Dev Biol Plant 48:107–112

Wang J, Gao W, Zuo B, Zhang L, Huang L (2013) Effect of methyl jasmonate on the ginsenoside content of *Panax ginseng* adventitious root cultures and on the genes involved in triterpene biosynthesis. Res Chem Intermed 39(5):1973–1980

Wang XD, Su GY, Zhao C et al (2018) Anticancer activity and potential mechanisms of 1C, a ginseng saponin derivative, on prostate cancer cells. J Ginseng Res 42:133–143

Wang Q, Wu X, Shi F, Liu Y (2019) Comparison of antidiabetic effects of saponins and polysaccharides from *Momordica charantia* L. in STZ-induced type 2 diabetic mice. Biomed Pharmacother 109:744–750

Watcharatanon K, Ingkaninan K, Putalun W (2019) Improved triterpenoid saponin glycosides accumulation in in vitro culture of *Bacopa monnieri* (L.) Wettst with precursor feeding and LED light exposure. Ind Crop Prod 134:303–308

Wiktorowska E, Długosz M, Janiszowska W (2010) Significant enhancement of oleanolic acid accumulation by biotic elicitors in cell suspension cultures of *Calendula officinalis* L. Enzym Microb Technol 46:14–20

Włodarczyk M, Pasikowski P, Osiewała K et al (2020) In search of high-yielding and single-compound-yielding plants: new sources of pharmaceutically important saponins from the Primulaceae family. Biomol Ther 10(3):376

Wu J, Lin L, Chau FT (2001) Ultrasound-assisted extraction of ginseng saponins from ginseng roots and cultured ginseng cells. Ultrason Sonochem 8:347–352

Xu HJ, Liu C, Xue Y, Xue CH, Wang Y (2011) The effect of dietary sea cucumber saponin on hyperuricemia in mice. Chin Pharmacol Bull 27:1064–1067

Xu M-Y, Lee SY, Kang SS, Kim YS (2014) Antitumor activity of jujuboside B and the underlying mechanism via induction of apoptosis and autophagy. J Nat Prod 77:370–376

Xu J, Wang S, Feng T et al (2018) Hypoglycemic and hypolipidemic effects of total saponins from *Stauntonia chinensis* in diabetic db/db mice. J Cell Mol Med 22:6026–6038

Yamaguchi Y, Higashi M, Kobayashi H (1997) Effects of ginsenosides on maze performance and brain choline acetyltransferase activity in scopolamine-treated young rats and aged rats. Eur J Pharmacol 329:37–41

Yang X, Lei Z, Yu Y et al (2019) Phytochemical characteristics of callus suspension culture of *Helicteres angustifolia* L. and its in vitro antioxidant, antidiabetic and immunomodulatory activities. S Afr J Bot 121:178–185

Yang J, Yang X, Li B et al (2020) Establishment of in vitro culture system for transgenic hairy roots. 3 Biotech 10:137

Yao L, Lu J, Wang J, Gao W-Y (2020) Advances in biosynthesis of triterpenoid saponins in medicinal plants. Chin J Nat Med 18:417–424

Yu KW, Hanh EJ, Paek KY (2003) Ginsenoside production by hairy root cultures of *Panax ginseng* C. A. Meyer in bioreactors. Acta Hortic 597:237–243

Yuan B, Dinssa FF, Simon JE, Wu Q (2019) Simultaneous quantification of polyphenols, glycoalkaloids and saponins in African nightshade leaves using ultra-high-performance liquid chromatography-tandem mass spectrometry with acid assisted hydrolysis and multivariate analysis. Food Chem:126030

Zhang H-S, Wang S-Q (2006) Notoginsenoside R1 from *Panax notoginseng* inhibits TNF-alpha-induced PAI-1 production in human aortic smooth muscle cells. Vasc Pharmacol 44:224–230

Zhang WJ, Wojta J, Binder BR (1997) Effect of notoginsenoside R1 on the synthesis of components of the fibrinolytic system in cultured smooth muscle cells of the human pulmonary artery. Cell Mol Biol 43:581–587

Zhang Y, Liu C, Qi Y, Li S, Wang J (2013) Application of accelerated solvent extraction coupled with counter-current chromatography to extraction and online isolation of saponins with a broad range of polarity from *Panax notoginseng*. Sep Purif Technol 106:82–89

Zhang CJ, Zhang Y, Wei Z et al (2021) Characterisation of the variation of total saponin content in the Medicago species using an optimised extraction technique. N Z J Crop Hortic Sci:1–13

Zhou Y, Garcia-Prieto C, Carney DA, Xu RH, Pelicano H, Kang Y, Yu W, Lou C, Kondo S, Liu J, Harris DM, Estrov Z, Keating MJ, Jin Z, Huang P (2005) OSW-1: a natural compound with potent anticancer activity and a novel mechanism of action. J Natl Cancer Inst 97(23): 1781–1785

Zhu X, Li C, Zhu Y (2020) Saponins extracted from *Asparagus officinalis* L. by products exerts hypoglycemic effect in streptozotocin induced type 2 diabetic rats. IOP Conf Ser Earth Environ Sci 559(1):012002

In Vitro Production of Steroids

11

Ehab M. B. Mahdy, Sherif F. El-Sharabasy,
and Maiada M. El-Dawayati

Abstract

Steroids are a cluster of secondary metabolites owning a range in their structure and biological activities. Natural products are often linked with harmful effects on health and have many medicinal applications. Secondary metabolites can potentially drive drug design or discovery. The prime information about such steroids has risen systematically. Advancement in biotechnological approaches affects the disciplines of biochemistry, botany, conservation biology, and toxicology. The genetic markers are a beneficial tool that provides more insight into pathways. Genetic markers accompanied by PTC can supply more insight into the path of steroid products. We will here highlight the biological effects of steroids and the biosynthesis pathways. Studies on steroid extraction and in vitro production will also be included. Besides, the bioreactor of steroids will be scaled up.

Keywords

Steroid · Biological effects · Gonane · Biosynthesis pathway · HPLC · Ecdysterone

E. M. B. Mahdy (✉)
National Gene Bank (NGB), Agricultural Research Centre (ARC), Giza, Egypt
e-mail: ehab.mahdy@arc.sci.eg

S. F. El-Sharabasy · M. M. El-Dawayati
Central Laboratory of Date Palm Research and Development (CLDPRD), Agricultural Research Center (ARC), Giza, Egypt
e-mail: sheriff.sharabasy@arc.sci.eg; maiada.eldawayati@arc.sci.eg

T. Belwal et al. (eds.), *Nutraceuticals Production from Plant Cell Factory*,
https://doi.org/10.1007/978-981-16-8858-4_11

11.1 Introduction

Steroids exist as a class of cholesterol-derived lipophilic, low-molecular-weight constituents, small molecules, and are soluble in organic solvents. The basic unit contains gonane (perhydro-1,2-cyclopentanoperhydrophenanthrene) structure and a minor diverse in this structure or having functional groups results in various groups. Steroid group is composed of different natural and synthetic sources (Aav et al. 2005; Stanczyk 2009; Sultan and Rauf Raza 2015). Steroids play a prime role in the living organisms' biochemical and physiological characteristics (Aav et al. 2005; Sultan and Rauf Raza 2015). The family of steroids contains hormones, sterols, and bile acids (Balandrin and Klocke 1998; Pattenden et al. 2004; De Boeck et al. 2005; Marion et al. 2006; Lopez et al. 2006; Jovanović-Šanta et al. 2015). A divergence of synthetic steroids is employed for a broad sense of medicine and pharmacology (Lopez et al. 2006; Emmanuel et al. 2011; Thao et al. 2015; Jovanović-Šanta et al. 2015). These natural compounds can be convened based on their chemical structure, biochemical effects, biological function, molecular actions, and biosynthesis tissues (Morgan and Melinda 1997; Sultan and Rauf Raza 2015). Plant steroids are of two broad groups: phytosterols (sterols) and brassinosteroids.

11.1.1 Chemical Structure of Steroids

Steroid compounds contain gonane structure (perhydro-1,2-cyclopentanoperhydrophenanthrene; Fig. 11.1), and a minor diverse in this structure or owing functional groups causes various classes of steroids. Skeleton of steroids varies in the number of carbons; C17 in gonanes, C18 in estranes, C19 in androstanes, C21 in pregnanes and cortisol, and C24 of testosterone. Except for vitamin D, steroids have a skeleton of cyclopenta (α) phenanthrene or one/more bond scissions, ring expansions, or contractions. Methyl-side is normal at C10 and C13, and mostly an alkyl-side chain is at C17 (Abd El-Bahaman 1991).

Modifications in steroid rings are expressed by a prefix: inclusion of methylene group (e.g., 4a-*Homo*-7-nor-5α-androstane), addition of a hydrogen atom at terminal group (e.g., 2,3-*seco*-5a-cholestane), loss of an unsubstituted methylene group (e.g., 4-*nor*-5a-androstane), or addition of a hydrogen atom at junction atom with the adjacent ring (e.g., *des*-A-androstane).

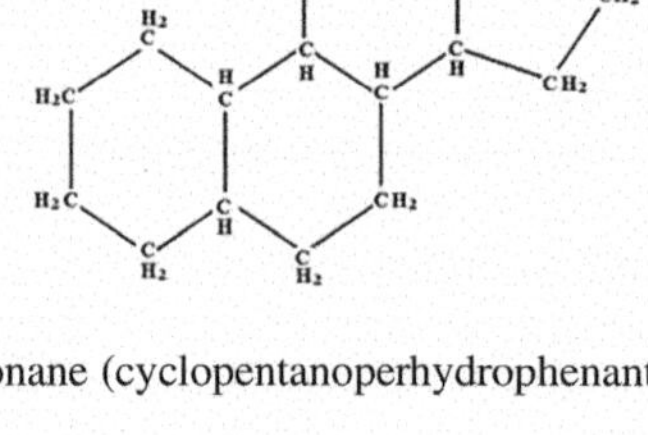

Fig. 11.1 The structure of gonane (cyclopentanoperhydrophenanthrene)

Cardiac glycoside is classified according to the five-membered lactone ring or six-membered lactone ring. (1) Cardenolides (so-called digitalis, Lanoxin, Digitek, Lanoxicaps) have a five-membered lactone ring at C17. Cardenolides have C23 and consist of steroids with methyl groups at C10 and C13. (2) Bufadienolides have the same assembly except for a six-membered lactone ring and two double bonds (Laurie et al. 2012). Stereochemistry shows steroids to be a chair form rather than a boat form through hexagonal C rings. These are oriented in equatorial or axial spot and bonded in trans-configuration. Steroids are identified generally via their trivial names (e.g., cortisol, testosterone), which often leads to confusion in some cases, and through the use of IUPAC, systematic nomenclature is recommended. IUPAC listed a complete description of steroid nomenclature, including (pregnane, androstane or estrange, cortisol, testosterone) number, position, and orientation (α or ß) of all functional groups.

We will here focus on the biological effects of steroids and the pathways of biosynthesis. Studies on steroid extraction and in vitro production will also be included. Also, the bioreactor of steroids will be scaled up.

11.2 Biosynthesis Pathway

Steroids are a class of derived lipophilic, low molecular weight, small molecules, and are soluble in organic solvents. The basic unit is gonane structure (perhydro-1,2-cyclopentanoperhydrophenanthrene), and a minor diverse in this structure or having functional groups results in various classes of steroids. Steroidal skeleton varies based on the number of carbons; C17 in gonanes, C18 in estranes, C19 in androstanes, C21 in pregnanes and cortisol, and C24 of testosterone. Steroids that have C17-C29 skeletons are not triterpenes. Nevertheless, steroid compounds come from the same C30. Squalene comes from the triterpenes. Squalene was isolated first from shark liver oil and found in almost all organisms (Melcangi et al. 2003). The squalene skeleton forms the union of two trans-farnesyl units joined in a head-to-head fashion (St-Onge et al. 2003). The pathway of squalene was mostly found by (Bucourt et al. 1969). The transform of the squalene sequence to the steroid nucleus is with an acid-catalyzed ring-opening of the mono-epoxide derivative. These then involve the formation of carbocationic-mediates series (Abd El-Bahaman 1991; Aav et al. 2005). The steric structures of steroids could streamline from the folding of polyprenyl-chain on the enzyme surface (St-Onge et al. 2003; Melcangi et al. 2003).

We will here highlight the synthesis pathway of ecdysterone (Fig. 11.2), a class of ecdysteroids. Ecdysteroids were discovered first in *Podocarpus nakaii* Hayata called ponasterones A, B, and C, and found later in *Podocarpus elatus* R.Br., *Polypodium vulgare* L. and *Achyranthes faurieri* H. Lev. and Vaniot (Baltaev 2000). Ecdysteroids exist in a broad sense of angiosperms belonging to the families Fabaceae, Caryophyllaceae, Amaranthaceae, Chenopodiaceae, Asteraceae, Lamiaceae Polypodiaceae, Pteridaceae, and Blechneaceae (Cocquyt et al. 2011; Wu et al. 2010; Vanyolos et al. 2012; Nowak et al. 2012, 2013). Ecdysteroids are a sort of sterols chemically associated with triterpenoids and have a direct precursor

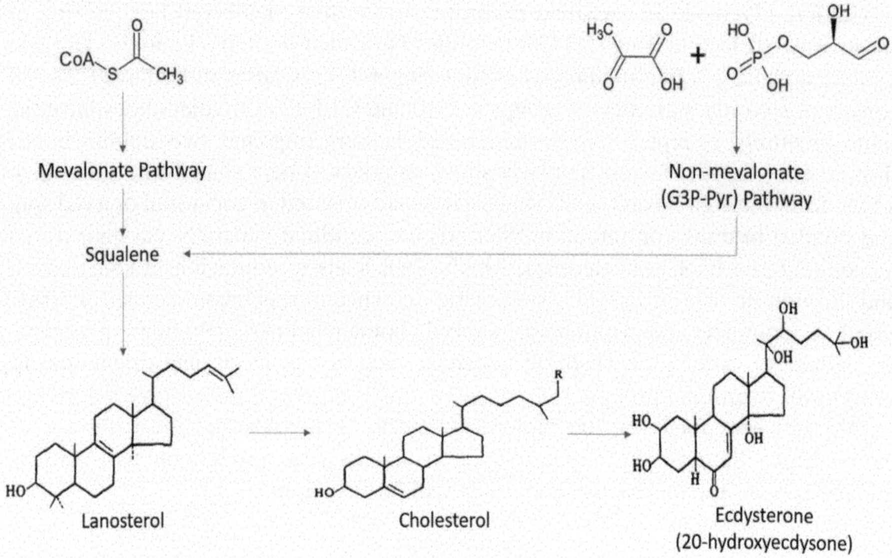

Fig. 11.2 The biosynthesis pathway of ecdysterone (20-E)

in most pathway stages of cholesterol biosynthesis. Ecdysteroids biosynthesized (Fig. 11.2) starting from the acetyl-CoA, through the mevalonate pathway, isoprenoids from pyruvate, and glycerylaldehyde-3-phosphate (via G3P-Pyr or non-mevalonate pathways).

The pathway passes isopentenyl pyrophosphate (IPP) and dimethylallyl pyrophosphate (DMAPP). Ecdysteroids depend on C27 cholest-7-en-6-one backbone, follow steps of their biosynthesis including conversion to cholesterol, dehydrogenation to 7-dehydrocholesterol, and more molecular modifications, and, most importantly, hydroxylation at various carbon atoms (Festucci-Buselli et al. 2008; Ikekawa et al. 2013). The aliphatic side chain, attached to the D ring, as in rubrosterone, is cleaved. 20-hydroxyecdysone (20E) and polypodine B (polB) are typical to ecdysteroids. The biosynthesis of ecdysteroids is performed via PTCs (Fujimoto et al. 2000; Okuzumi et al. 2003).

11.3 In Vitro Production Methodology

Due to the importance of steroid derivatives and profitable biosynthesis pathway, in vitro cultures offer rich biomass production. Generally, plant tissue culture (PTC) is highly a biotechnological technique in concern to its utilization of homogeneity explants, clonal propagation, the possibility of rapid, and the regardless of environmental factors (Rout et al. 2000; Karuppusamy 2009; Thiem et al. 2017). Homogeneity of explants also guarantees the integrity of their biometabolite composition. Plant tissue cultures have become an attractive methodology in the production of biometabolites of importance. Plant tissue culture is a potentially novel approach to obtaining various substances, especially those with complicated structures, of relatively high efficiency, and of low cost (Filova 2014). Therefore, it is predominantly to get a constant source of chemicals with biological activity. Less known biometabolites produced by rare or endangered species can be obtained from plant tissue cultures, as well as their effects can be scanned (Karuppusamy 2009).

Various PTC approaches could be utilized increasingly to produce in vitro bioactive steroids in a plant. The composition of media could be adjusted using variable ratios of plant growth regulators. Besides, all biotechnological strategies, including temperature, use of various biotic/abiotic elicitors, precursor feeding, and photoperiod, must be optimized due to their efficacy having a significant effect on bioactive metabolites yield. Induction of interest constituent production through elicitation is one of the few approaches which have been commercially applied (Smetanska 2008).

Various strategies may enhance steroid production through PCTs. Production of steroids such as ecdysteroids via PTCs has become of increased interest, though it makes available a stable source of these derivatives. Steroids may be extremely variable based upon the origin of tissue, time of excise, and various environmental factors that cannot be controlled. These methods permit in vitro cultures to yield safely bioactive metabolites. In vitro proliferation method can be utilized to overcome natural plant heterogeneity to improve and optimize yielding secondary metabolites (Smetanska 2008; Karuppusamy 2009; Filova 2014). Nowadays, the supply of steroids is based commercially on their extraction. It goes without the obvious obstacles regarding the agriculture, unstable environment, resource consumption, and less in the yield. Various factors, including environmental agents (e.g., temperature, seasonal variation, and geographic location) and biological agents (e.g., source of explants, uniformity of explant), may reflect the steroid-rich plant yields. It makes sure that the steroid-rich plants are an ecological source of biomass for research and supply commercially on steroids (Filova 2014; Thiem et al. 2017).

In vitro proliferation has been applied (Table 11.1) for various important steroid-rich species (Ahmad and Anis 2007; Smetanska 2008; Cheng et al. 2008; Parveen et al. 2008; Flores et al. 2010; Thanonkeo et al. 2011; Ahmad and Anis 2011; Chamnipa et al. 2012; Gnanaraj et al. 2012; Duan et al. 2012; Wang et al. 2013; Kaul et al. 2013; Thiem et al. 2013; Ahmad et al. 2013; Maliński et al. 2014; Vasconcelos et al. 2014; Zand et al. 2014; Wang et al. 2014; Sen et al. 2014; Skała et al. 2015). This procedure may be an alternative for biomass production. Various physicochemical agents may reflect on the in vitro production of bioactive steroids. This starts successfully from (1) the choice of a proper explant with a suitable technique, (2) optimizing tissue culture conditions with the high biomass growth and bioactive constituents, and (3) the selection of a high-performance line. Some modifications of the nutrition media composition or application of biotechnological strategies (e.g., elicitation or precursor feeding) may accumulate steroids (Karuppusamy 2009; Thanonkeo et al. 2011; Thiem et al. 2013; Skała et al. 2015; Thiem et al. 2017).

Some reviews showed that the in vitro production of steroids has a significant effect on future commercial utilization. However, it should recommend that the

Table 11.1 Best methods used for in vitro steroid-produced species

Species	Treatment	Method	Reference
Amaranthaceae			
Achyranthes bidentate Blume			
	MS + 1.5 μM NAA + 1.5 μM BAP	Cell suspension	Wang et al. 2013
	MS + 5.0 mg l^{-1} BAP + 1.0 mg l^{-1} IBA	Shoot tips	Gnanaraj et al. 2012
	MS + 0.5 mg l^{-1} 2,4-D + 1.0 mg l^{-1} NAA + 0.1 mg l^{-1}IBA + 0.1 mg l^{-1} ZT	Callus	Duan et al. 2012
A. aspera L.			
	MS + 5.0 mg l^{-1} BAP + 1.0 mg l^{-1} IBA	Shoot tips	Gnanaraj et al. 2012
	MS + 2.0 mg l^{-1} BAP + 1.0 mg l^{-1} IAA	Shoot tips	Parveen et al. 2008
	MS + BAP 2.0 mg l^{-1} + 0.5 mg l^{-1} NAA/ 3.0 mg l^{-1}IBA	Callus	Sen et al. 2014
Lamiaceae			
Ajuga bracteosa wall ex Benth			
	MS + 5.0 mg l^{-1} BAP + 2.0 mg l^{-1} IAA/0.5 mg l^{-1} IBA	Callus	Kaul et al. 2013
A. multiflora Bunge			
	MS only	Hairy roots	Kim et al. 2005
A. turkestanica (regel) Briq.			
	B_5 + 2.3 μM 2,4-D	Cell suspension	Cheng et al. 2008
	MS + 2.2 μM BAP + 2.5 μM IBA	Cell suspension	Cheng et al. 2008
Caryophyllaceae			
Lychnis flos-cuculi L.			
	MS + 100% BAP + 100% IAA/NAA	Shoot tips	Thiem et al. 2013; Maliński et al. 2014
Commelinaceae			
Cyanotis arachnoidea C.B. Clarke.			
	MS + 3 mg l^{-1} BAP + 0.2 mg l^{-1} NAA	Cell suspension	Wang et al. 2014
Asteraceae			
Rhaponticum carthamoides (Willd.) Iljin.			
	SH only	Hairy roots	Skała et al. 2015
	MS + 0.25 mg l^{-1} 2,4-D + 1.5 mg l^{-1} BAP	Callus	Zand et al. 2014
Lamiaceae			
Vitex glabrata R.Br.			
	B_5; 2.0 mg l^{-1} BA + 1.0 mg l^{-1} 2,4-D	Cell suspension	Sinlaparaya et al. 2007; Thanonkeo et al. 2011
	50:100% MS; 2.0 mg l^{-1} BA + 1.0 mg l^{-1} 2,4-D	Cell suspension	Chamnipa et al. 2012

(continued)

Table 11.1 (continued)

Species	Treatment	Method	Reference
Vitex negundo L.			
	MS + BAP 5.0 lM + 0.5 lM MS? IBA 10 lM	Shoot tips	Ahmad and Anis 2011
Amaranthaceae			
Pfaffia glomerate (Spreng.) Pedersen			
	MS + BAP 2.22 lM + NAA 2.68 lM + 0.1 lM glucose	Shoot tips	Vasconcelos et al. 2014
	MS only	Shoot tips	Flores et al. 2010
P. tuberosa (Spreng.) Hicken			
	MS + TDZ 1.0 lM (10.3 shoots)	Nodal segments	Flores et al. 2010

2,4-D 2,4-dichlorophenoxyacetic acid, *B5* Gamborg's medium, *BAP* 6-benzyloaminopurine, *IAA* indole-3-acetic acid, *IBA* indole-3-butyric acid, *MS* Murashige and Skoog medium, *NAA* a-naphthalene acetic acid

potential for biotechnological strategies of PTCs has not been examined due to their limitations (Filova 2014). Some biometabolites produced on scale-up production despite the PTCs have been highlighted in the first half of the twentieth century (Verpoorte et al. 2002; Karuppusamy 2009; Filova 2014).

11.4 Scale-up Techniques and Bioreactors

A bioreactor-based plant cell suspension process has the economic-potential advantages for large-scale production in plant pharmaceutical industry (Isah et al. 2018; Xu and Zhang 2014).

Bioreactor-based and scaling-up of desired SMs production is the final stage for the in vitro biosynthesis strategy. The plant in vitro system has the potential to provide a low-cost production of various plant-derived products. However, the scale-up production across bioreactors may take on many biologic obstacles, including low metabolite yield, cell size, heterogeneity, and genetic instability. Also, cell aggregation, shear stress sensitivity, aeration, and intensive mixing, besides adhesion of the cells and foaming, are considered the other technological obstacles of the bioreactors system. The bioreactor design and cultivation parameters must be optimized by maintaining temperature, light irradiation, adequate agitation, and ensuring gas exchange and homogeneity to enhance the biosynthetic capacity of secondary metabolites derived by plant cells (Georgiev 2020). There are many difficulties in obtaining plant steroid compounds since the amount of these compounds in raw materials is low; their chemical synthetics are not economically gainful for commercial investment. Therefore, industrial biotechnology is an alternative strategy for cost-effective large-scale production by using microbial (Shao et al. 2015; Kreit 2017; Mancilla et al. 2018) or plant cell cultures (Miras-Moreno et al. 2016). There are earlier successful attempts to produce well-known steroidal bioactive compounds in relatively large quantities by cell suspension systems (Kaul and Staba 1968; Kaul et al. 1969; Khanna et al. 1975; Marshall and Staba 1976). Kaul et al. (1969) have reported that diosgenin, a prime raw material in the commercial production of corticosteroids and steroid contraceptives, can be produced by suspension cultures of *Dioscorea* with a 1.5% dry weight content (Kaul et al. 1969). Kreis and Reinhard (1990) succeeded in the production of deacetyllanatoside C in 20-liter airlift bioreactors from two-stage cultivation of *Digitalis lanata* cells (Kreis and Reinhard 1990).

Monitoring the physiological case of cells is an important task. When scaling apparatus cultivation of cell cultures to industrial volumes according to this consideration, it reported that bubble bioreactors (20 and 630 L) with the annular aerator (K La 7.1–8.0 h − 1) showed higher biosynthetic rates (7.7–13.9%) of dry cell weight of steroidal glycosides production from *Dioscorea deltoidea* cell cultures (Titova et al. 2015). The biotic elicitors, *E. coli* (1.5%), proved best with a 9.1-fold increase in diosgenin content of *Helicteres isora*. The scaling-up across the suspension culture was performed for the production of diosgenin. It is worth mentioning that, in the scaling-up stages, the diosgenin yield obtained was in the range between 7.91 and 8.64 mg /l, where diosgenin content increased as the volume of the medium increased. The quantitative real-time PCR (qRT-PCR) analysis exhibited that biotic elicitors increased the expression levels of regulatory genes in the diosgenin biosynthetic pathway, which correlate with elicited diosgenin (Shaikh et al. 2020). Three configurations of bioreactors are optimized for extraction conditions of the steroidal glycosides of *Dioscorea deltoidea* wall cell suspension culture for LC-MS determination (Sarvin et al. 2018). Also, withanolides production during cell/organ culture, as important medicinal steroidal derivatives compounds have been cultured through bioreactors system. Sivanandhan et al. (2014) cultured *W. somnifera* on a medium consisting of 1 mg/l picloram, 0.5 mg/l kN, 200 mg/l glutamine, and 5% sucrose along with chitosan (100 mg/l) and squalene (6 mM) for 4–48 h exposure time, respectively. Withanolides compounds detected, after 28 days of culture, 7606.75 mg of withanolide-A, 4826.05 mg of withanolide B, 3732.81 mg of withaferin-A, 6538.65 mg of withanone, 3176.63 mg of 12-deoxy withanstramonolide, 2623.21 mg of withanoside IV, and 2861.18 mg of withanoside V (Sivanandhan et al. 2014). Ahlawat et al. (2017) elicited the cell suspension culture of Withania somnifera with cell homogenate of *P. indica* fungal elicitor at 3.0% (v/v)/7 days of exposure time to check the gene expression of withanolides biosynthesis pathways via quantitative PCR. Withanolides production was analyzed as withanolide A, withaferin A, and withanone. The bioreactor system enhanced the biomass to (1.13 folds), which is the upregulation of withanolide genes of biosynthetic pathways (Ahlawat et al. 2017). Steroids biosynthesis during date palm cell suspensions induced by adding pyruvic acid as a precursor and the bioreactor system scale-up their production 14 times. The date palm steroids could commercially be produced by using bioreactor systems (El-Sharabasy 2004; El-Sharabasy and El-Dawayati 2017). The evidence that the cell biomass grew by bioreactors enhances the production of steroid compounds and encourages the pharmaceutical industry of steroid derivatives. The cell culture is considered a promised goal for the advancement of innovative biotechnologies.

11.5 Extraction and Detection Techniques

Several detectors are now available to run through different apparatuses for isolating metabolites, including anion and cation exchange, ligand exchange, reverse phase, and size exclusion. Water is not often the best for extraction of lipophilic compounds (Colegate and Molyneux 2008). However, most extraction methods preferably used MeOH or ethanol after powdering the biomass using liquid nitrogen (LN). Treatment with LN, and then storing at low temperatures, should be directly done after the collection to lessen or inhibit the possibility of enzyme-reaction of the interest; Fig. 11.3 illustrates the general steps to the isolation of steroids.

Modern technologies play a significant role in the different bioactive progress of natural products. Several techniques have separated the mixtures of natural products (Colegate and Molyneux 2008). The extracts are very complex compounds in most cases and have mixtures of neutral, acidic, basic, lipophilic, hydrophilic, or amphiphilic compounds, and consequently, there will not be one method that uses all eventualities. The characterization of bioproducts necessitates sophisticated methodologies. These methodologies provide the best selective, sensitive, and structural information and determination on the interested product (Hostettmann and Wolfender 2004; Colegate and Molyneux 2008; Shakhmurova et al. 2012; Vanyolos et al. 2012).

High-throughput screening programs are in crucial need to be more sensitive methodologies. High-performance liquid chromatography (HPLC) is utilized habitually in preparative isolation and purification of natural products (Hostettmann et al.

Fig. 11.3 Illustration of general step for isolation of steroids

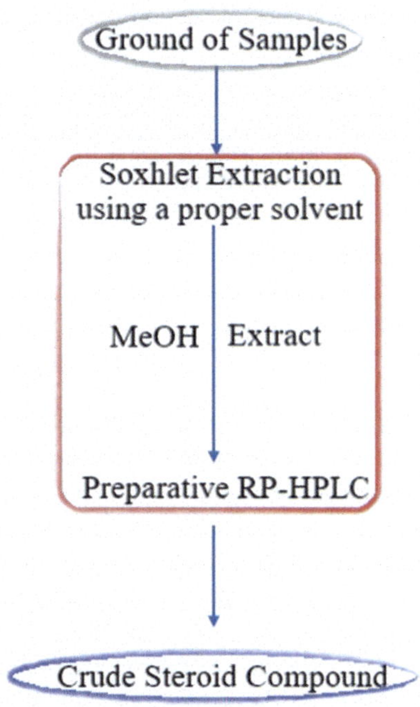

1998). This detector developed new powerful tools including LC/mass spectrometry (LC/MS; Niessen and Tinke 1995; Niessen 1999), LC/nuclear magnetic resonance (LC/NMR; Albert 1995; Sudmeier et al. 1996; Lindon et al. 1997; Wolfender et al. 1998a; Spraul et al. 1993), and LC/UV-photodiode array detection (LC/UV-DAD; Huber and George 1993). Combination with these detectors has achieved complementary spectroscopic data on an LC-peak of a target, which has often permitted their unambiguous structure determination. The LC-detectors integrated swiftly for the scan of crude extracts (Wolfender et al. 1998b). Other techniques, including IR or X-ray crystallography, have been utilized less when the other spectroscopic methods failed to give complete structure determination. The combination of different LC-techniques to determine metabolites in an extract ought to empower the full structure description at the corresponding LC-peak. The inherent sensitivity of used spectroscopic techniques is the core-one of the identification problems for LC (Hostettmann and Wolfender 2004). The LC/NMR allows structural data more than LC/MS, but it has much order of the magnitudes with minimal sensitivity. Several limitations have elucidated the use of potential LC-techniques.

LC/UV-DAD (Huber and George 1993) was utilized, in several labs, detecting metabolites (Bramley 1992; Yoshimura et al. 1994). LC/UV-DAD gives complementary information for different metabolites owing to strong chromophores. The selection of the proper LC-solvents should perform through their inherent UV limit to avoid interfering. Nonetheless, the only limitation of its application in plant extract analysis is that. New apparatuses allow the record of UV-spectral libraries, and automatic computer search can be achieved provided that UV metabolite-databases have built up. The UV-spectra measurements depend on the structure of solvent systems used, and matching with spectra measured in other solvents shows differences.

Mass spectrometry is right now one of the most sensitive molecular analyses. Moreover, it has probability to build up data on the molecular weight and structure of the test. Numerous LC/MS has been utilized and are available nowadays. These LC/MS crossing-points necessarily achieve nebulization and vaporization of a liquid, ionization of sample, removal of the excess solvent vapor, and extraction of the ions into the mass analyzer. Nowadays, the most commonly used LC/MS interfaces including electrospray (ES; Whitehouse et al. 1985), atmospheric pressure chemical ionization (APCI; Bruins et al. 1987), thermospray (TSP; Blakley and Vestal 1983), and continuous flow fast atom bombardment (CFFAB; Caprioli 1990) are very efficient for metabolites ionization (Wolfender et al. 1995).

Several analyzers, including the time of flight (TOF), ion traps (IT), and quadrupole (Q), could be utilized, which have, in terms, resolution, mass accuracy, and MS/MS possibilities (Mosi and Eigendorf 1998). The prime problems of LC/MS are the response for buffer and solvent used, nature of constituents, flow rate, and type of interfaces. Natural extracts have various physicochemical properties, which make it difficult to find optimum LC/MS conditions (Wolfender et al. 1994). Nevertheless, LC/MS/MS are extra reproducible, and the use of MS/MS databases of natural products can be considered for dereplication. Therefore, the LC/MS configuration can enormously be an influential system for screening crude extracts, but the precise

Table 11.2 Solvent used for steroid using TLC

Solvent	References
Benzene/benzene: ehyl acetate	Hunyadi et al. 2007); Nowak et al. 2013
Chloroform: ethanol	Lafont et al. 2000
Ethyl acetate: cyclohexane	Hunyadi et al. 2007); Nowak et al. 2013
Methanol: water	Lafont et al. 2000
Cyclohexane: heptane	Hunyadi et al. 2007); Nowak et al. 2013
Cyclohexane: ethyl acetate	Hunyadi et al. 2007); Nowak et al. 2013
Benzene: Chloroform	Hunyadi et al. 2007); Nowak et al. 2013
Chloroform: methanol: water	Cheng et al. 2008
Ethanol: water	Lafont et al. 2000
Acetonitrile: water	Lafont et al. 2000

conditions of ionization will have to be sensibly optimized. With the LC/MS appropriate configuration, the analysis of nonpolar to polar constituents, including the possibility of proteins will mostly determine the optimal ionization method and polarity used (Niessen 1999).

NMR provides the most useful evidence for the identification of natural products (Albert 1995). The combination of NMR and HPLC gives a piece of important information for direct metabolite identification. The rapid progress, including pulse field gradients and probe technology, has been given an impulse to LC/NMR (Spraul et al. 1993). Difficulty in detecting resonances existing in large resonances of the mobile phase is the prime problem in LC/NMR (Spraul et al. 1993; Albert 1995). Water suppression enhanced through T1 (WET) produces high-quality spectra in the procedure of on-flow and stop-flow (Albert 1995; Smallcombe et al. 1995). On-flow LC/NMR has been limited directly to the direct measurement of interest constituents and frequently loaded LC-conditions. One way to improve the exposure limit is to work in the stop-flow mode, which requires the retention times of interest or a sensitive method of the scan including LC/NMR, LC/UV, and LC/MS to activate the scan (Holt et al. 1998).

Thin-layer chromatography (TLC) is a widely used method for the isolation of natural and synthetic products. Norm- and reversed-phase TLC on silica plates with a solvent system (Table 11.2), such as chloroform: methanol: water (Cheng et al. 2008), chloroform: ethanol, methanol: water, ethanol: water, acetonitrile: water (Lafont et al. 2000) and others (Hunyadi et al. 2007); Nowak et al. 2013), has been described being an efficient way to separate steroids. The TLC plates are observed by UV lamps (254 nm) and performed visualization by nonspecific color reactions with anisaldehyde spray reagent (Cheng et al. 2008; Nowak et al. 2013), vanillin/sulfuric acid spray reagent, sulfuric (VI) acid or "specific" reactions with ammonium carbonate (fluorescence induction), 2,4-dinitrophenylhydrazine, triphenyltetrazolium chloride, and Folin-Ciocialteu reagent (color reactions) (Dinan 2001).

The TLC with mass spectrometry (TLC/MS; TLC/MS/MS) is a potential technique for identifying steroids from the crude plant extracts (Wilson et al. 1990).

Likewise, many TLC can permit more accurate quantitative measurements such as automatic multiple developments of TLC (AMDTLC), over-pressurized TLC (OPTLC), and high-performance TLC (HPTLC) (Read et al. 1990; Wilson et al. 1990; Lafont et al. 2000).

11.6 Biological Activities (Major)

Steroids play a prominent part in plant development and growth. Many steroids are isolated from plants, and the literature concerning this matter has been surveyed (Sultan and Rauf Raza 2015). Characteristics of the plant sterol biosynthesis and their role in cells are reviewed. Steroids play prime roles in several biochemical and physiological properties in all living organisms in which these are found (Abd El-Bahaman 1991; Sultan and Rauf Raza 2015). A broad sense of synthetic steroids is used as antihormones (Jovanović-Šanta et al. 2015), contraceptive drugs (Lopez et al. 2014), anticancer agents (Thao et al. 2015), cardiovascular agents (Rattanasopa et al. 2015), osteoporosis drugs (Emmanuel et al. 2011), antibiotics, anesthetics, anti-inflammatories, antidiabetic (Graf et al. 2014), and antiasthmatics (Aav et al. 2005).

Steroids in plants are of two main groups: brassinosteroids and phytosterols. Brassinosteroids, which are applied in several plants, tend to respond to biotic and abiotic stresses (Xia et al. 2009). Brassinosteroids show a broad sense of biological responses such as the inhibition of wilt, stimulating cell division, suppression of stress-regulated genes, and light-regulated genes in the darkness (Schwarz et al. 2003). Brassinosteroids have shown remarkable benefits for various responses through application in several plants. Previous reports pointed to an increase in metabolism and the elimination of pesticides. Therefore, it lessens the human ingestion of residual pesticides (Sondhi et al. 2008). The 24-epibranssinolide (EBL), isolated from *Aegle marmelos* Correa (family: Rutaceae), has pointedly diminished the maleic hydrazide-induced genotoxicity in chromosomal aberrations of *Allium cepa* (Howell and Buzdar 2005) and protected neuronal PC12 cells from 1-methyl-4-phenylpyridinium-(MPP+-)-induced oxidative stress and apoptosis in dopaminergic neurons (Julie et al. 2011). Analogs of branssinosteroid plant growth regulators such as 5α-hydroxy-6-ketopregnanes have been surveyed (Biggadike and Morton 2003). The 17-substituted pregnadienes have been made (Mellon et al. 2004) as potential inhibitors of testosterone-5α-reductase. An unusual Δ20-pregnene got from an octacoral has been inhibited by the mitochondrial respiratory chain (Meggers 2007).

Plant sterols are called phytosterols, which are a cluster of steroid alcohol that biosynthesize in plants. Phytosterols, which are white powders with mild and odor, solubilize in alcohol. Phytosterols are used in medicine, cosmetics, and food additive and fight cancer (Delvin 2002). Plant sterols sell as a dietary supplement (Delvin 2002). Ergosterol (Fig. 11.4) lowers the cholesterol level in human up to 15% (Banthorpe 1994; Vieira et al. 2005). Dietary guidelines for health call for lessening the cholesterol intake. For doing that, one strategy takes in the plant sterols, e.g., sitosterol, stigmasterol, stigmastanol, and campesterol. Despite their structural

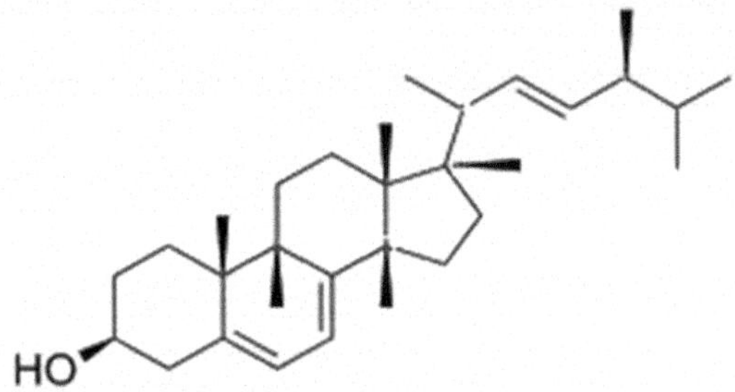

Fig. 11.4 Structure of ergosterol

similarity to cholesterol, minor isomeric modifications in the present/absence of methyl/ethyl groups in the side chains cause their less absorption by intestinal mucosal cells. Plant sterols block the absorption of cholesterol very high-efficient through intestinal cells despite not their absorbed by the body. Plant sterol drugs progressed to subject to the structures of sterols and the form of administered factor. For example, sitosterol is unsaturated and absorbed poorly in the human intestine, while sitostanol is a saturated analog and practically fully unabsorbable. The evidence of the plant sterols running in a soluble, micellar procedure effect in the inhibition of cholesterol absorption than those in a solid-crystalline form (Sharma et al. 2008).

Some miscellaneous steroids have the same vital roles. Derivatives of androstanes are separated for the treatment and inhibition of allergy (Sanjiv et al. 2010); others are effective against cardiovascular disease, cerebral-degenerative disorder, cartilage degeneration, periodontal disease, osteoporosis, metastatic bone disease, Paget's disease, endometriosis, uterine fibroid disease, restenosis, vascular smooth muscle cell proliferation, obesity, inflammatory bowel disease, hypertension, retinal degeneration, and cancer especially against breast cancer, uterus, and prostate (Chermnykh et al. 1988). The treatment by steroid multiple-course can increasingly and effectively get a better chance against preterm and Respiratory Distress Syndrome (RDS) babies and help the fragile babies to survive (Chohan et al. 2006). Antenatal steroids lessen the chances of hyaline membrane disease and intraventricular hemorrhage and chronic lung disease or death in very low-birth-weight infants (Singh et al. 2006). Steroids, e.g., stigmasterol, show antimutagenic, anxiolytic, analgesic, anticonvulsant, sedative, hypnotic, and anesthetic properties (Paschke et al. 2003; Chohan et al. 2004). By enhancing GABA receptor function in a non-genomic manner, some steroids produce an opponent for treatment of CNS abnormalities such as stress, PMS seizures, the anxiety produced by epilepsy and block muscle tension, depression, and induce anesthesia (Hollman 1996).

Cardiac glycosides (Fig. 11.5) are classified into cardenolide and bufadienolides (Laurie et al. 2012). Those have high potent cardiotoxins. Cardiac glycosides are used in the treatment of cardiac arrhythmia and congestive heart failure. Also, those have anticancer properties (Laurie et al. 2012).

Digoxin isolated from *Digitalis lanata*, ouabain (g-strophanthin) isolated from ripe seeds of *Strophanthus gratus,* and the bark of *Acokanthera ouabaio* are cardenolides. Cardenolides have a vital role in the treatment of various heart

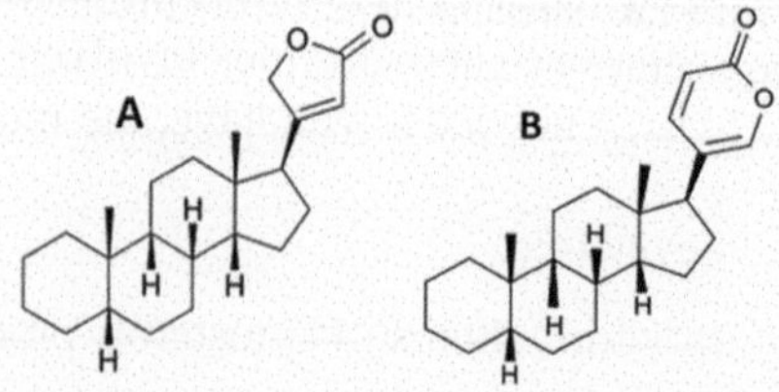

Fig. 11.5 Structures of cardiac glycosides: (**a**) cardenolide and (**b**) bufadienolide

conditions and various health risks such as increased blood pressure (BP) and tissue perfusion, anorexia, nausea, hallucinations, disorientation, insomnia, impaired yellow color perception, and gynecomastia in males (Gao et al. 2002; Pierog et al. 2009). In this trend, ouabains (g-strophanthin) are used by Somali tribesmen to poison hunting arrows due to their toxic effect, thus blocking the Na + pump in a fashion similar to one of steroidal alkaloid batrachotoxins isolated from poison arrow frogs (Wang and Sun 1994; Schneider et al. 1998).

Bufodienodiles, e.g., cinobufagin isolated from Chusan island toad (*Bufo gargarizans*) and Bufagins (a constituent of bufotoxins) from the secretion of the Cane toad (*Bufo marinus*)—so-called Toad's milk, have been utilized. Treatment with low doses of cinobufagin used traditional Chinese medicines against atrial fibrillation (Cardarelli and Kanakkanatt 1983). Bufagins have a characteristic of cardiotoxins. Bufagins derivatives have anesthetic action, anticancer properties against leukemia, melanoma, and prostate cancer cells (Jiun et al. 2002; Bick et al. 2002) nuclei.

Steroidal alkaloids, e.g., a perhydro-1,2-cyclopentanophenanthrene nucleus, are biosynthesized in the various plants as a glycosidic combination with carbohydrate moieties. Dihydroplakinamine K from a marine sponge (*Corocium niger*) is used for cytotoxic activity (Enev et al. 1998). Batrachotoxins isolated from *Phyllobates* spp., *Pitohui* spp., and *Iflita* spp. are very efficient neurotoxins. Samandarin isolated from *Salamandra salamandra* causes muscle convulsions, high blood pressure, and hyperventilation in vertebrates (Philomin et al. 1993).

Some steroidal derivatives have been used as fluorescent detectors for polycyclic aromatic hydrocarbon (PAH), like that results from cholic acid, which has a tweezer-like structure (Simon et al. 1999). The cholesteryl benzoate forms cholesteric liquid crystals with helical structures. Cholesteryl benzoate is used in thermochromic liquid crystals, hair colors, and some cosmetic preparations (Sevillano et al. 2002).

11.7 Commercial Utilization and Prospects

Steroids could be employed in several disciplines, e.g., multidrug resistance. The problematic obstacle of steroid application is the lipophilic compounds acting as inhibitors of vital bioactive macroconstituents (Martins et al. 2013). Several reports (Martins et al. 2013; Sultan and Rauf Raza 2015; Thiem et al. 2017) on multidrug resistance reported that dioxolane may lead reasonably to the invention of an effective MDR inhibitor (Martins et al. 2013). Antioxidant, free radical scavenging, and effects of 20-hydroxyecdysone neuroprotective were evidenced in in vitro oxidative damage and in vivo ischemic injury models. 20-hydroxyecdysone showed a protective effect in the adrenal gland and pheochromocytoma of rats against cobalt chloride-induced cell damage (Hu et al. 2010).

The effect of 20E on NF-κB and JNK signaling pathways and inhibition of the caspase-3 activity responsible for apoptosis reported that all destructive intracellular characteristics, e.g., the disruption of the mitochondrial membrane and inducing calcium level, have decreased via 20-hydroxyecdysone (Hu et al. 2010, 2012).

11.8 Conclusions and Recommendations

Despite the progression in enhancing bioactive metabolites production, the automation of production is still restricted. The most natural products biosynthesize in small quantities and disordered tissues. Economic production is the central bottleneck. Only a small number of plant kingdom around the globe has been economically and practically submitted to the various screening of bioactivities. There is currently a renaissance of interest as steroid-rich sources for introduction into different bioactivities programs. The rapid rate of loss of plant species has meant that it needs to report new methods leading to the rapid isolation and identification of bioproducts. The approach assumed to acquire utilizable pure constituent is interdisciplinary in botany, pharmacognosy, pharmacology, chemistry, toxicology, conservation biology, and genetics. The risk of pointless isolation with known activity is by following the guide of fractional processes. Besides, it needs to accompany constituent isolation of interest with its activities. Avoiding the time length in extraction, the automation of hyphenated techniques is used at the earliest stage of extraction. It is valuable to detect bioproducts with interest structures and to aim at their segregation.

Many studies on steroids reported raising interest in having beneficially interdisciplinary activities. The prime problems with steroids' commercial utilization are that the extraction processed hardly with a lower yield and is mostly based on numerous environmental factors. So, PTCs are a valuable alternative approach because plant biomass proliferated under the control of the whole conditions. The automation of production procedures may be improved using a broad sense of recent biotechnological methods. Progression in biotechnological tools may affect the advancement of other disciplines, including biochemistry, botany, conservation biology, and toxicology. The approach of genetic markers is a very beneficial tool that provides more insight into pathways. Till now, mini reviews of genetic markers are carried out. It should combine genetic markers with other methods. Besides, more research using genetic markers must investigate more into the pathway systems of steroids.

References

Aav R, Kanger T, Pehk T, Lopp M (2005) Unexpected reactivity of ethyl 2-(diethylphosphono) propionate toward 2,2-disubstituted-,3-cyclopentanediones. Phosphorus Sulfur Silicon Relat Elem 180:1739–1748

Abd El-Bahaman AA (1991) Biochemical studies on steroids., PhD Thesis. Fac of Agric, Moshtohor Zagazig Univ

Ahlawat S, Saxena P, Ali A et al (2017) Comparative study of withanolide production and the related transcriptional responses of biosynthetic genes in fungi elicited cell suspension culture of *Withania somnifera* in shake flask and bioreactor. Plant Physiol Biochem 114:19–28

Ahmad N, Anis M (2007) Rapid clonal multiplication of a woody tree, Vitex negundo L. through axillary shoots proliferation. Agrofor Syst 71:195–200

Ahmad N, Anis M (2011) An efficient in vitro process for recurrent production of cloned plants of Vitex negundo L. Eur J For Res 130:135–144

Ahmad N, Bin Javed S, Khan MI et al (2013) Rapid plant regeneration and analysis of genetic fidelity in micropropagated plants of Vitex trifolia: an important medicinal plant. Acta Physiol Plant 35:2493–2500

Albert K (1995) On-line use of NMR detection in separation chemistry. J Chromatogr A 703:123–147

Balandrin MF, Klocke JA (1998) Medicinal, aromatic and industrials products from plants: biotechn. in agric. and forest. In: Bajaj YPS (ed) Medicinal and aromatic plants I, vol 4. Springer, Berlin

Baltaev UA (2000) Phytoecdysteroids: structure, sources, and biosynthesis in plants. Russ J Bioorg Chem 26:799–831

Banthorpe DV (1994) Natural products: their chemistry and biological significance, 1st edn. Longman scientific and technical, Essex, p 289

Bick RJ, Poindexter BJ, Sweney RR, Dasgupta A (2002) Effects of Chan Su, a traditional Chinese medicine, on the calcium transients of isolated cardiomyocytes: cardiotoxicity due to more than Na, K-ATPase blocking. Life Sci 72:699–709

Biggadike K, Morton GE (2003). WO 0362,259

Blakley CR, Vestal ML (1983) Thermospray interface for liquid chromatography/mass spectrometry. Anal Chem 55:750–754

Bramley PM (1992) Analysis of carotenoids by high performance liquid chromatography and diode-array detection. Phytochem Anal 3:97–104

Bruins AP, Covey TR, Henion JT (1987) Ion spray for combined liquid chromatography mass spectrometry/atmospheric pressure chemical ionisation mass spectrometry. Anal Chem 59: 2642–2646

Bucourt R, Nedlec L, Gasc JC, Weill-Raynal (1969) J Bull Soc Chim Fr 1967:561

Caprioli RM (1990) Continuous flow fast atom bombardment mass spectrometry. Anal Chem 58: 949–954

Cardarelli NF, Kanakkanatt SV (1983). US 4,634,693

Chamnipa N, Thanonkeo S, Thanonkeo P (2012) Enhance production of 20-hydroxyecdysone in cell suspension cultures of Vitex glabrata R.Br. by elicitor feeding. J Med Plants Res 6(17): 3317–3323

Cheng DM, Yousef GG, Grace MH et al (2008) In vitro production of metabolism-enhancing phytoecdysteroids from Ajuga turkestanica. Plant Cell Tissue Organ Cult 93:73–83

Chermnykh NS, Shimanovski ÄNL, Shutko GV, Syrov VN (1988) The action of methandrostenolone and ecdysterone on the physical endurance of animals and on protein metabolism in the skeletal muscles. Farmakol Toksikol 51:57–60

Chohan ZH, Supuran CT, Scozzafava A (2004) Metalloantibiotics: synthesis and antibacterial activity of cobalt (II), copper (II), nickel (II) and zinc (II) complexes of kefzol. J Enzyme Inhib Med Chem 19:79–84

Chohan ZH, Arif M, Akhtar MA, Supuran CT (2006) Metal-based antibacterial and antifungal agents: synthesis, characterization, and in vitro biological evaluation of Co(II), Cu(II), Ni(II), and Zn(II) complexes with amino acid- derived compounds. Bioinorg Chem Appl 2006:83131

Cocquyt K, Cos P, Herdewijn P et al (2011) *Ajuga remota* Benth: from ethnopharmacology to phytomedical perspective in the treatment of malaria. Phytomedicine 18:1229–1237

Colegate SM, Molyneux RJ (2008) Bioactive natural products: detection, isolation, and structural determination, 2nd edn. CRC Press, Boca Raton

De Boeck B, Harrington-Frost NM, Pattenden G (2005) Tandem cyclisations involving α-ketenyl alkyl radicals. New syntheses of the natural triquinanes pentalenene and modhephene. Org Biomol Chem 3:340–347

Delvin MT (2002) Text book of biochemistry with clinical correlations, 5th edn. Wiley, Hoboken

Dinan L (2001) Phytoecdysteroids: biological aspects. Phytochemistry 57:325–339

Duan H, Ding X, Xing J et al (2012) Effects of plant growth substances on callus re-differentiation of medicinal plant Achyranthes bidentata. Plant Omics 5(3):261–265

El-Sharabasy SF (2004) Effects of some precursors on development of secondary products in tissues and media of embryogenic callus of date palm cv. Sewi Arab J Biotechnol 7(1):83–90

El-Sharabasy S, El-Dawayati M (2017) Bioreactor steroid production and analysis of date palm embryogenic callus. In: Date palm biotechnology protocols, vol 1. Humana Press, New York, pp 309–318

Emmanuel H, Gauthier M, Emmanuel B, Severine B, Florence C et al (2011) Joint Bone Spine 78(2):S191–S196

Enev VS, Mohr J, Harre M, Nikisch K (1998) Tetrahedron-Asymmetry 9:2693

Festucci-Buselli RA, Contim LSA, Barbosa LCA et al (2008) Biosynthesis and potential functions of the ecdysteroid 20-hydroxyecdysone—a review. Botany 86:978–987

Filova A (2014) Production of secondary metabolites in plant tissue cultures. Res J Agric Sci 46(1): 236–245

Flores R, Brondani D, Cezarotto V et al (2010) Micropropagation and beta-ecdysone content of the Brazilian ginsengs Pfaffia glomerata and Pfaffia tuberosa. In Vitro Cell Dev Biol Plant 46:210–217

Fujimoto Y, Ohyama K, Nomura K et al (2000) Biosynthesis of sterols and ecdysteroids in Ajuga hairy roots. Lipids 35:279–288

Gao J, Wymore RS, Wang Y, Gaudette GR, Krukenkamp IB et al (2002) Isoform-specific stimulation of cardiac Na/K pumps by nanomolar concentrations of glycosides. J Gen Physiol 119:297–312

Gnanaraj WE, Antonisamy JM, Monahamathi RB et al (2012) In vitro clonal propagation of Achyranthes aspera L. and Achyranthes bidentata Blume using nodal explants. Asian Pac J Trop Biomed 1:1–5

Graf BL, Poulev A, Kuhn P et al (2014) Quinoa seeds leach phytoecdysteroids and other compounds with anti-diabetic properties. Food Chem 163:178–185

Hollman A (1996) Drugs for atrial fibrillation. Digoxin comes from Digitalis lanata. Br Med J 312: 912

Holt RM, Newman MJ, Pullen FS et al (1998) High-performance liquid chromatography/NMR spectrometry/mass spectrometry: further advances in hyphenated technology. J Mass Spectrom 32:64–70

Hostettmann K, Wolfender J-L (2004) Applications of liquid chromatography/UV/MS and liquid chromatography/NMR for the on-line identification of plant metabolites. In: Tringali C (ed) Bioactive compounds from natural sources: isolation, characterisation and biological properties. CRC Press, Boca Raton

Hostettmann K, Marston A, Hostettmann M (1998) Preparative chromatography techniques: applications in natural product isolation, 2nd edn. Springer, Berlin

Howell A, Buzdar A (2005) Are aromatase inhibitors superior to antiestrogens? J Steroid Biochem Mol Biol 93:237–247

Hu J, Zhao TZ, Chu WH et al (2010) Protective effects of 20-hydroxyecdysone on CoCl2-induced cell injury in PC12 cells. J Cell Biochem 111:1512–1521

Hu J, Luo CX, Chu WH et al (2012) 20-hydroxyecdysone protects against oxidative stress-induced neuronal injury by scavenging free radicals and modulating NF-κB and JNK pathways. PLoS One 7(12):e50764

Huber L, George SA (1993) Diode array detection in HPLC. Marcel Dekker, New York

Hunyadi A, Gergely A, Simon A et al (2007) Preparative-Scale Chromatography of Ecdysteroids of Serratula wolffii Andrae. J Chromatogr Sci 45:76–86

Ikekawa N, Fujimoto Y, Ishiguro M (2013) Reminiscences of research on the chemistry and biology of natural sterols in insects, plants and humans. Proc Jpn Acad Ser B 89(8):349–369

Isah T, Umar S, Mujib A et al (2018) Secondary metabolism of pharmaceuticals in the plant in vitro cultures: strategies, approaches, and limitations to achieving higher yield. Plant Cell Tiss. Org. Cult. 132(2):239–265

Jiun Y, William JH, Shu K, Paulus SW (2002) Effects of bufalin and cinobufagin on the proliferation of androgen dependent and independent prostate cancer cells. Prostate 54:112–124

Jovanović-Šanta SS, Petri ET, Klisuric OR, Szecsi M, Kovacevic R et al (2015) Antihormonal potential of selected D-homo and D-seco estratriene derivatives. Steroids 97:45–53

Julie C, Fanny L, Benoit D, Maria-Grazia MJ (2011) 24-epibrassinolide, a phytosterol from the brassinosteroid family, protects dopaminergic cells against MPP+-induced oxidative stress and apoptosis. J Toxicol 2011:392859

Karuppusamy S (2009) A review on trends in production of secondary metabolites from higher plants by in vitro tissue, organ and cell cultures. J Med Plants Res 3(13):1222–1239

Kaul B, Staba EJ (1968) Dioscorea tissue cultures. I. Biosynthesis and isolation of diosgenin from *Dioseorea deltoidea* callus and suspension cells. Lloydia 31:171–179

Kaul B, Stohs SJ, Staba EJ (1969) Dioscorea tissue cultures. III. Influence of various factors on diosgenin production by *Dioseorea deltoidea* callus and suspension cultures. Lloydia 32:347–359

Kaul S, Das S, Srivastava PS (2013) Micropropagation of Ajuga bracteosa, a medicinal herb. Physiol Mol Biol Plants 19(2):289–296

Khanna P, Jain SC, Bansal R (1975) Effect of cholesterol on growth and production of diosgenin, gitogenin, tigogenin and sterols in suspension cultures. Indian J Exp Biol 1:211–213

Kim OT, Manickavasagm M, Kim YJ et al (2005) Genetic transformation of Ajuga multiflora Bunge with Agrobacterium rhizogenes and 20-hydroxyecdysone production in hairy roots. J Plant Biol 48(2):258–262

Kreis W, Reinhard E (1990) Two-stage cultivation of Digitalis lanata cells: semicontinuous production of deacetyllanatoside C in 20-litre airlift bioreactors. J Biotechnol 16(1–2):123–135

Lafont R, Blais C, Harmatha J et al (2000) Ecdysteroids: chromatography. In: Wilson ID, Adlard ER, Cooke M, Poole CF (eds) Encyclopedia of separation science. Academic Press, London, pp 2631–2641

Kreit J (2017) Microbial catabolism of sterols: focus on the enzymes that transform the sterol 3β-hydroxy-5-en into 3-keto-4-en. FEMS Microbiol Lett 364(3):7

Laurie M, Erika V, Sandy A, Isabelle M, Yuting M et al (2012) Sci Transl Med 4:143–199

Lindon JC, Nicholson JK, Sidelmann UG, Wilson ID (1997) Directly coupled HPLC-NMR and its application to drug metabolism. Drug Metab Rev 29:705–746

Lopez LM, Grimes DA, Shulz KF, Curtis KM (2006) Cochrane Database Syst Rev 2:7

Lopez LM, Grimes DA, Schulz KF (2014) Steroidal contraceptives: effect on carbohydrate metabolism in women without diabetes mellitus. Cochrane Database Syst Rev 2014(4): CD006133. https://doi.org/10.1002/14651858.CD006133.pub5

Maliński MP, Michalska AD, Tomczykowa M et al (2014) Ragged Robin (Lychnis flos-cuculi)—a plant with potential medicinal value. Rev Bras Farmacogn 24(6):722–730

Mancilla RA, Little C, Amoroso A (2018) Efficient bioconversion of high concentration phytosterol microdispersion to 4-androstene-3, 17-dione (AD) by mycobacterium sp. B3805. Appl Biochem Biotechnol 185(2):494–506

Marchev AS, Georgiev MI (2020) Plant in vitro systems as a sustainable source of active ingredients for cosmeceutical application. Molecules 25(9):2006

Marion F, Coulomb J, Servais A, Courillon C, Fensterbank C et al (2006) Radical cascade cyclizations and platinum(II)-catalyzed cycloisomerizations of ynamides. ChemInform 62: 3856–3871

Marshall JG, Staba EJ (1976) Hormonal effects on diosgenin biosynthesis and growth in Dioseorea deltoidea tissue cultures. Phytochemistry 12:53–55

Martins A, Csa´bi J, Bala´zs A et al (2013) Synthesis and structure-activity relationships of novel ecdysteroid dioxolanes as MDR modulators in cancer. Molecules 18:15255–15275

Meggers E (2007) Exploring biologically relevant chemical space with metal complexes. Curr Opin Chem Biol 11:287–292

Melcangi RC, Azcoitia I, Ballabio M, Cavarretta I, Gonzalez LC et al (2003) Neuroactive steroids influence peripheral myelination: a promising opportunity for preventing or treating age-dependent dysfunctions of peripheral nerves. Prog Neurobiol 71:57–66

Mellon S, Gong W, Griffin LD (2004) Niemann pick type C disease as a model for defects in neurosteroidogenesis. Endocr Res 30:727–735

Miras-Moreno B, Sabater-Jara AB, Pedreño MA, Almagro L (2016) Bioactivity of phytosterols and their production in plant in vitro cultures. J Agric Food Chem 64(38):7049–7058

Morgan P, Melinda SM (1997) Kirk Othmer encyclopedia of chemical technology by Wiley, Hoboken 4:2000

Mosi AA, Eigendorf GK (1998) Current mass spectrometric methods in organic chemistry. Curr Org Chem 2:145–172

Niessen WMA (1999) Liquid chromatography-mass spectrometry, 2nd edn. Dekker, New York

Niessen WMA, Tinke AP (1995) Liquid chromatography-mass spectrometry, general principles and instrumentation. J Chromatogr A 703:37–57

Nowak G, Moroch A, Urban´ska M et al (2012) Ekdysony roślinne. Plant ecdysones. Post Fitoter 13(1):15–21

Nowak G, Urbanska M, Nawrot J et al (2013) Color and chemical reactions of selected sesquiterpene lactones and ecdysone from Asteraceae on TLC plates. J Plant Chromatogr 26:289–293

Okuzumi K, Hara N, Fujimoto Y et al (2003) Biosynthesis of phytoecdysteroids in Ajuga hairy roots: clerosterol as a precursor of cyasterone, isocyasterone and 29-norcyasterone. Tetrahedron Lett 44:323–326

Parveen F, Ara KZG, Khan M et al (2008) In vitro propagation of Achyranthes aspera: effect of cytokinins and auxins. Abstracts. J Biotechnol 136S:S147–S169

Paschke R, Paetz C, Mueller T, Schmoll HJ, Mueller H et al (2003) Biomolecules linked to transition metal complexes—new chances for chemotherapy. Curr Med Chem 10:2033–2044

Pattenden G, Gonzalez MA, McCulloch S, Walter A, Woodhead SJ (2004) A total synthesis of estrone based on a novel cascade of radical cyclizations. Proc Natl Acad Sci U S A 101:12024–12029

Philomin V, Vessieres A, Gruselle M, Jaouen G (1993) Synthesis of cobalt carbonyl complexes of cortisol and testosterone. Study of their recognition by specific polyclonal antibodies. Bioconjug Chem 4:419–424

Pierog JE, Kane KE, Kane BG, Donovan JW, Helmick T (2009) Tricyclic antidepressant toxicity treated with massive sodium bicarbonate. Am J Emerg Med 27:1168

Rattanasopa C, Phungphong S, Wattanapermpool J, Bupha-Intr T (2015) Significant role of estrogen in maintaining cardiac mitochondrial functions. J Steroids Biochem Mol Biol 147:1–9

Read H, Wilson JD, Lafont R (1990) Overpressure thin-layer chromatography of ecdysteroids. In: McCaffery R, Wilson ID (eds) Chromatography and isolation of insect hormones and pheromones. Plenum Press, London, p 127

Rout GR, Samantatay S, Das P (2000) In vitro manipulation and propagation of medicinal plants. Biotechnol Adv 18:91–120

Sanjiv BA, Mark O, Ann E, Matthew M, Christy M et al (2010) J Pediatrics 156:377

Sarvin B, Fedorova E, Shpigun O et al (2018) LC-MS determination of steroidal glycosides from *Dioscorea deltoidea* wall cell suspension culture: optimization of pre-LC-MS procedure parameters by Latin Square design. J Chromatogr B 1080:64–70

Schneider R, Ralf V, Manfred N, Wolf DL, Ulrike K et al (1998) Bovine adrenals contain, in addition to ouabain, a second inhibitor of the sodium pump. J Biol Chem 273:784–792

Schwarz S, Mueller G, Osemund D, Richter M, Petus O, et al. (2003). US 65,83,130

Sen MK, Nasrin S, Rahman S et al (2014) In vitro callus induction and plantlet regeneration of Achyranthes aspera L., a high value medicinal plant. Asian Pac J Trop Biomed 4(1):40–46

Sevillano LG, Caballero E, Tome F, Medarde M, San Feliciano A (2002) Tetrahedron 58:10103

Shaikh S, Shriram V, Khare T, Kumar V (2020) Biotic elicitors enhance diosgenin production in *Helicteres isora* L. suspension cultures via up-regulation of CAS and HMGR genes. Physiol Mol Biol Plants 26(3):593–604

Shakhmurova GA, Mamadalieva NZ, Zhanibekov AA et al (2012) Effect of total ecdysteroid preparation from *Silene viridiflora* on the immune state of experimental animals under normal and secondary immunodeficiency conditions. Pharm Chem J 46:222–224

Shao M, Zhang X, Rao Z et al (2015) Enhanced production of androst-1, 4-diene-3, 17-dione by mycobacterium neoaurum JC-12 using three-stage fermentation strategy. PLoS One 10(9): e0137658

Sharma P, Bhardwaj R, Arora HK, Arora N (2008) Biol Plant 52:767

Simon M, Harwood KJ, Toyne J, Goodby W, Michael P (1999) The synthesis of 3α–/3β-cholesteryl and cholestanyl esters and ethers—an assessment of their mesogenicity. Mol Cryst Liq Cryst 332:485–495

Singh K, Singh DP, Barwa MS, Tyagi P, Mirza Y (2006) Some bivalent metal complexes of Schiff bases containing N and S donor atoms. J Enzyme Inhib Med Chem 21:749–755

Sinlaparaya D, Duanghaklang P, Panichajakul S (2007) Enhancement of 20-hydroxyecdysone production in cell suspension cultures of Vitex glabrata R.Br. by precursors feeding. Afr J Biotechnol 6(14):1639–1642

Sivanandhan G, Selvaraj N, Ganapathi A, Manickavasagam M (2014) Enhanced biosynthesis of withanolides by elicitation and precursor feeding in cell suspension culture of *Withania somnifera* (L.) Dunal in shake-flask culture and bioreactor. PLoS One 9(8):e104005

Skała E, Grąbkowska R, Sitarek P et al (2015) Rhaponticum carthamoides regeneration through direct and indirect organogenesis, molecular profiles and secondary metabolite production. Plant Cell Tissue Organ Cult 123:83–98

Smallcombe SH, Patt SL, Keiffer PA (1995) WET solvent suppression and its application to LC-NMR and high-resolution NMR spectroscopy. J Magn Reson Series A 117:295–303

Smetanska I (2008) Production of secondary metabolites using plant cell cultures. Adv Biochem Eng Biotechnol 111:187–228

Sondhi N, Bhardwaj R, Kaur S, Singh B, Kumar N (2008) Plant Growth Regul 54:217

Spraul M, Hoffmann M, Lindon JC et al (1993) Liquid chromatography coupled with high field proton nuclear magnetic resonance spectroscopy: Current status and future prospects. Anal Proc 30:390–392

Stanczyk FZ (2009) Production, clearance, and measurement of steroid hormones. Glob Libr Women's Med. https://doi.org/10.3843/GLOWM.10278

St-Onge MP, Lamarche B, Mauger JF, Jones PJ (2003) Consumption of a functional oil rich in phytosterols and medium-chain triglyceride oil improves plasma lipid profiles in men. J Nutr 133:1815–1820

Sudmeier JL, Günther UL, Alber K, Bachovchin WW (1996) Sensitivity optimization in continuous-flow FTNMR. J Magn Reson A 118:145–156

Sultan A, Rauf Raza A (2015) Steroids: a diverse class of secondary metabolites. Med Chem 5(5): 310–317. https://doi.org/10.4172/2161-0444.1000279

Thanonkeo S, Chamnipa N, Thanonkeo P (2011) Induced accumulation of 20-hydroxyecdysone in cell suspension cultures of Vitex glabrata R.Br. Afr J Biotechnol 10(52):10612–10617

Thao NP, Luyen BT, Kim EJ, Kang JI, Kang HK et al (2015) Steroidal constituents from the edible sea urchin Diadema savignyi Michelin induce apoptosis in human cancer cells. J Med Food 18: 45–53

Thiem B, Malin´ski M, Tomczyk M (2013) In vitro propagation of Lychnis flos-cuculi L., a plant with potential medicinal value. Acta Biol Cracov Ser Bot 55(2):66

Thiem B, Kikowska M, Maliński MP et al (2017) Ecdysteroids: production in plant in vitro cultures. Phytochem Rev 16:603–622. https://doi.org/10.1007/s11101-016-9483-z

Titova MV, Shumilo NA, Kulichenko IE et al (2015) Features of respiration and formation of steroidal glycosides in *Dioscorea deltoidea* cell suspension culture grown in flasks and bioreactors. Russ J Plant Physiol 62(4):557–563

Vanyolos A, Beni Z, Dekany M et al (2012) Novel ecdysteroids from *Serratula wolffii*. Sci World J. https://doi.org/10.1100/2012/651275

Vasconcelos JM, Saldanha GW, Dias LLC et al (2014) In vitro propagation of Brazilian ginseng [Pfaffia glomerata (Spreng.) Pedersen] as affected by carbon sources. In Vitro Cell Dev Biol Plant 50:746–751

Verpoorte R, Contin A, Memelink J (2002) Biotechnology for the production of plant secondary metabolites. Phytochem Rev 1:13–25

Vieira AT, Pinho V, Lepsch LB, Scavone C, Ribeiro IM et al (2005) Mechanisms of the anti-inflammatory effects of the natural secosteroids physalins in a model of intestinal ischaemia and reperfusion injury. Br J Pharmacol 146:244–251

Wang G, Sun G (1994) J Tradit Chin Med 14:132

Wang QJ, Zheng LP, Sima YH et al (2013) Methyl jasmonate stimulates 20-hydroxyecdysone production in cell suspension cultures of *Achyranthes bidentata*. Plant Omnics 6(2):116–120

Wang QJ, Zheng LP, Zhao PF et al (2014) Cloning and characterization of an elicitor-responsive gene encoding 3-hydroxy-3-methylglutaryl coenzyme A reductase involved in 20-hydroxyecdysone production in cell cultures of *Cyanotis arachnoidea*. Plant Physiol Biochem 84:1–9

Whitehouse RC, Dreyer RN, Yamashita M, Fenn JB (1985) Electrospray interface for liquid chromatographs and mass spectrometers. Anal Chem 57:675–679

Wolfender J-L, Maillard M, Hostettmann K (1994) Thermospray liquid chromatography-mass spectrometry in phytochemical analysis. Phytochem Anal 5:153–182

Wolfender J-L, Abe F, Nagao T et al (1995) Liquid chromatography combined with thermospray and continuous-flow fast atom bombardment mass spectrometry of glycosides in crude plant extracts. J Chromatogr A 712:155–168

Wolfender J-L, Ndjoko K, Hostettmann K (1998a) LC/NMR in natural products chemistry. Curr Org Chem 2:575–596

Wolfender J-L, Rodriguez S, Hostettmann K (1998b) Liquid chromatography coupled to mass spectrometry and nuclear magnetic resonance spectroscopy for the screening of plant constituents. J Chromatogr A 794:299–316

Wilson ID, Lafont R, Porter CJ et al (1990) Thin-layer chromatography of ecdysteroids—detection and identification. In: McCaffery AR, Wilson ID (eds) Chromatography and isolation of insect hormones and pheromones. Plenum Press, New York, pp 117–126

Wu P, Xie H, Tao W et al (2010) Phytoecdysteroids from the rhizomes of *Brainea insignis*. Phytochemistry 71:975–981

Xia XJ, Zhang Y, Wu JX, Wang JT, Zhou YH et al (2009) Brassinosteroids promote metabolism of pesticides in cucumber. J Agric Food Chem 57:8406–8413

Xu J, Zhang N (2014) On the way to commercializing plant cell culture platform for biopharmaceuticals: present status and prospect. Pharm Bioprocess 2(6):491–499

Yoshimura I, Yasuhiro K, Yoshikazu Y et al (1994) Analysis of secondary metabolites from lichen by high performance liquid chromatography with a photodiode array detector. Phytochem Anal 5:197–205

Zand A, Babaei A, Omidbaigi R et al (2014) Study on callus induction and plant regeneration of *Leuzea carthamoides* via tissue culture system. J Med Plants Res 8(5):260–268

In Vitro Production of Tocopherols

12

Vasantha Veerappa Lakshmaiah, Biljo Vadakkekudiyil Joseph, Rakesh Bhaskar, Rutwick Surya Ulhas, Jameel Mohamed Al-Khayri, and Praveen Nagella

Abstract

Tocopherols are an essential dietary nutrient for mammals and photosynthetic products produced by green plants. Tocopherols commonly referred to as vitamin E exist in four forms (α-, β-, γ-, and δ-tocopherol). Synthetic α-tocopherol is a mixture of eight racemic forms and is less effective than natural tocopherol, thus the demand for plant-derived tocopherols is high. Tocopherols are lipophilic antioxidant and extensively used as therapeutic agents such as anti-inflammatory, anti-infection, anticancer, immune-stimulant, and nephro-protectant. They are also used as food additives and nutraceuticals. Plant cell and tissue culture is one of the promising techniques for mass production of tocopherols to meet the commercial demand. Optimizing physical and chemical factors for in vitro culture system has resulted in better accumulation of the product. Moreover, using bioreactors, precursor feeding, elicitation, biotransformation, and metabolic engineering approaches have resulted in enhanced yield of tocopherols from in vitro cultures. The present chapter deals with various important aspects of tocopherol in vitro production such as biosynthesis of tocopherol with special emphasis on key enzymes involved in the pathway whose modulation in expression can increase the yield of the product. Topics discussed include production of tocopherol from callus, cell and organ culture, metabolic engineering for mass

V. V. Lakshmaiah · B. V. Joseph · R. Bhaskar · R. S. Ulhas · P. Nagella
Department of Life Sciences, CHRIST (Deemed to be University), Bengaluru, Karnataka, India
e-mail: vasantha.vl@christuniversity.in; biljo.joseph@christuniversity.in; rakesh.b@res.christuniversity.in; rutwick.u@science.christuniversity.in; praveen.n@christuniversity.in

J. M. Al-Khayri (✉)
Department of Agricultural Biotechnology, College of Agriculture and Food Sciences, King Faisal University, Al-Ahsa, Saudi Arabia
e-mail: jkhayri@kfu.edu.sa

T. Belwal et al. (eds.), *Nutraceuticals Production from Plant Cell Factory*,
https://doi.org/10.1007/978-981-16-8858-4_12

production, different methods employed for extraction and quantification of tocopherols, and their biological activities and commercial applications.

Keywords

Tocopherol · Plant cell and organ culture · Elicitation strategies · Metabolic engineering · Biological activity · Biosynthesis

12.1 Introduction

Vitamin E complex, which consists of tocopherols (T) and tocotrienols (T3) collectively called as tocols, was first discovered in 1922 by Herbert McLean Evans and Katherine Scott Bishop (Niki and Traber 2012) (Fig. 12.1). It was isolated in 1935 by Gladys Emerson, structurally elucidated by Erhard Fernholz in 1938, and synthesized artificially in 1939 by group of scientists headed by Paul Karrer (Babura et al. 2017). Since this vitamin plays a predominant role in fertilization, it was named tocopherol which in Greek means 'birth and to bear' (Evans et al. 1936; Evans and Bishop 1922). Tocopherols are membrane-soluble amphipathic antioxidants with polar chromanol ring and hydrophobic-saturated phytyl side chains, while tocotrienols have an unsaturated farnesyl tail. Four homologues exist within

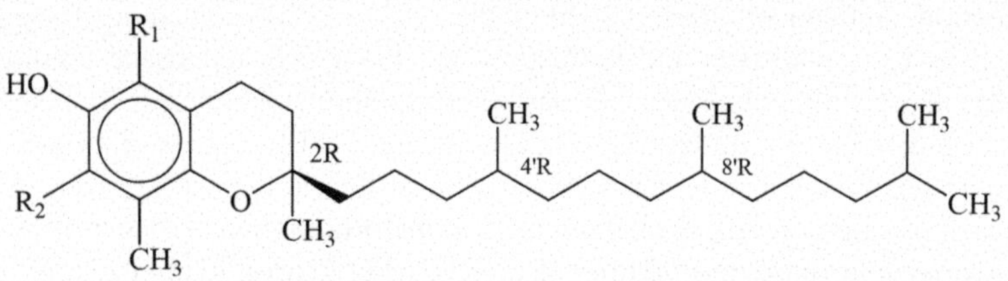

Tocopherol

R1	R2	
CH_3	CH_3	α
CH3	CH3	β
H	CH_3	γ
H	H	δ

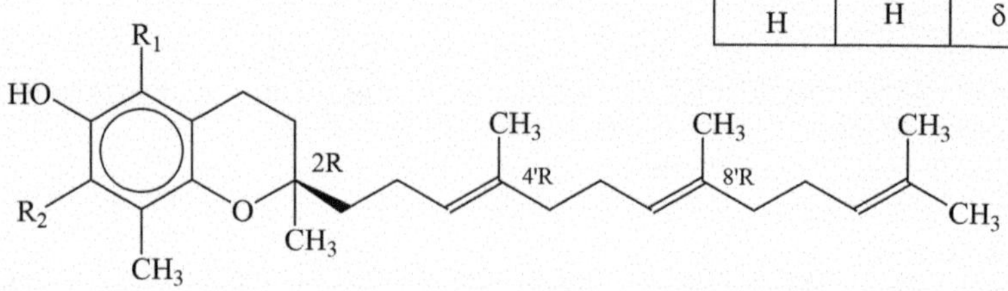

Tocotrienol

Fig. 12.1 Structural representation of tocopherols and tocotrienols (Source: Ricciarelli et al. 2002)

tocopherols and tocotrienols, namely alpha (α), beta (β), gamma (γ), and delta (δ), which differ in the position and number of methyl groups attached to the aromatic ring (Jukić Špika et al. 2016). Out of all the eight isomers of the tocols, α-tocopherol has the highest vitamin E activity (Ehrenbergerová et al. 2006). α-Tocopherol is produced in high quantities in green tissues, while γ-tocopherol and tocotrienols are produced in seeds, which prevents lipid peroxidation of polyunsaturated fatty acids and lipids in seeds (Collakova and DellaPenna 2003a, b).

Vitamin E is synthesized only in the plastids of plants and certain photosynthetic bacteria. Animals are unable to synthesize vitamin E, so they depend on the plant sources (Yasar and Sevket 2006). Vitamin E possesses high antioxidant activity which prevents the destructive oxidation in the cell membranes. Tocopherols act as free radical scavengers (Caretto et al. 2002). They act on reactive oxygen species (ROS) including singlet oxygen, superoxide radicals, and alkyl peroxy radicals that accumulate in plant cells in response to the environmental stress conditions (Sandorf and Holländer-Czytko 2002). Apart from their role as antioxidants, they also play a photoprotective role maintaining the membrane stability and are involved in cell signalling (Munné-Bosch and Falk 2004). The National Institute of Health (NIH) recommends 15 mg of vitamin E for adults and 10 mg for children per day. High doses of vitamin E reduce the risk of cardiovascular disorders, neurological disorders, and cataracts, reduce serum cholesterol levels, and inhibit the growth of cancer cells (Cahoon et al. 2003). In addition, they are required to maintain proper muscular and immune system. Tocopherols are used in cosmetics, animal feed, and as food additives to increase their shelf life (Karunanandaa et al. 2005). According to the USDA-ARS Nutrient Data Laboratory, the highest α-tocopherols are produced in sunflower seeds, almonds, hazelnuts, and pine nuts, while the highest sources of γ-tocopherols are black walnuts, sesame seeds, pecans, pistachios, flaxseeds, and pumpkin seeds. Oilseeds have a vitamin E in the range of 330–2000 μg/g of oil. However, the oilseeds accumulate γ-tocopherols which have a lower activity as compared to the α-tocopherols. However, the green tissues produce only 20–50 μg/g of α-tocopherols. Hence, the naturally produced Vitamin E is not sufficient to satisfy the growing demand of the vitamin, owning to its pharmaceutical health benefits and prevention of diseases like Parkinson's, Alzheimer's, atherosclerosis, diabetes, and coronary heart disease. Therefore, many strategies like increasing the biosynthesis of vitamin E through targeting their genes, breeding, and genetic engineering have been investigated to produce high α-tocopherol content (Ajjawi and Shintani 2004). The α-tocopherol yield of genetically engineered plants has increased by 20-fold. However, it still depends on constraints such as seed maturation, growth season, and climatic and edaphic factors (Caretto et al. 2010). One of the most promising approaches to enhance production lies in the plant tissue culture technology which provides continuous supply of uniform quality of these antioxidants (Caretto et al. 2004). In elicited cell suspension culture, a particular cell or organ with high α-tocopherol yield can be targeted and produced in high quantities in a shorter duration in a sophisticated environment which favors vitamin E biosynthesis (Almagro et al. 2016; Cetin et al. 2014). The present chapter deals with the production of tocopherols from cell and organ culture and metabolic engineering strategies for enhanced metabolite content. Also, emphasis has been given on the biosynthesis of tocopherols and the key genes involved in the biosynthetic pathway have been discussed. Furthermore, the different extraction and quantification methods employed for tocopherol and their potential biological activities have been mentioned.

12.2 Biosynthesis of Tocopherols

Tocopherol biosynthesis (Fig. 12.2) occurs in the plastids of higher plants. The biosynthesis involves conjugation of two precursors: namely, Homogentistate, which is an intermediate of degradation of aromatic amino acids, formed in cytosolic shikimate pathway, as its head (the aromatic ring of tocopherols); and Phytyl diphosphate that arises from methylerythritol phosphate pathway, as its tail.

The shikimate pathway has seven steps that conjugate Erythrose-4-phosphate (an intermediate of pentose phosphate pathway) and phosphoenol pyruvate (an intermediate of glycolysis) to form shikimate through a series of steps, which further produces the end product Chorismate. Quinones, tocopherols, and aromatic amino acids (Phe, Tyr.Trp) use Chorismate as a precursor. There is no correlation known between the pool of syntheses, dehydratases, and kinases of the Shikimate pathway and tocopherol biosynthesis. The limiting step of this pathway is the reversible formation of 5-Enolpyruvylshikimate-3-phosphate (EPSP) catalyzed by EPSP synthase. Chorismate upon the action of chorismate mutase forms Prephenate that further diverges from the pathway to give raise to Phe and Tyr from arogenate upon the action of arogenate dehydratase or arogenate dehydrogenase, respectively. But, prephenate, upon the action of prephenate dehydrogenase, produces the first intermediate in the tocopherol and tocotrienols biosynthesis which is p-Hydroxyphenyl pyruvate (HPP). A small fraction of carbon from Tyr is also incorporated in HPP formation by tyrosine aminotransferase (Lopukhina et al. 2001; Sandorf and Holländer-Czytko 2002), while Phe will be incorporated in synthesis of lignin, flavonoids, and other phenylpropanoids, and HPP by the action of HPP dioxygenase forms Homogentisate which is another precursor for tococromanols (Rippert 2004).

The isoprenyl side chain of tocopherol is derived from Phytyl diphosphate (PDP). Isopentenyldiphosphate (IPP) and its isomer dimethylallyldiphosphate (DMAPP) are the precursors of PDP; they are synthesized from Mevalonate (MVA) and methyl erythritol phosphate (MEP) pathways, respectively. The cytoplasmic MVA pathway converts C15 sesquiterpenes into C30 triterpenes. On the other hand, plastidic MEP pathway produces isopentenyl diphosphate which is a precursor of isoprenes, C10 Monoterpenes, C20 diterpenes, Carotenoids, and Plastoquinones along with phytol conjugates like chlorophyll and tocopherols.

The MEP pathway begins with the transketolase-type condensation of pyruvate and glycerldehyde-3-phosphate yielding 1-deoxy-D-xylulose-5-phosphate (DOXP) by the action of DOXP synthase, which is a key enzyme in plastidic IPP synthesis.

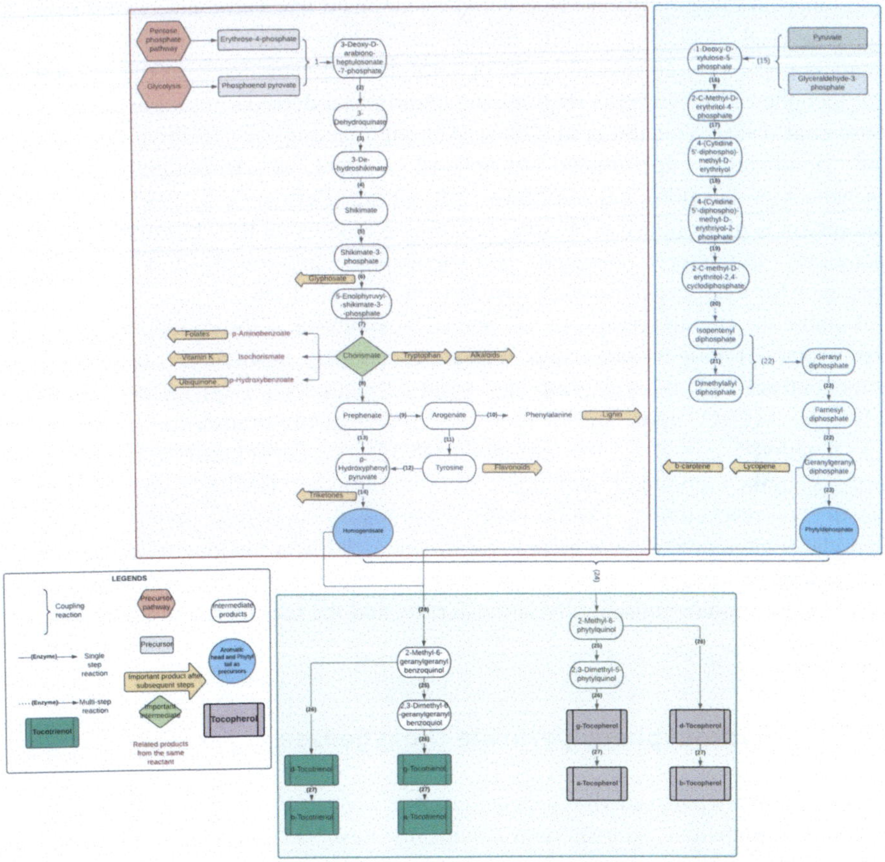

Fig. 12.2 A brief overview on the biosynthetic pathway of Tocopherols. The legend used in the figure is explained in the bottom left corner. The numbers in the reaction correspond the enzymes as follows: (1) 3-deoxy-D-arabino-heptulosonate 7-phosphate synthase; (2) 3-dehydroquinate synthase; (3) 3-dehydroquinate dehydratase; (4) shikimate dehydrogenase; (5) shikimate kinase; (6) 5-enolpyruvylshikimate 3-phosphate synthase; (7) chorismate synthase; (8) chorismate mutase; (9) prephenate aminotransferase; (10) arogenate dehydratase; (11) arogenate dehydrogenase; (12) tyrosine aminotransferase; (13) prephenate dehydrogenase; (14) p-hydroxyphenylpyruvate dioxygenase; (15) 1-deoxy-D-xylulose-5-phosphate (DOXP) synthase; (16) DOXP reductoisomerase; (17) 4-(cytidine 50-diphospho)-methyl-D-erythritol (CDP-ME) synthase; (18) CDP-ME kinase; (19) ME cyclodiphosphate synthase; (20) multistep reaction catalyzed by two reductases and two dehydratases; (21) isomerase; (22) geranylgeranyl reductase; (23) chlorophyll synthase; (24) HPT (VTE2); (25) MOBQMT (VTE3); (26) TC (VTE1); (27) g-TMT (VTE4); (28) HST

The IPP that is formed after subsequent steps is either isomerized to DMAPP or retained as the final product as seen in higher plants (Lichtenthaler 2000).

Geranylgeranyl diphosphate (GGPP) is a 20C intermediate formed by 4 molecules of IPP (Munné-Bosch and Alegre 2002) and is a precursor for tocotrienols along with other products of plastid isoprenoid metabolism like carotenoids, chlorophyll, and PDP (Ischebeck et al. 2006). Studies have shown that chlorophyll content in plants is negatively correlated by tocopherol content. This leads to the conclusion that tocopherol biosynthesis requires the chlorophyll-derived phytol as a precursor (Rise et al. 1989; Valentin et al. 2006). Although feeding of free Phytol has shown that significantly upregulated tocopherol content is *Arabidopsis* seedlings (Ischebeck et al. 2006) and Sunflower cell culture (Furuya et al. 1987), unless the free phytol is phosphorylated by two successive kinases on the chlorophyll envelop membrane into phytyl phosphate and diphosphate previously, it is not a direct precursor for tocopherol biosynthesis (Ischebeck et al. 2006).

Finally, the step to commit to tocopherol biosynthesis is the condensation for the precursors, Homogentisate, and PDP, respectively, which is catalyzed by homogentisate phytyltransferase to yield 2-methyl-6-phytyl-1,4-benzoquinone (MPBQ). MPBQ upon the action of MPBQ methyltransferase converts to 2,3-Dimethyl-5-phytyl-1,4-benzoquinol (DMPBQ). Tocopherol cyclase finally forms gamma tocopherol and delta tocopherol using DMPBQ and MPBQ as substrates, respectively. Additionally, gamma tocopherols methyltransferase methylates gamma and delta tocopherol to alpha and beta tocopherol, respectively. Throughout this process, the key regulatory enzymes are p-hydroxyphenylpyruvate dioxygenase, homogentisate phytyltransferase, and the two methyl transferases.

12.2.1 p-Hydroxyphenylpyruvate Dioxygenase

p-Hydroxyphenylpyruvate dioxygenase enzyme was isolated from plant, bacterial as well as mammalian sources and found to have a homodimer or rarely a homotetramer active region, with each of the subunit weighing 40–48 kDA (Endo et al. 1992; Lindstedt and Odelhog 1987; Roche et al. 1982; Garcia et al. 1997; Wada et al. 1975). It catalyzes a key step producing the first intermediate of the tocopherol biosynthesis, Homogentisate from p-Hydroxyphenyl pyruvate. Along with tocopherol biosynthesis, Homogentisate is also a precursor to a large variety of plastoquinons and other elements involved in photosynthetic electron transport chain, components of antioxidant system. The enzyme causes oxidative decarboxylation in the 2-oxoacid side chain of p-Hydroxyphenylpyruvate. The mechanism of this reaction involves hydroxylation of aromatic ring followed by 1,2-migration of carboxymethyl group (Jefford and Cadby 1981).

Inhibition of p-Hydroxyphenylpyruvate dioxygenase will inhibit tocopherol and plastoquinone biosynthesis, which inhibits carotenoid desaturation leading to phytoene accumulation which causes photooxidation of plastids. As a result, this enzyme is often targeted in bleaching herbicides. A study involving inactivation of Arabidopsis gene coding for p-Hydroxyphenylpyruvate dioxygenase, At1g06590, at PDS1 locus leads to inability to synthesize both tocopherol and plastoquinone (Norris et al. 1998). However, a similar inhibition of Hydroxyphenylpyruvate dioxygenase homolog in Synechocystis only led to impaired tocopherol biosynthesis and didn't affect plastoquinone biosynthesis, which indicates that this enzyme is not involved in plastoquinone biosynthesis in Synechocystis (Dahnhardt et al. 2002).

Additionally, a study involving increased expression of p-Hydroxyphenylpyruvate dioxygenase didn't result in increased tocopherol biosynthesis, which indicates that either this gene alone is not sufficient to enhance tocopherol biosynthesis or since this enzyme only produces homogentisate, an enhanced tocopherol yield must also require a mechanism to enhance homogentisate degradation which can push the flux forward (Tsegaye et al. 2002a, b). However, this result varied on a case to case basis; overexpression of p-Hydroxyphenylpyruvate dioxygenase in tobacco plant, for instance, showed no tocopherol yield enhancement in leaves, but a twofold enhancement in seeds (Falk et al. 2003), a seven-fold increase when this gene was overexpressed in Synechocystis, and a ten-fold increase when combined with overexpression of Eh-tyrA in the same species (Karunanandaa et al. 2005).

12.2.2 Homogentisate Phytyltransferase

Homogentisate phytyltransferase performs the committing step towards tocopherol biosynthesis, by condensation of Homogentisate ring and Phytyl diphosphate tail. Additionally, Homogentisate phytyltransferase also catalyzes condensation of Homogentisate with Geranylgeranyl diphosphate to form tocotrienols and Homogentisate with solanesyl diphosphate to form plastoquinone-9, respectively (Dellapenna 2005; Dellapenna and Pogson 2006). As a result, this enzyme shows a high flux coefficient (Collakova and Dellapenna 2003a). Although Homogentisate phytyltransferase uses both Phytyl diphosphate and Geranylgeranyl diphosphate as its substrates, in Arabidopsis and Synechocystis, Homogentisate phytyltransferase tends to preferentially use Phytyl diphosphate as a substrate (Karunanandaa et al. 2005; Collakova and DellaPenna 2001).

Studies on *Synechocystis sp.* PCC6803 and *Arabidopsis thaliana* have shown that homogentisate phytyltransferase is a limiting enzyme for tocopherol biosynthesis (Collakova and Dellapenna 2001; Collakova and Dellapenna 2003a; Schledz et al. 2001). Synechocystis showed a complete absence of tocopherol production in photoautotrophic environment and a four-fold decreased production in photoheterotrophic environment when its Homogentisate phytyltransferase was disrupted (Collakova and Dellapenna 2001; Savidge et al. 2002; Schledz et al. 2001). However, in all these experimental conditions, plastoquinone-9 synthesis remained unchanged, which indicates that different unrelated polyprenyltransferases are involved in tocopherol biosynthesis and plastoquinone biosynthesis in Synechocystis.

Tobacco has shown a reduction of tocopherol by up to 98% due to homogentisate phytyltransferase silencing (Abbasi et al. 2007). Additionally, a study involving HPT1 over expression led to enhancement of homogentisate phytyltransferase activity by ten-fold and consequently increasing the tocopherol content by 4.4-fold in leaves (majorly α-tocopherol and by a small extent c-tocopherol) and 1.4-fold in seeds (majorly γ-tocopherol) of Arabidopsis (Collakova and DellaPenna 2003a). Interestingly, when homogentisate phytyltransferase was co expressed with c-Tocopherol methyltransferase, the excess c-tocopherol converted to a-tocopherol as a consequence of almost complete methylation. Additionally, overexpression of these co-expressed enzymes led to enhanced tocopherol synthesis by 3.2-fold in leaves and 12-fold in seeds of *Brassica napus* (Lassner et al. 2001).

12.2.3 Tocopherol Cyclase

Tocopherol cyclase is an enzyme present in chloroplast and catalyzes the conversion of 2-methyl-6-phytyl-1,4-benzoquinone (MPBQ) or 2,3-dimethyl-6-phytyl-1,4-benzoquinone (DMPBQ) to δ-tocopherol and γ-tocopherol, respectively; this is the penultimate step of tocopherol biosynthesis (Zbierzak et al. 2010). The activity of cyclase is with regard to the chromanol ring, an oxygen containing heterocycle at the junction between the phytyl tail and homogentisate head (Sattler et al. 2003). Tocopherol cyclase was isolated, identified, and annotated for the first time from *Anabaena variabilis* (Stocker et al. 1993).

It was further noted that tocopherol cyclase originated from the locus At4g32770 of Chr 4 holding the gene VTE1, *A. thaliana,* VTE1-deficient mutants showed deficient Tocopherol cyclase activity and *E. coli* VTE1 overexpression enhanced tocopherol cyclase activity, which led to the conclusion that VTE1 encoded tocopherol cyclase (Hofius and Sonnewald 2003). Before the characterization of VTE1, sxd1, a tocopherol cyclase mutant maize, was already identified as a cause of carbohydrate accumulation in leaves due to disruption in symplastic movement of the molecules as a result of callose deposition between bundle sheath and vascular parenchyma cells (Ma et al. 2008; Russin et al. 1996; Provencher et al. 2001). However, recent studies have concluded that Arabidopsis VTE1 and Maize SXD1 along with stSXD1 from *Solanum tuberosum* are single copy orthologues coding for Tocopherol cyclase, accounting to 62% amino acid sequence identity (Hofius and Sonnewald 2003; Porfirova et al. 2002; Sattler et al. 2003; Hofius et al. 2004). It was observed that downregulation of all these genes coding for Tocopherol cyclase led to identical primary biochemical phenotypes characterized by accumulation of the substrate DMPBQ (Sattler et al. 2003). A notable difference was that, in contrast to SXD1 of Maize, the Arabidopsis VTE1 downregulation didn't show any sucrose export phenotype indicating that carbohydrates and anthocyanins didn't accumulate in Arabidopsis. SXD1 phenotype, thus, is an early evidence of tocopherols functioning as a signaling molecule in C4 plant metabolism (Sattler et al. 2003). Additionally, disruption of stSXD1 by RNAi showed defective photoassimilate export in addition to tocopherol deficiency in the C3 crop, potato (Hofius et al. 2004).

Another study conducted against Tocopherol cyclase activity and temperature as parameters concluded that at lower temperature, sugar and starch accumulation took place in Arabidopsis HPT mutant vte2 (Maeda et al. 2006, 2008). When tocopherol deficiency was induced in the same mutant line at the same cold temperature, lipid composition of endoplasmic reticulum was observed. This suggests that the disruption of photoassimilate transport at lower temperature was a result of tocopherol deficiency; thus indicating the extra plastidic function of tocopherol in endoplasmic reticulum lipid metabolism (Mène-Saffrané and DellaPenna 2010).

12.2.4 Methyltransferases

γ-tocopherol methyl transferase and 2-methyl-6-phytyl-1,4-benzoquinone methyltransferases (MPBQ MT) are the 2 major methyltransferases that determine the composition of tocopherol in the plant tissue which consequently decides the vitamin E activity, unlike the other key enzymes that primarily regulated net tocopherol flux. γ-Tocopherol methyl transferase catalyzes the conversion of γ-tocopherol and δ-tocopherol to α-tocopherol and β-tocopherol; this is the final step in the tocopherol biosynthesis pathway. MPBQ MT catalyzes the formation of 2,3-dimethyl-6-phytyl-1,4-benzoquinone by methylation of the first prenylquinone intermediate (Shintani et al. 2002). MPBQ MT also participates in the final step of plastoquinone biosynthesis by catalyzing the methylation of 2-methyl-6-solanyl-1,4-benzoquinone at its third carbon.

It was observed that the characteristic S-adenosylmethionine binding domain of γ-tocopherol methyltransferase among others was present in VTE4 of *A. thaliana* and slr0089 of *Synechocystis sp.* PCC6803; thus concluding that these genes encoded for γ-tocopherol methyltransferase (Shintani and Dellapenna 1998; Bergmüller et al. 2003). Similarly, VTE3 of *A. thaliana* and Sll0418 gene of the *Synechocystis* sp. PCC6803 were found to code for MPBQ MT based on its homology with γ-tocopherol methyl transferase. However, these VTE3 and sll0418 are not orthologous and only share an identity of 18% amino acids, unlike the orthologous genes coding for the other key enzymes. This was verified experimentally by altering its expression which led to change in tocopherol composition, with accumulation of β-tocopherol, while the net tocopherol content remains unchanged.

The partial VTE3 disruption led to higher amounts of β-tocopherol and δ-tocopherol, but consequently a lower amount of α-tocopherol and γ-tocopherol. A complete VTE3 disruption, on the other hand, caused complete accumulation of β-tocopherol and δ-tocopherol (Cheng et al. 2003). VTE3 overexpression in transgenic Soybean led to reduced amount of β-tocopherol and δ-tocopherol and a consequent increase in amount of α-tocopherol and γ-tocopherol (Van Eenennaam et al. 2003).

Further, γ-tocopherol accumulation and very small amounts of δ-tocopherol with complete absence of α-tocopherol and β-tocopherol were observed in leaves of vte4 mutant *A.* thaliana. However, the net tocopherol content remained unchanged in this mutant; this indicated a disruption of γ-tocopherol methyl transferase (Bergmüller et al. 2003; Semchuk et al. 2009). Similar results were observed in *Nicotiana tabacum* where disruption of γ-tocopherol methyl transferase led to 95% decline in α-tocopherol content in the leaves which was compensated by higher amount of γ-tocopherol (Abbasi et al. 2007). Additionally, higher conversion rates of γ-tocopherol were reported to α-tocopherol accounting to higher γ-tocopherol methyl transferase activity in *Lactuca sativa* with overexpressed *Arabidopsis* γ-tocopherol methyl transferase gene (Cho et al. 2005).

Studies involving overexpression of *B. napus* VTE4 gene in *A. thaliana* have reported a raise in the amount of α-tocopherol by 50-fold while keeping the net tocopherol amount constant (Endrigkeit et al. 2009). Similarly, overexpression of γ-tocopherol methyl transferase in soybean led to a ten-fold increase in α-tocopherol amount and a 15-fold increase in β-tocopherol amount, consequently increasing the effective vitamin E activity by five-fold. Another similar study involving simultaneous overexpression of VTE1 (which led to accumulation of Plastochromanol-8), VTE4, caused the conversion of a fraction of the accumulated Plastochromanol-8 into α-tocopherol and 5-methyl-plastochromanol-8; clearly indicating that although γ-tocopherol methyl transferase could use plastochromanol-8 as a substrate, it preferentially uses γ-tocopherol (Zbierzak et al. 2010). Finally, there have also been reports of almost complete tocopherol content in a plant converting to just α-tocopherol, resulting in 95% α-tocopherol when both VTE3 and VTE4 were expressed simultaneously in soybean (Van Eenennaam et al. 2003).

12.3 Production of Tocopherol From Cell and Organ Cultures

Callus cultures consist of an undifferentiated, proliferating mass of cells usually arising on wounds of differentiated tissues and cells. Callus can be initiated either from the source tissue synthesizing the secondary compound of interest or from other tissues like embryo. Generally, the plants which accumulate relatively high yields of specific secondary metabolites give rise to tissue cultures producing high levels of secondary metabolites and vice versa. This is because the capacity for the biosynthesis of secondary metabolites is genetically determined (Lindsey and Yeoman 1985). Callus can be used to establish single cell suspension culture by using batch and continuous fermentation techniques and can be easily scaled up for the production of biomass and metabolite content in a short period of time. Furuya et al. (1987) established the callus and cell suspension cultures from petal explants of *Carthamus tinctorius* (Safflower) cultured on MS medium supplemented with 1 ppm 2,4-D and 0.1 ppm KN and further cell lines were analyzed in terms of both growth rate and amounts of tocopherols produced. Furthermore, tocopherol production was effectively stimulated by addition of biosynthetic precursors, such as phytol administration as precursor increased the total tocopherol content by 18-fold and α-tocopherol content by some 11-fold on dry weight basis (Furuya et al. 1987). Studies have shown that media supplemented with precursors like phytol and homogentistic acid had improved the production of α-tocopherol by safflower callus culture. In recent times, the focus has shifted to analyze the conditioning factor and dissolved oxygen in media as a limiting factor in production of α-tocopherol. A batch culture system was established by Takeda et al. (1993) to study these factors. The results showed that by increasing dissolved oxygen up to 9 mg/L showed a linearly increased yield of product by the culture. Additionally, the conditioning factor released by the cells had an impact on the product yield and its presence in the early stages of batch culture was significant in product production (Takeda et al. 1993). A structured model study was designed to examine the respiratory intermediates as rate limiting factor for plant cell growth and production of secondary metabolites like tocopherol. Cell structure components and storage carbohydrates were also considered as constituting compartments. Both batch and semi-batch fermenters were set with safflower cell culture to study the growth kinetics and analyzed the products formed. Studies confirm that the production of tocopherol was limited by the intracellular concentrations of the respiratory intermediates and compounds involved in casamino acid active transport from the medium (Takeda et al. 1998).

Culture conditions were optimized for α-Tocopherol using callus and suspension cultures of sunflower (*Helianthus annuus*) as bio-factories for large-scale production; wherein MS medium supplemented with 0.5 mg/l BAP and 0.5 mg/l NAA (MS-N5B5) in combination with casaminoacids (CA) and myo-inositol (MI) both at concentration of 0.1% were effective for production of α-tocopherol. Further, addition of biosynthetic precursors homogentisic acid and phytol into culture media had enhanced the in vitro production of α-tocopherol by approximately 30%. Longer subculture intervals were also found to be more efficient for production of α-tocopherol by 78% (Caretto et al. 2004). Sunflower cell suspension culture was established in 5 L bioreactor and in shaker flask to analyze the metabolite content. In both culture mode, the amount of α-tocopherol produced was high during initial culture period of 2 days. Later on, the concentration reduced and this may be attributed to the stress condition the cells experience because of inoculation procedure. α-tocopherol production can be increased in culture on a continuous basis by subjecting the cells to different stress factors (Gala et al. 2005). However, bioactive metabolite production was high with bioreactor compared to shaker flask as the aeration was high in bioreactor, resulting in more ROS production in turn inducing antioxidant tocopherol synthesis by cells (Haas et al. 2008). Two cell lines [High yielding (HT) and low yielding (LT)] of sunflower (*Helianthus annuus* L. cv. Gloriasol) cultures were cultured in in vitro and their growth rate and production of antioxidant molecules such as α-tocopherol, ascorbate (ASC), and the tripeptide glutathione (GSH) were studied. HT cell line was reported to produce higher amount of all these antioxidant metabolites compared to LT cell line. The results obtained can be correlated to better efficiency to withstand the adverse stress effect by HT cell lines by producing antioxidants. HL cell line also produced ASC and GSH contents more than that can increase tocopherol production efficiency, while LT cells respond to the stress effect by producing ascorbate peroxidase and catalase (Caretto et al. 2002). Earlier studies have identified that two well-established sunflower cell lines, namely HL and LT, were capable of synthesizing tocopherol with varying efficiency where HL was more potent than LT (Caretto et al. 2002). In a study conducted by Fachechi et al. (2006), a new cell line was derived from HS3 from HL cell line by limited sucrose supply in medium. This Photomixotrophic cell line was compared to HL and TL with respect to various parameters such as photosynthetic activity, chlorophyll content, chloroplast number, photosynthesis-related enzymes (Rubisco), and α-tocopherol production. The lack of exogenous supply of carbon source had improved their photosynthetic ability and also showed three-fold increase in tocopherol synthesis in comparison to HL and LT. Further studies also revealed higher expression of gene coding for enzyme geranyl-geranylpyrophosphate synthase (GGPPS), which is involved in chloroplast metabolism for the biosynthesis of α-tocopherol. This clearly explains that the cells with high photosynthetic activity are always a better candidate for natural Vitamin E production (Fachechi et al. 2006).

Nisi et al. (2010) studied the effect of aeration on production of α-tocopherol using *Arabidopsis thaliana* cell suspension culture, where the culture system was maintained on MS medium supplemented with 30 g/l sucrose, 0.5 mg/l NAA, 0.05 mg/l Kinetin, and under continuous light (125 μmol photons m^{-2} s^{-1}). When the suspension culture was maintained as static culture, there was 62% increase in production of α-tocopherol as compared to culture in shaker after 16 to 24 h. The main reason for this difference in production is the lack of oxygen transfer in static cultures which acts as stress factor and impairs the cellular redox balance, and in turn, induces the production of antioxidant metabolite leading to increase in α-tocopherol content. Harish et al. (2012) also studied the effect of intermediates/precursors (tyrosine, p-hydroxyphenyl pyruvic acid, homogentisic acid (HGA), and phytol in different concentrations and combinations) supplementation on the biosynthesis of α-tocopherol using tobacco cell suspension cultures. The results showed that combination of intermediates/precursors (150 μM HGA + 100 μM phytol) was beneficial and enhanced α-tocopherol production by approximately 36%. Overall feeding the cell suspension cultures with intermediates/precursors has been found to show a positive response in increasing the α-tocopherol content. Geipel et al. (2013) studied the effect of photomixotrophic and heterotrophic conditions on callus culture of sunflower (*Helianthus annuus L.)* for production of α-tocopherol. Callus was induced using different explants like stem, hypocotyls, cotyledons, and leaves; and good callus induction was obtained from stem on MS media supplemented with 0.5 mg/l NAA and 0.5 mg/l 6-BAP. The callus so obtained was further used for tocopherol production. Photomixotrophic culture was maintained with callus on solid MS at 26 °C and high light intensity of 165 μmol m^{-2} s^{-1} with varying concentration of sucrose (30, 15 and 3 g/l), in order to enhance photosynthetic activity for a better yield of tocopherol. The results proved that culture with low concentration of sucrose (3 g/l) showed poor growth, while the callus on MS media with higher amounts of carbon source (15 and 30 g/l) expanded well, initially pale yellow and, in later stages, callus showed green pigmentation under photomixotrophic conditions. A considerable increase of around 230% of α-tocopherol production in the cells was obtained within the photomixotrophic cell culture (40 μg/g DW) compared to a heterotrophic cell culture (15 μg/g DW). The high yield of α-tocopherol in photomixotrophic culture is due to photoactive stimulation of α-tocopherol biosynthetic pathway (Geipel et al. 2013). Application of UV-C radiation to increase the production of α-tocopherol has been worked out by optimizing certain parameters like irradiation duration, distance from samples, and incubation time for grape (*Vitis vinifera* L. Öküzgözü) calli. The results have shown that α-tocopherol content increased with 10 min UV-C radiation exposure when maintained at 30 cm after 24 and 48 h of incubation time. However, effects of UV irradiation on tocopherol contents can change depending on the genotypes (Cetin 2014). *Arabidopsis* cells suspension culture was maintained in stirred tank bioreactor in order to study the effect of oxygen concentration over its ability to produce secondary metabolite. Culture was initially maintained in anoxia state without aeration for 8 h followed by reoxygenation. This varied oxygen concentration disturbs the balance of redox state of cells. In response to this, plant cells produce ROS, H_2O_2, and NO, which were observed using Confocal Laser Scanning Microscopy and fluorescent probes. Cells also trigger their antioxidant defense system by producing several molecules, both hydrophilic and lipophilic such as ascorbic acid, glutathione, and α-tocopherol in order to reach redox homeostasis. Thus, when plant cells are maintained under oxidative stress conditions, anoxia or hypoxia had resulted in significant raise in α-tocopherol in medium (Blokhina et al. 2003; Paradiso et al. 2016).

Callus obtained from petiole tissue of grape was used to establish cell suspension culture in order to study the effect of cadmium chloride on production of α-tocopherol. 1.0 mM concentration of $CdCl_2$ was optimal to enhance the production of tocopherol. When this metabolite was measured at regular interval of 2 days, there was gradual increase in its concentration until 6 days, after which its concentration reduced. Further studies concluded that when concentration was increased to 1.5 mM $CdCl_2$, an increase in exposure time drastically reduced the efficiency of culture to produce the product. This is due to the toxic effect of $CdCl_2$ affecting the cell viability. However, at lower concentration they serve as stress inducer in culture, resulting in production of ROS and antioxidant metabolites like α-tocopherol and antioxidant defense system such as catalase, superoxide dismutase, and glutathione reductase to cope up with stress and also as potent defense mechanism (Cetin et al. 2014).

Elicitation strategy has been employed to study the effect of methyl jasmonate on the cell cultures for production of α-tocopherol (Antognoni et al. 2009), where callus culture was established from *Amaranthus caudatus* and *Chenopodium quinoa* using different explants such as hypocotyls, cotyledon, and leaves. Best results were obtained with hypocotyl explants and callus was initiated best on MS medium supplemented with 2.0 mg/L 2,4-D and 0.75 mg/L Kinetin. α-tocopherol produced by these callus cultures was low when compared to greenhouse grown plant. Production of α-tocopherol improved when methyl jasmonate (MJ) was used as elicitor, but the growth hormones used in media and duration of exposure had an effect on α-tocopherol production. For *A. caudatus,* callus grown on media with BA as the cytokinin and when exposed to MJ (100 μM) for 72 h showed a good response in production of product compared to media with KIN. The increase in α-tocopherol content is associated with a proportional increase in tyrosine aminotransferase (TAT) activity, which is one of the key enzymes involved in biosynthesis of tocopherol. In case of *C. quinoa,* callus cultures did not show any elicitor response towards α-tocopherol production with MJ. This varied response to elicitor can be attributed to species-specific sensitivity to elicitors (Antognoni et al. 2009). Improved production of 49% and 66% of α-tocopherol has been reported in cell cultures of sunflower and Arabidopsis, respectively, elicited with jasmonic acid concentration of 5 mM for 72-h treatment (Gala et al. 2005). *Carthamus tinctorius* cell cultures were established and examined the effects of plant growth regulators, incubation period, and biotic and abiotic elicitors on cell growth, α-tocopherol, and pigment (red and yellow). Significant improvement in the cell growth, α-tocopherol, and pigment content was observed on MS liquid medium supplemented with 50.0 μM NAA and 2.5 μM BA on 28 days of incubation period. Addition of *T. versicolor* at 50 mg/l significantly increased the production of α-tocopherol (12.7-fold) and red pigment (4.24-fold) among other biotic elicitors used. Similarly, supplementation of 30 mg/l *T. versicolor* (7.54-fold) and 70 mg/l *Mucor* sp. (7.40-fold) significantly increased the production of yellow pigment. Among the abiotic elicitors studied, NaCl (50–70 mg/l) and $MgSO_4$ (10–30 mg/l) notably improved production of α-tocopherol (1.24-fold) and red pigment (20-fold). Increase in the yellow pigment content was found with all the abiotic elicitor treatments. These fungi can bind to receptors on plant cell, and thus trigger signal transduction leading to expression of genes that participate in α-tocopherol synthesis pathway (Chavan et al. 2011). The effect of elicitors on *Linum usitatissimum L.* (Flax) cell suspension culture over its potential in production of tocopherol production has been evaluated. Three different elicitors were used in the study, mainly the chemical derivatives of β-cyclodextrins (CD) such as hydroxypropylated-β-cyclodextrins (CDH) or methylated-β-cyclodextrins (CDM) were used as they were analogue to alkyl-derived oligosaccharides that are released by plant cells during fungal infections. The second elicitor was (Z)-3-hexenol (Hex) which is a leaf C6-volatiles produced by plants during natural attack by enemies. β-Glucans (Glu) was the third elicitor which was fungi-based oligosaccharide produced during plant infection. These elicitors' effect was tested individually as well as in combination. The studies revealed that CDM was not effective in increasing the production of tocopherol by the suspension culture when elicited individually or in combination with Glu and Hex. On the other hand, CDH proved to be an effective elicitor which could increase target metabolite production at two different concentrations, 25 mM or 50 mM, when used with other elicitor combination of Glu (174 mg/g DW) or Hex (257 mg/g DW), which proved to be excellent elicitor combination and resulted in high extracellular release and intracellular accumulation of tocopherol. Synergistic effect of these elicitors on flax cell suspension culture proved beneficial (Almagro et al. 2016). p-hydroxyphenylpyruvate dioxygenase (HPPD) obtained from *A. thaliana* (At-HPPD) was overexpressed in *H. annuus* plant cell lines. The established cell line showed enhanced antioxidant activity and resistance to Sulcotrione which is an enzyme inhibitor of p-Hydroxyphenylpyruvate dioxygenase (HPPD). Studies also revealed this to be the top ranked enzyme in biosynthetic pathway of α-tocopherol as the yield of this product was ten-fold higher than untransformed plant cells. HPPD can be considered as a crucial enzyme for metabolic engineering for production of α-tocopherol by plant cells. (Srinivasan et al. 2019).

Hairy and adventitious root cultures are extensively used for mass production of plant-derived bioactive molecules. Adventitious root culture can easily adopt to stress factors and serve as feasible method for mass culture using bioreactors for commercially important secondary metabolite. Hairy root cultures are transformed root by *Agrobacterium rhizogenes* infection and are considered to be one of the best in vitro culture methods for production of secondary metabolites. This kind of in vitro culture system is highly recommended due to its chromosomal stability and high secondary metabolite production efficiency (Murthy et al. 2008; Korde et al. 2016). *Corylus avellana L.* (hazelnut) is a good source for vitamin E, a potent antioxidant molecule used in treatment of various diseases such as Alzheimer's and cardiovascular diseases. Adventitious roots were induced from hazelnut and the cultures were established for production of tocopherol. Addition of elicitor molecules into the culture system influences/triggers the biosynthetic pathway, leading to higher metabolite yields. The hazelnut adventitious root cultures were elicited with methyl jasmonic acid before 5 days of harvest of the biomass and the cultures elicited with 100 μM/L were found to be more beneficial as they were able to produce α-tocopherol content 4 times higher than that of the unelicited cultures (Bacchetta et al. 2009). Ginseng (*Panax ginseng* CA Mayer) adventitious roots were established and studies were carried out to evaluate the effects of different light spectra, where the ginseng roots were irradiated with (red-630 nm, blue-465 nm) LED light or fluorescent lamp (FL) light. The quantitative results showed that ginseng adventitious roots irradiated with blue LED light had enhanced concentrations of α-tocopherol and β-amyrin, as well as phenolic acids compared with FL-irradiated ginseng adventitious roots (Park et al. 2013).

12.4 Metabolic Engineering Studies for Enhanced Production of Tocopherol

Plants are the excellent source for plethora of beneficial metabolites which are produced by complex multistep enzymatic reactions. Studies have proved that by identifying the key genes involved in a particular pathway and by genetic engineering, one can manipulate the metabolism of plant cells and induce them to serve as factory producing bioactive molecules, which are commercially important. Stable and upregulation of these genes can improve the efficiency of cells to enhance the yield of interested products. (Verpoorte et al. 2000; Ebrahimi and Mokhtari 2017).

Tsegaye et al. (2002a, b) produced transgenic Arabidopsis plant overexpressing the enzyme p-hydroxyphenylpyruvate dioxygenase (HPPD), which catalyzes the conversion of p-hydroxyphenylpyruvate to homogentisic acid (HGA), the aromatic precursor for the biosynthesis of vitamin E (α-tocopherol) and plastoquinone. Analysis of tocopherol from leaf and seed material revealed that increased expression levels of HPPD led to an increase of 37% tocopherol in leaf and 28% in seed when compared with control. The results showed that levels of HPPD and HGA are at least a single factor limiting the tocopherols' production in photosynthetic and non-photosynthetic plant tissues. HGA phytyltransferase (HPT) is one of the committed steps in biosynthesis of tocopherol. Studies have been conducted to evaluate whether HPT activity is limiting for tocopherol synthesis in plants; the gene encoding Arabidopsis HPT, HPT1, was constitutively overexpressed in

Arabidopsis. In leaves, overexpression of HPT1 showed a ten-fold increase in HPT-specific activity and, in turn, total tocopherol content also increased to 4.4-fold compared to wild type. In seeds, overexpression of HPT1 resulted in a four-fold increase in HPT-specific activity and 40% higher total seed tocopherol content was observed than wild type, which is primarily because of an increase in γ-tocopherol content. The pool of γ-tocopherol so obtained was almost converted to α-tocopherol by crossing HPT1 overexpressing plants with lines constitutively overexpressing γ-tocopherol methyltransferase. The seed of the lines which was overexpressed twice had a 12-fold increase in vitamin E activity relative to wild type (Collakova and DellaPenna 2003a, b). cDNAs encoding homogentisic acid geranylgeranyltransferase (HGGT), which catalyzes the committed step of tocotrienol biosynthesis, were isolated from barley, wheat, and rice. Gene was expressed in *A. thaliana* and tobacco callus under the control of cauliflower mosaic virus (CaMV) 35S promoter. Tobacco callus showed high amount of α-tocopherol, while transgenic expression of HGGT in *A. thaliana* resulted in accumulation of tocotrienols which were absent in non-transformed plants and a total of 10 to 15-fold increase of vitamin E antioxidants (tocotrienols + tocopherols) was evident. Six-fold increase in the content of tocotrienol and tocopherol was observed when the barley HGGT overexpressed in corn seeds. This provides clear evidence that HGGT-catalyzed pathway produces tocotrienols (Cahoon et al. 2003). Rippert (2004) reported the production of transgenic tobacco plants overexpressing Prephenate dehydrogenase (PDH) gene from yeast and HPPD gene of tobacco plant in order to increase Hydroxyphenylpyruvate(HPP) and homogentisic acid (HGA) influxes, the aromatic precursors for vitamin E. Thus, the co-expression of PDH and HPPD resulted in 10 to 11-fold increase in production of Vitamin E in transgenic plant compared to wild type. *P. frutescens* γ-TMT gene was expressed in soya seed which was considered to be one of the limiting factors for the production of α-tocopherol. Somatic embryos derived from the cotyledon of soya seeds were bombarded with transgene construct for expression of γ-TMT gene under the control of seed-specific promoter for seed-specific expression of γ-TMT gene. Transgenic soyabean plants, so obtained, showed a higher yield of α-tocopherol when compared to non-transgenic plants (Tavva et al. 2006). Similar kind of studies was performed to express tomato γ-TMT gene in canola cultivar for better yield of α-tocopherol through *Agrobacterium tumefaciens*-mediated transformation (Babaei and Hosseini 2019).

Transformation studies were initiated for stable expression of γ-TMT (-γ-tocopherol methyltransferase) that catalyzes the conversion of γ-tocopherol into α-tocopherol in *Perilla frutescens Britt* with *Agrobacterium tumefaciens*. The gene was cloned under the control of seed-specific vicillin promoter, and the confirmation of the transgene was performed through molecular characterization. Regenerated plants showed an enhanced yield of α-tocopherol and also high ratio of α-tocopherol: γ-tocopherol (α-tocopherol and γ-tocopherol contents ranged from 0.9–11.4 mg/100 g and 199.4–272.8 mg/100 g, respectively, in untransformed plants to α-tocopherol and γ-tocopherol contents were 153.2–200.4 mg/100 g, and 37.1–59.6 mg/100 g, respectively, in transformed plants). This conversion was observed because of the overexpression of γ-tocopherolmethyl transferase (Lee et al. 2007). Similar overexpression studies of γ-tocopherolmethyl transferase gene driven by CaMV 35S promoter for enhancement of α-tocopherol have been reported in *Perilla frutescens* (Ghimire et al. 2011). Transgenic approach was applied for increasing the content of α-tocopherol in tobacco where constitutive overexpression of the genes [Arabidopsis homogentisate phytyltransferase (HPT) and tocopherol cyclase (TC)] involved in α-tocopherol biosynthesis has been expressed in tobacco through Agrobacterium-mediated transformation. The confirmation of the transgenes was performed through molecular analyzes of PCR, RT-PCR, and Southern hybridization and the α-tocopherol content in transgenic plants expressing HPT and TC increased by 5.5 and 4.1, respectively, over the wild type. The activities of these 2 genes were found to be beneficial in increasing the levels of α-tocopherol in tobacco plants (Harish et al. 2012). Genetic engineering of chloroplast was employed to improve the tocopherol production ability in tobacco and lettuce plants. Transplastomic tobacco and lettuce with Toc cyclase (TC) or γ-Toc methyltransferase (γ-TMT) gene and the TC plus γ-TMT genes in chloroplast genome were obtained for studies. In pTTC-TMT plants, total Tocopherol content was higher than wild plants and majorly composed of α-tocopherol. In pTTC plants, there was significant increase in tocopherol content and mostly comprised of γ-tocopherol isoform. However, pTMT plants reported a decrease in γ-tocopherol, but total phenol content and α-tocopherol were low compared to wild type (Yabuta et al. 2012).

2-methyl-6-phytyl-1, 4-benzoquinol methyltransferase (MPBQ-MT) is an important enzyme catalyzing methylation step in both α/γ-tocopherol and plastoquinone biosynthetic pathways. Gene-encoding MPBQ-MT was isolated from lettuce (*Lactuca sativa*) and overexpressed in lettuce which brought a significant enhancement of α- and γ-tocopherol contents and also reduction in phylloquinone content was evident. This suggested a competition for a common substrate PDP between the two biosynthetic pathways. Further increase in the concentration of tocopherol and plastoquinone levels by overexpression of LsMT also showed improvement of plants' tolerance capacity and photosynthesis under light stress (Tang et al. 2016). Expression of homogentisate phytyltransferase (HPT) and tocopherolcyclase (TC) isolated from *Arabidopsis thaliana* was transiently expressed in *Nicotiana benthamiana* by Agrobacterium-mediated method for industrial production of α-tocopherol. The expression of the gene was done individually and in combination. Molecular approaches were used to confirm the presence and expression of these transgenes. HPLC analysis was used to quantify the tocopherol content in leaves where 4.2, 5.9, and 11.3-fold increase was observed in transgenic leaves expressing homogentisate phytyltransferase (HPT), tocopherolcyclase (TC), and HPT + TC, respectively. Thus, combined expression of HPT and TC can greatly influence the plant's efficiency to produce the product (Sathish et al. 2018).

12.5 Extraction and Detection Techniques of Tocopherols From Plant Samples

Tocopherols and Tocotrienols can be extracted from plant samples by soxhlet method, cold pressing, or Bligh-Dyer method (Bligh and Dyer 1959), which uses a combination of solvents (Chloroform, methanol, and water) with a vigorous homogenization process. Though the oil content from Faveleira seeds (*Cnidoscolus quercifolius*) was reported to be high in soxhlet extraction, the tocopherol content was found to be highest in cold-pressed method (21.56 mg/100 g) followed by Bligh-Dyer method (9.01 mg/100 g) and completely absent in soxhlet extraction due to the exposure and drying of samples at high temperature for longer durations (Riberio et al. 2019). Ching et al. studied α-tocopherol content in sixty-two edible plants by subjecting them to saponification and extracting the sample by Konings method (Konings et al. 1996). The common solvents used for the tocopherol extraction are ethanol for cereal grains (Panfili et al. 2003), hexane for legumes (Kalogeropoulos et al. 2010), methanol: dichloromethane (1:2) for leafy vegetables (Cruz and Casal 2013), acetone for wheat (Ziegler et al. 2015), acetonitrile for barley (Tsochatzis and Tzimou-Tsitouridou 2015), and hexane: isopropanol (3:2) for lentils (Zhang et al. 2014). Some of the common extraction methods used for the extraction of tocopherols from plant samples have been mentioned in Table 12.1.

After the extraction of tocopherols with suitable techniques, its separation and confirmation can be performed by using any one of the several commonly used analytical techniques like Thin Layer Chromatography (TLC), High Performance Liquid Chromatography (HPLC), Gas chromatography (GC-MS), Capillary electrochromatography (CEC), Supercritical fluid chromatography (SFC), and Nano-liquid chromatography (Nano-LC). The diluted samples are subjected to Liquid Chromatography coupled with detectors like photodiodes, ultraviolet, fluorescence, evaporative light scattering, flame ionization, charged aerosol detectors, and mass spectrometer (Saini and Keum 2016). The commonly used liquid chromatography is a Reverse Phase High Performance Liquid Chromatography (RP-HPLC) which uses a C-18 column. The mobile phase may be of 100% methanol (Riberio et al. 2019), Methanol: water (94:6) (Ching and Mohamed 2001) or a mixture of 85% methanol, 7.5% ethanol, and 7.5% acetonitrile (Abu-Fayyad and Nazzal 2017). The solvent is eluted in an isocratic manner with the flow rate of 1–1.5 mL/min and the tocopherol is detected at 292 nm. Column chromatography packed with silica gel and a mobile phase with a gradient elution of n-hexane and increasing amount of ethyl acetate (0–12%) were used to extract Vitamin-E isomers from oil palm (*Elaeis guineensis*). It was noted that at 1% ethyl acetate, α-T3 fraction was eluted followed by γ-T3 at 2% and a mixture of δ and γ-T3 at 3%. As the ethyl acetate concentration increased, pure fractions of δ-T3 were eluted. These fractions were collected separately and further confirmed by TLC evaluation under UV light at 254 and 366 nm by spraying 4-anisaldehyde reagent followed by HPLC and H-NMR studies (Abu-Fayyad and Nazzal 2017). Apart from these commonly used methods, α-tocopherol was determined in palm olein using sodium chloride window utilization by Fourier transform-infrared spectroscopy (FT-IR) (Che Man et al. 2005).

Table 12.1 Common extraction methods for tocopherols from plant samples

Extraction methods	Advantages	Disadvantages
Solvent extraction	• Simple and widely used method • Vortexing, sonication improves the quality • Saponification, acid hydrolysis reduces the impurities	• Suitable only for hydrophobic compound extraction • Longer duration with the usage of more toxic solvents
Ultrasonic-assisted extraction	• Low cost, most effective, fast, and alternative method • More samples can be analyzed simultaneously • Ultrasound helps in solvent penetration into the cell • Low temperature facilitates extraction of heat-sensitive compounds • Less sample and solvent requirement	• Efficiency depends on nature of plant matrix • High amount of heat generated, hence cooling of vessel is required • No uniformity in ultrasound energy distribution
Matrix solid-phase dispersion extraction	• Useful in selection and extraction of single compound from a mixture • Extraction from solid, semi-solid, and viscous samples are possible • Small amount of solvent is used	• Laborious and time-consuming
Supercritical fluid extraction	• Offers selective extractions • Physicochemical properties of supercritical fluid (CO_2) can be manipulated • Extraction prevents damage to thermolabile compounds	• High cost • Due to presence of water in sample, blockage of system occurs
Pressurized liquid extraction	• High temperature and high pressure for enhanced solubility and extraction • Samples required in lesser quantity • Harmful organic solvents are avoided	• Costlier process • Not suitable for heat-sensitive compounds

Source: Saini and Keum (2016)

Tocopherols, tocotrienols, and plastochromanol-8 estimated from the seed oil of canola, flax, sunflower, and soybean (Ahmed et al. 2005) and α-tocopherol from thirteen vegetable oils (Silva et al. 2009) were quantified using ZnSe attenuation by FT-IR. Raman microscopy was also used to map vitamin E in biological samples (Beattie et al. 2007). Molecular Imprinting Polymer (MIP), a detection technique based on the fluorescence nano-sensing principle using CdSe or ZnS detects the presence of tocols in vegetable oils and rice. The intensity decreases as the concentration increases (Liu et al. 2012). Tocopherols in edible oils were characterized and its quality over storage was monitored by using synchronous fluorescence spectroscopy (Sikorska et al. 2008). Tocopherol content present in various plants are tabulated and represented in Table 12.2.

Table 12.2 Tocopherol contents in different plants

Sl. No.	Scientific name of the plant	Common name	α-Tocopherol	γ-Tocopherol	δ-Tocopherol
1	*Triticum aestivum* L.	Wheat	149.40	–	–
2	*Corylus avellana* L.	Hazelnut	47.20	–	–
3	*Helianthus annuus* L.	Sunflower	41.08	–	–
4	*Prunus amygdalus* (mill.)	Almond	39.20	–	–
5	*Oryza sativa* L.	Rice bran	32.30	–	–
6	*Vitis vinifera* L.	Grapeseed	28.80	–	–
7	*Arachis hypogaea* L.	Peanut	15.69	15.95	1.37
8	*Zea mays* L. and *Brassica napus* L.	Corn and canola	14.82	35.37	1.28
9	*Olea europaea* L.	Olive	14.35	0.83	–
10	*Glycine max* (L.) Merr	Soybean	8.18	64.26	21.30
11	*Capsicum frutescens* L.	Chili	38.14	3.41	–
12	*Capsicum annuum* L.	Paprika	29.10	3.54	0.25
13	*Origanum vulgare* L.	Oregano	18.23	24.42	0.92
14	*Ocimum basilicum* L.	Basil	10.70	0.77	–
15	*Petroselinum crispum* (mill.) fuss	Parsley	8.96	1.53	–
16	*Curcuma longa* L.	Turmeric	4.43	0.72	–
17	*Cuminum cyminum* L.	Cumin seed	3.33	–	–
18	*Brassica nigra L.*	Mustard seed	5.07	19.82	0.81
19	*Cinnamomum verum* L.	Cinnamon	3.32	10.44	0.26
20	*Papaver somniferum* L.	Poppy seed	1.77	8.82	0.23
21	*Pistacia vera* L.	Nuts, pistachio	2.86	20.41	0.80
22	*Juglans regia L.*	Walnuts	2.08	28.78	1.51
23	*Daucus carota subsp. Sativus*	Carrot	5.45	–	–
24	*Taraxacum officinale L.*	Dandelion	3.44	–	–
25	*Brassica rapa var. rapa*. L.	Turnip	2.86	–	–
26	*Coriandrum sativum* L.	Coriander (cilantro)	2.50	–	–
27	*Persea americana* Mill.	Avocados	2.66	–	–
28	*Actinidia deliciosa* L.	Kiwifruit	1.46	–	–

(continued)

Table 12.2 (continued)

Sl. No.	Scientific name of the plant	Common name	α-Tocopherol	γ-Tocopherol	δ-Tocopherol
29	*Vaccinium macrocarpon* L.	Cranberries	1.32	–	–
30	*Cucurbita pepo* L.	Pumpkin and squash	2.18	35.10	0.44
31	*Linum Usitatissimum* L.	Flaxseed	3.31	19.95	0.35
32	*Spirulina* sp.	Seaweed, spirulina	5.00	–	–

Source: Saini and Keum (2016)

12.6 Biological Activity of Tocopherols

The first observation on the biological activity was related to reproduction. Many groups have done extensive research on this activity using fetal resorption assay in female rats and Horwitt studied the effect of vitamin E in human male (Horwitt et al. 1956; Horwitt 1960).

The main activity of tocopherol in the body is to act as a lipid-soluble antioxidant. It is a chain breaking antioxidant and inhibits the production of reactive oxygen species that occur during the oxidation of fats and also the propagation of other free radical reactions, thereby protecting the cellular and subcellular membranes and lipoproteins (Burton et al. 1983a, b). α-Tocopherol which is incorporated to the membrane helps to stabilize the lipoprotein structure, thereby stabilizing the membrane integrity. Antioxidant property of tocopherol can protect the lipids and lipoprotein, thereby increasing the orderliness and, in turn, stabilizing the membranes. Apart from the stabilization of the membrane, it can also promote the high levels of plasma membrane repair as shown in the cultured monocytes (Howard et al. 2011; Szczeklik et al. 1985). This is achieved by the prevention of generation of oxidized phospholipids in the membrane. It is also noted that the presence of α-tocopherol is relatively high in the membranes of mitochondria and endoplasmic reticulum of heart and lungs where the formation of free radicals is more compared to other tissue types. Though the ROS scavenging capacity of both alpha and gamma tocopherols is the same, majority of the ROS scavenging in the tissue is done by the alpha tocopherol compared to the gamma tocopherol because of the higher concentration of the former (Huang et al. 2013; Nishio et al. 2013; Yoshida et al. 2007). Unlike alpha tocopherols, gamma tocopherols react with the reactive nitrogen species also (Patel et al. 2007). Antioxidant properties of tocopherols provide protection from developing Alzheimer's disease. It reduces the oxidative stress and blocks the production of hydrogen peroxide, which in turn prevents the oxidation of protein, and β-amyloid protein-induced cytotoxicity-mediated neuronal death is prevented (Mangialasche et al. 2010; Sano et al. 1997). To act as an antioxidant, it will donate a proton [H^+] to the lipid peroxyl radical to convert it into lipid hydroperoxide and the tocopherol will be converted to tocopheroxyl radical as a comparatively stable compound. Being a relatively more stable and less reactive species, tocopheroxyl radical will now stop the continuation of the chain reaction and prevent the oxidation of other molecules in the cell or membrane. This radical may get converted back to tocopherol by the help of coenzyme Q or ascorbate. In the cell, the relative antioxidant activity of α-tocopherol is the highest followed by β-tocopherol, γ-tocopherol, and the least antioxidant activity is observed in δ-tocopherol, respectively (Burton and Ingold 1981).

While all the free radical mechanisms have been studied in vitro on tissue samples and cell cultures, there is still an existing argument over the existence of antioxidant activity of tocoperols in vivo (Azzi and Stocker 2000; Munteanu et al. 2004; Zingg and Azzi 2004). Additionally, some clinical trials have also concluded tocopherols to be ineffective against any disorders due to oxidative stress such as cancer and cardiovascular disorders, contrary to the previously assumed bioactivity based on its in vitro antioxidant activity (Genkinger et al. 2004; Lonn et al. 2005; Poston et al. 2006). There are also several reports and clinical studies showing toxicity of tocopherols on human subjects (Vivekananthan et al. 2003; Miller III et al. 2005).

Many cardiovascular complications can be reduced by increasing the nitric oxide synthase activity with the use of γ-tocopherol along with reduction of platelet aggregation and cholesterol (Ricciarelli et al. 1999). Additionally, supplementation of Vitamin E is attributed to the reduced risk of cardiovascular diseases in some of the epidemiological studies (Rimm et al. 1993; Stampfer et al. 1993). A recent study from Boston, USA, shows the level of tocopherol in the body and the risk of tuberculosis infection in a high-risk group based on household contact is inversely proportional (Aibana et al. 2018). Tocopherols are reported to have stimulation activity of wild-type p53; downregulation of mutated p53 activities. α-tocopherol is known to inhibit protein kinase C and collagenase, thereby inhibiting cancer cell growth. By preventing the accumulation of proteins and hence reducing the age-related lens opacification, tocopherols can act as cataract protective agent. (Behl et al. 1992; Goodman and Mattson 1994).

In physiological concentration, alpha tocopherol can inhibit PKC (Protein Kinase C) activity in vascular smooth muscle cells, and as a consequence, inhibit the proliferation of smooth muscle cell (Boscoboinik et al. 1991). The supplementation of tocopherol can decrease monocyte superoxide production by downregulating the PKC pathway by a dephosphorylation that inhibits PKC. It can also reduce expression of ICAM-1 and VCAM-1, the adhesion molecules (Azzi and Stocker 2000). Additionally, α-tocopherol has also shown inhibitory activity against various inflammatory signals, molecules, and cytokines such as IL-1β, IL-8, monocyte chemoattractant protein-1, and platelet aggregation, among others (Brigelius-Flohé et al. 2002).

It is interesting to note that although β-tocopherols show in vitro antioxidant activity similar to alpha tocopherols, they either do not show any of the consequent bioactivity mentioned previously or its intensity is negligible. As a result, there are very few reports and clinical studies held on the bioactivity of beta tocopherols. This could possibly further indicate the independence of antioxidant activity from other studied bioactivities (Brigelius-Flohe 2006). Unlike the alpha and beta tocopherols, gamma tocopherols lack a methyl group on the chroman ring which results in two changes in its antioxidant activity. First, the antioxidant activity itself is lower than the former. Second, with reactive nitrogen species, gamma tocopherol gets nitrated to form 5-nitro-gamma tocopherol (Kamal-Eldin and Appelqvist 1996; Christen et al. 1997; Hoglen et al. 1997). As a consequence of its activity against ROS and RNS, gamma tocopherol has been tested for activity against oxidative inflammatory disorders, predominantly cancer and neurodegenerative diseases. It was observed that gamma tocopherol downregulated prostaglandin E2 production by inhibition of Cyclo-oxygenase 2 (COX2) activity (Jiang et al. 2000). Various studies have shown that anticancer activity of gamma Tocopherol is relatively more effective than other classes of tocopherols. The most responsive cancer cell lines for tocopherols are prostate (LNCaP and DU-145) and colon cancer (CaCo2) cell lines by induction of apoptosis, downregulation of Cyclin D1 and cyclin E1, inhibition of p21, and upregulation of receptor gamma which is activated during proliferation (Gysin et al. 2002). Delta tocopherol is by far one of the least studied tocopherols due to its relatively lesser bioactivity. However, it has been reported that delta tocopherol is the most toxic of all classes of tocopherols and is reported to induce apoptosis in short-term exposure, but induce resistance to cytotoxicity with long-term exposure (McCormick and Parker 2004).

12.7 Commercial Applications of Tocopherol

A good number of commercial applications have been attributed for tocopherols; where tocopherol as a potent antioxidant is incorporated into polymers such as-polylactic acid (PLA) to produce ecofriendly biodegradable packing films. This serves as an excellent packaging material with better mechanical properties such as stiffness and transparency. Producers and consumers also demand for natural anti-oxidant, instead of artificial antioxidant. Such packing films are replacing plastic to pack and preserve food products (Marcos et al. 2014). α–Tocopherol (vitamin E) when incorporated in chitosan film has improved its antioxidant property and as good barrier of UV rays. This modified chitosan film preserves the quality of food packed and extends its shelf life (Martins et al. 2012). Advanced packaging materials are used to preserve the quality and prolong the shelf life of food. Studies have shown that supercritical carbon dioxide has been used to impregnate α-tocopherol in single and multilayer of polyethylene terephthalate (PET)/polypropylene (PP) films used in packing food. This active film gradually releases the antioxidant towards food surface (Franco et al. 2019).

Innovative inulin (INU)–vitamin E succinate (VITE) bioconjugates (INVITE) was synthesized and characterized. Amphiphilic INU-based polymers self-assemble to form nano-micelle system. This can serve as potential drug delivery system (DDS) to fight against infection and inflammation in urinary tract. Clinical studies have revealed that α-tocopherol can greatly benefit the patients suffering from chronic diseases such as chronic kidney diseases (CKD), hemodialysis (HD) patients, focal segmental glomerulosclerosis (FSGS), and nephropathy by reducing oxidative stress (Mandracchia et al. 2014). Three-dimensional (3D) hydrogels-based scaffolds of chitosan and alginate were prepared with α–tocopherol (vitamin E). The developed hydrogel was used to test its efficiency in wound healing in rat model. Results revealed that the hydrogel had exhibited greater wound closure percentage and can be used as dressing to threat skin injuries (Ehteramia et al. 2019). Topical application of α-tocopherol 24 h before exposure to UV radiation in hairless mice had reduced cutaneous oxidative damaged caused by UV radiation and also unregulated network of enzymatic and nonenzymatic antioxidants (Lopez-Torres et al. 1998).

Foliar applications of alpha-tocopherol to *Vigna radiate* cultivar had significantly enhanced the accumulation of chlorophyll, proline, and total phenolic contents in fresh green pods (Sadiq et al. 2017). Nanotechnology provides new dimension in the field of therapeutics. Nanoemulsion with bioactive compounds Astaxanthin-alpha tocopherol was found to be very effective in fast wound healing with broad antimicrobial activity and anticancer properties, which was proved with different cancer cell lines (CT26, HeLa, Panc1, and T24 cells) (Karuppusamy et al. 2018).

RNA interference is one of the powerful molecular tools to specifically knockout the gene expression. α–Tocopherol (vitamin E) was used as carrier to deliver and knock out the expression of targeting apolipoprotein B (apoB) in liver. Toc-siRNA reduces the mRNA of apoB. This confirms that α–tocopherol (vitamin E) is effective and safe carrier of siRNA in gene silencing (Nishina et al. 2008).

12.8 Conclusions and Future Prospects

Plants are the excellent source for plethora of beneficial metabolites which are produced by complex multistep enzymatic reactions. Studies have proved that identifying genetically engineered key genes involved in a particular pathway facilitates the manipulation of plant cell metabolism to serve as a factory producing bioactive molecules of commercially important. Stable upregulation of these genes can improve the efficiency of cells to enhance the yield of the products of interest. In addition, some of the plants are seasonal and geographically restricted. Due to all these factors, conventional methods of isolation and purification of secondary metabolites from plants are tedious. Plant cell culture is one of the most recommended sustainable production systems for metabolites from plants. By choosing the suitable bioreactor and optimizing the in vitro culture condition, one can easily eliminate the limiting factors such as soil or climatic conditions and can easily scale up to extract the interested metabolites from plant cells. Furthermore, by using genetic engineering techniques, one can upregulate the enzymes involved in the production of α–tocopherol for best yield. Applying metabolomics and metabolic engineering can further add new dimension for mass production of α–tocopherol.

To meet the growing demand for tocopherol and for mass production, different strategies can be adopted. Conventional methods can be replaced with modern biotechnological approaches like plant cell and organ culture, which is free from climatic and geographical conditions. Future potential is towards screening of more plants for interested tocopherol and to isolate high yielding cell lines for commercial production. Further research in optimizing the crucial physical and chemical factors can enhance their yield by in vitro production. Another effective approach for better productivity of SM is by elicitation. Many secondary metabolites are triggered by both biotic and abiotic elicitors, and hence, research in this field to determine the ideal combination of elicitors can remarkably enhance the yield of SM. Transformed hairy root culture is most reliable method for easy genetic manipulation for steady and stable production; when coupled to bioreactor, these can serve as successful alternate approach.

Bioreactor is the key for commercial production of metabolites, and hence, there is need for designing bioreactor and optimizing its operational conditions which can enhance the yield of metabolite as there is limited bioreactor model currently available for tocopherol production. Furthermore, in-depth knowledge of complex metabolic pathway and their regulatory mechanism has paved a new way to metabolic engineering, leading to overexpression of crucial regulatory gene for enhanced yield of product. Metabolic engineering aims at channelizing the carbon source to interested product. This can be achieved by specific mutation in catalytic activities or by genetic manipulation of regulatory protein genes. Another emerging prospective is application of nanotechnology which has been proved in certain plant cell culture; when exposed to various nanoparticles, their efficiency in production of SM has drastically improved.

References

Abbasi AR, Hajirezaei M, Hofius D, Sonnewald U, Voll LM (2007) Specific roles of α- and γ-tocopherol in abiotic stress responses of transgenic tobacco. Plant Physiol 143:1720–1738

Abu-Fayyad A, Nazzal S (2017) Extraction of vitamin E isomers from palm oil: methodology, characterization, and *in vitro* anti-tumor activity. J Am Oil Chem Soc 94:1209–1217. https://doi.org/10.1007/s11746-017-3025-8

Ahmed MK, Daun JK, Przybylski R (2005) FT-IR based methodology for quantitation of total tocopherols, tocotrienols and plastochromanol-8 in vegetable oils. J Food Compos Anal 18: 359–364. https://doi.org/10.1016/j.jfca.2003.12.008

Aibana O, Franke MF, Huang CC, Galea JT, Calderon R, Zhang Z, Becerra MC, Smith ER, Contreras C, Yataco R, Lecca L, Murray MB (2018) Vitamin E status is inversely associated with risk of incident tuberculosis disease among household contacts. J Nutr 148:56–62

Ajjawi I, Shintani D (2004) Engineered plants with elevated vitamin E: a nutraceutical success story. Trends Biotechnol 22:103–104. https://doi.org/10.1016/j.tibtech.2004.01.001

Almagro L, García-Pérez P, Belchí-Navarro S, Sánchez-Pujante PJ, Pedreño MA (2016) New strategies for the use of *Linum usitatissimum* cell factories for the production of bioactive compounds. Plant Physiol Biochem 99:73–78. https://doi.org/10.1016/j.plaphy.2015.12.009

Antognoni F, Faudale M, Poli F, Biondi S (2009) Methyl jasmonate differentially affects tocopherol content and tyrosine amino transferase activity in cultured cells of *Amaranthus caudatus* and *Chenopodium quinoa*. Plant Biol 11(2):161–169

Azzi A, Stocker A (2000) Vitamin E: non-antioxidant roles. Prog Lipid Res 39:231–255

Babaei M, Hosseini R (2019) Increasing vitamin E content of canola (Brassica napus L.) by transferring γ-tmt gene. Iran J Genet Plant Breed 8(1):55–63

Babura SR, Abdullah SNA, Khaza'ai H (2017) Advances in genetic improvement for tocotrienol production: a review. J Nutr Sci Vitaminol 63:215–221. https://doi.org/10.3177/jnsv.63.215

Bacchetta L, Aramini M, Bernardini C, Sivakumar LG (2009) High-tech production of bioactive α-tocopherol from *Corylus avellana* adventitious roots by bioreactor culture. Acta Hortic 845: 713–716. https://doi.org/10.17660/ActaHortic.2009.845.112

Beattie JR, Maguire C, Gilchrist S, Barrett LJ, Cross CE, Possmayer F, Ennis M, Elborn JS, Curry WJ, McGarvey JJ, Schock BC (2007) The use of Raman microscopy to determine and localize vitamin E in biological samples. FASEB J 21:766–776. https://doi.org/10.1096/fj.06-7028com

Behl C, Davis J, Cole GM, Schubert D (1992) Vitamin E protects nerve cells from amyloid beta protein toxicity. Biochem Biophys Res Commun 186:944–950

Bergmüller E, Porfirova S, Dörmann P (2003) Characterization of an Arabidopsis mutant deficient in γ-tocopherol methyltransferase. Plant Mol Biol 52:1181–1190

Bligh EG, Dyer WJ (1959) A rapid method of total lipid extraction and purification. Can J Biochem Physiol 37(1):911–917

Blokhina O, Virolainen E, Fagerstedt KV (2003) Antioxidants, oxidative damage and oxygen deprivation stress: a review. Ann Bot 91(2):179–194

Boscoboinik D, Szewczyk A, Azzi A (1991) α-Tocopherol (vitamin E) regulates vascular smooth muscle cell proliferation and protein kinase C activity. Arch Biochem Biophys 286:264–269

Brigelius-Flohe R (2006) Bioactivity of vitamin E. Nutr Res Rev 19:174–186

Brigelius-Flohé R, Kelly FJ, Salonen J, Neuzil J, Zingg J-M, Azzi A (2002) The European perspective on vitamin E: current knowledge and future research. Am J Clin Nutr 76:703–716

Burton GW, Ingold KU (1981) Autoxidation of biological molecules. 1. The antioxidant activity of vitamin E and related chain breaking phenolic antioxidants in vitro. J Am Chem Soc 103:6472–6477

Burton GW, Cheeseman KH, Doba T, Ingold KU, Slater TF (1983a) Vitamin E as an antioxidant in vitro and in vivo. CIBA Found Symp 101:4–18

Burton GW, Joyce A, Ingold KU (1983b) Is vitamin E the only lipid-soluble, chain-breaking antioxidant in human blood plasma and erythrocyte membranes? Arch Biochem Biophys 221: 281–290

Cahoon EB, Hall SE, Ripp KG, Ganzke TS, Hitz WD, Coughlan SJ (2003) Metabolic redesign of vitamin E biosynthesis in plants for tocotrienol production and increased antioxidant content. Nat Biotechnol 21:1082–1087. https://doi.org/10.1038/nbt853

Caretto S, Bray Speth E, Fachechi C, Gala R, Zacheo G, Giovinazzo G (2004) Enhancement of vitamin E production in sunflower cell cultures. Plant Cell Rep 23(3):174–179. https://doi.org/10.1007/s00299-004-0799-6

Caretto S, Nisi R, Paradiso A, de Gara L (2010) Tocopherol production in plant cell cultures. Mol Nutr Food Res 54:726–730. https://doi.org/10.1002/mnfr.200900397

Caretto S, Paradiso A, D'Amico L, De Gara L (2002) Ascorbate and glutathione metabolism in two sunflower cell lines of differing α-tocopherol biosynthetic capability. Plant Physiol Biochem 40: 509–513. https://doi.org/10.1016/S0981-9428(02)01419-5

Cetin E, Babalik Z, Hallac-Turk F, Gokturk-Baydar N (2014) The effects of cadmium chloride on secondary metabolite production in *Vitis vinifera* cv. cell suspension cultures. Biol Res 47(1):47. https://doi.org/10.1186/0717-6287-47-47

Cetin ES (2014) Induction of secondary metabolite production by UV-C radiation in *Vitis vinifera* L. Öküzgözü callus cultures. Biol Res 47:37

Chavan SP, Lokhande VH, Nitnaware KM et al (2011) Influence of growth regulators and elicitors on cell growth and α-tocopherol and pigment productions in cell cultures of *Carthamus tinctorius* L. Appl Microbiol Biotechnol 89:1701–1707. https://doi.org/10.1007/s00253-010-3014-4

Che Man YB, Ammawath W, Mirghani MES (2005) Determining α-tocopherol in refined bleached and deodorized palm olein by Fourier transform infrared spectroscopy. Food Chem 90:323–327. https://doi.org/10.1016/j.foodchem.2004.05.059

Cheng Z, Sattler S, Maeda H, Sakuragi Y, Bryant DA, DellaPenna D (2003) Highly divergent methyltransferases catalyze a conserved reaction in tocopherol and plastoquinone synthesis in cyanobacteria and photosynthetic eukaryotes. Plant Cell 15:2343–2356

Ching LS, Mohamed S (2001) Alpha-tocopherol content in 62 edible tropical plants. J Agric Food Chem 49:3101–3105. https://doi.org/10.1021/jf000891u

Cho EA, Lee CA, Kim YS, Back SH, de los Reyes BG, Yun SJ (2005) Expression of γ-tocopherol methyltransferase transgene improves tocopherol composition in lettuce (*Latuca sativa* L.). Mol Cell 19:16–22

Christen S, Woodall AA, Shigenaga MK, Southwell-Keely PT, Duncan MW, Ames BN (1997) g-Tocopherol traps mutagenic electrophiles such as NO(X) and complements a-tocopherol: physiological implications. Proc Natl Acad Sci U S A 94:3217–3222

Collakova E, DellaPenna D (2001) Isolation and functional analysis of homogentisate phytyltransferase from Synechocystis sp. PCC 6803 and Arabidopsis. Plant Physiol 127: 1113–1124

Collakova E, DellaPenna D (2003a) Homogentisate phytyltransferase activity is limiting for tocopherol biosynthesis in *Arabidopsis thaliana*. Plant Physiol 131:632–642. https://doi.org/10.1104/pp.015222

Collakova E, DellaPenna D (2003b) The role of homogentisate phytyltransferase and other tocopherol pathway enzymes in the regulation of tocopherol synthesis during abiotic stress. Plant Physiol 133:930–940. https://doi.org/10.1104/pp.103.026138

Cruz R, Casal S (2013) Validation of a fast and accurate chromatographic method for detailed quantification of vitamin E in green leafy vegetables. Food Chem 141:1175–1180. https://doi.org/10.1016/j.foodchem.2013.03.099

Dahnhardt D, Falk J, Appel J, van der Kooij TA, Schulz-Friedrich R, Krupinska K (2002) The hydroxyphenylpyruvate dioxygenase from Synechocystis sp. PCC 6803 is not required for plastoquinone biosynthesis. FEBS Lett 523:177–181

DellaPenna D (2005) Progress in the dissection and manipulation of vitamin E synthesis. Trends Plant Sci 10:574–579

DellaPenna D, Pogson BJ (2006) Vitamin synthesis in plants: tocopherols and carotenoids. Annu Rev Plant Biol 57:711–738

Ebrahimi M, Mokhtari A (2017) Engineering of secondary metabolites in tissue and cell culture of medicinal plants: an alternative to produce beneficial compounds using bioreactor technologies. In: Abdullah SNA et al (eds) Crop improvement. Springer, New York, pp 137–167. https://doi.org/10.1007/978-3-319-65079-1_7

Ehrenbergerová J, Belcrediová N, Prýma J, Vaculová K, Newman CW (2006) Effect of cultivar, year grown, and cropping system on the content of tocopherols and tocotrienols in grains of hulled and hulless barley. Plant Foods Hum Nutr 61:145–150. https://doi.org/10.1007/s11130-006-0024-6

Ehteramia A, Salehib M, Farzamfard S, Samadiane H, Vaezd A, Ghorbanif S, Aid J, Sahrapeyma H (2019) Chitosan/alginate hydrogels containing alpha-tocopherol for wound healing in rat model. J Drug Delivery Sci Technol 51:204–213

Endo F, Awata H, Tanoue A, Ishiguro M, Eda Y, Titani K, Matsuda I (1992) Primary structure deduced from complementary DNA sequence and expression in cultured cells of mammalian 4-hydroxyphenylpyruvic acid dioxygenase: evidence that the enzyme is a homodimer of identical subunit homologous to rat liver-specific alloantigen F. J Biol Chem 267:24235–24240

Endrigkeit J, Wang X, Cai D, Zhang C, Long Y, Meng J, Jung C (2009) Genetic mapping, cloning, and functional characterization of the BnaX.VTE4 gene encoding a c-tocopherol methyltransferase from oilseed rape. Theor Appl Genet 119:567–575

Evans HM, Bishop KS (1922) On the existence of a hitherto unrecognized dietary factor essential for reproduction. Science 56:650–651

Evans HM, Emerson OH, Emerson GA (1936) The isolation from wheat germ oil of an alcohol, tocopherol, having the properties of vitamin E. J Biol Chem 113:319–332

Fachechi C, Nisi R, Gala R, Leone A, Caretto S (2006) Tocopherol biosynthesis is enhanced in photomixotrophic sunflower cell cultures. Plant Cell Rep 26(4):525–530. https://doi.org/10.1007/s00299-006-0268-5

Falk J, Andersen G, Kernebeck B, Krupinska K (2003) Constitutive overexpression of barley 4-hydroxyphenylpyruvate dioxygenase in tobacco results in elevation of the vitamin E content in seeds but not in leaves. FEBS Lett 540:35–40

Franco P, Incarnato L, De Marco I (2019) Supercritical CO_2 impregnation of α-tocopherol into PET/PP films for active packaging applications. J CO2 Util 34:266–273. https://doi.org/10.1016/j.jcou.2019.06.012

Furuya T, Yoshikawa T, Kimura T, Kaneko H (1987) Production of tocopherols by cell culture of safflower. Phytochemistry 26:2741–2747

Gala R, Mita G, Caretto S (2005) Improving α-tocopherol production in plant cell cultures. J Plant Physiol 162(7):782–784

Garcia I, Rodgers M, Lenne C, Roland A, Saillant A, Matringe M (1997) Subcellular localization and purification of a p-hydroxyphenylpyruvate dioxygenase from cultured carrot cells and characterization of the corresponding cDNA. Biochem J 325:761–769

Geipel K, Song X, Socher ML, Kümmritz S, Püschel J, Bley T et al (2013) Induction of a photomixotrophic plant cell culture of Helianthus annuus and optimization of culture conditions for improved α-tocopherol production. Appl Microbiol Biotechnol 98(5):2029–2040. https://doi.org/10.1007/s00253-013-7

Genkinger JM, Platz EA, Hoffman SC, Comstock GW, Helzlsouer KJ (2004) Fruit, vegetable, and antioxidant intake and all-cause, cancer, and cardiovascular disease mortality in a community-dwelling population in Washington County, Maryland. Am J Epidemiol 160:1223–1233

Ghimire BK, Seong ES, Lee CO, Lim JD, Lee JG, Yoo JH, Chung IM, Kim NY, Yu CY (2011) Enhancement of α-tocopherol content in transgenic *Perilla frutescens* containing the g-TMT gene. Afr J Biotechnol 10(13):2430–2439. https://doi.org/10.5897/AJB

Goodman Y, Mattson MP (1994) Secreted forms of beta-amyloid precursor protein protect hippocampal neurons against amyloid beta-peptide-induced oxidative injury. Exp Neurol 128:1–12

Gysin R, Azzi A, Visarius T (2002) γ-Tocopherol inhibits human cancer cell cycle progression and cell proliferation by downregulation of cyclins. FASEB J 16:1952–1954

Haas C, Weber J, Ludwig-Müller J, Deponte S, Bley T, Georgiev M (2008) Flow cytometry and phytochemical analysis of a sunflower cell suspension culture in a 5-L bioreactor. Z Für Naturforsch C 63(9–10):699–705. https://doi.org/10.1515/znc-2008-9-1015

Harish MC, Dachinamoorthy P, Balamurugan S, Bala Murugan S, Sathishkumar R (2012) Enhancement of α-tocopherol content through transgenic and cell suspension culture systems in tobacco. Acta Physiol Plant 35(4):1121–1130. https://doi.org/10.1007/s11738-012-1149-x

Hofius D, Hajirezaei MR, Geiger M, Tschiersch H, Melzer M, Sonnewald U (2004) RNAi-mediated tocopherol deficiency impairs photoassimilate export in transgenic potato plants. Plant Physiol 135:1256–1268

Hofius D, Sonnewald U (2003) Vitamin E biosynthesis: biochemistry meets cell biology. Trends Plant Sci 8:6–8

Hoglen NC, Waller SC, Sipes IG, Liebler DC (1997) Reactions of peroxynitrite with g-tocopherol. Chem Res Toxicol 10:401–407

Horwitt MK (1960) Vitamin E and lipid metabolism in man. Am J Clin Nutr 8:451–461

Horwitt MK, Harvey CC, Duncan GD, Wilson WC (1956) Effects of limited tocopherol intake in man with relationships to erythrocyte hemolysis and lipid oxidations. Am J Clin Nutr 4:408–419

Howard AC, McNeil AK, McNeil PL (2011) Promotion of plasma membrane repair by vitamin E. Nat Commun 2:597

Huang PH, Chuang HC, Chou CC, Wang H, Lee SL, Yang HC, Chiu HC, Kapuriya N, Wang D, Kulp SK, Chen CS (2013) Vitamin E facilitates the inactivation of the kinase Akt by the phosphatase PHLPP1. Sci Signal 6:19

Ischebeck T, Zbierzak AM, Kanwischer M, Dörmann P (2006) A salvage pathway for phytol metabolism in Arabidopsis. J Biol Chem 281:2470–2477

Jefford CW, Cadby PA (1981) Evaluation of models for the mechanism of 4-hydroxyphenylpyruvate dioxygenase. Exp Dermatol 37:1134–1137

Jiang Q, Elson-Schwab I, Courtemanche C, Ames BN (2000) γ-Tocopherol and its major metabolite, in contrast to a-tocopherol, inhibit cyclooxygenase activity in macrophages and epithelial cells. Proc Natl Acad Sci U S A 97:11494–11499

Jukić Špika M, Kraljić K, Škevin D (2016) Tocopherols: chemical structure, bioactivity, and variability in Croatian virgin olive oils. In: Products from olive tree. https://doi.org/10.5772/64658

Kalogeropoulos N, Chiou A, Ioannou M, Karathanos VT, Hassapidou M, Andrikopoulos NK (2010) Nutritional evaluation and bioactive microconstituents (phytosterols, tocopherols, polyphenols, triterpenic acids) in cooked dry legumes usually consumed in the Mediterranean countries. Food Chem 121:682–690. https://doi.org/10.1016/j.foodchem.2010.01.005

Kamal-Eldin A, Appelqvist LA (1996) The chemistry and antioxidant properties of tocopherols and tocotrienols. Lipids 31:671–701

Karunanandaa B, Qi Q, Hao M, Baszis SR, Jensen PK, Wong YHH, Jiang J, Venkatramesh M, Gruys KJ, Moshiri F, Post-Beittenmiller D, Weiss JD, Valentin HE (2005) Metabolically engineered oilseed crops with enhanced seed tocopherol. Metab Eng 7:384–400. https://doi.org/10.1016/j.ymben.2005.05.005

Karuppusamy S, Kim H, Saravana PS, Chun BS, Kang HW (2018) Astaxanthin-alpha tocopherol nanoemulsion formulation by emulsification methods: investigation on anticancer, wound healing, and antibacterial effects. Colloids Surf B Biointerfaces. https://doi.org/10.1016/j.colsurfb.2018.08.042

Konings EJM, Roomans HHS, Beljaars PR (1996) Liquid chromatographic determination of tocopherols and tocotrienols in margarine, infant foods, and vegetables. J AOAC Int 79:902–906. https://doi.org/10.1093/jaoac/79.4.902

Korde VV, Dhas SS, Gurave NA (2016) Hairy root culture: a promising approach in biotransformation. Asian J Plant Sci Res 6(4):6–11

Lassner MW, Savidge B, Weiss JD, Mitsky TA, Post-Beittenmiller MA, Valentin HE (2001) Nucleic acid sequences to proteins involved in tocopherol synthesis. WO 02/33060 (Monsanto)

Lee BK, Kim SL, Kim KH, Yu SH, Lee SC, Zhang Z et al (2007) Seed specific expression of perilla γ-tocopherol methyltransferase gene increases α-tocopherol content in transgenic perilla (*Perilla frutescens*). Plant Cell Tissue Organ Cult 92(1):47–54. https://doi.org/10.1007/s11240-007-9301-9

Lichtenthaler HK (2000) The non-mevalonate isoprenoid biosynthesis: enzymes, genes and inhibitors. Biochem Soc Trans 28:787–792

Lindstedt S, Odelhog B (1987) 4-Hydroxyphenylpyruvate dioxygenase from pseudomonas. Meth Enzymol 142:143–148

Lindsey K, Yeoman MM (1985) Dynamics of plant cell cultures. Cell Cult Somatic Cell Genet Plants 2:61–101

Liu H, Fang G, Li C, Pan M, Liu C, Fan C, Wang S (2012) Molecularly imprinted polymer on ionic liquid-modified CdSe/ZnS quantum dots for the highly selective and sensitive optosensing of tocopherol. J Mater Chem A 22:19882–19887. https://doi.org/10.1039/c2jm33522k

Lonn E, Bosch J, Yusuf S et al (2005) Effects of long-term vitamin E supplementation on cardiovascular events and cancer: a randomized controlled trial. J Am Med Assoc 293:1338–1347

Lopez-Torres M, Thiele JJ, Shindo Y, Han D, Packer L (1998) Topical application of α-tocopherol modulates the antioxidant network and diminishes ultraviolet-induced oxidative damage in murine skin. Br J Dermatol 138(2):207–215. https://doi.org/10.1046/j.1365-2133.1998.02062.x

Lopukhina A, Dettenberg M, Weiler EW, Hollander-Czytko H (2001) Cloning and characterization of a coronatine-regulated tyrosine aminotransferase from Arabidopsis. Plant Physiol 126:1678–1687

Ma Y, Baker RF, Magallanes-Lundback M, DellaPenna D, Braun DM (2008) Tie-dyed1 and sucrose export defective1 act independently to promote carbohydrate export from maize leaves. Planta 227:527–538

Maeda H, Sage TL, Isaac G, Welti R, DellaPenna D (2008) Tocopherols modulate extra plastidic polyunsaturated fatty acid metabolism in Arabidopsis at low temperature. Plant Cell 20:452–470

Maeda H, Song W, Sage TL, DellaPenna D (2006) Tocopherols play a crucial role in low-temperature adaptation and phloem loading in Arabidopsis. Plant Cell 18:2710–2732

Mandracchia D, Tripodo G, Latrofa A, Dorati R (2014) Amphiphilic inulin-d-α-tocopherol succinate (INVITE) bioconjugates for biomedical applications. Carbohydr Polym 103:46–54. https://doi.org/10.1016/j.carbpol.2013.11.056

Mangialasche F, Kivipelto M, Mecocci P, Rizzuto D, Palmer K, Winblad B, Fratiglioni L (2010) High plasma levels of vitamin E forms and reduced Alzheimer's disease risk in advanced age. J Alzheimers Dis 20:1029–1037

Marcos B, Sárraga C, Castellari M, Kappen F, Schennink G, Arnau J (2014) Development of biodegradable films with antioxidant properties based on polyesters containing α-tocopherol and olive leaf extract for food packaging applications. Food Packag Shelf Life 1(2):140–150. https://doi.org/10.1016/j.fpsl.2014.04.002

Martins JT, Cerqueira MA, Vicente AA (2012) Influence of α-tocopherol on physicochemical properties of chitosan-based films. Food Hydrocoll 27(1):220–227. https://doi.org/10.1016/j.foodhyd.2011.06.011

McCormick CC, Parker RS (2004) The cytotoxicity of vitamin E is both vitamer- and cell-specific and involves a selectable trait. J Nutr 134:3335–3342

Mène-Saffrané L, DellaPenna D (2010) Biosynthesis, regulation and functions of tocochromanols in plants. Plant Physiol Biochem 48:301–309

Miller ER III, Pastor-Barriuso R, Dalal D, Riemersma RA, Appel LJ, Guallar E (2005) Meta-analysis: high-dosage vitamin E supplementation may increase all-cause mortality. Ann Intern Med 142:37–46

Munné-Bosch S, Alegre L (2002) The function of tocopherols and tocotrienols in plants. Crit Rev Plant Sci 21:31–57

Munné-Bosch S, Falk J (2004) New insights into the function of tocopherols in plants. Planta 218: 323–326. https://doi.org/10.1007/s00425-003-1126-0

Munteanu A, Zingg JM, Azzi A (2004) Anti-atherosclerotic effects of vitamin E–myth or reality? J Cell Mol Med 8:59–76

Murthy H, Hahn E, Paek K (2008) Adventitious roots and secondary metabolism. Chin J Biotechnol 24(5):711–716. https://doi.org/10.1016/s1872-2075(08)60035-7

Niki E, Traber MG (2012) A history of vitamin e. Ann Nutr Metab 61:207–212. https://doi.org/10.1159/000343106

Nishina K, Unno T, Uno Y, Kubodera T, Kanouchi T, Mizusawa H, Yokota T (2008) Efficient in vivo delivery of siRNA to the liver by conjugation of α-tocopherol. Mol Ther 16(4):734–740. https://doi.org/10.1038/mt.2008.14

Nishio K, Horie M, Akazawa Y, Shichiri M, Iwahashi H, Hagihara Y, Yoshida Y, Niki E (2013) Attenuation of lipopolysaccharide (LPS)-induced cytotoxicity by tocopherols and tocotrienols. Redox Biol 1:97–103

Nisi R, Paradiso A, De Gara L, D'Amico L, Caretto S (2010) Cultivation of Arabidopsis cell cultures in a stirred bioreactor at variable oxygen levels: influence on tocopherol production. Plant Biosyst 144(3):721–724. https://doi.org/10.1080/11263501003755572

Norris SR, Shen X, DellaPenna D (1998) Complementation of the Arabidopsis pds1 mutation with the gene encoding p-hydroxyphenylpuruvate dioxygenase. Plant Physiol 117:1317–1323

Panfili G, Fratianni A, Irano M (2003) Normal phase high-performance liquid chromatography method for the determination of tocopherols and tocotrienols in cereals. J Agric Food Chem 51: 3940–3944. https://doi.org/10.1021/jf030009v

Paradiso A, Caretto S, Leone A, Bove A, Nisi R, De Gara L (2016) ROS production and scavenging under anoxia and reoxygenation in Arabidopsis cells: a balance between redox signaling and impairment. Front Plant Sci 7:1803

Park SY, Lee JG, Cho HS, Seong ES, Kim HY, Yu CY, Kim JK (2013) Metabolite profiling approach for assessing the effects of colored light-emitting diode lighting on the adventitious roots of ginseng (*Panax ginseng* CA Mayer). Plant Omics 6:224

Patel A, Liebner F, Netscher T, Mereiter K, Rosenau T (2007) Vitamin E chemistry. Nitration of non-alpha-tocopherols: products and mechanistic considerations. J Org Chem 72:6504–6512

Porfirova S, Bergmuller E, Tropf S, Lemke R, Dormann P (2002) Isolation of an Arabidopsis mutant lacking vitamin E and identification of a cyclase essential for all tocopherol biosynthesis. Proc Natl Acad Sci U S A 99:12495–12500

Poston L, Briley AL, Seed PT, Kelly FJ, Shennan AH (2006) Vitamin C and vitamin E in pregnant women at risk for preeclampsia (VIP trial): randomised placebo-controlled trial. Lancet 367: 1145–1154

Provencher LM, Miao L, Sinha N, Lucas WJ (2001) Sucrose export defective1 encodes a novel protein implicated in chloroplast-to nucleus signaling. Plant Cell 13:1127–1141

Riberio PPC, de Lima e Silva DM, Dantas MM, da Silva Ribeiro KD, Dimenstein R, da Silva Chaves Damasceno KSF (2019) Determination of tocopherols and physicochemical properties of faveleira (*Cnidoscolus quercifolius*) seed oil extracted using different methods. Food Sci Technol 39:280–285. https://doi.org/10.1590/fst.24017

Ricciarelli R, Maroni P, Ozer N, Zingg JM, Azzi A (1999) Age-dependent increase of collagenase expression can be reduced by alpha-tocopherol via protein kinase C inhibition. Free Radic Biol Med 27:729–737

Ricciarelli R, Zingg JM, Azzi A (2002) The 80th anniversary of vitamin E: beyond its antioxidant properties. Biol Chem 383:457–465. https://doi.org/10.1515/BC.2002.048

Rimm EB, Stampfer MJ, Ascherio A, Giovannucci E, Colditz GA, Willett WC (1993) Vitamin E consumption and the risk of coronary heart disease in men. N Engl J Med 328:1450–1456

Rippert P (2004) Engineering plant shikimate pathway for production of tocotrienol and improving herbicide resistance. Plant Physiol 134(1):92–100. https://doi.org/10.1104/pp.103.032441

Rise M, Cojocaru M, Gottlieb HE, Goldschmidt EE (1989) Accumulation of a-tocopherol in senescing organs as related to chlorophyll degradation. Plant Physiol 89:1028–1030

Roche PA, Moorehead TJ, Hamilton GH (1982) Purification and properties of hog liver 4-hydroxyphenylpyruvate dioxygenase. Arch Biochem Biophys 216:62–73

Russin WA, Evert RF, Vanderveer PJ, Sharkey TD, Briggs SP (1996) Modification of a specific class of plasmodesmata and loss of sucrose export ability in a sucrose export defective1 maize mutant. Plant Cell 8:645–658

Sadiq M, Akram NA, Ashraf M (2017) Foliar applications of alpha-tocopherol improves the composition of fresh pods of *Vigna radiata* subjected to water deficiency. Turk J Bot 41:244–252

Saini RK, Keum YS (2016) Tocopherols and tocotrienols in plants and their products: a review on methods of extraction, chromatographic separation, and detection. Food Res Int 82:59–70. https://doi.org/10.1016/j.foodres.2016.01.025

Sandorf I, Holländer-Czytko H (2002) Jasmonate is involved in the induction of tyrosine aminotransferase and tocopherol biosynthesis in *Arabidopsis thaliana*. Planta 216:173–179. https://doi.org/10.1007/s00425-002-0888-0

Sano M, Ernesto C, Thomas RG, Klauber MR, Schafer K, Grundman M, Woodbury P, Growdon J, Cotman CW, Pfeiffer E, Schneider LS, Thal LJ (1997) A controlled trial of selegiline, alpha-tocopherol, or both as treatment for Alzheimer's disease. The Alzheimer's disease cooperative study. N Engl J Med 336:1216–1222

Sathish S, Preethy KS, Venkatesh R, Sathishkumar R (2018) Rapid enhancement of α-tocopherol content in *Nicotiana benthamiana* by transient expression of *Arabidopsis thaliana* tocopherol cyclase and homogentisate phytyl transferase genes. 3 Biotech s(12):485. https://doi.org/10.1007/s13205-018-1496-4

Sattler SE, Cahoon EB, Coughlan SJ, DellaPenna D (2003) Characterization of tocopherol cyclases from higher plants and cyanobacteria: evolutionary implications for tocopherol synthesis and function. Plant Physiol 132:2184–2195

Savidge B, Weiss JD, Wong Y-HH, Lassner MW, Mitsky TA, Shewmaker CK, Post-Beittenmiller-D, Valentin HE (2002) Isolation and characterization of homogentisate phytyltransferase genes from Synechocystis sp. PCC 6803 and Arabidopsis. Plant Physiol 129:321–332

Schledz M, Seidler A, Beyer P, Neuhaus G (2001) A novel phytyltransferase from Synechocystis sp. PCC 6803 involved in tocopherol biosynthesis. FEBS Lett 499:15–20

Semchuk N, Lushchak OV, Falk J, Krupinska K, Lushchak VI (2009) Inactivation of genes, encoding tocopherol biosynthetic pathway enzymes, results in oxidative stress in outdoor grown *Arabidopsis thaliana*. Plant Physiol Biochem 47:384–390

Shintani D, DellaPenna D (1998) Elevating the vitamin E content of plants through metabolic engineering. Science 282(5396):2098–2100

Shintani DK, Cheng Z, DellaPenna D (2002) The role of 2-methyl-6- phytylbenzoquinone methyltransferase in determining tocopherol composition in Synechocystis sp. PCC 6803. FEBS Lett 511:1–5

Sikorska E, Khmelinskii IV, Sikorski M, Caponio F, Bilancia MT, Pasqualone A, Gomes T (2008) Fluorescence spectroscopy in monitoring of extra virgin olive oil during storage. Int J Food Sci Technol 43:52–61. https://doi.org/10.1111/j.1365-2621.2006.01384.x

Silva SD, Rosa NF, Ferreira AE, Boas LV, Bronze MR (2009) Rapid determination of α-tocopherol in vegetable oils by fourier transform infrared spectroscopy. Food Anal Methods 2:120–127. https://doi.org/10.1007/s12161-008-9047-y

Srinivasan A, Vijaykumar S, Raman K, Srivastava S (2019) Rational metabolic engineering for enhanced alpha-tocopherol production in *Helianthus annuus* cell culture. Biochem Eng J 151: 107256. https://doi.org/10.1016/j.bej.2019.107256

Stampfer MJ, Hennekens CH, Manson JE, Colditz GA, Rosner B, Willett WC (1993) Vitamin E consumption and the risk of coronary disease in women. N Engl J Med 328:1444–1449

Stocker A, Ruttimann A, Woggon WD (1993) Identification of the tocopherol cyclase in the blue–green algae Anabaena variabilis Kutzing (cyanobacteria). Helv Chim Acta 76:1729–1738

Szczeklik A, Gryglewski RJ, Domagala B, Dworski R, Basista M (1985) Dietary supplementation with vitamin E in hyperlipoproteinemias: effects on plasma lipid peroxides, antioxidant activity, prostacyclin generation and platelet aggregability. Thromb Haemost 54:425–430

Takeda T, Seki M, Furusaki S, Furuya T (1993) Factors affecting vitamin E production using plant cell culture of *Carthamus tinctorius*. J Chem Eng Japan 26(5):470–474

Takeda T, Takeuchi Y, Seki M, Furusaki S, Matsuoka H (1998) Kinetic analysis of cell growth and vitamin e production in plant cell culture of *Carthamus tinctorius* using a structured model. Biochem Eng J 1(3):233–242. https://doi.org/10.1016/s1369-703x(98)00007-2

Tang Y, Fu X, Shen Q, Tang K (2016) Roles of MPBQ-MT in promoting α/γ-tocopherol production and photosynthesis under high light in lettuce. PLoS One 11(2):e0148490. https://doi.org/10.1371/journal.pone.0148490

Tavva VS, Kim YH, Kagan IA, Dinkins RD, Kim KH, Collins GB (2006) Increased α-tocopherol content in soybean seed overexpressing the *Perilla frutescens* γ-tocopherol methyltransferase gene. Plant Cell Rep 26(1):61–70. https://doi.org/10.1007/s00299-006-0218-2

Tsegaye Y, Shintani DK, DellaPenna D (2002a) Overexpression of the enzyme p-hydroxyphenylpyruvate dioxygenase in Arabidopsis and its relation to tocopherol biosynthesis. Plant Physiol Biochem 40:913–920

Tsegaye Y, Shintani DK, DellaPenna D (2002b) Overexpression of the enzyme p-hydroxyphenolpyruvate dioxygenase in Arabidopsis and its relation to tocopherol biosynthesis. Plant Physiol Biochem 40(11):913–920. https://doi.org/10.1016/s0981-9428(02)01461-4

Tsochatzis ED, Tzimou-Tsitouridou R (2015) Validated RP-HPLC method for simultaneous determination of tocopherols and tocotrienols in whole grain barley using matrix solid-phase dispersion. Food Anal Methods 8:392–400. https://doi.org/10.1007/s12161-014-9904-9

Valentin HE, Lincoln K, Moshiri F, Jensen PK, Qi Q, Venkatesh TV, Karunanandaa B, Baszis SR, Norris SR, Savidge B, Gruys KJ, Last RL (2006) The Arabidopsis vitamin E pathway gene5-1 mutant reveals a critical role for phytol kinase in seed tocopherol biosynthesis. Plant Cell 18: 212–224

Van Eenennaam AL, Lincoln K, Durrett TP, Valentin HE, Shewmaker CK, Thorne GM, Jiang J, Baszis SR, Levering CK, Aasen ED, Hao M, Stein JC, Norris SR, Last RL (2003) Engineering vitamin E content: from Arabidopsis mutant to soy oil. Plant Cell 15:3007–3019

Verpoorte R, van der Heijden R, Memelink J (2000) Engineering the plant cell factory for secondary metabolite production. Transgenic Res 9(4/5):323–343. https://doi.org/10.1023/a:1008966404981

Vivekananthan DP, Penn MS, Sapp SK, Hsu A, Topol EJ (2003) Use of antioxidant vitamins for the prevention of cardiovascular disease: meta-analysis of randomised trials. Lancet 361:2017–2023

Wada GH, Fellman JH, Fujita FS, Roth E (1975) Purification and properties of avian liver p-hydroxyphenylpyruvate hydroxylase. J Biol Chem 250:6720–6726

Yabuta Y, Tanaka H, Yoshimura S, Suzuki A, Tamoi M, Maruta T, Shigeoka S (2012) Improvement of vitamin E quality and quantity in tobacco and lettuce by chloroplast genetic engineering. Transgenic Res 22(2):391–402. https://doi.org/10.1007/s11248-012-9656-5

Yasar D, Sevket G (2006) Tocopherol and fatty acids of *Spirulina platensis* biomass in glass panel bioreactor. Pak J Biol Sci 9:2901–2904

Yoshida Y, Saito Y, Jones LS, Shigeri Y (2007) Chemical reactivities and physical effects in comparison between tocopherols and tocotrienols: physiological significance and prospects as antioxidants. J Biosci Bioeng 104:439–445

Zbierzak AM, Kanwischer M, Wille C, Vidi PA, Giavalisco P, Lohmann A, Briesen I, Porfirova S, Brehelin C, Kessler F, Dormann P (2010) Intersection of the tocopherol and plastoquinol metabolic pathways at the plastoglobule. Biochem J 425:389–399

Zhang B, Deng Z, Tang Y, Chen P, Liu R, Ramdath DD, Liu Q, Hernandez M, Tsao R (2014) Fatty acid, carotenoid and tocopherol compositions of 20 Canadian lentil cultivars and synergistic contribution to antioxidant activities. Food Chem 161:296–304. https://doi.org/10.1016/j.foodchem.2014.04.014

Ziegler JU, Schweiggert RM, Carle R (2015) A method for the simultaneous extraction and quantitation of lipophilic antioxidants in Triticum sp. by HPLC-DAD/FLD-MSn. J Food Compos Anal 39:94–102. https://doi.org/10.1016/j.jfca.2014.11.011

Zingg JM, Azzi A (2004) Non-antioxidant activities of vitamin E. Curr Med Chem 11:1113–1133

In Vitro Production of Phytosterols

13

Mostafa M. Hegazy and Wahidah H. Al-Qahtani

Abstract

Phytosterols are plant-derived metabolites which are present in edible and non-edible plants and proven to have many health promoting effects. β-sitosterol, campesterol, and stigmasterol are the major phytosterols present in plants. The positive impact on the health came from their ability to reduce plasma cholesterol levels and anti-inflammatory, antidiabetic, and anticancer activities. The beneficial effects of these phytotherapeutic molecules create market need for pharmaceutical and enriched food products based on phytosterols. However, phytosterols production is encountered by many difficulties, since phytosterols levels produced by plants are low and the chemical synthesis offered no actual solution for commercial production, neither practically nor economically. In vitro culture production of phytosterols is an alternative technique with bright future for these biomolecules which includes many successful strategies such as elicitation and metabolic engineering through DNA technologies for overexpression of phytosterols. Phytosterols production by this technique needs to have knowledge with genes and enzymes involved in phytosterols biosynthesis pathway, and the more knowledge you have, the more successful production you get.

Keywords

In vitro cultures · Phytosterols production · Biosynthesis

M. M. Hegazy (✉)
Department of Pharmacognosy, Faculty of Pharmacy, Al-Azhar University (Boys), Cairo, Egypt
e-mail: MostafaHegazy@azhar.edu.eg

W. H. Al-Qahtani
Department of Food Sciences and Nutrition, College of Food and Agriculture Sciences, King Saud University, Riyadh, Saudi Arabia
e-mail: wahida@ksu.edu.sa

T. Belwal et al. (eds.), *Nutraceuticals Production from Plant Cell Factory*,
https://doi.org/10.1007/978-981-16-8858-4_13

13.1 Introduction

Phytosterols are plant-derived lipophilic steroid compounds that are structurally related to cholesterol (Fig. 13.1) which gives them many health benefits when consumed as functional foods and/or nutraceutical preparations (Mackay and Jones 2011). β-Sitosterol, campesterol, and stigmasterol are the main bioactive dietary phytosterols and all of them have double bonds in the ring B of the steroid nucleus at C-5 position. However, C-5 double bond hydrogenation yields the corresponding saturated β-sitostanol and campestanol which are called stanols (Normén et al. 2001). The dominant phytosterols are β-sitosterol and campesterol which constitute about 95% of total plant sterols (Mccarthy et al. 2005). Health benefits of phytosterols can be obtained through many biological effects like plasma

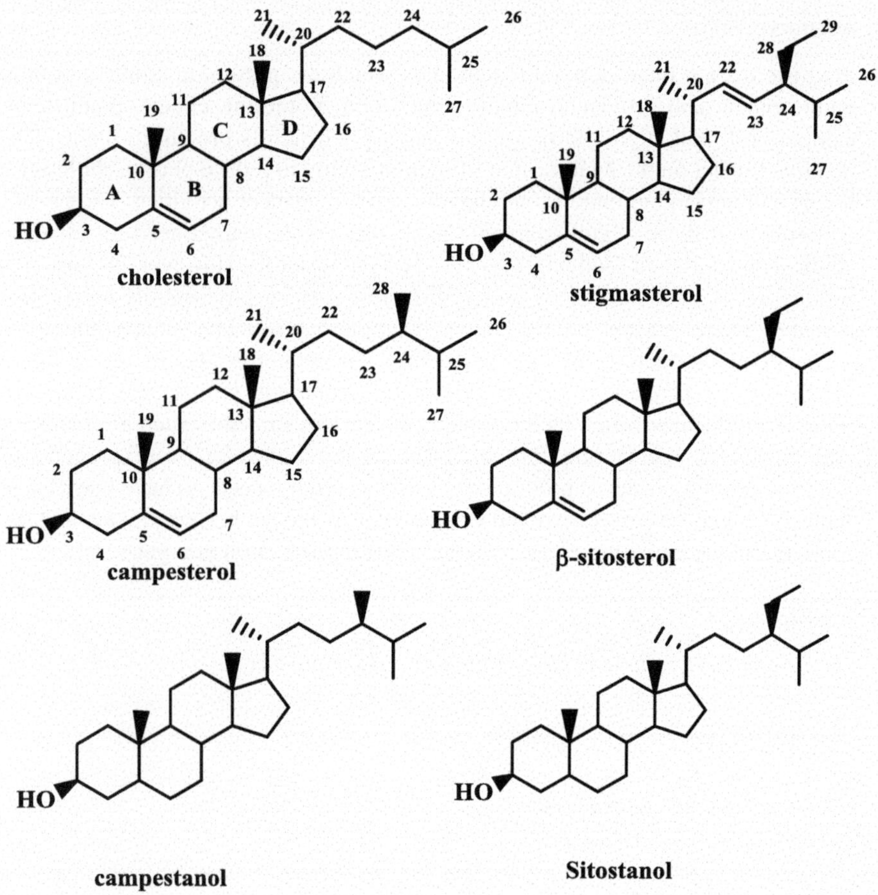

Fig. 13.1 Structures of animal sterol; cholesterol (C27), plant sterols; β-sitosterol (C29), campesterol (C28) and stigmasterol (C29) and plant stanols; Sitostanol (C29) and campestanol (C28)

cholesterol levels reduction, anti-inflammatory, antidiabetic, and anticancer activities (Miras-Moreno et al. 2016). Elevated blood cholesterol level is a risky factor for development of atherosclerosis, which in turn is responsible for cardiovascular and cerebrovascular diseases (Śliż et al. 2019). β-sitosterol was approved by FDA for treatment of coronary heart disease, alopecia, and Benign prostatic hyperplasia which can be found in many nutraceutical preparations in the global market (Bin Sayeed et al. 2016). Phytosterols are secondary metabolites which are biosynthesized by all plants as a cellular structural component and involved in many physiological functions of the plants. Phytosterols are essential for plant cell membrane cohesion, embryonic growth, and formation of liquid-ordered lipid domains (lipid rafts), which play key role in important biological processes like signal transduction and infectious diseases (Dufourc 2008). Vegetable oils are considered richest source of phytosterols, corn (0.91% w/v), sunflower (0.41% w/v), soybean (0.32% w/v), and olive (0.30% w/v); cereals like wheat germ (0.34%) and wheat bran (0.2%); in addition, fruits and vegetables like passion fruit (0.044%), orange (0.024%), and cauliflower (0.04%) (Normén et al. 2001). Phytostanols present as a minor constituent in most of the plants except in cereal grains which are relatively abundant (Vriet et al. 2013). Clinical studies demonstrated that the intake of bioactive phytosterols (2 g/day) can cause a significant reduction (8–10%) in the plasma levels of low-density lipoprotein cholesterol. So, the nutraceutical preparations and food enriched with phytosterols were recommended in several guidelines in dose of 2 g/day in order to reduce LDL-cholesterol levels. Phytosterols consumed in the western diet are about 300 mg/day which do not supply the body with the effective daily dose (Cabral and Klein 2017) that is because of low plant production (Normén et al. 2001). Replacement of cholesterol-rich foods with phytosterols in diet was the message of many health authorities which was accepted by consumers which creates market need. Chemical synthesis of phytosterols provides no actual solution since the starting material is another less prevalent sterol like stigmasterol and final product contains isomerization by-products (Mccarthy et al. 2005). The market demand for natural phytosterols based on nutraceutical preparations and enriched foods results in major challenge to find alternative ways for production of these highly valuable molecules.

13.2 Phytosterols Biosynthetic Pathway

The mevalonate pathway, squalene formation, and cyclization are the main three stages (Fig. 13.2) involved in phytosterol biosynthesis in the plant cell (Valitova et al. 2016). Phytosterols like β-sitosterol, campesterol, and stigmasterol, also produced by microalgae in pathway, contain both features of plant and fungal biosynthetic pathways (Jaramillo-Madrid et al. 2019). The second stage provides squalene as a central precursor for production of phytosterols in plant cell and cholesterol in the animal cell (Zhang et al. 2020).

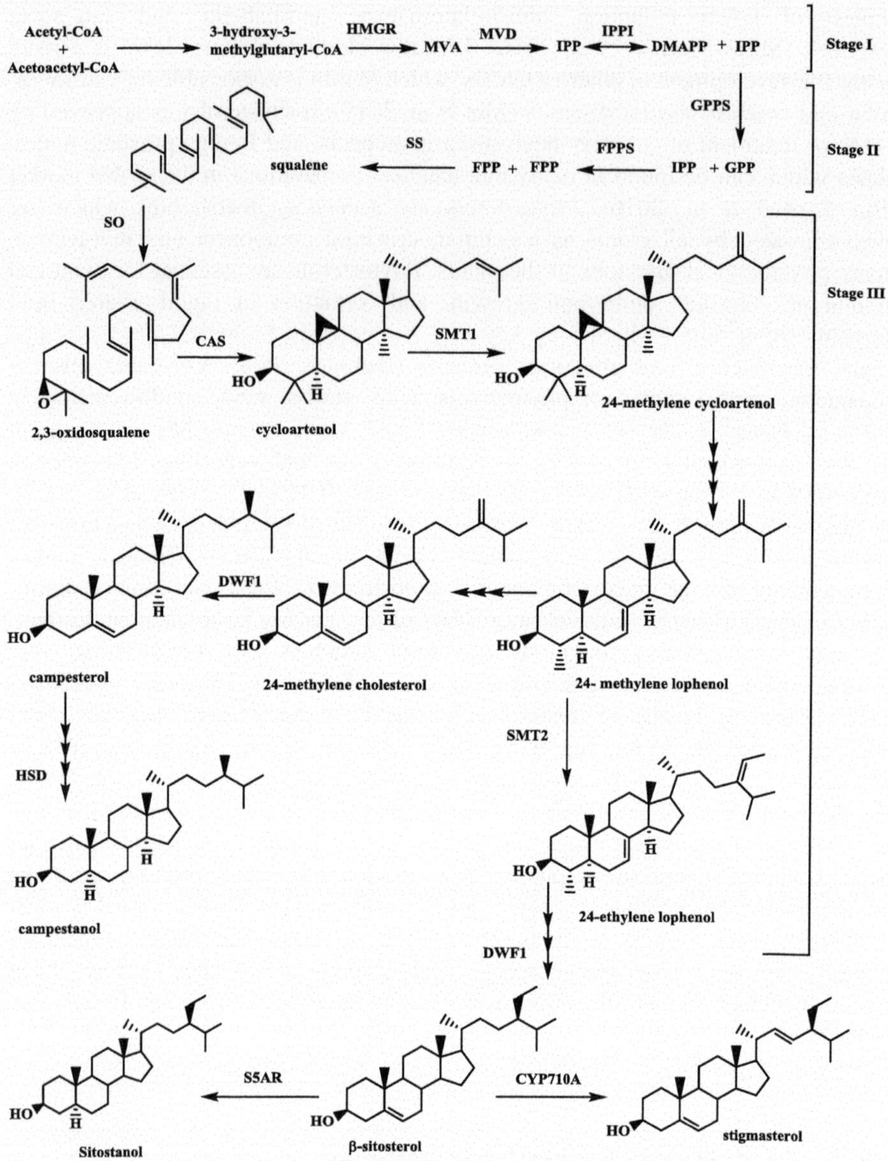

Fig. 13.2 Scheme for the main pathway of phytosterols and phytostanols biosynthesis in plants

13.2.1 Stage I: Mevalonate Pathway

One unit of acetyl-CoA and acetoacetyl-CoA were condensed together forming 3-hydroxy-3-methylglutaryl-CoA which in turn is converted into mevalonate (MVA) by the enzyme 3-hydroxy-3-methylglutaryl-coenzyme A reductase

(HMGR) (Vranová et al. 2013). In the cytoplasmic MVA pathway, HMGR is the rate-limiting enzyme (Zhang et al. 2020) and very essential to maintain necessary levels of phytosterols (Valitova et al. 2016). After transformation of MVA to mevalonate 5-diphosphate, isopentenyl diphosphate (IPP) is produced by the catalysis of mevalonate-5-pyrophosphate decarboxylase (MVD) (Vranová et al. 2013). Although IPP, C5 isoprene unit, is also produced by the plastid methylerythritol phosphate pathway, the cytoplasmic MVA pathway is the predominant supplier for phytosterols biosynthesis (Opitz et al. 2014). IPP produced by MVA pathway is reversibly isomerized to dimethylallyl diphosphate (DMAPP) by the action of the enzyme Isopentenyl diphosphate Δ-isomerase (IPPI) (Vranová et al. 2013).

13.2.2 Stage II: Formation of Squalene

A pair of active isoprenyl diphosphate units, IPP and DMAPP, interact to form C10 geranyl diphosphate (GPP) by the action of Geranyl diphosphate synthase (GPPS) (Bin Sayeed et al. 2016). Then another IPP unit is added by the enzyme farnesyl diphosphate synthase (FPPS) to produce C15 farnesyl diphosphate (FPP). After that, two molecules of FPP combined tail to tail forming C30 squalene by the catalysis of squalene synthase (SS) (Miras-Moreno et al. 2016).

13.2.3 Stage III: Squalene Cyclization

Cyclization of squalene starts with 2,3-oxidosqualene formation by the action of enzyme squalene epoxidase (SO) which is followed by production of tetra-ring structure of cycloartenol and lanosterol by catalysis of cycloartenol synthase (CAS) and lanosterol synthase (LAS), respectively. Formation of cycloartenol or lanosterol is thought to be the difference between plant and fungal/animal cells since lanosterol continues to form cholesterol and ergosterol in animal and fungal cells, respectively (Zhang et al. 2020). Cycloartenol is the main source of plant phytosterols' accumulation, but on the other hand lanosterol contributes a small fraction as a supplementary pathway in response of the plant to stresses (Ohyama et al. 2009). Cycloartenol is methylated via C24-sterol methyltransferase 1 (SMT1) to produce 24-methylene cycloartenol (Zhang et al. 2020). After that, many intermediates are formed in series of oxidative demethylation reactions at C4 and C14 to produce 24-methylene lophenol, which is a precursor for both 24-methylsterol and 24-ethylsterol (Valitova et al. 2016). Then 24-methylene lophenol is transformed to 24-methylene cholesterol which produces campesterol via Dimunito/Dwarf1 (DWF1) (Zhang et al. 2020). Conversion of campesterol to campestanol is the first step in brassinosteroids through three intermediates which involves catalysis of 3β-HydroxySteroid Dehydrogenase (HSD) (Nomura et al. 2004; Vriet et al. 2013). For β-sitosterol, another methyl is added to 24-methylene lophenol for production of 24-ethylene lophenol by SMT2 followed by action of DWF1. Then β-sitosterol turns into stigmasterol by action of C22-sterol desaturase

(CYP710A) (Zhang et al. 2020). Sitostanol formation from β-sitosterol is catalyzed by the action of steroid 5α-reductase (S5AR) (Nomura et al. 2004).

13.2.4 Crucial Importance of Phytosterols Biosynthesis

Using in vitro plant culture for production of secondary metabolites needs all information about biosynthetic pathway, enzymes, precursors, intermediates, and mechanisms involved in their production. This knowledge is very important for development of effective production strategies and techniques (Bhatia et al. 2015). Although a large number of studies are devoted to find out the full biosynthetic pathway of phytosterols, some enzymes are still unknown and needed to be discovered (Miras-Moreno et al. 2016).

13.3 In Vitro Plant Culture for Phytosterols Production

Phytosterols are membrane components and they are important for many vital functions such as signal transduction, resistance of environmental stress conditions, and maintaining of membrane homeostasis. Dramatic changes in the chemical composition of plant cells occur to adapt to the effect of environmental stress factors like low/high temperatures, UV radiation, drought, and salinity. Some observations were correlated between changing in phytosterols' total content/profile and the response to stress (Rogowska and Szakiel 2020). In vitro plant culture production of phytosterols is considered an alternative way and sustainable process for the whole plant extraction. Many strategies are using plant cells' ability to acclimatize various stresses by production of secondary metabolites (Bhatia et al. 2015); also genetic modification can greatly affect in vitro culture production of phytosterols (Mottaki et al. 2019). Strategies used for enhancement of phytosterols production are the selection of highly productive cell lines, determination of the differentiation stage and culture age of high production, and optimization of the cultivation condition, using biotic and abiotic elicitors and metabolic engineering of phytosterol biosynthesis. Production of phytosterols by in vitro culture systems uses summation of all knowledge about biosynthetic pathway and influencing factors which can be useful for development of new strategies that can maximize phytosterols accumulation (Bhatia et al. 2015).

13.3.1 Phytosterols Production Enhancement Strategies

13.3.1.1 Selection of Highly Producing Cell Lines

Plant cells are the production machines of phytosterols, so careful selection must be considered by screening of many cell lines for the highly producing one. There are great variations of production capacities between in vitro plants of different taxonomy, varieties of the same species, and different organs and genotypes of the same

Table 13.1 Cell line selection effect on phytosterols productivity

Plant culture	Medium	Phytosterols production	Ref.
Cell suspension of *T. divaricate*	MS with 10 mg/L NAA	3.170 mg/g DW for total phytosterols	Dyas et al. (1994)
Cell suspension of *S. tuberosum*	MS with 10 mg/L NAA	1.190 mg/g DW for total phytosterols	Dyas et al. (1994)
Callus of *M. pumila* cv. Golden delicious	MS and B5 enriched with 30.0 g/L sucrose and supplemented with 2.0 mg/L BAP and 0.2–2.0 mg/L NAA or 0.2–2.0 mg/L 2,4-D	715.6 μg/g DW for β-Sitosterol	Verardo et al. (2017)
Callus of *M. pumila* cv. Mela Rosa Marchigiana	MS and B5 enriched with 30.0 g/L sucrose and supplemented with 2.0 mg/L BAP and 0.2–2.0 mg/L NAA or 0.2–2.0 mg/L 2,4-D	467.4 μg/g DW for β-Sitosterol	Verardo et al. (2017)
Callus of *E. alba* stem-derived	MS supplemented by 3 mg/L BAP and 1 mg/L NAA	0.32025 mg/g DW for stigmasterol	Khurshid et al. (2018)
Callus of *E. alba* leaf-derived	MS supplemented by 3 mg/L BAP and 1 mg/L NAA	0.20758 mg/g DW for stigmasterol	(Khurshid et al. 2018)
Cell suspension of *C. roseus* (CRPP) cell line	B5 containing 30 g/L sucrose and 1.86 mg/L NAA for 21 days	604 μg/g DW for Campesterol	Saiman et al. (2014)
Cell suspension of *C. roseus* (A12A2) cell line	B5 containing 30 g/L sucrose and 1.86 mg/L NAA for 21 days	124 μg/g DW for Campesterol	Saiman et al. (2014)

plant. Cell suspension cultures of *Tabernaemontana divaricate* (Apocynaceae) and *Solanum tuberosum* (Solanaceae) were grown in Murashige and Skoog media (MS) under the same conditions which were harvested after reaching the stationary phase and the total phytosterols levels were (3.170 mg/g Dry Weight (DW)) and (1.190 mg/g DW) (Table 13.1), respectively (Dyas et al. 1994). Callus Cultures from fruit pulp of two apple varieties *Malus pumila* cv. Golden Delicious and cv. Mela Rosa Marchigiana were cultivated on MS and Gamborg B5 media enriched with 30.0 g/L of sucrose and supplemented with 2.0 mg/L 6-benzylaminopurine (BAP) and 0.2–2.0 mg/L 1-naphthalene acetic acid (NAA) or 0.2–2.0 mg/L 2,4-dichlorophenoxyacetic acid (2,4-D). β-Sitosterol maximal production was detected after 28 days for Golden Delicious (715.6 μg/g DW) and Mela Rosa Marchigiana (467.4 μg/g DW) varieties (Verardo et al. 2017). Leaf and stem explants of *Eclipta alba* were grown on MS medium supplemented by 3 mg/L BAP and 1 mg/L NAA to form callus cultures and the highest estimated concentrations of stigmasterol in stem and leaf-derived callus were (0.32025 mg/g DW) and (0.20758 mg/g DW), respectively (Khurshid et al. 2018). Cell suspension of nine *Catharanthus roseus* cell lines is grown for 21 days in B5 medium containing

30 g/L sucrose and 1.86 mg/L NAA with varying production capacities of campesterol ranging from (124–604 μg/g DW). Saiman et al. suggest that differences in the metabolite profiles were influenced by genotypic variation (Saiman et al. 2014).

13.3.1.2 Effect of Cell Differentiation Stage and Cell Age of Culture

Cunha and Ferreira observe that the phytosterols' total content of *Linum usitatissimum* cultures developed from flax hypocotyl in the same medium is dependent on cell differentiation level. They found that the lower level of phytosterol was in dry flax seeds (0.94 mg/g DW) which increased in non-organogenic callus (1.32 mg/g DW), shoot organogenic callus (1.48 mg/g DW), embryogenic callus (1.45 mg/g DW), and regenerated shoots (2.1 mg/g DW) and somatic embryos (2.13 mg/g DW) (Table 13.2). Stigmasterol to β-sitosterol ratio decreased in non-organogenic callus during growth due to drastic increase of β-sitosterol which increased in derived embryos and shoots induced by somatic embryogenesis and shoot organogenesis (Cunha and Ferreira 1997). On the other hand, phytosterols' production in *Chenopodium rubrum* cell suspension cultures was affected by the cell age of cultures and the total phytosterols' amounts of younger and older cultures were (402.5 μg/g DW) and (260.1 μg/g DW), respectively. This reduction of amounts was due to increase of sterol oxidation products which may be used as a measure for culture aging. However, the main phytosterols' ratio (β-sitosterol, stigmasterol, and campesterol) remained approximately unchanged (Meyer and Spiteller 1997). Callus cultures of *Cissus quadrangularis* were increased by aging, since the cultures cultivated for 2 weeks produced (257 μg/g DW and 150 μg/g DW), for 4 weeks produced (262 μg/g DW and 157 μg/g DW), and for 6 weeks produced (272 μg/g DW and 167 μg/g DW) of β-sitosterol and stigmasterol, respectively. However, after 8 weeks, the contents of β-sitosterol and stigmasterol start to decline (267 μg/g DW and 162 μg/g DW). Thus, screening phytosterols in different ages should be considered to find out the most appropriate age of high production (Sharma et al. 2011).

13.3.1.3 Manipulation of Medium

Different media, growth regulators, carbon source sugars, and dark/light can alter phytosterols' content of in vitro culture. *Euphorbia characias* callus culture was grown on 20 media for screening the effect of media composition on phytosterols productivity. Fernandes-Ferreira et al. observed variation in amounts of β-sitosterol and campesterol in 20 media; the highest produced levels were in Gautheret medium (1033 μg/g DW) and (174 μg/g DW) and the lowest produced were in half-strength MS (322 μg/g DW) and (75 μg/g DW), respectively (Table 13.3) (Fernandes-Ferreira et al. 1992). Phytosterols production and culture growth of *C. quadrangularis* callus culture were altered by changing the growth regulator and its concentration. The best conditions for callus induction were the use of MS medium with 3% sucrose, 2.0 mg/L NAA, and 0.5 mg/L BAP after 6 weeks. However, the maximal levels of phytosterols were (4.839 mg/g DW), (2.239 mg/g DW), and (0.439 mg/g DW), which were identified in the presence of 5 mg/L

Table 13.2 Effect of cell differentiation stage and cell age of culture

Plant culture	Medium	Phytosterols production	Ref.
Non-organogenic callus of *L. usitatissimum*	MS supplemented with 0.4 mg/L 2,4 D and 1.6 g/L zeatin (ZEA) for 8 weeks	1.32 mg/g DW for total phytosterols	Cunha and Ferreira (1997)
Shoot organogenic callus of *L. usitatissimum*	MS supplemented with 0.4 mg/L 2,4 D and 1.6 g/L ZEA for 8 weeks	1.48 mg/g DW for total phytosterols	Cunha and Ferreira (1997)
Embryogenic callus of *L. usitatissimum*	MS supplemented with 0.4 mg/L 2,4 D and 1.6 g/L ZEA for 8 weeks	1.45 mg/g DW for total phytosterols	Cunha and Ferreira (1997)
Regenerated shoots of *L. usitatissimum*	MS supplemented with 0.4 mg/L 2,4 D and 1.6 g/L ZEA for 8 weeks	2.1 mg/g DW for total Phytosterols	Cunha and Ferreira (1997)
Regenerated somatic embryos of *L. usitatissimum*	MS supplemented with 0.4 mg/L 2,4 D and 1.6 g/L ZEA for 8 weeks	2.13 mg/g DW for total phytosterols	Cunha and Ferreira (1997)
Cell suspension of *C. rubrum*	MS supplemented with 1% 2,4 D for 11 days	402.5 μg/g DW for total phytosterols	Meyer and Spiteller (1997)
Cell suspension of *C. rubrum*	MS supplemented with 1% 2,4 D for 8 weeks	260.1 μg/g DW for total phytosterols	Meyer and Spiteller (1997)
Callus of *C. quadrangularis*	MS supplemented with 3% sucrose, 2 mg/L NAA, and 0.5 mg/L BAP for 2 weeks	257 μg β-sitosterol/g DW and 150 μg stigmasterol/g DW	Sharma et al. (2011)
Callus of *C. quadrangularis*	MS supplemented with 3% sucrose, 2 mg/L NAA, and 0.5 mg/L BAP for 4 weeks	262 μg β-sitosterol/g DW and 157 μg stigmasterol/g DW	Sharma et al. (2011)
Callus of *C. quadrangularis*	MS supplemented with 3% sucrose, 2 mg/L NAA, and 0.5 mg/L BAP for 6 weeks	272 μg β-sitosterol/g DW and 167 μg stigmasterol/g DW	Sharma et al. (2011)
Callus of *C. quadrangularis*	MS supplemented with 3% sucrose, 2 mg/L NAA, and 0.5 mg/L BAP for 8 weeks	267 μg β-sitosterol/g DW and 162 μg stigmasterol/g DW	Sharma et al. (2011)

indole-3-acetic acid (IAA), 5 mg/L 2,4 D, and in control conditions without hormones, respectively (Sharma et al. 2011). Moreover, Mangas et al. tested the influence of different concentrations of N-(2-chloro- 4-pyridyl)-N′-phenylurea (4PU-30) on the total content of phytosterols in callus culture of *Centella asiatica.* The comparison of phytosterols' content in *C. asiatica* calli grown for 4 weeks

Table 13.3 Effect of media manipulation on phytosterols levels

Plant culture	Medium	Phytosterols production	Ref.
Callus of *E. characias*	Half-strength MS supplemented with 0.5 mg/L benzyl adenine (BA), 0.25 mg/L 2, 4-D, 5 mg/L ascorbic acid, 1 g/L casein hydrolysate (CH), and 30 g/L sucrose for 30 days	322 μg β-sitosterol/g DW and 75 μg campesterol/g DW	Fernandes-Ferreira et al. (1992)
Callus of *E. characias*	Gautheret medium supplemented with 0.5 mg/L benzyl adenine (BA), 0.25 mg/liter 2, 4-D, 5 mg/L ascorbic acid, 1 g/L casein hydrolysate (CH), and 30 g/L sucrose for 30 days	1033 μg β-sitosterol/g DW and 174 μg campesterol/g DW	Fernandes-Ferreira et al. (1992)
Callus of *C. quadrangularis*	MS supplemented with 3% sucrose	0.439 mg/g DW for total phytosterols	Sharma et al. (2011)
Callus of *C. quadrangularis*	MS supplemented with 3% sucrose with 5 mg/L 2,4 D	2.239 mg/g DW for total phytosterols	Sharma et al. (2011)
Callus of *C. quadrangularis*	MS supplemented with 3% sucrose with 5 mg/L IAA	4.839 mg/g DW for total phytosterols	Sharma et al. (2011)
Non-organogenic callus of *C. asiatica*	MS supplemented with 0.1 mg/L indole-3-butyric acid (IBA) and 1 mg/L 4PU-30 for 4 weeks	Approximately 150 μg/g DW for total phytosterols	Mangas et al. (2008)
Non-organogenic callus of *C. asiatica*	MS supplemented with 0.1 mg/L IBA and 3 mg/L 4PU-30 for 4 weeks	Approximately 190 μg/g DW for total phytosterols	Mangas et al. (2008)
Hairy root of *A. thaliana* ecotype Columbia	Half-strength MS supplemented with nicotinic acid, pyridoxine, thiamine, inositol, and 3% glucose for 42 days	1476 μg/ g DW for total phytosterols	Jozwiak et al. (2013)
Hairy root of *A. thaliana* ecotype Columbia	Half-strength MS supplemented with nicotinic acid, pyridoxine, thiamine, inositol, and 0.5% glucose for 42 days	2129 μg/g DW for total phytosterols	Jozwiak et al. (2013)
Hairy root of *A. thaliana* ecotype Columbia	Half-strength MS supplemented with nicotinic acid, pyridoxine, thiamine, inositol, and 3% sucrose for 42 days	2234 μg/g DW for total phytosterols	Jozwiak et al. (2013)
Callus of *B. diffusa*	MS fortified with 18 μM 2,4-D, 2 μM BAP, and 3% galactose for 60 days	Approximately 215 μg β-sitosterol/g DW	Patil and Bhalsing (2016)
Callus of *B. diffusa*	MS fortified with 18 μM 2,4-D, 2 μM BAP, and 3% sucrose 60 days	Approximately 460 μg β-sitosterol/g DW	Patil and Bhalsing (2016)

(continued)

Table 13.3 (continued)

Plant culture	Medium	Phytosterols production	Ref.
Callus of *B. diffusa*	MS fortified with 18 μM 2,4-D, 2 μM BAP, and 3% glucose 60 days	575 μg β-sitosterol/g DW	Patil and Bhalsing (2016)
Callus of *P. dactylifera* cv. Hayani	MS supplemented with 10 mg/L 2,4-D and 3 mg/L 2-isopentyladenine under dark condition 0 μmol/m2/s PPFD	0.1 mg β-sitosterol/g DW	El-Dawayati et al. (2020)
Callus of *P. dactylifera* cv. Hayani	MS supplemented with 10 mg/L 2,4-D and 3 mg/L 2-isopentyladenine The light intensity was 14 μmol/m2/s PPFD with 16 h photoperiod for 16 weeks	Approximately 0.29 mg β-sitosterol/g DW	El-Dawayati et al. (2020)
Callus of *P. dactylifera* cv. Hayani	MS supplemented with 10 mg/L 2,4-D and 3 mg/L 2-isopentyladenine The light intensity was 42 μmol/m2/s PPFD with 16 h photoperiod for 16 weeks	0.52 mg β-sitosterol/g DW	El-Dawayati et al. (2020)

reveals that the higher phytosterols' concentration (approximately 190 μg/g DW) was present in calli grown with 3 mg/L 4PU-30, while the lowest (approximately 150 μg/g DW) was present in calli grown with 1 mg/L 4PU-30. In fact, the results were confirming the direct relationship between the cytokinin concentration and phytosterols levels (Mangas et al. 2008). On the other hand, sugar type and concentration can affect amount of phytosterols produced by *Arabidopsis thaliana* hairy root culture containing either sucrose (3%) or glucose (0.5, 3.0%) as carbon source. The total phytosterols produced in half-strength MS medium were (2234 μg/g DW), (2129 μg/g DW), and (1476 μg/g DW) for media containing 3% sucrose, 0.5% glucose, and 3.0% glucose, respectively (Jozwiak et al. 2013). β-sitosterol accumulated in *Boerhaavia diffusa* callus culture was affected by the used sugar and the maximal detected levels were (575 μg/g DW), (approximately 460 μg/g DW), and (approximately 215 μg/g DW) in the media containing 3% glucose, 3% sucrose, and 3% galactose, respectively (Patil and Bhalsing 2016). El-Dawayati et al. observed that the production of β-sitosterol in *Phoenix dactylifera* cv. Hayani Callus Cultures was increased by increasing light intensity starting from incubation in dark condition. The callus culture which was cultivated under dark condition (0 μmol/m2/s photosynthetic photon flux density (PPFD)) was the lowest in production (0.1 mg/g DW), and by increasing intensity of the light (14 and 42 μmol/m2/s PPFD), the β-sitosterol was accumulated to approximately 2.9 to 5.2-fold (El-Dawayati et al. 2020). Conditions' optimization of the media is an essential element for the maximum in vitro production of phytosterols through testing various conditions for reaching to the most appropriate.

13.3.1.4 Elicitation for Enhancement of Phytosterols Accumulation

Elicitation is one of the used strategies to improve the phytosterols production in plant cell cultures. Thus, using of elicitors like methyl jasmonate, pectins, benzothiadiazole, β-glucan, (Z)-3-hexenol, chitosan, cadmium chloride, silver nitrate, and cyclodextrins (CDs) can modify the produced amount of phytosterols after using either separately or in combinations. Elicitation of *C. asiatica* cell suspension culture with methyl jasmonate (MJ) increases centelloside production. However, the total phytosterols amounts were decreased after 30 days of elicitation of 100 or 200 μM methyl jasmonate from (approximately 0.240 mg/g DW) of control to (approximately 0.22 and 0.17 mg/g DW), respectively (Bonfill et al. 2011). In a similar way, the phytosterol accumulation of *Uncaria tomentosa* cell suspension cultures wasn't affected (approximately 120 μg/g DW) after the treatment with 0.1% pectin w/v and four fungal strains *Trichoderma* sp., *Pestalotia* sp., *Epicoccum nigrum,* and *Alternaria tenuis* (Flores-Sánchez et al. 2002). The concentrations of phytosterols in *Calendula officinalis* hairy root cultures (cotyledon-derived CC16 line, 460.23, 546.98, 613.02, 625.88 and 685.44 μg/g DW) increased in a dose depending manner in half-strength MS medium after 7 days of elicitation with cadmium chloride (0, 25, 50, 100 and 150 μM $CdCl_2$), respectively (Alsoufi et al. 2019). However, phytosterols were produced in high level at 144 h when *Daucus carota* cell suspension cultures were treated with methylated-β-cyclodextrins (CDM, 10.8 mg/g DW), being approximately 2.7-fold higher than cultures treated with hydroxypropylated-β-cyclodextrins (CDH, approximately 4 mg/g DW, Table 13.4) (Sabater-Jara and Pedreño 2013). In a similar way, biosynthesis of phytosterols was enhanced in flax cell suspension cultures by elicitation with β-glucan, (Z)-3-hexenol, hydroxypropylated-β-cyclodextrins (CDH), and methylated-β-cyclodextrins (CDM), which were added in the media either separately or in combinations. In fact, elicitation with 1 mg/L β-glucan and 40 μM (Z)-3-hexenol enhances intracellular accumulation of phytosterols (192.58 or 231.61 μg/g DW), which represent 13 and 16-fold higher than the detected amount in control culture (14.53 μg/g DW) after 6 days, respectively. Moreover, addition of 50 mM CDH alone increases phytosterols production (259.47 μg/g DW) than 1 mg/L β-glucan and 40 μM (Z)-3-hexenol, each of them alone, while combination of 50 mM CDH with 1 mg/L β-glucan or 40 μM (Z)-3-hexenol showed more elevated levels of production (503.88 and 530.81 μg/g DW), respectively. Elicitation with 50 mM CDM alone showed five-fold increase in total phytosterols content (1325 μg/g DW) than same concentration of CDH; also combination of CDM with (Z)-3-hexenol showed synergistic effect (1507 μg/g DW) and antagonistic activity with 1 mg/L β-glucan (1278 μg/g DW) (Almagro et al. 2016). Therefore, β-cyclodextrins not only induce the biosynthesis of phytosterols, but also increase their extracellular secretion and accumulation in the culture medium which make it easy for direct recovery from the medium without destruction of biomass (Belchí-Navarro et al. 2012). Elicitor activity of β-cyclodextrins is due to their chemical structure similarity with the alkyl-derived oligosaccharides which are produced from plant cell walls during fungal infections (Bru et al. 2006). Phytosterols accumulation by stressor agent like cyclodextrins is one of the induced defense responses which lead to

Table 13.4 phytosterols enhancement by elicitation

Plant culture	Medium	Phytosterols production	Ref.
Cell suspension of *C. asiatica*	MS supplemented with 30 g sucrose, 2,4-D (2 mg/L), and BA (0.1 mg/L) for 30 days as a control	Approximately 0.240 mg/g DW for total phytosterols	Bonfill et al. (2011)
Cell suspension of *C. asiatica*	MS supplemented with 30 g sucrose, 2,4-D (2 mg/L), and BA (0.1 mg/L) for 30 days of elicitation with 100 μM MJ	Approximately 0.22 mg/g DW for total phytosterols	Bonfill et al. (2011)
Cell suspension of *C. asiatica*	MS supplemented with 30 g sucrose, 2,4-D (2 mg/L), and BA (0.1 mg/L) for 30 days of elicitation with 200 μM MJ	Approximately 0.17 mg/g DW for total phytosterols	Bonfill et al. (2011)
Cell suspension of *U. tomentosa*	Modified Nitsch–Nitsch supplemented with 2 mg/L 1 2.4-D, 2 mg/L kinetin, and 20 g/L sucrose as a control	Approximately 120 μg/g DW for total phytosterols	Flores-Sánchez et al. (2002)
Cell suspension of *U. tomentosa*	Modified Nitsch–Nitsch supplemented with 2 mg/L 1 2.4-D, 2 mg/L kinetin and 20 g/L sucrose with combination of 0.1% pectin w/v and four fungal strains *Trichoderma* sp., *Pestalotia* sp., *Epicoccum nigrum*, and *Alternaria tenuis* as elicitors	Approximately 120 μg/g DW for total phytosterols	Flores-Sánchez et al. (2002)
Cell suspension of *D. carota*	MS supplemented with Morel vitamins, 0.25 g/L casein hydrolysate, 30 g/L sucrose, 1 mg/L 2, 4-D, and CDH as elicitor for 144 h	Approximately 4 mg/g DW for total phytosterols	Sabater-Jara and Pedreño (2013)
Cell suspension of *D. carota*	MS supplemented with Morel vitamins, 0.25 g/L casein hydrolysate, 30 g/L sucrose, 1 mg/L 2, 4-D, and CDM as elicitor for 144 h	10.8 mg/g DW for total phytosterols	Sabater-Jara and Pedreño (2013)
Cell suspension of *L. usitatissimum*	MS supplemented with 0.4 mg/ L thiamin, 100 mg/L myo-inositol, and 30 g/ L sucrose as a control	14.53 μg/g DW for total phytosterols	Almagro et al. (2016)
Cell suspension of *L. usitatissimum*	MS supplemented with 0.4 mg/L thiamin, 100 mg/L myo-inositol, 30 g/L sucrose, and 1 mg/L β-glucan as elicitor for 6 days	192.58 μg/g DW for total phytosterols	Almagro et al. (2016)
Cell suspension of *L. usitatissimum*	MS supplemented with 0.4 mg/L thiamin, 100 mg/L myo-inositol, 30 g/L sucrose, and 40 μM (Z)-3-hexenol as elicitor for 6 days	231.61 μg/g DW for total phytosterols	Almagro et al. (2016)
Cell suspension of *L. usitatissimum*	MS supplemented with 0.4 mg/ L thiamin, 100 mg/L myo-inositol, 30 g/ L sucrose, and 50 mM CDH as elicitor for 6 days	259.47 μg/g DW for total phytosterols	Almagro et al. (2016)

(continued)

Table 13.4 (continued)

Plant culture	Medium	Phytosterols production	Ref.
Cell suspension of *L. usitatissimum*	MS supplemented with 0.4 mg/L thiamin, 100 mg/L myo-inositol, 30 g/L sucrose and 50 mM CDH, and 1 mg/L β-glucan as elicitors for 6 days	503.88 μg/g DW for total phytosterols	Almagro et al. (2016)
Cell suspension of *L. usitatissimum*	MS supplemented with 0.4 mg/ L thiamin, 100 mg/L myo-inositol, 30 g/L sucrose, and 50 mM CDH and 40 μM (Z)-3-hexenol as elicitors for 6 days	530.81 μg/g DW for total phytosterols	Almagro et al. (2016)
Cell suspension of *L. usitatissimum*	MS supplemented with 0.4 mg/L thiamin, 100 mg/L myo-inositol, 30 g/L sucrose, and 50 mM CDM as elicitor for 6 days	1325 μg/g DW for total phytosterols	Almagro et al. (2016)
Cell suspension of *L. usitatissimum*	MS supplemented with 0.4 mg/L thiamin, 100 mg/L myo-inositol, 30 g/L sucrose, and 50 mM CDM and 1 mg/L β-glucan as elicitors for 6 days	1278 μg/g DW for total phytosterols	Almagro et al. (2016)
Cell suspension of *L. usitatissimum*	MS supplemented with 0.4 mg/L thiamin, 100 mg/L myo-inositol, 30 g/L sucrose, and 50 mM CDM and 40 μM (Z)-3-hexenol as elicitors for 6 days	1507 μg/g DW for total phytosterols	Almagro et al. (2016)
Cell suspension of *C. annum*	MS supplemented with Morel vitamins, 300 mg/L CH, 2.5% sucrose, 0.2 mg/L BA, 2 mg/L NAA, and 50 mM CDM as elicitor for 96 h	776 μg β-sitosterol/ g DW	Sabater-Jara et al. (2010)
Cell suspension of *C. annum*	MS supplemented with Morel vitamins, 300 mg/L CH, 2.5% sucrose, 0.2 mg/L BA, 2 mg/L NAA, and 100 μM MJ as elicitor for 96 h	951 μg β-sitosterol/ g DW	Sabater-Jara et al. (2010)
Cell suspension of *C. annum*	MS supplemented with Morel vitamins, 300 mg/L CH, 2.5% sucrose, 0.2 mg/L BA, 2 mg/L NAA, and 50 mM CDM and 100 μM MJ as elicitors for 96 h	1943 μg β-sitosterol/g DW	Sabater-Jara et al. (2010)

modification and reinforcement to the architecture of the cell wall. β-sitosterol production in *Capsicum annum* cell suspension cultures was increased due to the additive effect of combination of both 50 mM CDM and 100 μM MJ (1943 μg/g DW) than when the elicitation was performed separately for each of them (CDM, 776 μg/g DW and MJ, 951 μg/g DW) (Sabater-Jara et al. 2010). Taking all results into account, elicitation is a successful strategy for phytosterols in vitro production and elicitors' joint action can be a useful tool for scaling up.

13.3.1.5 Metabolic Engineering as a Biotechnological Alternative Strategy

Transgenic hairy root cultures are used for phytosterols production, since these cultures are developed with high growth rate in hormones-free medium and these transformed cell lines are showing genetic and biosynthetic stability for successive generations than undifferentiated cells. Two genetically transformed *Chlorophytum borivilianum* hairy root culture lines (2364a and 2364b) are produced by *Agrobacterium rhizogenes* MTCC *2364* strain. Stigmasterol produced in control untransformed root culture treated with 3 mg/L IBA (46.4 mg/g) and without IBA (3.937 mg/g) was enhanced in two cell lines (2364b, 83.952 mg/g) with 1.8 and 21-fold and (2364a, 72.098 mg/g) with 1.5 and 18-fold than control culture with 3 mg/L IBA and without IBA, respectively (Table 13.5) (Bathoju et al. 2017). Overexpressing of FPPS in transgenic hairy roots culture of *C. asiatica* resulted in

Table 13.5 Metabolic engineering strategy for phytosterols production

Plant culture	Medium	Phytosterols production	Ref.
Untransformed root culture of *C. borivilianum*	MS for 25 days	3.937 mg Stigmasterol/g	Bathoju et al. (2017)
Untransformed root culture of *C. borivilianum*	MS supplemented with 3 mg/L IBA for 25 days	46.4 mg Stigmasterol/g	Bathoju et al. (2017)
Hairy root culture 2364a line of *C. borivilianum*	MS for 25 days	72.098 mg Stigmasterol/g	Bathoju et al. (2017)
Hairy root culture 2364b line *of C. borivilianum*	MS for 25 days	83.952 mg Stigmasterol/g	Bathoju et al. (2017)
Hairy root culture transformed by empty vector harboring *A. rhizogenes* of *C. asiatica*	MS supplemented with 3% sucrose for 28 days as a control	Approximately 235 μg/g DW for total sterols	Kim et al. (2010)
Hairy root culture transformed by *Panax ginseng* FPPS vector harboring *A. rhizogenes* strain R1000 of *C. asiatica*	MS supplemented with 3% sucrose for 28 days of FPPS gene overexpression	Approximately 640 μg/g DW for total sterols	Kim et al. (2010)
Non-transgenic root lines of *P. ginseng*	MS for 4 weeks as control	452 μg stigmasterol/g DW	Kim et al. (2014)
Hairy root culture of *P. ginseng*	MS for 4 weeks MVD overexpression	1389 μg stigmasterol/g DW	Kim et al. (2014)
Non-transgenic root lines of *P. ginseng*	MS for 4 weeks as control	4436 μg β-sitosterol/g DW	Kim et al. (2014)
Hairy root culture of *P. ginseng*	MS for 4 weeks MVD overexpression	3076 μg β-sitosterol/g DW	Kim et al. (2014)

production of phytosterols precursor squalene in high levels which in turn showed elevated amounts of total phytosterols (approximately 640 μg/g DW) than those detected in control lines (approximately 235 μg/g DW). Thus, more than two-fold greater elevation indicates the regulatory function of FPPS in biosynthetic pathway of phytosterols (Kim et al. 2010). Phytosterols accumulation was 1.6-fold higher in transgenic hairy root lines due to HMGR overexpression in *Platycodon grandiflorum* cultures than the detected amounts in control lines (Kim et al. 2013). Overexpressing MVD in transgenic lines of *P. ginseng* hairy root cultures resulted in higher production levels of stigmasterol (1389 μg/g DW) and β-sitosterol (4436 μg/g DW) by 3 and 1.4-fold than wild-type control lines (452 μg/g DW) and (3076 μg/g DW), respectively (Kim et al. 2014). In addition, *P. ginseng* adventitious transgenic roots-mediated transformation by *Agrobacterium tumefaciens* resulted in overexpression of SS gene which had a positive effect on phytosterols production (approximately 600 μg/g DW) compared to those detected in non-transgenic (approximately 200 μg/g DW) (Lee et al. 2004). In a similar way, overexpression of SS gene in *Eleutherococcus senticosus* transgenic somatic embryos also showed increased levels of stigmasterol and β-sitosterol (approximately 2.1 and 0.9 mg/g DW, respectively), compared to non-transgenic (approximately 1.3 and 0.4 mg/g DW), respectively (Seo et al. 2005). Thus, all these results indicate that metabolic-engineered cultures could be a powerful form of cellular factory for enhancing phytosterols bioproduction.

13.4 Extraction and Detection Techniques

Extraction method of phytosterols from plant matrices depends on its physical state (solid or liquid such as culture medium and vegetable oil) and chemical form (glycosylated, free, or esterified). Acid hydrolysis and saponification are included in many extraction procedures to liberate free phytosterols from glycosylated and esterified molecules, respectively (Almeida et al. 2020). In this sense, many solvents and solvent system could be used for extraction such as hexane, methylene chloride, ethyl acetate, chloroform–methanol, and chloroform–methanol–water (Lagarda et al. 2006). Efficient extraction of these valuable molecules is an important step which influences purity, functionality, and product final cost. Developing green methods and technologies for extraction should be considered to minimize health and environmental problems caused by organic solvents. Green phytosterols extraction such as direct citric acid hydrolysis could be used as efficient and eco-friendly method (Feng et al. 2020b). In similar way, the supercritical fluid extraction for phytosterols is considered to be efficient and selective with shorter time and eco-friendly technology. However, the disadvantage of using supercritical CO2 is the expensive high pressure equipment (Uddin et al. 2015). One of production limitations is the cost of large quantity used from plant material to increase the produced amount of phytosterols. However, flax cultures elicitation with β-cyclodextrins increased accumulation and secretion of phytosterols into the medium which are extracted with ethyl acetate. Phytosterols production in the medium is one of the strategies that may be the key for sustainable industrial production of phytosterols without affecting cell culture viability (Almagro et al. 2016). Nowadays, commercially produced phytosterols came from extraction of vegetable oils and tall oil which are obtained from coniferous trees as a by-product during wood pulp manufacture. Phytosterols are extracted with hexane followed by refining process involving high temperatures and pressures to remove impurities, which is considered to be very effective for obtaining 90% of vegetable oil total phytosterols. However, refining process increases the cost of the final product; also nonfood source should be used as alternative for edible oils. On contrary, refining process of tall oil caused degradation to more than half of β-sitosterol and decreased the yield to 8 and 20% of the oil total phytosterols (Fernandes and Cabral 2007). The most common analysis for phytosterols detection includes the following steps: (1) extraction; (2) acid hydrolysis and/or saponification; (3) unsaponifiable matter extraction; (4) purification/enrichment using separation techniques such as crystallization, solid phase extraction, and column chromatography; and (6) detection using chromatographic techniques such as gas chromatography (GC), high pressure liquid chromatography (HPLC), and ultra-pressure liquid chromatography-mass spectrometry (UPLC-MS) (Almeida et al. 2020). Gas chromatography, with flame ionization or mass spectrometer detectors, is the most common method for qualitative and quantitative analysis of phytosterols which should be derivatized to steryl acetates or trimethylsilyl ethers before analysis (Moreau et al. 2018). HPLC and UPLC coupled to Mass spectrometry (MS or MS/MS) have more sensitivity and precision than ordinary chromatographic techniques which make ease of identification and quantitation of phytosterols through intense fragment ions $[M + H\text{-}H_2O]^+$ (397, 383 and 395 m/z) for β-sitosterol, campsterol, and stigmasterol, respectively (Lagarda et al. 2006).

13.5 Phytosterols Biological Activities

Phytosterols have captured scientific and market interests primarily through their cholesterol-lowering activity which leads to their protective effect against cardiovascular diseases (García-Llatas and Rodríguez-Estrada 2011), and also for other important pharmacological properties such as anticancer (Woyengo et al. 2009), anti-inflammatory, and antidiabetic activities (Santas et al. 2013).

13.5.1 Hypocholesterolemic Activity of Phytosterols

Reduction of plasma cholesterol levels of phytosterols is mainly for their structural similarity with cholesterol which give them the ability to reduce small intestine absorption of cholesterol. In fact, phytosterols displace cholesterol from the phospholipid micelles which is considered an important vehicle for cholesterol absorption and this is because of the more lipophilic properties of phytosterols than cholesterol (Jones and Abumweis 2009; Rozner and Garti 2006). Cholesterol absorption decreased by 30 to 40% after intaking of 1.5–2.0 g of phytosterols/day and the dose of 2.0 g/day is proposed to be the optimal daily dose which can reduce plasma LDL cholesterol by 10% (Rocha et al. 2011; Santas et al. 2013). However, guidelines don't recommend any doses higher than 2.0 g/day since phytosterols in higher doses cause no further reduction in cholesterol levels, but cause a reduction in plasma levels of carotenoid and α-tocopherol (Santas et al. 2013). This phytosterols-related problem could be solved out by intake of α-tocopherol and β-carotene-rich foods (Noakes et al. 2002).

13.5.2 Immune System Modulation and Anti-Inflammatory Properties

Phytosterols consumption also involved immune system regulation and inflammatory disorders reduction. Enhanced activity of T lymphocytes and natural killer cells was clear evidence of immune system modulating activity. Moreover, clinical findings of phytosterols consumption for pulmonary-tuberculosis patients showed improvement of recovery of those patients (Santas et al. 2013). Vitamin D effect on the immune function of macrophages was increased with β-sitosterol (Alappat et al. 2010), and in dose-dependent manner, showed decreased in vitro production of mitogen-induced IL-2 (Aherne and O'brien 2008). In similar way, proinflammatory cytokines levels were decreased by phytosterols (Devaraj et al. 2011).

13.5.3 Antidiabetic Effects of Phytosterols

Fasting plasma insulin levels were increased with orally administered β-sitosterol which was accompanied by decrease in fasting glycemia in a way comparable to Glibenclamide as standard anti-hyperglycemic drug (Bin Sayeed et al. 2016). Although phytosterols as compounds have been reported with antidiabetic activity, further clinical studies are needed to assess this effect.

13.5.4 Anticancer Effects of Phytosterols

Campesterol isolated from *Chrysanthemum coronarium* can reduce the proliferation of human umbilical vein endothelial cells and inhibition of capillary differentiation antiangiogenic activity of campesterol (Choi et al. 2007). β-sitosterol showed reduction in the growth of prostate cancer cell lines by increasing prostaglandins levels and arresting cell cycle at the G2/M phase (Awad et al. 2005). Moreover, cytotoxicity and apoptosis were triggered in human monoblastic leukemia with β-sitosterol in a dose-dependent manner. The apoptosis enhancement triggered by β-sitosterol was mediated by the degradation of poly-(ADP-ribose) polymerase, the activation of caspase-3, and downregulation of Bcl-2 (Park et al. 2007). Phytosterols also can protect against carcinogenesis caused by reactive oxygen species after DNA damage. Antioxidant activity of macrophages enzymes like glutathione peroxidase and superoxide dismutase is enhanced with β-sitosterol in macrophages cultures which is induced with oxidative stress (Vivancos and Moreno 2005).

13.6 Phytosterols Commercial Utilization and Prospects

Plant sterols are phytotherapeutic compounds of high value which are used commercially to protect and treat cardiovascular diseases due to high plasma cholesterol levels. Cardiovascular diseases are the major noncommunicable diseases, accounting for about 50% of the world's noncommunicable-related deaths (Jagannathan et al. 2019). More than 35 estimated phytosterols-based pharmaceutical products in global market are used for treatment of hypercholesteremia and more 40 products containing β-sitosterol for treatment of benign prostatic hyperplasia which are listed in Martindale: the complete drug reference, edition 38 (Bayer 2015). Phytosterols' health promoting effects caused increased consumption of phytosterol-enriched foods. Nowadays, many functional food products in European markets contain phytosterols like margarine which has cholesterol-lowering effects (Piironen et al. 2000). Many foods are enriched with these bioactive compounds, especially those of lipid matrix such as spreads, dairy, bakery, and meat products and low-lipid matrix, like orange juice (Tolve et al. 2020). Monitoring survey on customers' purchase behavior for phytosterols-enriched food in six European countries revealed that the highest consumption was in United Kingdom by 736 kg and lowest was in Greece by 65 kg. Majority of the phytosterols were consumed as spreads which was followed by yogurt drinks (Feng et al. 2020a). Customer interests with natural ingredients of cosmetics and phytosterols have skin moisturizing, anti-inflammatory, and regenerating properties which made them suitable additive for dry, sun-damaged, and aged skin care cosmetics (Doering et al. 2005). In the same way, antidiabetic and anticancer activities of phytosterols need more clinical studies which may open new prospects for treatment of diabetic and cancer patients which may help in relieving their suffering.

13.7 Conclusions

Phytosterols are proven to be essential biomolecules for protection and treatment of cardiovascular diseases, treatment of benign prostatic hypertrophy, and skin care preparations. Phytosterols became an ingredient of many commercial nutraceutical preparations and enriched foods in the global market. Currently, their commercial production is still based on vegetable oils which decrease market share of these edible oils; in times, some parts of world suffer from hunger and tall oil which is produced as a by-product of paper industry also affect environment through deforestation. In fact, in vitro production of these phytotherapeutic molecules is sustainable and of eco-friendly technique. Enhancement of production by technique is optimized by many strategies such as selection of high production cell lines in appropriate age and conditions, elicitation, and genetic modification through hairy root in vitro cultures. In the future, I am suggesting that the key to scale-up of phytosterols to industrial production is through elicitors that can accumulate these valuable molecules in media like β-cyclodextrin.

References

Aherne SA, O'brien NM (2008) Modulation of cytokine production by plant sterols in stimulated human Jurkat T cells. Mol Nutr Food Res 52(6):664–673

Alappat L, Valerio M, Awad AB (2010) Effect of vitamin D and β-sitosterol on immune function of macrophages. Int Immunopharmacol 10(11):1390–1396

Almagro L, García-Pérez P, Belchí-Navarro S et al (2016) New strategies for the use of *Linum usitatissimum* cell factories for the production of bioactive compounds. Plant Physiol Biochem 99:73–78

Almeida CAS, Baggio SR, Mariutti LRB et al (2020) One-step rapid extraction of phytosterols from vegetable oils. Food Res Int 130:108891

Alsoufi ASM, Pączkowski C, Długosz M et al (2019) Influence of selected abiotic factors on triterpenoid biosynthesis and saponin secretion in marigold (*Calendula officinalis* L.) in vitro hairy root cultures. Molecules 24(16):2907

Awad AB, Burr AT, Fink CS (2005) Effect of resveratrol and β-sitosterol in combination on reactive oxygen species and prostaglandin release by PC-3 cells. Prostaglandins Leukot Essent Fat Acids 72(3):219–226

Bathoju G, Rao K, Giri A (2017) Production of sapogenins (stigmasterol and hecogenin) from genetically transformed hairy root cultures of *Chlorophytum* borivilianum (Safed musli). Plant Cell Tissue Organ Cult 131(3):369–376

Bayer G (2015) Martindale: the complete drug reference. Aust Prescr 38(2):59

Belchí-Navarro S, Almagro L, Lijavetzky D et al (2012) Enhanced extracellular production of trans-resveratrol in *Vitis vinifera* suspension cultured cells by using cyclodextrins and methyljasmonate. Plant Cell Rep 31(1):81–89

Bhatia S, Bera T, Dahiya R et al (2015) Classical and nonclassical techniques for secondary metabolite production in plant cell culture. In: Modern applications of plant biotechnology in pharmaceutical sciences. Elsevier, London, pp 231–291

Bin Sayeed MS, Karim SMR, Sharmin T et al (2016) Critical analysis on characterization, systemic effect, and therapeutic potential of beta-sitosterol: a plant-derived orphan phytosterol. Medicines 3(4):29

Bonfill M, Mangas S, Moyano E et al (2011) Production of centellosides and phytosterols in cell suspension cultures of *Centella asiatica*. Plant Cell Tissue Organ Cult 104(1):61–67

Bru R, Sellés S, Casado-Vela J et al (2006) Modified cyclodextrins are chemically defined glucan inducers of defense responses in grapevine cell cultures. J Agric Food Chem 54(1):65–71

Cabral CE, Klein MRST (2017) Phytosterols in the treatment of hypercholesterolemia and prevention of cardiovascular diseases. Arq Bras Cardiol 109(5):475–482

Choi JM, Lee EO, Lee HJ et al (2007) Identification of campesterol from *Chrysanthemum coronarium* L. and its antiangiogenic activities. Phytother Res 21(10):954–959

Cunha A, Ferreira MF (1997) Differences in free sterols content and composition associated with somatic embryogenesis, shoot organogenesis and calli growth of flax. Plant Sci 124(1):97–105

Devaraj S, Jialal I, Rockwood J et al (2011) Effect of orange juice and beverage with phytosterols on cytokines and PAI-1 activity. Clin Nutr 30(5):668–671

Doering T, Holtkötter O, Schlotmann K et al (2005) Cutaneous restructuration by apple seed phytosterols: from DNA chip analysis to morphological alterations. Int J Cosmet Sci 27(2): 142–142

Dufourc EJ (2008) The role of phytosterols in plant adaptation to temperature. Plant Signal Behav 3(2):133–134

Dyas L, Threlfall DR, Goad LJ (1994) The sterol composition of five plant species grown as cell suspension cultures. Phytochemistry 35(3):655–660

El-Dawayati MM, El-Sharabasy S, Gantait S (2020) Light intensity-induced morphogenetic response and enhanced β-Sitosterol accumulation in date palm (*Phoenix dactylifera* L. cv. Hayani) callus culture. Sugar Tech 22(6):1122–1129

Feng S, Belwal T, Li L et al (2020a) Phytosterols and their derivatives: potential health-promoting uses against lipid metabolism and associated diseases, mechanism, and safety issues. Compr Rev Food Sci Food Saf 19(4):1243–1267

Feng S, Wang L, Belwal T et al (2020b) Phytosterols extraction from hickory (*Carya cathayensis* Sarg.) husk with a green direct citric acid hydrolysis extraction method. Food Chem 315:126217

Fernandes P, Cabral J (2007) Phytosterols: applications and recovery methods. Bioresour Technol 98(12):2335–2350

Fernandes-Ferreira M, Pais MS, Novais J (1992) The effects of medium composition on biomass, sterols and triterpenols production by in-vitro cultures of *Euphorbia characias*. Bioresour Technol 42(1):67–73

Flores-Sánchez IJ, Ortega-López J, Montes-Horcasitas MDC et al (2002) Biosynthesis of sterols and triterpenes in cell suspension cultures of *Uncaria tomentosa*. Plant Cell Physiol 43(12): 1502–1509

García-Llatas G, Rodríguez-Estrada MT (2011) Current and new insights on phytosterol oxides in plant sterol-enriched food. Chem Phys Lipids 164(6):607–624

Jagannathan R, Patel SA, Ali MK et al (2019) Global updates on cardiovascular disease mortality trends and attribution of traditional risk factors. Curr Diab Rep 19(7):44

Jaramillo-Madrid AC, Ashworth J, Fabris M et al (2019) Phytosterol biosynthesis and production by diatoms (Bacillariophyceae). Phytochemistry 163:46–57

Jones PJ, Abumweis SS (2009) Phytosterols as functional food ingredients: linkages to cardiovascular disease and cancer. Curr Opin Clin Nutr Metab Care 12(2):147–151

Jozwiak A, Ples M, Skorupinska-Tudek K et al (2013) Sugar availability modulates polyisoprenoid and phytosterol profiles in *Arabidopsis thaliana* hairy root culture. Biochim Biophys Acta Mol Cell Biol Lipids 1831(2):438–447

Khurshid R, Khan T, Zaeem A et al (2018) Biosynthesis of precious metabolites in callus cultures of *Eclipta alba*. Plant Cell Tissue Organ Cult 135(2):287–298

Kim OT, Kim SH, Ohyama K et al (2010) Upregulation of phytosterol and triterpene biosynthesis in *Centella asiatica* hairy roots overexpressed ginseng farnesyl diphosphate synthase. Plant Cell Rep 29(4):403–411

Kim Y-K, Kim JK, Kim YB et al (2013) Enhanced accumulation of phytosterol and triterpene in hairy root cultures of *Platycodon grandiflorum* by overexpression of Panax ginseng 3-hydroxy-3-methylglutaryl-coenzyme a reductase. J Agric Food Chem 61(8):1928–1934

Kim Y-K, Kim YB, Uddin MR et al (2014) Enhanced triterpene accumulation in *Panax ginseng* hairy roots overexpressing mevalonate-5-pyrophosphate decarboxylase and farnesyl pyrophosphate synthase. ACS Synth Biol 3(10):773–779

Lagarda M, García-Llatas G, Farré R (2006) Analysis of phytosterols in foods. J Pharm Biomed Anal 41(5):1486–1496

Lee M-H, Jeong J-H, Seo J-W et al (2004) Enhanced triterpene and phytosterol biosynthesis in *Panax ginseng* overexpressing squalene synthase gene. Plant Cell Physiol 45(8):976–984

Mackay DS, Jones PJ (2011) Phytosterols in human nutrition: type, formulation, delivery, and physiological function. Eur J Lipid Sci Technol 113(12):1427–1432

Mangas S, Moyano E, Osuna L et al (2008) Triterpenoid saponin content and the expression level of some related genes in calli of *Centella asiatica*. Biotechnol Lett 30(10):1853

Mccarthy FO, Chopra J, Ford A et al (2005) Synthesis, isolation and characterisation of β-sitosterol and β-sitosterol oxide derivatives. Org Biomol Chem 3(16):3059–3065

Meyer W, Spiteller G (1997) Oxidized phytosterols increase by ageing in photoautotrophic cell cultures of Chenopodium rubrum. Phytochemistry 45(2):297–302

Miras-Moreno B, Sabater-Jara ABN, Pedreño M et al (2016) Bioactivity of phytosterols and their production in plant in vitro cultures. J Agric Food Chem 64(38):7049–7058

Moreau RA, Nyström L, Whitaker BD et al (2018) Phytosterols and their derivatives: structural diversity, distribution, metabolism, analysis, and health-promoting uses. Prog Lipid Res 70:35–61

Mottaki Z, Rezayian M, Niknam V et al (2019) Using hairy roots for production of secondary metabolites in Artemisia. Plant Biotechnol Rep 13(3):263–271

Noakes M, Clifton P, Ntanios F et al (2002) An increase in dietary carotenoids when consuming plant sterols or stanols is effective in maintaining plasma carotenoid concentrations. Am J Clin Nutr 75(1):79–86

Nomura T, Jager CE, Kitasaka Y et al (2004) Brassinosteroid deficiency due to truncated steroid 5α-reductase causes dwarfism in the lk mutant of pea. Plant Physiol 135(4):2220–2229

Normén AL, Brants HA, Voorrips LE et al (2001) Plant sterol intakes and colorectal cancer risk in the Netherlands cohort study on diet and cancer. Am J Clin Nutr 74(1):141–148

Ohyama K, Suzuki M, Kikuchi J et al (2009) Dual biosynthetic pathways to phytosterol via cycloartenol and lanosterol in Arabidopsis. Proc Natl Acad Sci 106(3):725–730

Opitz S, Nes WD, Gershenzon J (2014) Both methylerythritol phosphate and mevalonate pathways contribute to biosynthesis of each of the major isoprenoid classes in young cotton seedlings. Phytochemistry 98:110–119

Park C, Moon D-O, Rhu C-H et al (2007) β-Sitosterol induces anti-proliferation and apoptosis in human leukemic U937 cells through activation of caspase-3 and induction of Bax/Bcl-2 ratio. Biol Pharm Bull 30(7):1317–1323

Patil KS, Bhalsing SR (2016) Effect of sugars on production of β-sitosterol from in vitro callus culture of *Boerhaavia diffusa* L. Acta Biol Szeged 60(2):99–104

Piironen V, Lindsay DG, Miettinen TA et al (2000) Plant sterols: biosynthesis, biological function and their importance to human nutrition. J Sci Food Agric 80(7):939–966

Rocha M, Banuls C, Bellod L et al (2011) A review on the role of phytosterols: new insights into cardiovascular risk. Curr Pharm Des 17(36):4061–4075

Rogowska A, Szakiel A (2020) The role of sterols in plant response to abiotic stress. Phytochem Rev 2020:1–14

Rozner S, Garti N (2006) The activity and absorption relationship of cholesterol and phytosterols. Colloids Surf A Physicochem Eng Asp 282:435–456

Sabater-Jara AB, Pedreño MA (2013) Use of β-cyclodextrins to enhance phytosterol production in cell suspension cultures of carrot (*Daucus carota* L.). Plant Cell Tissue Organ Cult 114(2): 249–258

Sabater-Jara AB, Almagro L, Belchí-Navarro S et al (2010) Induction of sesquiterpenes, phytoesterols and extracellular pathogenesis-related proteins in elicited cell cultures of *Capsicum annuum*. J Plant Physiol 167(15):1273–1281

Saiman MZ, Mustafa NR, Pomahočová B et al (2014) Analysis of metabolites in the terpenoid pathway of *Catharanthus roseus* cell suspensions. Plant Cell Tissue Organ Cult 117(2):225–239

Santas J, Codony R, Rafecas M (2013) Phytosterols: beneficial effects. In: Natural products. Springer, New York, pp 3437–3464

Seo J-W, Jeong J-H, Shin C-G et al (2005) Overexpression of squalene synthase in *Eleutherococcus senticosus* increases phytosterol and triterpene accumulation. Phytochemistry 66(8):869–877

Sharma N, Nathawat R, Gour K et al (2011) Establishment of callus tissue and effect of growth regulators on enhanced sterol production in *Cissus quadrangularis* L. Int J Pharmacol 7(5): 653–658

Śliż D, Marcinkiewicz A, Olejniczak D et al (2019) Hypercholesterolemia and prevention of cardiovascular diseases in the light of preventive medical examinations of employees in Poland. Int J Occup Med Environ Health 32(6):865–872

Tolve R, Cela N, Condelli N et al (2020) Microencapsulation as a tool for the formulation of functional foods: the phytosterols' case study. Foods 9(4):470

Uddin MS, Sarker MZI, Ferdosh S et al (2015) Phytosterols and their extraction from various plant matrices using supercritical carbon dioxide: a review. J Sci Food Agric 95(7):1385–1394

Valitova J, Sulkarnayeva A, Minibayeva F (2016) Plant sterols: diversity, biosynthesis, and physiological functions. Biochem Mosc 81(8):819–834

Verardo G, Gorassini A, Ricci D et al (2017) High triterpenic acids production in callus cultures from fruit pulp of two apple varieties. Phytochem Anal 28(1):5–15

Vivancos M, Moreno JJ (2005) β-Sitosterol modulates antioxidant enzyme response in RAW 264.7 macrophages. Free Radic Biol Med 39(1):91–97

Vranová E, Coman D, Gruissem W (2013) Network analysis of the MVA and MEP pathways for isoprenoid synthesis. Annu Rev Plant Biol 64:665–700

Vriet C, Russinova E, Reuzeau C (2013) From squalene to brassinolide: the steroid metabolic and signaling pathways across the plant kingdom. Mol Plant 6(6):1738–1757

Woyengo T, Ramprasath V, Jones P (2009) Anticancer effects of phytosterols. Eur J Clin Nutr 63(7):813–820

Zhang X, Lin K, Li Y (2020) Highlights to phytosterols accumulation and equilibrium in plants: biosynthetic pathway and feedback regulation. Plant Physiol Biochem 155:637–649

In Vitro Production of Quinones

14

Ehab M. B. Mahdy, Sherif F. El-Sharabasy, and Maiada M. El-Dawayati

Abstract

Quinones as natural derivatives of phenolic constituents have a broad sense of biological activities, including anticancer, antimicrobial, anti-inflammatory, and purgative effects. Quinones have multiple applications such as dyes, pigments, cosmetics, food additives, flavors, agrochemicals, and pharmaceuticals. Those molecules have existed across the whole of organisms; plants are a prime natural source. Concerning the bio-based economy (bio-economy), a renewed push has been intended in the plant biotechnology area to crop high-value bio-ingredients for various downstream applications. Quinones divide into four classes along with a number of benzene rings: phenanthrenequinone, naphthoquinone, anthraquinone, and benzoquinone. The quinone molecules have to be interestingly investigated through their chemistry, classification, nomenclature, biosynthesis, biological effects, and distribution naturally in plants. We will here give an update on account of reports and studies done on bioactive quinones. The chemical structure with biological effects and biosynthesis will be the focus. Besides, we provide the progress on some significant plant-derived quinones and their mode of action to endorse more beneficial biologic features.

E. M. B. Mahdy (✉)
National Gene Bank (NGB), Agricultural Research Centre (ARC), Giza, Egypt
e-mail: ehab.mahdy@arc.sci.eg

S. F. El-Sharabasy · M. M. El-Dawayati
Central Laboratory of Date Palm Research and Development (CLDPRD), Agricultural Research Center (ARC), Giza, Egypt
e-mail: sheriff.sharabasy@arc.sci.eg; maiada.eldawayati@arc.sci.eg

T. Belwal et al. (eds.), *Nutraceuticals Production from Plant Cell Factory*,
https://doi.org/10.1007/978-981-16-8858-4_14

Keywords

Quinones · Phenanthrenequinone · Naphthoquinone · Anthraquinone · Benzoquinone

14.1 Introduction

Plants biosynthesize a wide range of natural products that use dyes, pigments, cosmetics, food additives, flavors, agrochemicals, and pharmaceuticals (Malik et al. 2014a, b, c). These secondary metabolites are heterogeneous compounds and can generally survive and biosynthesize from a bewildering array of the environment (Lubbe and Verpoorte 2011) under specific conditions, viz., specific nutrition and abiotic stresses (Malik et al. 2010a, b, 2013). These metabolites require primarily a renaissance of natural products (Lu et al. 2020; Raskin et al. 2002); so, the world market is annually rising at a rate of 15–18%.

Secondary metabolites are divided primarily into three key groups: (1) polyphenols, (2) alkaloids, and (3) terpenes. Quinones, as derivatives from oxidization of hydroquinones or polyphenols (Eyong et al. 2013a, b; Gong et al. 2014), are biosynthesized across the whole of kingdoms (Hillion and Antelmann 2015; Widhalm and Rhodes 2016; Yamamoto et al. 2018). Plants are the prime source of quinones, which exist in a wide range of families, including Vitaceae, Arecaceae, Verbenaceae, Bignoniaceae, Fabaceae, Caesalpiniaceae, Rhamnaceae, Polygonaceae, Rubiaceae, Orobanchaceae, Boraginaceae, and Lamiaceae (Oni et al. 2015; Soladoye and Chukwuma 2012). The role of quinones in plants is still mostly mysterious. Quinones are often negligent in the research of signal agents (Widhalm and Rhodes 2016). Plants do not have homologs of recognized quinones sensor as in animals and bacteria that own cysteine amendments of receptor proteins, which affect cytoprotective gene expression (Yamamoto et al. 2018). They consequently are probable to use sensing strategies different from those of animals (Farmer and Davoine 2007). Plastoquinones, α-tocopherol quinones, and phylloquinone are mainly metabolites existing in all photosynthesizing cells. Ubiquinones are made in most species of plantae kingdom. The majority of quinones are simple benzoquinones, naphthoquinones, or anthraquinones. However, less common skeletal structures occur, for instance terpenoid quinones and higher polycyclic quinones. In nature, most quinones are *p*-quinones, nonetheless *o*-quinones also occur (Fig. 14.2).

Anthraquinones originate in a broad sense of families such as Rubiaceae, Fabaceae, Rhamnaceae, Polygonaceae, Liliaceae, and Scrophulariaceae. Most naphthoquinones accumulate in Bignoniaceae, Verbenaceae, Juglandaceae, Plumbaginaceae, Boraginaceae, Lythraceae, Balsaminaceae, Ebenaceae, and Droseraceae. Myrsinaceae and Borraginaceae that are most families accumulate benzoquinones.

Quinones are divided into four sorts according to the number of benzene rings: phenanthrenequinone, naphthoquinone, anthraquinone, and benzoquinone

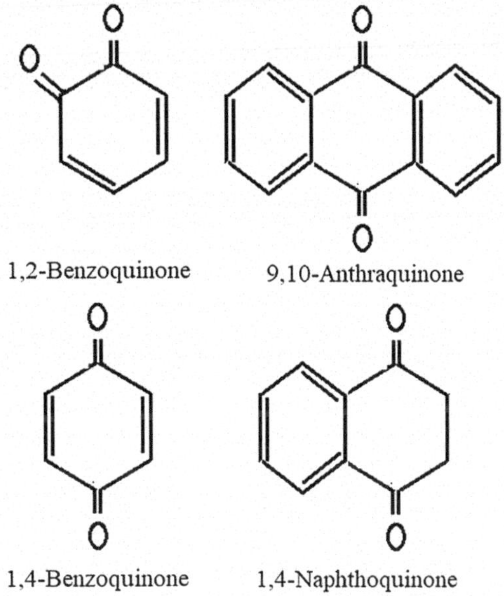

Fig. 14.1 Structure of four classes of quinones

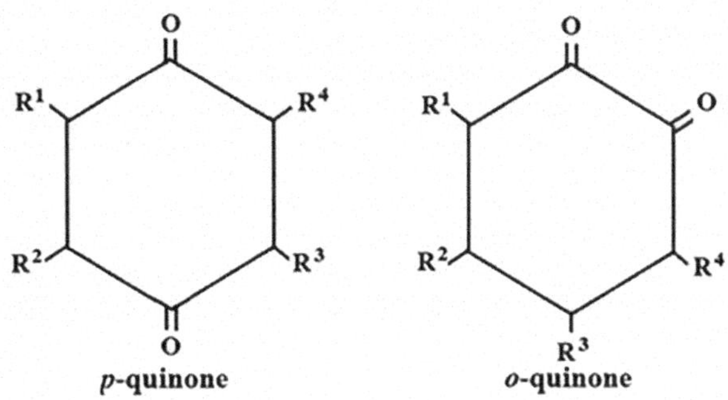

Fig. 14.2 Structure of *para*-quinones (1) and *ortho*-quinones (2)

(Fig. 14.1). Quinones play a prime role, as a signal agent, of biological activities, i.e., anticancer activities, antibacterial activities, food additives, and purgative effects (Aithal et al. 2009, 2011; Padhye et al. 2012). The compound 2,6-dimethoxy-1,4-benzoquinone (DMBQ) is only well-documented in the induction of haustoria across parasitic roots. Nevertheless, the sense of DMBQ is still blurred.

Focusing on the bio-economy (biological-based economy) has increasingly meant a modern drive in the biotechnology approach to yield high-value biomaterials for several downstream applications. Bio-economy emphasizes maintainable ('green') manufacture of renewable bio-resources and their change into value-added products in food, feed, chemicals, energy, and healthcare and wellness. Secondary metabolites have always been the prime source of biomaterial for many medicinal and manufacturing applications (Chakraborty 2018; Misra 2014; Nomura et al. 2018; Owen et al. 2017; Xin et al. 2017; Yang et al. 2016).

We will highlight the progression of some significant quinone derivatives, their chemistry, biosynthesis pathways, mode of action, and in vitro production. Besides, biological activities and scale-ups have considered encouraging more valuable biologic properties.

14.2 Biosynthesis Pathway

Biosynthesis is a dynamic method to get quinones and it is obligatory to reconnoiter the pathway and sense the mechanism of the mode of actions. Numerous quinone derivatives were reported (Camelio et al. 2015; Eyong et al. 2013a, b; Gong et al. 2014). The complementary aspect in the biosynthesis of bioactive products is to clarify their pathways. Quinones are naturally among the most miscellaneous biometabolites; however, they are almost all biosynthesized through the mevalonate pathway or 2-C-methylerythritol 4-phosphate (Nowicka and Kruk 2010; Pina et al. 2016; Skorupinska-Tudek et al. 2008; Socaciu 2007; Widhalm and Rhodes 2016). Two main pathways regulate the biosynthesis of natural products. By C13 labeling experiments, isopentenyl diphosphate (IPP) could switch between both pathways and contribute to the consequent biosynthesis pathways (Skorupinska-Tudek et al. 2008). The acetyl-coenzyme A (Co-A) units are a start step of the biosynthesis pathway, which catalyze acetyl-CoA C-acetyltransferase (AACT) to produce acetoacetyl-Co-A. The gene encoding AACT was cloned successfully from *Tripterygium wilfordii* (Zhao et al. 2015a, b). Liu et al. (2014) cloned the gene encoding 3-hydroxy-3-methylglutaryl-CoA synthase. Wu et al. (2012a, b) characterized the gene encoding NADPH-dependent HMG-CoA reductase (HMGR). Mevalonate pathway is gone down into IPP through subsequent chains of enzymes (Zhao et al. 2013). The IPP catalyzes to dimethylallyl pyrophosphate (DMAPP) using IPP isomerase. Both substrates are commonly for two pathways (Drummond et al. 2019). Pyruvate substrates are significant in the pathway of 2-C-methylerythritol 4-phosphate (Lichtenthaler 1999). Two genes, encoding 1-deoxy-D-xylulose-5-phosphate synthase (DXS) and 1-deoxy-d-xylulose-5-phosphate reductoisomerase (DXR), can be expressed highly in roots of *T. wilfordii* cloned (Tong et al. 2015a, b, c; Zhang et al. 2020). IPP isomerase catalyzed the conversion of IPP and DMAPP and was joined by protonation and deprotonation reactions (Xiang et al. 2010; Zhang et al. 2015; Zhao et al. 2015a, b; Tong et al. 2016). Cloning and expression of the MEP pathway were screened by the enzyme-encoding genes in *Osmanthus fragrans* (Xu et al. 2016). Zhou et al. (2018) identified the functions of five squalene epoxidase (SE) genes of *T. wilfordii* by *erg*1 mutant yeast constructed using CRISPR/Cas9.

Quinones properties are based on the structure of the chemical and aromatic ring, side-chain groups. Quinones, which have the aromatic di-one or di-ketone system in all creatures, can be substituted both in para- or ortho-patterns (R1 – R4) of its hydroquinone precursor (Fig. 14.2) (Weaver and Pettus 2014). So, quinones come from the corrosion of hydroquinones or polyphenols (Eyong et al. 2013a, b; Gong et al. 2014). Based on ring structure, quinones are classified into (1) benzoquinone

(1-ring structure), (2) naphthoquinone (2-ring structure), (3) anthraquinones, and (4) anthraquinones (Eyong et al. 2013a, b; Gong et al. 2014).

More intensive efforts are required to expand and progress the insight into biosynthetic processes. Biosynthesis pathway must be clear by illustrating the pathway of enzymes with their information and defining each intermediate. Biosynthesis pathways of quinones have stated various features of biosynthesis research (Kristensen et al. 2020). Quinones come from acetate/malonate (polyketide), iso-chorismate/*o*-succinylbenzoic acid, and p-hydroxybenzoic acid (Nowicka and Kruk 2010; Socaciu 2007; Widhalm and Rhodes 2016).

Plastoquinones, α-tocopherolquinones, and phylloquinone are primary metabolites existing in all photosynthesizing cells. Ubiquinones are present in the most species of plantae kingdom. The majority of quinones are majority of simple benzoquinones, naphthoquinones, or anthraquinones, although less common structures occur, such as higher polycyclic quinones and terpenoid quinones. Quinones, in nature, are *p*-quinones, or *o*-quinones. Anthraquinones accumulate in Rubiaceae, Fabaceae, Rhamnaceae, Polygonaceae, Liliaceae, and Scrophulariaceae; naphthoquinones accumulate in Bignoniaceae, Verbenaceae, Juglandaceae, Plumbaginaceae, Boraginaceae, Lythraceae, Balsaminaceae, Ebenaceae, and Droseraceae; while, benzoquinones accumulate in Myrsinaceae and Boraginaceae. Benzoquinones biosynthesize by the acetate/malonate or shikimic acid pathways. The gentisic acid (quinol form) is a side product formed by the oxidation of gentisaldehyde in the biosynthesis of putalin. The quinol forms biosynthesize naturally and most isolated benzoquinones are artifacts. Few families have been exhaustively investigated and led to the isolation of a few naphthoquinones and their glycosides. The 7-methyljuglone coming from Droseraceae and Ebenaceae is biosynthesized by the acetate-malonate pathway. However, composites without methyl groups are biosynthesized by the shikimic acid pathways, a situation that parallels that of the acetate-derived 6-methyl-salicylic acid and shikimate-derived salicylic acid. 3-chloroplumbagin, the chlorinated compound isolated from Plumbago (Plambaginaceae) and *Drosera* species (Droseraceae), should initiate a polyketide.

Anthraquinones biosynthesized through two pathways, one of which is emodin, with substituents in both rings A and C, are usually acetate-derived plants (lower and higher plants). Alizarin and its derivatives, without substituents in ring-A, are biosynthesized by the shikimic acid pathway. However, pachybasin from the fungus is acetate-derived, despite the absence of substituents in ring A.

Naphthoquinones started from two different precursors of pathways summarized in Fig. 14.3. Naphthoquinones (shikonin) have a wide variety of metabolites starting from simple compounds like 5-hydroxy derivative juglone to pigments with isoprenyl attachment such as alkannin. The precursor of 4-hydroxybenzoic acid (4HB) is composed of the shikimate and the phenylpropanoid pathway. The precursor of isoprenoid, geranyldiphosphate (GPP), comes from the mevalonate pathway (Zhang et al. 2015). In the first intermediate in shikonin biosynthesis, the 4HB geranyltransferase reaction binds the 4HB precursor to the isoprenoid (GPP) precursor and gives 3-geranyl-4-hydroxybenzoate (GBA). Two enzymes that regulate

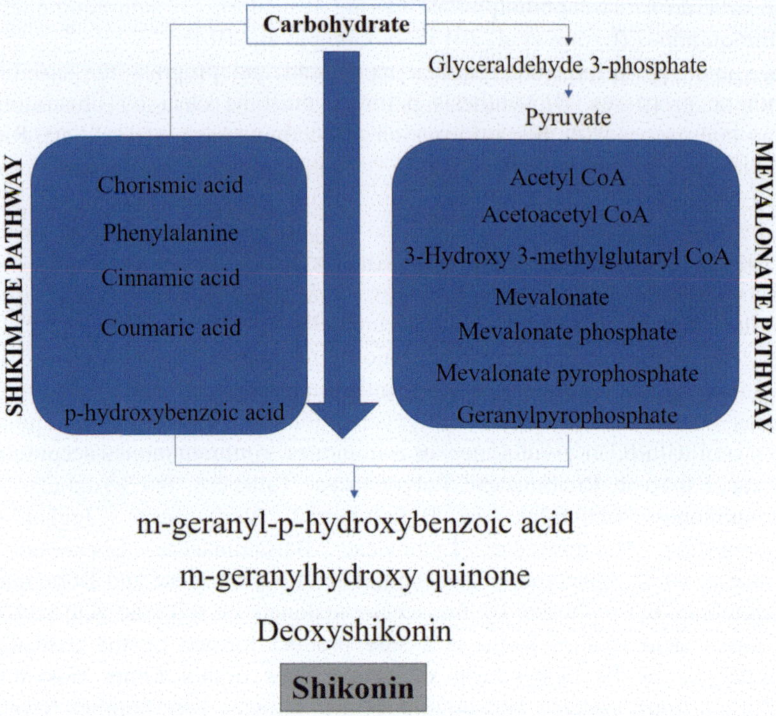

Fig. 14.3 Pathways in Shikonin biosynthesis

shikonin biosynthesis pathways are hydroxyl-3-methylglutaryl co-enzyme-A reductase (HMGR) and 4HB geranyltransferase (Wu et al. 2012a, b). Phenylalanine ammonia-lyase (PAL) and 3-hydroxyl-3-methylglutaryl-CoA-reductase own high activities on biosynthetic enzymes of shikonin from *L. erythrorhizon* (Srinivasan and Ryu 1992).

In eukaryotes, ubiquinone and plastoquinone function manner-oxygenic electron-transport in photosynthesis and the aerobic respiratory chain (Liu and Lu 2016; Malik et al. 2014a, b, c). The quinones function to go with reversible $2e^-$ redox reactions, which protect the cells against free radicals and other potentially harmful oxidants (Kristensen et al. 2020; Liu and Lu 2016; Malik et al. 2014a, b, c).

14.3 In Vitro Production Methodology

Recent plant biotechnological approaches may offer an opportunity to overwhelm variation obstacles against the biotic/abiotic stresses, genetic manipulation, and production of metabolites (Kumar et al. 2011a, b; Malik et al. 2011a, b, 2014a, b, c). Quinones production might be in vitro supportive to understand well the biosynthesis pathway of metabolites. Plant tissue culture (PTC) has totipotency

Table 14.1 Works on quinones production via PTCs

Species	Method	Reference
Lithospermum erythrorhizon		
	Cell suspension Callogenesis Organogenesis and embryogenesis	Yamamoto et al. (2000), Lin and Wu (2002), Yamamoto et al. (2002), Yu et al. (1997)
L. canescens		
	Organogenesis	Sykłowska-Baranek et al. (2012)
Echium lycopsis		
	Callus cultures	Fukui et al. (1983)
A. Euchroma		
	Cell culture Direct regeneration Cell suspension Callogenesis	Zakhlenjuk et al. (1993), Sokha et al. (1996), Ji and Wang (2001), Malik et al. (2010b), Zhang et al. (2013), Pietrosiuk et al. (1999), Sharma et al. (2008), Malik et al. (2010b), Bulgakov et al. (2001)
A. hispidissima		
	Cell culture	Jain et al. (1999), Shekhawat and Shekhawat (2011)
	Organogenesis	Singh et al. (2002), Chaudhury and Pal (2010)
	Callus suspension	Shekhawat and Shekhawat (2011)
Arnebia sp.		
	Cell suspension	Gupta et al. (2013)
Onosma paniculatum		
	Cell suspension	Wu et al. (2009)
E. italicum		
	Cell culture	Zare et al. (2010)
	Callogenesis	Zare et al. (2011)
Rheum emodi wall		
	Organogenesis	Malik et al. (2010a, b)

under complete-control conditions, in which exploiting is in the direction of interests. The PTC was exercised for multipurposes. Various reports used various explants on a broad sense of species (Malik et al. 2011a; Nosov 2012). The production of bioactive metabolites characterizes annually more consumption in different fields. PTCs take advantage of meeting the need for metabolites far away from the traditional extraction methods by intact plants (Nosov 2012). Many reports have accumulated and produced quinones through PTC. The chronicle investigations of quinone production via in vitro tissue cultures are highlighted in Table 14.1. They used different ways such as exposure to elicitors/inducers, precursors, plant growth regulators (PGR) concentrations, and basal media, e.g., Murashige and Skoog (1962) media and Linsmaier and Skoog (1965). Shikonin derivatives and pyrrolizidine alkaloids are isolated from *A. euchroma* (Pietrosiuk et al. 1999, 2006). Sharma et al. (2008) reported acetylshikonin and b-acetoxyisovalerylshikonin through cell suspension proliferation. Jain et al. (1999) produced arnebins from *the A. hispidissima* plant. Some compounds' isolation such as isohexenylnaphthazarin has been proved through cell suspension cultures of *A. euchroma* (Damianakos et al. 2012).

However, few bioactive compounds have been made marketable or got to industrial-scale via PTCs. Numerous efforts have been concentrated on increasing the scale-up production by separating the biosynthetic activities based on choosing proper explants, optimizing the nutrient media, and optimizing the PGRs' concentration and proliferation conditions (Bulgakov et al. 2001; Malik et al. 2008, 2009, 2011a, b; Zare et al. 2011 Zhang et al., 2013). The TDC1 and TDC3 isoforms are induced via abiotic and biotic stress factors in rice (Dharmawardhana et al. 2013).

Various genera belonging to the Boraginaceae family showed a 10% higher content of acetylshikonin and isobutrylshikonin (Pietrosiuk et al. 2006). Alkannin contents accumulated on half-strength agar-solidified MS medium from hairy roots of *A. hispidissima* induced with *A. rhizogenes* (Singh et al. 2002). The functions of guaiacol peroxidase (G-POD), ascorbate peroxidase (APX), and catalase (CAT) at quarter strength (0.25) of MS media-treated root tissue culture of *Morinda citrifolia* were efficient to remove the probable hazard of hydrogen peroxide (H2O2). Baque et al. (2010) demonstrated lowing the accumulation of H2O2 and the level of lipid peroxidation.

Photo-protection as free radical hunters and carotenoids are the precursors for quite a few physiological apocarotenoids (Auldridge et al. 2006; Bouvier et al. 2005) like the phytohormone abscisic acid (ABA), originating from 9-cis-violaxanthin, and 9-cis-neoxanthin (Qin and Zeevaart 2002).

The main complex problem is the impact of turbulence through the scale-up of PTC (Busto et al. 2008; Busto et al. 2013). The impact of unrest and light on cell feasibility, biomass, and anthraquinones production has been measured by Busto et al. (2013). The biomass concentration and anthraquinones production were similar to the achievement of Busto et al. (2008) on *R. tinctorum* suspension cultures by a stirred tank bioreactor. Scheme of simple and downscaled culture proliferation is a valid scale-up for inducing anthraquinones through PTCs.

14.4 Scale-up Techniques and Bioreactors

For long decades, many efforts have been commercially allowed to produce quinone derivatives using PTCs based on the optimizing of media, environment conditions, and downstream parameters. The Mitsui Petrochemical Co., Japan, is one of the first companies in the shikonin production on the commercial scales. The quinone production has fruitfully scaled up using bioreactors. Those productions by PTC and bioreactor technologies have highly accumulated quinones 20–30% rather than the intact plants, which yielded in a few years in only a few days 3–6% (Ge et al. 2006; Renneberg 2008). Several investigations were proceeded to understand well how quinone derivatives are biosynthesizing and their mode of action. Shikonins are produced by changing the biosynthesis pathway (Sommer et al. 1999). The biosynthesis has identified the path of shikonins (Singh et al. 2010). Zhang et al. (2011) characterized that the gene LeERF-1 in *L. erythrorhizon* downregulated shikonin by light optimization. Factors affecting low energy ultrasound (Lin and Wu 2002), physicochemical effects (Malik et al. 2011a, b), and selection of PFP-resistant cell lines (Zakhlenjuk et al. 1993) have been considered (Shekhawat and Shekhawat 2011; Sykłowska-Baranek et al. 2012; Zare et al. 2010, 2011). Zhang et al. (2013) studied the effects of fungal elicitor on accumulation of shikonins.

14.5 Extraction and Detection Techniques

Advancement in phytochemical approaches has witnessed a progression in isolating and detecting biometabolites. Phytochemical methods have been utilized (Mboso et al. 2013; Obadoni and Ochuko 2001; Sofowora 1993; Soladoye and Chukwuma 2012). HPLC is a universal tool to detect secondary metabolites in plants.

14.5.1 Extraction and Isolation of Quinone Compounds

The best tissue for isolating quinone derivatives could be extracted from any part of the tissue based on natural constituents, a species, age of tissue, and existing interest-accumulated tissues.

Box 14.1 Extraction of Quinone Derivatives

- Dry and fine-powder of tissues (200–300 g).
- Dissolve with 500 mL a proper solvent (hexane, chloroform, and ethyl-acetate) using Soxhlet; then filter.
- Subject to column chromatography over silica gel via the gradient elution.
- Read on high-performance liquid-chromatography (HPLC).

Standard quinones are isolated from samples (Box 14.1) that should be fine-powdered and dry well. Proper solvents, including hexane, chloroform, and ethyl acetate, are subjected to using Soxhlet extractions to obtain extracts via column chromatography over silica gel using gradient elution method (Singh et al. 2005). Ethyl acetate extracts on column purification with methanol–chloroform (20:80, v/v) produced emodin glycoside and chrysophanol glycoside. Chloroform extract on column purification with ethyl acetate-hexane (30:70, v/v) cropped emodin, while hexane extract on column purification with ethyl acetate-hexane (5:95, v/v) borne chrysophanol and physcion (Singh et al. 2005). Quinone derivatives scored via spectrophotometer, analyzed using ultra-performance liquid chromatography (UPLC), and compared with the reference values (Coskun et al. 1990; Kubo et al. 1992;Semple et al. 2001; Yang et al. 2016). Plastoquinones, α-tocopherol quinones, and phylloquinone are prime metabolites presenting in all photosynthesizing tissues. Ubiquinone is found in most genera. The quinones, which are prime in plants, are relatively simple benzoquinones, naphthoquinones, or anthraquinones. Skeletal structures of quinones are also found, for instance, terpenoid quinones and higher polycyclic quinones.

14.5.2 Determining Quinone

Box 14.2 Anthraquinone Determination

- Soak 50 mg of the sample in 50 mL of dH2O/16 h.
- Heat sample suspensions at 70oC/1 h in water bath; then cool.
- Add 50 mL of 50% to the sample; the filter.
- Read the spectrophotometric value at a 450 nm.
- Compare with a standard (1 mg/100 mL).

Anthraquinone derivatives could be spectrophotometrically determined as described in Box 14.2, according to Zenk et al. (1975) and Soladoye and Chukwuma (2012). Quinones scored via spectrophotometer, analyzed using high-performance liquid chromatography (HPLC), and compared with the reported values (Semple et al. 2001; Yang et al. 2016).

14.5.3 Genetic Approaches to Detect Quinones

The progression in genetic tools is utilized rapidly for a broad sense of biological disciplines. Low production of quinones by PTC is a chief bottleneck at commercial scales (Charlwood and Pletsch 2002). Therefore, genetic tools, e.g., genetic engineering, can be probable to be employed for regulatory steps of biosynthetic pathways to high-yield content of constituents' scales (Charlwood and Pletsch 2002). Quinone transformation via *Agrobacterium rhizogenes* to promote the bioactive compounds has been stated. Those have been very evidenced genetic stable with rather than 10–20% quinones production such as acetylshikonin and isobutrylshikonin (Boehm et al. 2000; Kohle et al. 2002; Pietrosiuk et al. 2006). Pietrosiuk et al. (2006) produced ca. at higher 10% acetylshikonin and isobutrylshikonin of *L. canescens* hairy roots than natural roots.

Several investigations on genetic and biochemical approaches had found that the 1,4-naphthalenoid ring started from the pathways of shikimate and *o*-succinylbenzoate. Genetic manipulation of various regulatory factors takes advantage of tolerance mechanisms of stress in species. Numerous genes were biosynthesized quinone pathways in a broad sense of species. Gene expression analysis of bioactive constituents may be an additive or inhibitory effect on the expression of PGR actions (Dharmawardhana et al. 2013; Pelagio-Flores et al. 2011). The family of *ubi* genes has a primary role in biosynthesizing quinone pathways (Kohle et al. 2002). The *ubi*A gene has not expressed the shikonin across *L. erythrorhizon* tissue culture (Boehm et al. 2000), whereas quinones have transformed with HMGR cDNA and shown high activity using callogenesis of *A. euchroma* (Malik et al. 2014a, b, c). The genetic manipulation of the biosynthesis pathway accumulated high quinone through PTCs (Kohle et al. 2002; Sommer et al. 1999).

Dihydroorotate:quinone oxidoreductases (DHOQO) are membrane-bound enzymes for oxidizing dihydroorotate (DHO) orotate with concomitant reduction of quinone to quinol. de Sousa et al. (2020) concluded the DHOQOs could be the single dihydroorotate dehydrogenase or could exist with other dihydroorotate dehydrogenases (DHODH), as the NAD+ or fumarate reducing enzymes. Stimulation of different overexpression of genes or suppression regulation could affect the quinones biosynthesis (Malik et al. 2014a, b, c). Freshly, Next Generation Sequencing (NGS) analysis applicable on tissue culture-infected Chardonnay propagules exhibited some genes responsible for photosynthesis process under both light and dark phases that were downregulated (Bertazzon et al. 2019).

14.6 Biological Activities (Major)

Investigations on various biological properties, e.g., medical, pharmaceutical, and other industrial activities, have been shown (Oliveira et al. 2020; Prateeksha et al. 2019). A broad sense of biological activities has been recognized as natural therapeutic and dying properties. Previous reviews reported numerous activities, including antioxidant, anti-ulcer, anti-inflammatory, anticancer, neuroprotective, antiaging, lung protective, and hepatoprotective properties (Choi et al. 2005; Prateeksha et al. 2019; Singh and Chauhan 2004). Plant extracts possess antineoplastic properties and hepatoprotective activity and own the Ayurvedic system using anti-PAF (platelet-activating factor), immunomodulant, blood purifier, and anti-inflammatory. Other extracts from *R. yunnanensis* enhance the ATP magnitude of both brain and heart, an increase of leukocytes, and the treatment of psoriasis (Singh and Chauhan 2004). Major biological investigations are detailed below. Various reports have generally shown the changes in metabolites due to the detrimental infection impacts at molecular and physiological levels (Margaria et al. 2014; Oliveira et al. 2020; Prezelj et al. 2016). Those are due to the extreme gathering of carbohydrates, phenolics, and chlorophyll deficiency (Oliveira et al. 2020; Prezelj et al. 2016). Accumulations such as high starch in leaf mesophylls are leaded to accompany through upregulation of prime genes with their biosynthesis (Hren et al. 2009; Margaria et al. 2014; Oliveira et al. 2020; Prezelj et al. 2016). Anthocyanins detected in micropropagation-infected cv. 'Barbera' leaf propagules and transcript genes of the flavonoid pathway and pro-anthocyanidins were higher in infected plants (Hren et al. 2009). Photosynthesis is the key sponsor of redox equivalents and energy that provides amino acids, secondary metabolites, and carbohydrates (Allahverdiyeva et al. 2015; Margaria et al. 2013; Oliveira et al. 2020; Vitali et al. 2013). Tocopherols perform protecting PSII from photoinactivation and membrane lipids from photooxidation (Allahverdiyeva et al. 2015; Havaux and García-Plazaola

2014). Isoprenoids are vital electron transporters acting in oxidative phosphorylation and photosynthesis (Ducluzeau et al. 2012; Havaux and García-Plazaola 2014).

14.6.1 Chrysophanol

Many quinones possess antidiabetic activity (Prateeksha et al. 2019; Singh and Chauhan 2004). Chrysophanol activity signalizes mainly insulin. It lowers postprandial hyperglycemia by approximately a half percent of the bodyweight of the albino rat (Arvindekar et al. 2015; Choi et al. 2005). This compound served as inactive in contradiction of intestinal alpha-glucosidase enzyme (Arvindekar et al. 2015; Lee and Sohn 2008); as well enhanced the phosphorylation of insulin receptor substrate-1 (IRS-1) through an aggressive impact on its inhibitor, protein-tyrosine phosphatase 1B (PTP-1B) (Arvindekar et al. 2015; Prateeksha et al. 2019). Chrysophanols could raise the phosphorylated-activated protein kinase-B (P-AKT) level in human-liver carcinoma cells (Onoda et al. 2016). Chrysophanol with the peroxisome proliferator-activated receptor (PPAR-γ) might perform against PPAR-γ expressing in adipose tissue and regulates the metabolism of lipid and glucose (Onoda et al. 2016). More articles at the scale of in vitro, in vivo, and in silico assays may provide a useful direction to confirm the agonistic effect on PPAR-γ and DPP-IV (Arvindekar et al. 2015;Onoda et al. 2016; Prateeksha et al. 2019).

Anticancer properties have been established by several investigations via necrosis, a caspase-independent phenomenon, which causes unalterable inflammatory sense against tumor cells (Onoda et al. 2016; Prateeksha et al. 2019). Chrysophanols have lowered the cell viability of liver cancer by necrosis through a proper concentration and incubation time (Lu et al. 2010). Apoptosis-associated signals, including the loss of mitochondrial membrane potential (MMP), raise the reactive oxygen species (ROS), cytosolic Ca + 2, and the release of cytochrome c from mitochondria. It delays the externalization of phosphatidylserine in chrysophanol-treated liver cancer and indicates apoptosis. Extra levels of apoptotic signal-associated proteins, for instance, apoptosis-inducing factor (AIF), were lowered in the treated cells (Lu et al. 2010). Correspondingly, chrysophanol affects cell viability in renal cancer and lung cancer (Choi 2016; Ni et al. 2014). Also, chrysophanol cytotoxicity detected contra human breast cancer and leukemia cells (Choi et al. 2007; Kang et al. 2008; Sun et al. 2014). Chrysophanol has a potential against antagonistic to MYLK4 and affects metastasis and tumor invasion (Demirezer et al. 2016; Lee et al. 2011).

The protective effects of quinones inspected (Lin et al. 2015; Prateeksha et al. 2019) on lipopolysaccharide (LPS)-induced inflammatory and the oxidative response of murine microglial cells were explored [95]. Likewise, the effect on the hippocampus of lead-poisoned neonatal mice was spotted, in which chrysophanol diminished the hippocampal injury and enhanced the learning memory through encouraging the cell-defense system of antioxidants. The lead-exposed neonatal rat has lowered dose-based means in the heart, spleen, liver, kidney, brain, and blood (Lin et al. 2015; Yan et al. 2014; Zhang et al. 2014). The action of chrysophanol on cerebral ischemic stroke through endoplasmic reticulum stress was illuminated (Zhang et al. 2014; Zhao et al. 2016). Chrysophanols have the probability to reduce retinitis pigmentosa (RP), an inherited photoreceptor-degenerative disease. It inhibited apoptosis, gliosis, activation of microglia, and matrix metallopeptidase 9 (MMP-9) expression in N-methyl-N-nitrosourea (MNU)-induced mouse model of RP (Lin et al. 2017; Prateeksha et al. 2019).

Chrysophanol has hepatoprotective effects (Jiang et al. 2016). It regulates inflammatory sensing brought in mice via hepatic injury. The block of caspases-associated apoptosis induction indicated to chrysophanol could shield in contradiction of liver injury by its antiapoptotic effect (Jiang et al. 2016). Hepatoprotective activity via chrysophanol lowered the gamma-glutamyl transpeptidase (GGT) activity that necessitates GSH homeostasis. The immunoblot analyses decrease in the expressions of cytochrome, GGT, and GSH (Jiang et al. 2016; Qian et al. 2011). Some reports suggested that the chrysophanol could save against the gastrointestinal effects of cold-resistant ulcers, alcohol, aspirin, and pyloric ligation-induced ulcer in mice (Suleyman et al. 2004). Chrysophanols have decreased the acids through the inhabitation of H+/K + -ATPase activity. Nevertheless, its activity shows less than the emodin molecule. Upregulation of mucin secretion, a protect mechanism of the ulcer, is also noted in chrysophanol-treated rats.

Regarding the activities of anti-inflammatory, various essays stated the mode of defense action via chrysophanol in numerous diseases (Malik and Muller 2016). The effects of chrysophanol on dextran sulfate sodium (DSS)-induced colitis and LPS-induced inflammatory were sensed by Kim et al. (2010) in rats' peritoneal macrophages. Furthermore, chrysophanol significantly removes atopic dermatitis that rises in phosphorylated-mitogen-activated protein kinase. This active constituent, AST2017-01, is a potent anti-inflammatory of konsentrasi hambat minimum (KHM) or functional food (Jeong et al. 2018). Carrageenan, histamine, dextran, serotonin formaldehyde-induced edema tests, cotton-pellet granuloma, and Kabak tests could be used to examine anti-inflammatory in mice (Jeong et al. 2018; Suleyman et al. 1999).

The activity of chrysophanol has shown anti-microorganisms, e.g., viruses (Bunluepuech et al. 2016; Chang et al. 2014; Ramana et al. 2017), fungi (Choi et al. 2004; Liu et al. 2009; Malik and Muller 2016; Ren et al. 2012), and bacteria (Rodrigues et al. 2017; Singh et al. 2017).

Numerous pharmacological activities on chrysophanols are recognized. This molecule is an antiprotozoal factor agonist chloroquine-resistant and sensitive to *Plasmodium falciparum* strain (Abdissa et al. 2017). The molecule chrysophanol is demonstrated as an anti-obesity response to recover pulmonary injury in mice through advanced anti-inflammatory and antioxidant retort to injured tissue (Li et al. 2016). Also, it inhibits triglyceride and cholesterol in the fish zebra, which is on a high cholesterol diet. Antituberculosis affects *Mycobacterium tuberculosis* and *M. bovis* (Schorkhuber et al. 1998). The antinomic activity of chrysophanol against root-knot nematode was also testified (Tripathi et al. 2014). Likewise, chrysophanols are the main constituents of various species such as Cassia and Aloe owning the laxative effect (Malik and Muller 2016).

14.6.2 Aloe-Emodin (AE)

Quinone of AE is a bioactive hydroxyanthraquinone, which exists in a broad spectrum of medicinal plants, having laxative, antifungal, antiviral, anticancer, and hepatoprotective activities (Agarwal et al. 2000; Arosio et al. 2000). The modes of human cancer contain various carcinoma cells which include nasopharyngeal carcinoma, lung cancer, leukemia, neuroectodermal tumor, colon cancer, and hepatocellular carcinoma (Lee et al. 2006; Lin et al. 2010, 2011; Pecere et al. 2000; Suboj et al. 2012). AE-induced cancer cell cultivating inhibition is produced by prolonged G1, S, or G2/M phase cell cycle capture or apoptosis based on the tissues and treatment procedure (Guo et al. 2007, 2008; Xiao et al. 2007). This remarkably prevents chorioallantoic membrane angiogenesis and inhibits tubule formation of endothelial cells on matrigel (Cardenas et al. 2006; Chiu et al. 2009). AE suppresses cancer cell migration, invasion, and metastasis (Chen et al. 2010a, b; Tabolacci et al. 2010). Moreover, antiproliferative effects come by the linkage of 5-fluorouracil, doxorubicin, tyrosine kinase, and the cisplatin inhibitor (Fenig et al. 2004).

AE encourages DNA damage and obstructs DNA repair gene expression in cancer cells akin to the promotion of reactive oxygen species (ROS) (Lee et al. 2006). It downregulates cyclin A, cyclin-dependent kinase 2 (CDK2), protein kinase C (PKC), and c-Myc as well as upregulates cyclin B1, CDK1, p53, and p21 (Chiu et al. 2009; Guo et al. 2007, 2008). The c-Jun N-terminal kinase (JNK) activation (Lu et al. 2007), p53 pathway, Fas pathway, and caspase activation are mechanistically involved in aloe-emodin-induced apoptosis (Chiu et al. 2009; Pecere et al. 2003). Aloe-emodin also performs the following activities: decreases the levels of urokinase (Cardenas et al. 2006); suppresses the expression of metalloproteinase-2 (MMP-2) and metalloproteinase-9 (MMP-9) (Chen et al. 2010a, b; Tabolacci et al. 2010); as well locks up translocation and the binding of DNA binding (Suboj et al. 2012).

Other hydroxy-anthraquinones, like emodin, rhein, and chrysophanol, own anticancer effects. Their mode of action is akin to those of AE (Cha et al. 2005; Chun-Guang et al. 2010; Lu et al. 2010; Shi et al. 2008; Tan et al. 2011). Further, some derivatives such as 1,8-Di-O-alkylaloe-emodin and 15-amino-emodin, 15-thiocyano-emodin, and 15-selenocyanochrysophanol initiated from AE might show potential to the cytotoxic effects on cancer cells (Cui et al. 2008).

14.6.3 Juglone

Juglone is isolated from *Juglans mandshurica* Maxim. Their bioactivities, including anticancer and antimicrobials, have been demonstrated in anti-growth, induction of apoptosis, and intermediate to G2-phase cell cycle (Aithal et al. 2009, 2011; Li et al. 2010). Juglones inhibit the intestinal carcinogenesis of mice treating with azoxymethane. They are hopefully chemo-defensive agents for intestinal neoplasia. Juglone shows to be possibly immune-stimulating for oncogenes (Polonik et al. 2003). Juglones have been activated synergistically by the cytotoxicity of the etoposide process. The toxicity of juglone could not be ignored (Mathur et al. 2011; Seshadri et al. 2011). The anticancer activity of juglone could be attributed to the reduction of glutathione (GSH), induction of oxidative stress, and cell death through apoptosis and necrosis due to the harm of cell membranes (Fila et al. 2008; Ji et al. 2011). In addition, the inhibition of peptidyl-prolyl isomerase overexpresses in several cancer cells (Chao et al. 2001; Xu et al. 2012). Juglones suppress the activator protein-2α (AP-2α) and the epidermal growth factor receptor-2 (HER-2) promoter activity, the changing of B-cell lymphoma 2 (Bcl-2) and Bcl-2-associated x protein (Bax), the activating of poly (ADP-ribose) polymerase (PARP), the release of cytochrome *c*, Smac, and apoptosis-inducing factor (AIF), and the induction of caspases activation (Fila et al. 2008; Ji et al. 2011; Khanal et al. 2010; Xu et al. 2010). Some analogs including 2,5-dihydroxy-3-(3-methylbut-2-enyl) naphthalene-1,4-dione and 2,3-dihydro-5-hydroxy-2-(prop-1-2-enyl) naphtho-[2,3-b]-furan-4,9-dione are most potential to anti-growth factors (Bonifazi et al. 2010).

14.6.4 β-Lapachone

A natural naphthoquinone, *β*-Lapachone, is isolated from *Tabebuia avellanedae* Lor. (Hussain et al. 2007). Their activities include anticancer, antimicrobials, and anti trypanocidal activities (Medeiros et al. 2010). *β*-lapachones activate selective death with a broad sense of cancers without eliminating unaffected tissues and anti-growth at the G1 transition (Li et al. 2003). Anticancer effects have been demonstrated in xenograft animals (; Dong et al. 2009). The *β*-lapachone constituent with radiotherapy has shown a potential treatment for agonist cancer (Suzuki et al. 2006). *β*-lapachone is utilized in monotherapy and other medicines (Miao et al. 2009). *β*-Lapachone activates topoisomerase II (Topo II) *α*-mediated DNA breaks instead of the Topo I-mediated DNA breaks. It inhibits the activity of Topo I contrary to camptothecin (Bentle et al. 2007; Frydman et al. 1997; Hori et al. 2011; Li et al. 1993). Therefore, *β*-lapachone inhibits DNA polymerase, the elicitation of endoplasmic reticulum stress, and induction of caspases activation (Lien et al. 2008). The increasing cytosolic Ca2+ is a vital mechanism for *β*-lapachone-induced apoptosis (Hori et al. 2011).

β-lapachone makes cells more sensitive to radiation through interrupting the contribution of quinone oxidoreductase (NQO1) in radiation-induced activation of NF-κB and inhibiting the repair of sublethal radiation damage (Bentle et al. 2007; Dong et al. 2010; Suzuki et al. 2006). However, *β*-lapachone has significant effects agonist antitumor; it was terminated as an antineoplastic in 1970 because of its toxicity (Almeida 2009).

14.6.5 Plumbagin

Plumbagins, an organic yellow dye, is isolated from *Plumbago* spp. L., and *Dyerophytum africanum* (Lam.) Kuntze. Their effects have been utilized in agonist antioxidant, anticancer, anti-inflammatory, analgesic, and antimicrobial activities (Luo et al. 2010; Padhye et al. 2012). Plumbagins prevent prostate tumor, cancer angiogenesis, growth of carcinoma, and cancer growth (Lai et al. 2012; Li et al. 2012; Sand et al. 2012; Subramaniya et al. 2011; Sun and McKallip 2011). This effect can be seen in solid tumors and Ehrlich ascites model, melanoma, and hormone-refractory prostate cancer (Lai et al. 2012). Furthermore, plumbagin sponsors micronuclei induction (Aziz et al. 2008; Kumar et al. 2011a, b; Lai et al. 2012). Besides, plumbagin encourages cell cycle arrest at the G2/M phase (Gomathinayagam et al. 2008; Wang et al. 2008). This mechanism is due to their apoptosis, including intracellular ROS induction (Xu and Lu 2010), upregulation of p53, downregulation of cyclooxygenase-2 (COX-2), encouragement of caspases, and Bcl-2 family proteins (Kawiak et al. 2007; Li et al. 2012; Powolny and Singh 2008; Subramaniya et al. 2011; Sun and McKallip 2011; Tian et al. 2012; Xu et al. 2010).

Plumbagin inactivates the NF-κB pathway that suppresses NF-κB-regulated gene products, such as IAP, Bcl-2, Bcl-xL, survivin, cyclin D1, COX-2, and MMP-9 (Sandur et al. 2006). This constituent prevents the invasion and evolution of cancer by downregulating the expression of chemokine receptor CXCR4 by NF-κB inhibition (Manu et al. 2011). Furthermore, plumbagin has multidrug resistance related to ATP binding cassette drug transporter (Shukla et al. 2007). Other bioactivities inhibit mycobacterial development and reduce insect feed. (Dandawate et al. 2012; Mathew et al. 2010; Sreelatha et al. 2009).

14.6.6 Shikonin

Shikonins are isolated from the roots of *Lithospermum erythrorhizon* Sieb. et Zucc., which are utilized for burns, sore throats, measles, carbuncles, macular eruptions, and allergic disease (Wang et al. 2010). This bioactive constituent as a naphthoquinone pigment has anticancer, antioxidant, and anti-inflammatory activities (Wang et al. 2010; Yang et al. 2009). Anticancer activities involving the induction of apoptosis, delay of invasion, and antiproliferation have been in vivo determined in homograft and heterograft animal models (Yang et al. 2009). Anticancer mechanisms inactivate the NF-κB and upregulate p53 and p21, ROS generation. Others downregulate the CDKs and ER-α and activate caspases (Chang et al. 2010; Min et al. 2008; Min et al. 2011; Rao et al. 2011; Wu et al. 2012a, b; Yao and Zhou 2010). Also, this bioactive quinone prevents Topo II activity and damage DNA (Kretschmer et al. 2012; Yang et al. 2006). Shikonin is a chosen estrogen enzyme modulator expressing steroid sulfatase (STS) for estrogen biosynthesis (Zhang et al. 2009). As well, shikonin analogs may inhibit pyruvate kinase-M2 (Chen et al. 2011). Natural analogs evade drug resistance mediated by multiple drug resistance protein 1 and breast cancer resistance protein 1. They wield antiangiogenesis effect by the hang-up of vascular endothelial growth factor receptor 2 (Komi et al. 2009; Xuan and Hu 2009).

14.6.7 Thymoquinone

Thymoquinone is isolated from *Nigella sativa* L. This bioactive formulation belonging to benzoquinones has been vitally utilized for antioxidant, anti-inflammatory, and anticancer activities (Banerjee et al. 2009; Chehl et al. 2009; Sayed-Ahmed et al. 2010). The anticancer effects refer to diverse modes of action with the inducing of apoptosis, antiproliferation, antiangiogenesis, anti-metastasis, and cell cycle seizure (Banerjee et al. 2009; El-Mahdy et al. 2005; El-Najjar et al. 2010; Kolli-Bouhafs et al. 2011). The antitumor was examined in xenograft mice models for tumors of prostate, pancreatic, and colon (Banerjee et al. 2009; Gali-Muhtasib et al. 2008; Jafri et al. 2010; Kaseb et al. 2007). Thymoquinone prevents doxorubicin-resistant human breast cancer MCF-7/DOX cell proliferation deprived of unveiling cytotoxicity to normal human colonic FHs74Int cells (Arafa et al. 2011; El-Najjar et al. 2010). Thymoquinone with orthodox chemotherapeutic medicines, including cisplatin, gemcitabine, 5-fluorouracil, doxorubicin, and oxaliplatin, affect various cancer families (Banerjee et al. 2009; Jafri et al. 2010; Lei et al. 2012; Woo et al. 2011). Doses of toxic thymoquinone make the effusion of lymphoma cells more sensitive to the TNF-related apoptosis-induced ligand. It is due to the upregulation of the death receptor-5 (Hussain et al. 2011).

The anticancer activity effect takes place via the modification of numerous molecular targets, for instance, NF-κB, p73, AKT, tubulin, peroxisome proliferator-activated receptor γ (PPARγ), the signal transducer and activator of transcription 3 (STAT3), phosphatase and tensin homolog (PTEN), polo-like kinase-1 (PLK1), androgen receptor (AR), E2F-1, Bcl-2, FAK, and MMPs (Alhosin et al. 2010; 2012; Arafa et al. 2011; Badr et al. 2011a, b; Banerjee et al. 2009; Hussain et al. 2011; Kaseb et al. 2007; Kolli-Bouhafs et al. 2011; Reindl et al. 2008; Sethi et al. 2008; Woo et al. 2011). Thymoquinone produces ROS. Thymoquinone damages DNA and inhibits telomerase activity (El-Najjar et al. 2010; Gurung et al. 2010; Hussain et al. 2011). Treatment by thymoquinone stops CXCL12-mediated chemotaxis in myeloma cells and lowers CXCR4 expression and CXCL12-mediated CXCR4/CD45 association (Badr et al. 2011a, b). Numerous analogs of thymoquinone are manufactured utilizing modulations at the benzenoid or carbonyl spots. Some formulations are more effective than thymoquinone and affect anti-proliferation in human pancreatic cancers (Banerjee et al. 2010). The 6-hencosahexaenyl conjugate is an effective antiproliferative of 518A2 melanoma, HL-60 leukemia, KB-V1/Vbl cervix carcinoma, and MCF-7/Topo breast adenocarcinoma cells (Breyer et al. 2009).

14.7 Commercial Utilization and Prospects

Quinones are utilized and served in various capacities. They manufactured dye fabrics (red-violet) and used them to transport energy and hydrogen atoms (Coenzyme Q10). Therefore, they provide essential nutrients (Vitamin K) and serve as cocatalysts (PQQ). In the pharmaceutical industry, quinones eradicate cancer (doxorubicin), HIV (streptonigrin, STN), and bacteria (prekinamycin).

Previous pharmacologic studies stated promising activities, including antipruritic, anti-dermatitic, antihistaminic, antimicrobial (bacteria and fungi), analgesic, antioxidant, anti-inflammatory, antirheumatic, anti-anaphylactic, antitumor, and anticancer activities. Various extracts are often associated with an existence of natural constituents, including quinones, glycosides, alkaloids, flavonoids/flavanols, anthocyanins, saponins, phenolics, and terpenoids.

Orobanchaceae plants' parasitizing roots that reason extensive losses in agro-production worldwide advance haustoria in the existence of the DMBQ. The host-derived quinone compound DMBQ is an effective haustorium-inducing factor in various Orobanchaceae species. The molecular processes that are basis of conception and go in advance of the formation of the haustorium are unclear. The DMBQ signaling components identification is challenging in parasitic plants, as key research tools, for instance, trans-generational genetic changes are unavailable. Understanding plant quinone identifying and the molecular proceedings that track DMBQ conception would sense a better comprehend of signaling and offer a molecular target to contest parasitic plants. This offers insights into the role of quinone signaling. Perception of CARD1 function helps to better realize the signaling pathways during the haustorium formation in parasitic plants and the immunity events in nonparasitic plants.

14.8 Conclusions and Recommendations

The importance of commercial and societal demand makes the scientific-gate always require more efforts towards biosynthesis pathway and its scale-ups of natural products. However, evident efforts on in vitro production are obviously from new works. Progression of plant biotechnology helps obviously to be utilized in various directions: (1) biodiversity conservation. The demand for raw materials of tissues, (2) alternative means. It offers numerous new strategies to justify industrial calls, (3) eating-up the time. The high and fast yield of metabolite products may reduce the time, (4) biosynthesis pathway. Help to sense well the complex pathways of biometabolites, (5) physiological and genetic correlation. Well, understand the relation between bearers of metabolic engineering and suppression/overexpression genes. (6) chemotaxonomic marker.

References

Abdissa D, Geleta G, Bacha K, Abdissa N (2017) Phytochemical investigation of Aloe pulcherrima roots and evaluation for its antibacterial and antiplasmodial activities. PLoS One 12:e0173882

Agarwal SK, Singh SS, Verma S, Kumar S (2000) Antifungal activity of anthraquinone derivatives from Rheum emodi. J Ethnopharmacol 72(1–2):43–46

Aithal BK, Kumar MR, Rao BN et al (2009) Juglone, a naphthoquinone from walnut, exerts cytotoxic and genotoxic effects against cultured melanoma tumor cells. Cell Biol Int 33(10): 1039–1049

Aithal KB, Kumar S, Rao BN et al (2011) Tumor growth inhibitory effect of juglone and its radiation sensitizing potential: *in vivo* and *in vitro* studies. Integr Cancer Ther 11(1):68–80

Alhosin M, Abusnina A, Achour M et al (2010) Induction of apoptosis by thymoquinone in lymphoblastic leukemia Jurkat cells is mediated by a p73-dependent pathway which targets the epigenetic integrator UHRF1. Biochem Pharmacol 79(9):1251–1260

Alhosin M, Ibrahim A, Boukhari A et al (2012) Anti-neoplastic agent thymoquinone induces degradation of alpha and beta tubulin proteins in human cancer cells without affecting their level in normal human fibroblasts. Invest New Drugs 30(5):1813–1819

Allahverdiyeva Y, Suorsa M, Tikkanen M, Aro EM (2015) Photoprotection of photosystems in fluctuating light intensities. J Exp Bot 66(9):2427–2436. https://doi.org/10.1093/jxb/eru463

Almeida ER (2009) Preclinical and clinical studies of lapachol and beta-lapachone. Open Nat Prod J 2:42–47

Arafa el SA, Zhu Q, Shah ZI et al (2011) Thymoquinone upregulates PTEN expression and induces apoptosis in doxorubicin resistant human breast cancer cells. Mutat Res 706(1–2):28–35

Arosio B, Gagliano N, Fusaro LM et al (2000) Aloe-Emodin quinone pretreatment reduces acute liver injury induced by carbon tetrachloride. Pharmacol Toxicol 87(5):229–233

Arvindekar A, More T, Payghan PV et al (2015) Evaluation of anti-diabetic and alpha glucosidase inhibitory action of anthraquinones from Rheum emodi. Food Funct 6:2693–2700

Auldridge ME, McCarty DR, Klee HJ (2006) Plant carotenoid cleavage oxygenases and their apocarotenoid products. Curr Opin Plant Biol 9(3):315–321

Aziz MH, Dreckschmidt NE, Verma AK (2008) Plumbagin, a medicinal plant-derived naphthoquinone, is a novel inhibitor of the growth and invasion of hormone-refractory prostate cancer. Cancer Res 68(21):9024–9032

Badr G, Lefevre EA, Mohany M (2011a) Thymoquinone inhibits the CXCL12-induced chemotaxis of multiple myeloma cells and increases their susceptibility to Fas-mediated apoptosis. PLoS One 6(9):e23741

Badr G, Mohany M, Abu-Tarboush F (2011b) Thymoquinone decreases F-actin polymerization and the proliferation of human multiple myeloma cells by suppressing STAT3 phosphorylation and Bcl2/Bcl-XL expression. Lipids Health Dis 10:236

Banerjee S, Kaseb AO, Wang Z et al (2009) Antitumor activity of gemcitabine and oxaliplatin is augmented by thymoquinone in pancreatic cancer. Cancer Res 69(13):5575–5583

Banerjee S, Azmi AS, Padhye S et al (2010) Structure-activity studies on therapeutic potential of Thymoquinone analogs in pancreatic cancer. Pharm Res 27(6):1146–1158

Baque MA, Lee E-J, Paek K-Y (2010) Medium salt strength induced changes in growth, physiology and secondary metabolite content in adventitious roots of *Morinda citrifolia*: the role of antioxidant enzymes and phenylalanine ammonia lyase. Plant Cell Rep 29:685–694. https://doi.org/10.1007/s00299-010-0854-4

Bentle MS, Reinicke KE, Dong Y et al (2007) Nonhomologous end joining is essential for cellular resistance to the novel antitumor agent, beta-lapachone. Cancer Res 67(14):6936–6945

Bertazzon N, Bagnaresi P, Forte V et al (2019) Grapevine comparative early transcriptomic profiling suggests that Flavescence dorée phytoplasma represses plant responses induced by vector feeding in susceptible varieties. BMC Genomics 20(1):526. https://doi.org/10.1186/s12864-019-5908-6

Boehm R, Sommer S, Li SM, Heide L (2000) Genetic engineering on shikonin biosynthesis: expression of the bacterial *ubi*A gene in *Lithospermum erythrorhizon*. Plant Cell Physiol 41: 911–919

Bonifazi EL, Rios-Luci C, Leon LG et al (2010) Antiproliferative activity of synthetic naphthoquinones related to lapachol. First synthesis of 5-hydroxylapachol. Bioorg Med Chem 18(7):2621–2630

Bouvier F, Rahier A, Camara B (2005) Biogenesis, molecular regulation and function of plant isoprenoids. Prog Lipid Res 44:357–429

Breyer S, Effenberger K, Schobert R (2009) Effects of thymoquinone-fatty acid conjugates on cancer cells. Chem Med Chem 4(5):761–768

Bulgakov VP, Kozyrenko MM, Fedoreyev SA et al (2001) Shikonin production by p-fluorophenylalanine resistant cells of *Lithospermum erythrorhizon*. Fitoterapia 72:394–401

Bunluepuech K, Tewtrakul S, Wattanapiromsakul C (2016) Anti-HIV-1 protease activity of compounds from *Cassia garrettiana*. Walailak J Sci Technol 13:827–835

Busto VD, Rodríguez-Talou J, Giulietti AM, Merchuk JC (2008) Effect of shear stress on anthraquinones production by Rubia tinctorum suspension cultures. Biotechnol Prog 24:175–181

Busto VD, Calabró-López A, Rodríguez-Talou J et al (2013) Anthraquinones production in Rubia tinctorum cell suspension cultures: down scale of shear effects. Biochem Eng J 77:119–128

Camelio AM, Johnson TC, Siegel D (2015) Total synthesis of celastrol, development of a platform to access celastroid natural products. J Am Chem Soc 137(37):11864–11867

Cardenas C, Quesada AR, Medina MA (2006) Evaluation of the antiangiogenic effect of aloe-emodin. Cell Mol Life Sci 63(24):3083–3089

Cha TL, Qiu L, Chen CT et al (2005) Emodin down-regulates androgen receptor and inhibits prostate cancer cell growth. Cancer Res 65(6):2287–2295

Chakraborty P (2018) Herbal genomics as tools for dissecting new metabolic pathways of unexplored medicinal plants and drug discovery. Biochim Open 6:9–16

Chang IC, Huang YJ, Chiang TI et al (2010) Shikonin induces apoptosis through reactive oxygen species/extracellular signal-regulated kinase pathway in osteosarcoma cells. Biol Pharm Bull 33(5):816–824

Chang SJ, Huang SH, Lin YJ et al (2014) Antiviral activity of *Rheum palmatum* methanol extract and chrysophanol against Japanese encephalitis virus. Arch Pharm Res 37:1117–1123

Chao SH, Greenleaf AL, Price DH (2001) Juglone, an inhibitor of the peptidyl-prolyl isomerase Pin1, also directly blocks transcription. Nucleic Acids Res 29(3):767–773

Charlwood BV, Pletsch M (2002) Manipulation of natural product accumulation in plants through genetic engineering. Int J Geogr Inf Syst 9(2–3):139–151

Chaudhury A, Pal M (2010) Induction of Shikonin production in hairy root cultures of *Arnebia hispidissima* via *agrobacterium rhizogenes*-mediated genetic transformation. J Crop Sci Biotechnol 13:99–106

Chehl N, Chipitsyna G, Gong Q et al (2009) Antiinflammatory effects of the Nigella sativa seed extract, thymoquinone, in pancreatic cancer cells. HPB 11(5):373–381

Chen YY, Chiang SY, Lin JG et al (2010a) Emodin, aloe-emodin and rhein inhibit migration and invasion in human tongue cancer SCC-4 cells through the inhibition of gene expression of matrix metalloproteinase-9. Int J Oncol 36(5):1113–1120

Chen YY, Chiang SY, Lin JG et al (2010b) Emodin, aloe-emodin and rhein induced DNA damage and inhibited DNA repair gene expression in SCC-4 human tongue cancer cells. Anticancer Res 30(3):945–951

Chen J, Xie J, Jiang Z et al (2011) Shikonin and its analogs inhibit cancer cell glycolysis by targeting tumor pyruvate kinase-M2. Oncogene 30(42):4297–4306

Chiu TH, Lai WW, Hsia TC et al (2009) Aloe-emodin induces cell death through S-phase arrest and caspase-dependent pathways in human tongue squamous cancer SCC-4 cells. Anticancer Res 29(11):4503–4511

Choi JS (2016) Chrysophanic acid induces necrosis but not necroptosis in human renal cell carcinoma Caki-2 cells. J Cancer Prev 21:81–87

Choi GJ, Lee SW, Jang KS et al (2004) Effects of chrysophanol, parietin, and nepodin of Rumex crispus on barley and cucumber powdery mildews. Crop Prot 23:1215–1221

Choi SZ, Lee SO, Jang KU et al (2005) Antidiabetic stilbene and anthraquinone derivatives from Rheum undulatum. Arch Pharm Res 28:1027–1030

Choi SG, Kim J, Sung ND et al (2007) Anthraquinones, Cdc25B phosphatase inhibitors, isolated from the roots of Polygonum multiflorum Thunb. Nat Prod Res 21:487–493

Chun-Guang W, Jun-Qing Y, Bei-Zhong L et al (2010) Anti-tumor activity of emodin against human chronic myelocytic leukemia K562 cell lines in vitro and in vivo. Eur J Pharmacol 627(1–3):33–41

Coskun M, Satake T, Hori K (1990) Anthraquinone glycosides from *Rhamnus libanoticus*. Phytochemistry 29:2018–2020

Cui XR, Takahashi K, Shimamura T et al (2008) Preparation of 1,8-di-O-alkylaloe-emodins and 15-amino-, 15-thiocyano-, and 15-selenocyanochrysophanol derivatives from aloe-emodin and studying their cytotoxic effects. Chem Pharm Bull 56(4):497–503

Damianakos H, Kretschmer N, Sykłowska-Baranek K et al (2012) Antimicrobial and cytotoxic isohexenylnaphthazarins from *Arnebia euchroma* (Royle) Jonst. (Boraginaceae) callus and cell suspension culture. Molecules 17:14310–14322

Dandawate P, Khan E, Padhye S et al (2012) Synthesis, characterization, molecular docking and cytotoxic activity of novel plumbagin hydrazones against breast cancer cells. Bioorg Med Chem Lett 22(9):3104–3108

de Sousa FM, Refojo PN, Pereira MM (2020) Investigating the amino acid sequences of membrane bound dihydroorotate: quinone oxidoreductases (DHOQOs): structural and functional implications. Biochim Biophys Acta – Bioenerg 1862(1):148321. https://doi.org/10.1016/j.bbabio.2020.148321

Demirezer LOO, Mine U, Yukselen O et al (2016) Molecular docking of Anthranoids on some targeted human proteins. FABAD J Pharm Sci 41:1–16

Dharmawardhana P, Ren L, Amarasinghe V et al (2013) A genome scale metabolic network for rice and accompanying analysis of tryptophan, auxin and serotonin biosynthesis regulation under biotic stress. Rice 6:15

Dong Y, Chin SF, Blanco E et al (2009) Intratumoral delivery of beta-lapachone via polymer implants for prostate cancer therapy. Clin Cancer Res 15(1):131–139

Dong GZ, Oh ET, Lee H et al (2010) Beta-lapachone suppresses radiation-induced activation of nuclear factor-kappaB. Exp Mol Med 42(5):327–334

Drummond L, Kschowak MJ, Breitenbach J et al (2019) Expanding the isoprenoid building block repertoire with an IPP methyltransferase from Streptomyces monomycini. ACS Synth Biol 8(6): 1303–1313

Ducluzeau AL, Wamboldt Y, Elowsky CG et al (2012) Gene network reconstruction identifies the authentic trans-prenyl diphosphate synthase that makes the solanesyl moiety of ubiquinone-9 in Arabidopsis. Plant J 69:366–375. https://doi.org/10.1111/j.1365-313X.2011.04796.x

El-Mahdy MA, Zhu Q, Wang QE et al (2005) Thymoquinone induces apoptosis through activation of caspase-8 and mitochondrial events in p53-null myeloblastic leukemia HL-60 cells. Int J Cancer 117(3):409–417

El-Najjar N, Chatila M, Moukadem H et al (2010) Reactive oxygen species mediate thymoquinone-induced apoptosis and activate ERK and JNK signaling. Apoptosis 15(2):183–195

Eyong KO, Kuete V, Eferth T (2013a) Quinones and benzophenones from the medicinal plants of Africa. Med plant res Afr Pharmacol Chem. Elsevier, pp 351–391. https://doi.org/10.1016/B978-0-12-405927-6.00010-2

Eyong KO, Puppala M, Kumar PS et al (2013b) A mechanistic study on the hooker oxidation: synthesis of novel indane carboxylic acid derivatives from lapachol. Org Biomol Chem 11(3): 459–468

Farmer EE, Davoine C (2007) Reactive electrophile species. Curr Opin Plant Biol 10:380–386

Fenig E, Nordenberg J, Beery E et al (2004) Combined effect of aloe-emodin and chemotherapeutic agents on the proliferation of an adherent variant cell line of Merkel cell carcinoma. Oncol Rep 11(1):213–217

Fila C, Metz C, van der Sluijs P (2008) Juglone inactivates cysteinerich proteins required for progression through mitosis. J Biol Chem 283(31):21714–21724

Frydman B, Marton LJ, Sun JS et al (1997) Induction of DNA topoisomerase II-mediated DNA cleavage by beta-lapachone and related naphthoquinones. Cancer Res 57(4):620–627

Fukui H, Tsukada M, Mizukami H, Tabata M (1983) Formation of stereoisomeric mixtures of naphthoquinone derivatives in *Echium lycopsis* callus cultures. Phytochemistry 22:453–456

Gali-Muhtasib H, Kuester D, Mawrin C et al (2008) Thymoquinone triggers inactivation of the stress response pathway sensor CHEK1 and contributes to apoptosis in colorectal cancer cells. Cancer Res 68(14):5609–5618

Ge F, Yuan X, Wang X et al (2006) Cell growth and shikonin production of *Arnebia euchroma* in a periodically submerged airlift bioreactor. Biotechnol Lett 28:525–529

Gomathinayagam R, Sowmyalakshmi S, Mardhatillah F et al (2008) Anticancer mechanism of plumbagin, a natural compound, on non-small cell lung cancer cells. Anticancer Res 28(2A):785–792

Gong H, He Z, Peng A, Zhang X, et al (2014) Efects of several quinones on insulin aggregation. Sci Rep 4:1–8. https://doi.org/10.1038/srep05648

Guo JM, Xiao BX, Liu Q et al (2007) Anticancer effect of aloe-emodin on cervical cancer cells involves G2/M arrest and induction of differentiation. Acta Pharmacol Sin 28(12):1991–1995

Guo J, Xiao B, Liu Q et al (2008) Suppression of C-myc expression associates with anti-proliferation of aloe-emodin on gastric cancer cells. Cancer Invest 26(4):369–374

Gupta K, Garg S, Singh J, Kumar M (2013) Enhanced production of napthoquinone metabolite (shikonin) from cell suspension culture of *Arnebia sp.* and its up-scaling through bioreactor. 3 Biotech 4:263–273

Gurung RL, Lim SN, Khaw AK et al (2010) Thymoquinone induces telomere shortening, DNA damage and apoptosis in human glioblastoma cells. PLoS One 5(8):e12124

Havaux M, García-Plazaola JI (2014) Beyond non-photochemical fluorescence quenching: the overlapping antioxidant functions of zeaxanthin and tocopherols. In: Non-photochemical quenching and energy dissipation in plants, algae and cyanobacteria. Springer, Dordrecht, pp 583–603

Hillion M, Antelmann H (2015) Thiol-based redox switches in prokaryotes. Biol Chem 396:415–444

Hori T, Kondo T, Lee H et al (2011) Hyperthermia enhances the effect of beta-lapachone to cause gammaH2AX formations and cell death in human osteosarcoma cells. Int J Hyperthermia 27(1): 53–62

Hren M, Nikolić P, Rotter A et al (2009) Bois noir'phytoplasma induces significant reprogramming of the leaf transcriptome in the field grown grapevine. BMC Genomics 10(1):460. https://doi.org/10.1186/1471-2164-10-460

Hussain H, Krohn K, Ahmad VU et al (2007) Lapachol: an overview. Rev Accounts Arkivoc 2: 145–171. https://doi.org/10.3998/ark.5550190.0008.204

Hussain AR, Ahmed M, Ahmed S et al (2011) Thymoquinone suppresses growth and induces apoptosis via generation of reactive oxygen species in primary effusion lymphoma. Free Radic Biol Med 50(8):978–987

Jafri SH, Glass J, Shi R et al (2010) Thymoquinone and cisplatin as a therapeutic combination in lung cancer: in vitro and in vivo. J Exp Clin Cancer Res 29:87

Jain SC, Singh B, Jain R (1999) Arnebins and antimicrobial activities of *Arnebia hispidissima* (Lehm.) DC cell cultures. Phytomedicine 6:474–476

Jeong HJ, Kim HY, Kim HM (2018) Molecular mechanisms of anti-inflammatory effect of chrysophanol, an active component of AST2017-01 on atopic dermatitis *in vitro* models. Int Immunopharmacol 54:238–244

Ji QL, Wang WG (2001) Asexual propagation of *Arnebia euchroma* and exploration of hereditary stability in regenerated plantlets. Plant Physiol Commun 37:499–502

Ji YB, Qu ZY, Zou X (2011) Juglone-induced apoptosis in human gastric cancer SGC-7901 cells via the mitochondrial pathway. Exp Toxicol Pathol 63(1–2):69–78

Jiang W, Zhou R, Li P et al (2016) Protective effect of chrysophanol on LPS/d-GalN-induced hepatic injury through the RIP140/NF-_B pathway. RSC Adv 6:38192–38200

Kang SC, Lee CM, Choung ES et al (2008) Anti-proliferative effects of estrogen receptor-modulating compounds isolated from *Rheum palmatum*. Arch Pharm Res 31:722–726

Kaseb AO, Chinnakannu K, Chen D et al (2007) Androgen receptor and E2F-1 targeted thymoquinone therapy for hormonerefractory prostate cancer. Cancer Res 67(16):7782–7788

Kawiak A, Piosik J, Stasilojc G et al (2007) Induction of apoptosis by plumbagin through reactive oxygen species-mediated inhibition of topoisomerase II. Toxicol Appl Pharmacol 223(3): 267–276

Khanal P, Namgoong GM, Kang BS et al (2010) The prolyl isomerase Pin1 enhances HER-2 expression and cellular transformation via its interaction with mitogen-activated protein kinase/extracellular signal-regulated kinase kinase 1. Mol Cancer Ther 9(3):606–616

Kim SJ, Kim MC, Lee BJ et al (2010) Anti-inflammatory activity of chrysophanol through the suppression of NF-kappaB/caspase-1 activation *in vitro* and *in vivo*. Molecules 15:6436–6451

Kohle A, Sommer S, Yazaki K et al (2002) High level expression of chorismate pyruvate-lyase (ubiC) and HMG-CoA reductase in hairy root cultures of *Lithospermum erythrorhizon*. Plant Cell Physiol 43:894–902

Kolli-Bouhafs K, Boukhari A, Abusnina A et al (2011) Thymoquinone reduces migration and invasion of human glioblastoma cells associated with FAK, MMP-2 and MMP-9 down-regulation. Invest New Drugs 30:2121–2131

Komi Y, Suzuki Y, Shimamura M et al (2009) Mechanism of inhibition of tumor angiogenesis by beta-hydroxyisovalerylshikonin. Cancer Sci 100(2):269–277

Kretschmer N, Rinner B, Deutsch AJ et al (2012) Naphthoquinones from Onosma paniculata induce cell-cycle arrest and apoptosis in melanoma cells. J Nat Prod 75(5):865–869

Kristensen SB, van Mourik T, Pedersen TB et al (2020) Simulation of electrochemical properties of naturally occurring quinones. Sci Rep 10:13571. https://doi.org/10.1038/s41598-020-70522-z

Kubo I, Murai Y, Soediro I et al (1992) Cytotoxic anthraquinones from *Rheum palmatum*. Phytochemistry 31:1063–1065

Kumar MR, Aithal BK, Udupa N (2011a) Formulation of plumbagin loaded long circulating pegylated liposomes: in vivo evaluation in C57BL/6J mice bearing B16F1 melanoma. Drug Deliv 18(7):511–522

Kumar R, Sharma N, Malik S et al (2011b) Cell suspension culture of *Arnebia euchroma* (Royle) Johnston—a potential source of naphthoquinone pigments. J Med Plant Res 5:6048–6054

Lai L, Liu J, Zhai D et al (2012) Plumbagin inhibits tumour angiogenesis and tumour growth through the Ras signalling pathway following activation of the VEGF receptor-2. Br J Pharmacol 165(4b):1084–1096

Lee MS, Sohn CB (2008) Anti-diabetic properties of chrysophanol and its glucoside from rhubarb rhizome. Biol Pharm Bull 31:2154–2157

Lee HZ, Lin CJ, Yang WH (2006) Aloe emodin induced DNA damage through generation of reactive oxygen species in human lung carcinoma cells. Cancer Lett 239(1):55–63

Lee MS, Cha EY, Sul JY et al (2011) Chrysophanic acid blocks proliferation of colon cancer cells by inhibiting EGFR/mTOR pathway. Phytother Res 25:833–837

Lei X, Lv X, Liu M et al (2012) Thymoquinone inhibits growth and augments 5-fluorouracilinduced apoptosis in gastric cancer cells both *in vitro* and *in vivo*. Biochem Biophys Res Commun 417(2):864–868

Li CJ, Averboukh L, Pardee AB (1993) Beta-Lapachone, a novel DNA topoisomerase I inhibitor with a mode of action different from camptothecin. J Biol Chem 268(30):22463–22468

Li Y, Sun X, LaMont JT et al (2003) Selective killing of cancer cells by beta-lapachone: direct checkpoint activation as a strategy against cancer. Proc Natl Acad Sci U S A 100(5):2674–2678

Li L, Chen P, Lian HK et al (2010) Juglone inhibits the proliferation of human esophageal carcinoma EC1 cells. World Chin J Digestol 18(11):1147–1151

Li J, Shen L, Lu FR et al (2012) Plumbagin inhibits cell growth and potentiates apoptosis inhuman gastric cancer cells in vitro through the NF-kappaB signaling pathway. Acta Pharmacol Sin 33(2):242–249

Li A, Liu Y, Zhai L et al (2016) Activating peroxisome proliferator-activated receptors (PPARs): a new sight for Chrysophanol to treat Paraquat-induced lung injury. Inflammation 39:928–937

Lichtenthaler HK (1999) The 1-deoxy-D-xylulose-5-phosphate pathway of isoprenoid biosynthesis in plants. Annu Rev Plant Biol 50(1):47–65

Lien YC, Kung HN, Lu KS et al (2008) Involvement of endoplasmic reticulum stress and activation of MAP kinases in beta-lapachone-induced human prostate cancer cell apoptosis. Histol Histopathol 23(11):1299–1308

Lin L, Wu J (2002) Enhancement of shikonin production in single and two-phase suspension cultures of *Lithospermum erythorhizon* cells using low-energy ultrasound. Biotechnol Bioeng 78:81–88

Lin ML, Lu YC, Chung JG et al (2010) Human nasopharyngeal carcinoma cells via caspase-8-mediated activation of the mitochondrial death pathway. Cancer Lett 291(1):46–58

Lin ML, Lu YC, Su H.L et al (2011) Destabilization of CARP mRNAs by aloe-emodin contributes to caspase-8-mediated p53-independent apoptosis of human carcinoma cells. J Cell Biochem 112(4):1176–1191

Lin F, Zhang C, Chen X et al (2015) Chrysophanol affords neuroprotection against microglial activation and free radical-mediated oxidative damage in BV2 murine microglia. Int J Clin Exp Med 8:3447–3455

Lin FL, Lin CH, Ho JD et al (2017) The natural retinoprotectant chrysophanol attenuated photoreceptor cell apoptosis in an N-methyl-N-nitrosourea-induced mouse model of retinal degeneration. Sci Rep 7:41086

Liu M, Lu S (2016) Plastoquinone and ubiquinone in plants: biosynthesis, physiological function and metabolic engineering. Front Plant Sci 7:1–18

Liu SY, Lo CT, Shibu MA et al (2009) Study on the anthraquinones separated from the cultivation of *Trichoderma harzianum* strain Th-R16 and their biological activity. J Agric Food Chem 57: 7288–7292

Liu Y-J, Zhao Y-J, Zhang M et al (2014) Cloning and characterisation of the gene encoding 3-hydroxy-3-methylglutaryl-CoA synthase in *Tripterygium wilfordii*. Molecules 19(12): 19696–19707

Linsmaier EM, Skoog F (1965) Organic growth factor requirements of tobacco tissue cultures. Physiol Plant 18:100–127

Lu GD, Shen HM, Chung MC, Ong CN (2007) Critical role of oxidative stress and sustained JNK activation in aloe-emodin mediated apoptotic cell death in human hepatoma cells. Carcinogenesis 28(9):1937–1945

Lu CC, Yang JS, Huang AC et al (2010) Chrysophanol induces necrosis through the production of ROS and alteration of ATP levels in J5 human liver cancer cells. Mol Nutr Food Res 54:967–976

Lu Y, Liu Y, Zhou J, Li D, Gao W (2020) Biosynthesis, total synthesis, structural modifications, bioactivity, and mechanism of action of the quinone-methide triterpenoid celastrol. Med Res Rev:1–39. https://doi.org/10.1002/med.21751

Lubbe A, Verpoorte R (2011) Cultivation of medicinal and aromatic plants for specialty industrial materials. Ind Crop Prod 34(1):785–801

Luo P, Wong YF, Ge L et al (2010) Anti-inflammatory and analgesic effect of plumbagin through inhibition of nuclear factor-kappaB activation. J Pharmacol Exp Ther 335(3):735–742

Malik EM, Muller CE (2016) Anthraquinones as pharmacological tools and drugs. Med Res Rev 36:705–748

Malik S, Bhushan S, Verma SC et al (2008) Production of naphthoquinone pigments in cell suspension cultures of *Arnebia euchroma* (Royle) Johnston: influence of pH on growth kinetics and acetyl shikonin. Med Aromat Plant Sci Biotechnol 2:43–49

Malik S, Kumar R, Vats SK et al (2009) Regeneration in *Rheum emodi* wall.: a step towards conservation of an endangered medicinal plant species. Eng Life Sci 9:130–134

Malik S, Sharma S, Sharma M, Ahuja PS (2010a) Direct shoot regeneration from intact leaves of *Arnebia euchroma* (Royle) Johnston using thidiazuron. Cell Biol Int 34:537–542

Malik S, Sharma N, Sharma UK et al (2010b) Qualitative and quantitative analysis of anthraquinone derivatives in rhizomes of tissue culture-raised Rheum emodi wall. Plants J Plant Physiol 167: 749–756. https://doi.org/10.1016/j.jplph.2009.12.007
Malik S, Bhushan S, Sharma M, Ahuja PS (2011a) Physico-chemical factors influencing the shikonin derivatives production in cell suspension cultures of *Arnebia euchroma* (Royle) Johnston, a medicinally important plant species. Cell Biol Int 35:152–157
Malik S, Cusido RM, Mirjalili MH et al (2011b) Production of the anticancer drug taxol in *Taxus baccata* suspension cultures: a review. Process Biochem 46:23–34
Malik S, Andrade SAL, Sawaya ACHF et al (2013) Root-zone temperature alters alkaloid synthesis and accumulation in *Catharanthus roseus* and *Nicotiana tabacum*. Ind Crop Prod 49:318–325
Malik S, Bhushan S, Sharma M, Ahuja PS (2014a) Biotechnological approaches to the production of shikonins: a critical review with recent updates. Crit Rev Biotechnol:1–14. https://doi.org/10.3109/07388551.2014.961003
Malik S, Bhushan S, Sharma M et al (2014b) Biotechnological approaches to the production of shikonins: a critical review with recent updates. Crit Rev Biotechnol:1–14. https://doi.org/10.3109/07388551.2014.961003
Malik S, Biba O, Gruz J et al (2014c) Biotechnological approaches for producing aryltetralin lignans from *Linum* species. Phytochem Rev. https://doi.org/10.1007/s11101-014-9345-5
Manu KA, Shanmugam MK, Rajendran P et al (2011) Plumbagin inhibits invasion and migration of breast and gastric cancer cells by downregulating the expression of chemokine receptor CXCR4. Mol Cancer 10:107
Margaria P, Abbà S, Palmano S (2013) Novel aspects of grapevine response to phytoplasma infection investigated by a proteomic and phospho-proteomic approach with data integration into functional networks. BMC Genomics 14(1):38. https://doi.org/10.1186/1471-2164-14-38
Margaria P, Ferrandino A, Caciagli P et al (2014) Metabolic and transcript analysis of the flavonoid pathway in diseased and recovered Nebbiolo and Barbera grapevines (*Vitis vinifera* L.) following infection by Flavescence dorée phytoplasma. Plant Cell Environ 37(9):2183–2200. https://doi.org/10.1111/pce.12332
Mathew R, Kruthiventi AK, Prasad JV et al (2010) Inhibition of mycobacterial growth by plumbagin derivatives. Chem Biol Drug Des 76(1):34–42
Mathur R, Chandna S et al (2011) Peptidyl prolyl isomerase, Pin1 is a potential target for enhancing the therapeutic efficacy of etoposide. Curr Cancer Drug Targets 11(3):380–392
Mboso OE, Eyong EU, Odey MO, Osakwe E (2013) Comparative phytochemical screening of Ereromastax speciosa and Ereromastax polysperma. J Nat Prod Plant Resour 3:37–41
Medeiros CS, Pontes-Filho NT, Camara CA et al (2010) Antifungal activity of the naphthoquinone beta-lapachone against disseminated infection with Cryptococcus neoformans var. neoformans in dexamethasone-immunosuppressed Swiss mice. Braz J Med Biol Res 43(4):345–349
Miao XS, Zhong C, Wang Y et al (2009) In vitro metabolism of beta-lapachone (ARQ 501) in mammalian hepatocytes and cultured human cells. Rapid Commun Mass Sp 23(1):12–22
Min R, Tong J, Wenjun Y et al (2008) Growth inhibition and induction of apoptosis in human oral squamous cell carcinoma Tca-8113 cell lines by shikonin was partly through the inactivation of NF-kappaB pathway. Phytother Res 22(3):407–415
Min R, Zun Z, Min Y et al (2011) Shikonin inhibits tumor invasion via down-regulation of NF kappa B-mediated MMP-9 expression in human ACC-M cells. Oral Dis 17(4):362–369
Misra BB (2014) An updated snapshot of recent advances in transcriptomics and genomics of phytomedicinals. Postdoc J 2:1–15
Murashige T, Skoog F (1962) A revised medium for rapid growth and bioassays with tobacco tissue cultures. Plant Physiol:15473497
Ni CH, Yu CS, Lu HF et al (2014) Chrysophanol-induced cell death (necrosis) in human lung cancer A549 cells is mediated through increasing reactive oxygen species and decreasing the level of mitochondrial membrane potential. Environ Toxicol 29:740–749

Nomura T, Ogita S, Kato Y (2018) Rational metabolic-flow switching for the production of exogenous secondary metabolites in bamboo suspension cells. Sci Rep 8:13203

Nosov AM (2012) Application of cell technologies for production of plant-derived bioactive substances of plant origin. Appl Biochem Microbiol 48(7):609–624

Nowicka B, Kruk J (2010) Occurrence, biosynthesis and function of isoprenoid quinones. Biochim. Biophys. Acta-Bioenergetics 1797:1587–1605

Obadoni BO, Ochuko PO (2001) Phytochemical composition, spoilage and shelf life extension. Studies and comparative efficacy of the crude extracts of some homeostatic plants in Edo and Delta states of Nigeria. Global J Pure Appl Sci 8:203–208

Oliveira MJ, Castro S, Paltrinieri S et al (2020) "Flavescence dorée" impacts growth, productivity and ultrastructure of *Vitis vinifera* plants in Portuguese "Vinhos Verdes" region. Sci Hortic 261: 108742. https://doi.org/10.1016/j.scienta.2019.108742

Oni OE, Schmidt F, Miyatake T et al (2015) Microbial communities and organic matter composition in surface and subsurface sediments of the Helgoland mud area. North Sea Front Microbiol 6:1290. https://doi.org/10.3389/fmicb.2015.01290

Onoda T, Li W, Sasaki T et al (2016) Identification and evaluation of magnolol and chrysophanol as the principle protein tyrosine phosphatase-1B inhibitory compounds in a Kampo medicine. Masiningan J Ethnopharmacol 186:84–90

Owen C, Patron NJ, Huang A, Osbourn A (2017) Harnessing plant metabolic diversity. Curr Opin Chem Biol 40:24–30

Padhye S, Dandawate P, Yusufi M et al (2012) Perspectives on medicinal properties of plumbagin and its analogs. Med Res Rev 32(6):1131–1158

Pecere T, Gazzola MV, Mucignat C et al (2000) Aloe-emodin is a new type of anticancer agent with selective activity against neuroectodermal tumors. Cancer Res 60(11):2800–2804

Pecere T, Sarinella F, Salata C et al (2003) Involvement of p53 in specific anti-neuroectodermal tumor activity of aloe-emodin. Int J Cancer 106(6):836–847

Pelagio-Flores R, Ortíz-Castro R, Méndez-Bravo A et al (2011) Serotonin, a tryptophan-derived signal conserved in plants and animals, regulates root system architecture probably acting as a natural auxin inhibitor in Arabidopsis thaliana. Plant Cell Physiol 52:490–508

Pietrosiuk A, Urmantseva V, Furmanowa M (1999) Some naphthoquinones and pyrrolizidine alkaloids in cell culture of *Arnebia euchroma* (Royle) Jonst. Herba Pol 65:354–361

Pietrosiuk A, Syklowska-Baranek K, Wiedenfeld H et al (2006) The shikonin derivatives and pyrrolizidine alkaloids in hairy root cultures of *Lithospermum canescens* (Michx.) Lehm. Plant Cell Rep 25:1052–1058

Pina ES, Silva DB, Teixeira SP et al (2016) Mevalonate-derived quinone-methide triterpenoid from in vitro roots of *Peritassa laevigata* and their localization in root tissue by MALDI imaging. Sci Rep 6:22627

Polonik SG, Prokof'eva NG, Agafonova IG, Uvarova NI (2003) Antitumor and immunostimulating activity of 5-hydroxy-1,4-naphthoquinone (juglone) o- and s-acetylglycosides. Pharm Chem J 37(8):397–398

Powolny AA, Singh SV (2008) Plumbagin-induced apoptosis in human prostate cancer cells is associated with modulation of cellular redox status and generation of reactive oxygen species. Pharm Res 25(9):2171–2180

Prateeksha YMA, Singh BN et al (2019) Chrysophanol: a natural Anthraquinone with multifaceted biotherapeutic potential. Biomol Ther 9:68. https://doi.org/10.3390/biom9020068

Prezelj N, Covington E, Roitsch T et al (2016) Metabolic consequences of infection of grapevine (*Vitis vinifera* L.) cv. "Modra frankinja" with Flavescence Dorée phytoplasma. Front. Plant Sci 7:711. https://doi.org/10.3389/fpls.2016.00711

Qian ZJ, Zhang C, Li YX et al (2011) Protective effects of emodin and chrysophanol isolated from marine fungus *aspergillus sp.* on ethanol-induced toxicity in HepG2/CYP2E1 cells. Evid Based Complement Alternat Med 2011:452621

Qin X, Zeevaart JA (2002) Overexpression of a 9-cis-epoxycarotenoid dioxygenase gene in *Nicotiana plumbaginifolia* increases abscisic acid and phaseic acid levels and enhances drought tolerance. Plant Physiol 128(2):544–551
Ramana LV, Apparao KC, Rao BN, Rao MS (2017) New phenolic constituents from bark of Walsura trifoliate. J Pharmacogn Phytochem 6:1314–1316
Rao Z, Liu X, Zhou W et al (2011) Synthesis and antitumour activity of beta-hydroxyisovalerylshikonin analogs. Eur J Med Chem 46(9):3934–3941
Raskin I, Ribnicky DM, Komarnytsky S et al (2002) Plants and human health in the twenty-first century. Trends Biotechnol 20(12):522–531
Reindl W, Yuan J, Kramer A et al (2008) Inhibition of polo-like kinase 1 by blocking polo-box domain-dependent protein-protein interactions. Chem Biol 15(5):459–466
Ren HM, Fan F, Cao KQ (2012) Ultrastructural changes of *phaerotheca fuliginea* (Schlechtend.: Fr.) pollacci in cucumber after treated by chrysophanol. J Integr Agric 11:970–977
Renneberg R (2008) Green biotechnology. In: Demain AL (ed) Biotechnology for beginners. Academic Press, New York, pp 210–213
Rodrigues CF, Rodrigues ME, Silva S, Henriques M (2017) Candida glabrata biofilms: how far have we come? J Fungi 3:11
Sand JM, Hafeez BB, Jamal MS et al (2012) Plumbagin (5-hydroxy-2-methyl-1,4-naphthoquinone), isolated from Plumbago zeylanica, inhibits ultraviolet radiation-induced development of squamous cell carcinomas. Carcinogenesis 33(1):184–190
Sandur SK, Ichikawa H, Sethi G et al (2006) Plumbagin (5-hydroxy-2-methyl-1,4-naphthoquinone) suppresses NF-kappaB activation and NF-kappaB-regulated gene products. J Biol Chem 281(25):17023–17033
Sayed-Ahmed MM, Aleisa AM, Al-Rejaie SS et al (2010) Thymoquinone attenuates diethylnitrosamine induction of hepatic carcinogenesis through antioxidant signaling. Oxid Med Cell Longev 3(4):254–261
Schorkhuber M, Richter M, Dutter A et al (1998) Effect of anthraquinone-laxatives on the proliferation and urokinase secretion of normal, premalignant and malignant colonic epithelial cells. Eur J Cancer 34:1091–1098
Semple SJ, Pyke SM, Reynolds GD (2001) Flower RLP. In vitro antiviral activity of the anthraquinone chrysophanic acid against poliovirus. Antiviral Res 49:169–178
Seshadri P, Rajaram A, Rajaram R (2011) Plumbagin and juglone induce caspase-3-dependent apoptosis involving the mitochondria through ROS generation in human peripheral blood lymphocytes. Free Radic Biol Med 51(11):2090–2107
Sethi G, Ahn KS, Aggarwal BB (2008) Targeting nuclear factor-kappa B activation pathway by thymoquinone: role in suppression of antiapoptotic gene products and enhancement of apoptosis. Mol Cancer Res 6(6):1059–1070
Sharma N, Sharma UK, Malik S et al (2008) Isolation and purification of acetylshikonin and b-acetoxyisovalerylshikonin from cell suspension cultures of *Arnebia euchroma* (Royle) Johnston using rapid preparative HPLC. J Sep Sci 31:629–635
Shekhawat MS, Shekhawat NS (2011) Micropropagation of *Arnebia hispidissima* (Lehm).DC. And production of alkannin from callus and cell suspension culture. Acta Physiol Plant 33:1445–1450
Shi P, Huang Z, Chen G (2008) Rhein induces apoptosis and cell cycle arrest in human hepatocellular carcinoma BEL-7402 cells. Am J Chin Med 36(4):805–813
Shukla S, Wu CP, Nandigama K, Ambudkar SV (2007) The naphthoquinones, vitamin K3 and its structural analogue plumbagin, are substrates of the multidrug resistance linked ATP binding cassette drug transporter ABCG2. Mol Cancer Ther 6:3279–3286
Singh R, Chauhan S (2004) 9,10-Anthraquinones and other biologically active compounds from the genus Rubia. Chem Biodivers 1:1241–1264. https://doi.org/10.1002/cbdv.200490088
Singh B, Sahu PM, Sharma MK et al (2002) Production and secretion of alkannin by hairy root cultures of *Arnebia hispidissima* (Lehm.) D.C. J Plant Biol 29:293–300

Singh NP, Gupta AP, Sinha AK, Ahuja PS (2005) High-performance thin layer chromatography method for the quantitative determination of four major anthraquinone derivatives in *Rheum emodi*. J Chromatogr A 1077:202–206

Singh RS, Gara RK, Bhardwaj PK et al (2010) Expression of 3-hydroxy-3-methylglutaryl-CoA reductase, p-hydroxybenzoate-m-geranyltransferase and genes of phenylpropanoid pathway exhibitspositive correlation with shikonins content in arnebia [Arnebiaeuchroma (Royle) Johnston]. BMC Mol Biol 11:88

Singh S, Singh SK, Chowdhury I, Singh R (2017) Understanding the mechanism of bacterial biofilms resistance to antimicrobial agents. Open Microbiol J 11:53–62

Skorupinska-Tudek K, Poznanski J, Wojcik J et al (2008) Contribution of the mevalonate and methylerythritol phosphate pathways to the biosynthesis of dolichols in plants. J Biol Chem 283(30):21024–21035

Socaciu C (2007) Food colorants. Chemical and Functional Properties (CRC Press, Boca Raton

Sofowora A (1993) Medicinal plants and traditional medicine in Africa. John Wiley and sons, pp 112–142

Sokha V, Nikolaeva L, Pank F (1996) The shikonin production in plant tissue culture: selection of cell lines with high productivity. Proc Int 2:322–326

Soladoye MO, Chukwuma EC (2012) Quantitative phytochemical profile of the leaves of Cissus populnea Guill. & Perr. (Vitaceae)—an important medicinal plant in Central Nigeria. Arch Appl Sci Res 4:200–206

Sommer S, Kohle A, Yazaki K et al (1999) Genetic engineering of shikonin biosynthesis hairy root cultures of *Lithospermum erythrorhizon* transformed with the bacterial ubiC gene. Plant Mol Biol 39:683–693

Sreelatha T, Hymavathi A, Babu KS et al (2009) Synthesis and insect antifeedant activity of plumbagin derivatives with the amino acid moiety. J Agric Food Chem 57(14):6090–6094

Srinivasan V, Ryu DDY (1992) Enzyme activity and shikonin production in Lithospermum erythrorhizon cell cultures. Biotechnol Bioeng 40:69–74

Suboj P, Babykutty S, Valiyaparambil Gopi DR et al (2012) Aloe emodin inhibits colon cancer cell migration/angiogenesis by downregulating MMP-2/9, RhoB and VEGF via reduced DNA binding activity of NF-kappaB. Eur J Pharm Sci 45(5):581–591

Subramaniya BR, Srinivasan G, Sadullah SS (2011) Apoptosis inducing effect of plumbagin on colonic cancer cells depends on expression of COX-2. PLoS One 6(4):e18695

Suleyman H, Demirezer LO, Kuruuzum A et al (1999) Anti-inflammatory effect of the aqueous extract from Rumex patientia L-roots. J Ethnopharmacol 65:141–148

Suleyman H, Demirezer LO, Kuruuzum-Uz A (2004) Effects of Rumex patientia root extract on indomethacine and ethanol induced gastric damage in rats. Die Pharm 59:147–149

Sun J, McKallip RJ (2011) Plumbagin treatment leads to apoptosis in human K562 leukemia cells through increased ROS and elevated TRAIL receptor expression. Leuk Res 35(10):1402–1408

Sun YL, Zhang XY, Zheng ZH et al (2014) Three new polyketides from marine-derived fungus Penicillium citrinum SCSGAF 0167. Nat Prod Res 28:239–244

Suzuki M, Amano M, Choi J et al (2006) Synergistic effects of radiation and beta-lapachone in DU-145 human prostate cancer cells in vitro. Radiat Res 165(5):525–531

Sykłowska-Baranek K, Pietrosiuk A, Gawron A et al (2012) Enhanced production of antitumour naphthoquinones in transgenic hairy root lines of *Lithospermum canescens*. Plant Cell Tiss Org Cult 108:213–219

Tabolacci C, Lentini A, Mattioli P et al (2010) Antitumor properties of aloe emodin and induction of transglutaminase 2 activity in B16-F10 melanoma cells. Life Sci 87(9–10):316–324

Tan W, Lu J, Huang M et al (2011) Anticancer natural products isolated from Chinese medicinal herbs. Chinas Med 6(1):27

Tian L, Yin D, Ren Y et al (2012) Plumbagin induces apoptosis via the p53 pathway and generation of reactive oxygen species in human osteosarcoma cells. Mol Med Rep 5(1):126–132

Tong YR, Su P, Zhang M et al (2015a) Cloning and expression analysis of 2-C-methyl-D-erythritol 4-phosphate cytidylyltransferase gene in *Tripterygium wilfordii*. Zhongguo Zhong Yao Za Zhi 40(22):4378–4383

Tong Y, Su P, Zhao Y et al (2015b) Molecular cloning and characterization of DXS and DXR genes in the terpenoid biosynthetic pathway of *Tripterygium wilfordii*. Int J Mol Sci 16(10): 25516–25535

Tong YR, Su P, Zhao YJ et al (2015c) Cloning and expression analysis of 4-(cytidine-5-diphospho)-2-C-methyl-D-erythritol kinase gene in *Tripterygium wilfordii*. Zhongguo Zhong Yao Za Zhi 40(21):4165–4170

Tong Y, Zhang M, Su P et al (2016) Cloning and functional characterization of an isopentenyl diphosphate isomerase gene from *Tripterygium wilfordii*. Biotechnol Appl Biochem 63(6): 863–869

Tripathi B, Bhatia R, Pandey A et al (2014) Potential antioxidant Anthraquinones isolated from Rheum emodi showing Nematicidal activity against Meloidogyne incognita. J Chem 2014: 652526

Vitali M, Chitarra W, Galetto L, Bosco D, Marzachì C, Gullino ML, Spanna F, Lovisolo C (2013) Ann Appl Biol 162(3):335–346. https://doi.org/10.1111/aab.12025

Wang CC, Chiang YM, Sung SC et al (2008) Plumbagin induces cell cycle arrest and apoptosis through reactive oxygen species/c-Jun N-terminal kinase pathways in human melanoma A375. S2 cells. Cancer Lett 259(1):82–98

Wang Z, Liu T, Gan L et al (2010) Shikonin protects mouse brain against cerebral ischemia/ reperfusion injury through its antioxidant activity. Eur J Pharmacol 643(2–3):211–217

Weaver MG, Pettus TRR (2014) Synthesis of Para- and ortho-quinones. Comprehensive organic synthesis II, vol vol. 7, pp 373–410. https://doi.org/10.1016/B978-0-08-097742-3.00717-5

Widhalm JR, Rhodes D (2016) Biosynthesis and molecular actions of specialized 1,4-naphthoquinone natural products produced by horticultural plants. Hortic Res 3:16046

Woo CC, Loo SY, Gee V et al (2011) Anticancer activity of thymoquinone in breast cancer cells: possible involvement of PPAR-gamma pathway. Biochem Pharmacol 82(5):464–475

Wu SJ, Qi JL, Zhang WJ et al (2009) Nitric oxide regulates shikonin formation in suspension-cultured *Onosma paniculatum* cells. Plant Cell Physiol 50:118–128

Wu Y, Li W, Chuan-shu W Y-h, Zhang X (2012a) Characterization and expression of gene encoding HMGR from *Tripterygium wilfordii*. J Northwest A & F Univ 40(11):103–111

Wu YY, Wan LH, Zheng XW et al (2012b) Inhibitory effects of beta, beta-dimethylacrylshikonin on hepatocellular carcinoma in vitro and in vivo. Phytother Res 26(5):764–771

Xiang L, Zhao K, Chen L (2010) Molecular cloning and expression of *Chimonanthus praecox* farnesyl pyrophosphate synthase gene and its possible involvement in the biosynthesis of floral volatile sesquiterpenoids. Plant Physiol Biochem 48:845–850

Xiao B, Guo J, Liu D, Zhang S (2007) Aloe-emodin induces in vitro G2/M arrest and alkaline phosphatase activation in human oral cancer KB cells. Oral Oncol 43(9):905–910

Xin J, Zhang RC, Wang L, Zhang YQ (2017) Researches on transcriptome sequencing in the study of traditional Chinese medicine. Evid Based Complement Alternat Med 2017:7521363

Xu KH, Lu DP (2010) Plumbagin induces ROS-mediated apoptosis in human promyelocytic leukemia cells in vivo. Leuk Res 34(5):658–665

Xu HL, Yu XF, Qu SC et al (2010) Anti-proliferative effect of Juglone from *Juglans mandshurica* maxim on human leukemia cell HL-60 by inducing apoptosis through the mitochondria-dependent pathway. Eur J Pharmacol 645(1–3):14–22

Xu HL, Yu XF, Qu SC et al (2012) Juglone, from *Juglans mandshruica* maxim, inhibits growth and induces apoptosis in human leukemia cell HL-60 through a reactive oxygen species-dependent mechanism. Food Chem Toxicol 50(3–4):590–596

Xu C, Li H, Yang X et al (2016) Cloning and expression analysis of MEP pathway enzyme-encoding genes in Osmanthus fragrans. Gene 7(10):78

Xuan Y, Hu X (2009) Naturally-occurring shikonin analogs—a class of necroptotic inducers that circumvent cancer drug resistance. Cancer Lett 274(2):233–242

Yamamoto H, Inouye K, Li SM, Heide L (2000) Geranylhydroquinone 3-hydroxylase, a cytochrome P-450 monooxygenase from *Lithospermum erythrorhizon* cell suspension cultures. Planta 210:312–317

Yamamoto H, Zhao P, Yazaki K, Inoue K (2002) Regulation of lithospermic acid B and shikonin production in *Lithospermum erythrorhizon* cell suspension cultures. Chem Pharm Bull 50: 1086–1090

Yamamoto M, Kensler TW, Motohashi H (2018) The KEAP1–NRF2 system: a thiol-based sensor-effector apparatus for maintaining redox homeostasis. Physiol Rev 98:1169–1203

Yan J, Zheng MD, Zhang DS (2014) Chrysophanol liposome preconditioning protects against cerebral ischemia-reperfusion injury by inhibiting oxidative stress and apoptosis in mice. Int J Pharm 10:55–68

Yang F, Chen Y, Duan W et al (2006) A new synthesized shikonin derivative, exerting its potent antitumor activities as a topoisomerase inhibitor. Int J Cancer 119(5):1184–1193

Yang H, Zhou P, Huang H et al (2009) Shikonin exerts antitumor activity via proteasome inhibition and cell death induction in vitro and in vivo. Int J Cancer 124(10):2450–2459

Yang L, Yang C, Li C et al (2016) Recent advances in biosynthesis of bioactive compounds in traditional Chinese medicinal plants. Sci Bull 61(1):3–17

Yao Y, Zhou Q (2010) A novel antiestrogen agent shikonin inhibits estrogen-dependent gene transcription in human breast cancer cells. Breast Cancer Res Tr 121(1):233–240

Yu HJ, Oh SK, Oh MH et al (1997) Plant regeneration from callus cultures of *Lithospermum erythrorhizon*. Plant Cell Rep 16:261–266

Zakhlenjuk OV, Vidmachenko TV, Kunakh VA et al (1993) Comparative characteristics of shikonin accumulating *Arnebia euchroma* suspension culture and p-fluorophenylalanine-resistant variant. Acta Hortic 330:293–298

Zare K, Nazemiyeh H, Movafeghi A et al (2010) Bioprocess engineering of *Echium italicum* L.: induction of shikonin and alkannin derivatives by two-liquid-phase suspension cultures. Plant Cell Tiss Org Cult 100:157–164

Zare K, Khosrowshahli M, Nazemiyeh H et al (2011) Callus culture of *Echium italicum* L. towards production of a shikonin derivative. Nat Prod Res 25:1480–1487

Zenk MH, Shagi HEL, Schulte U (1975) Anthraquinone production by cell suspension cultures of M. citrifolia. Planta Med 28:79–101

Zhang Y, Qian RQ, Li PP (2009) Shikonin, an ingredient of *Lithospermum erythrorhizon*, down-regulates the expression of steroid sulfatase genes in breast cancer cells. Cancer Lett 284(1): 47–54

Zhang W, Zou A, Miao J et al (2011) LeERF-1, a novel AP2/ERF family gene within the B3 subcluster, is down-regulated by light signals in *Lithospermum erythrorhizon*. Plant Biol 13: 343–348

Zhang P, Wang F, Zhu C (2013) Influence of fungal elicitor and macroporous resin on shikonin accumulation in hairy roots of Arnebia euchroma (Royle) Johnst. Chin J Biotechnol 29:214–223

Zhang J, Yan C, Wang S et al (2014) Chrysophanol attenuates lead exposure-induced injury to hippocampal neurons in neonatal mice. Neural Regen Res 9:924–930

Zhang M, Su P, Zhou YJ et al (2015) Identification of geranylgeranyl diphosphate synthase genes from *Tripterygium wilfordii*. Plant Cell Rep 34(12):2179–2188

Zhang Y, Zhao Y, Wang J et al (2020) The expression of TwDXS in the MEP pathway specifically affects the accumulation of triptolide. Physiol Plant 169(1):40–48

Zhao L, Chang W-c, Xiao Y, Liu H-w, Liu P (2013) Methylerythritol phosphate pathway of isoprenoid biosynthesis. Annu Rev Biochem 82:497–530

Zhao YJ, Chen X, Zhang M et al (2015a) Molecular cloning and characterisation of farnesyl pyrophosphate synthase from *Tripterygium wilfordii*. PLoS One 10(5):e0125415

Zhao YJ, Zhang M, Liu YJ et al (2015b) Cloning and expression analysis of a acetyl-CoA U-acetyltransferase gene (TwAACT) from *Tripterygium wilfordii*. China J Chin Mater Med 40(5):847–852

Zhao YM, Fang YL, Li JC et al (2016) Neuroprotective effects of chrysophanol against inflammation in middle cerebral artery occlusion mice. Neurosci Lett 630:16–22

Zhou J, Zhang Y, Hu T et al (2018) Functional characterization of squalene epoxidase genes in the medicinal plant Tripterygium wilfordii. Int J Biol Macromol 120:203–212

Part III

Strategic Advances and Challenges

Optimization of In Vitro Cell Culture Conditions for Increasing Biomass and Nutraceutical Production

15

Deepika Tripathi, Arti Bisht, Mithilesh Singh, and I. D. Bhatt

Abstract

Plants are one of the most reliable resources of nutrient supplements and pharmaceutical compounds, but rapidly losing from the wild due to intensified harvesting for the bioactive secondary metabolites of nutraceutical interest. Recent resurgence in the nutraceutical and its importance in the human health has aroused worldwide researchers to find efficient techniques for nutraceutical production. Among others, in vitro plant cell culture technique is a viable alternative for large-scale propagation of germplasm with the high content of nutraceutical compounds. However, for successful in vitro plant cell culture, selection of a suitable plant material and optimization of several physical and chemical culture conditions such as nutrient media content, mineral composition, phytohormone supplementation, temperature, pH, light, humidity, aeration, agitation, etc. is imperative. The optimization of these parameters accelerates the growth rate of plant cells under in vitro conditions along with the improvement of production of nutraceuticals. Therefore, the present chapter provides details on the optimization of in vitro plant cell culture methods with respect to sustainable production of nutraceutical compounds.

Keywords

Nutraceuticals · In vitro cell culture · Optimization · Secondary metabolites · Suspension culture

D. Tripathi · A. Bisht · M. Singh (✉) · I. D. Bhatt
G. B. Pant National Institute of Himalayan Environment, Kosi-Katarmal, Almora, Uttarakhand, India
e-mail: singh.mithilesh@gbpihed.nic.in

T. Belwal et al. (eds.), *Nutraceuticals Production from Plant Cell Factory*,
https://doi.org/10.1007/978-981-16-8858-4_15

15.1 Introduction

Plants are rich source of bioactive compounds which have the miraculous ability to cure human health problems and used in numerous industrial areas like cosmetics, pharmaceuticals, nutraceuticals, etc. According to the world health organization (WHO), approximately 80% human population depends upon plants and their related bioactive compounds for nutritional and medicinal purposes. Nutraceuticals are naturally plant-derived bioactive compounds/secondary metabolites that possess health and nutritional benefits (Sampaio et al. 2020). The term 'nutraceutical' was given by DeFelice (director of the Foundation for Innovation in Medicine) in 1989, which is derived from the confluence of two words 'nutrients' and 'pharmaceutical'. Nutraceuticals can be grouped into the following three main categories: (a) Nutrients: Substances with established nutritional functions, such as vitamins, minerals, amino acids, and fatty acids; (b) Herbals: Herbs or botanical products as concentrates and extracts; (c) Dietary supplements: Reagents derived from other sources serving specific functions, such as sports nutrition, weight-loss supplements, and meal replacements (Dureja et al. 2003). Nowadays, globally nutraceuticals have received much attention due to potential nutrient value, safety, and pharmacological efficiency. Along with this, nutraceuticals are becoming a part of people's daily diet due to increasing prevalence of diseases and increasing consciousness of the consumers towards preventive healthcare measures. According to market analysis report (2020), world nutraceuticals' market size was estimated to be US$233.9 billion at the time of COVID-19 crisis. It has been observed that the demand of nutraceuticals is increasing all across the globe and especially in developing countries. In India, ASSOCHAM report (2018) predicted that "the Indian nutraceuticals market is expected to grow from $4 billion in 2017 to $18 billion in 2025." In upcoming time, India can be the largest source of nutraceuticals production with high biodiversity and robust research infrastructure.

Nutraceuticals are generally obtained from naturally grown plants, but their commercial production is limited due to environmental, seasonal, and regional constraints. The spontaneous supply of raw materials can't be maintained from natural plant population. For homogenous production of nutraceuticals, old conventional methods are time-consuming because plant takes several years to grow and reach the suitable age for desired bioactive compounds production. Also, one of the main problems with the commercial supply of plant-derived molecules is the limited availability of the bioactive compounds in plants. The content of secondary metabolites in field grown plants is generally less than 1% dry weight. Therefore, an alternative method is requisite to surpass such situation for the production of higher plant biomass and nutraceuticals. In this context, in vitro plant cell culture technique is well-known to efficiently produce maximum plant biomass along with high content of nutraceutical compounds within a short duration and cost-effective manner for commercial applications. In vitro cell culture is more appropriate for optimizing the biomass and bioactive metabolites production, including nutraceuticals. For instance, attempts have been made on optimization of in vitro culture method for nutraceutical compounds productions in *Celastrus paniculatus*,

Bacopa monnieri, *Aegle marmelos*, *Cucumis melo*, etc. (Anusha et al. 2016; Rahman et al. 2002; Sainulabdeen and Sulekha 2008; Kintzios and Taravira 1997). However, the detail documentation of different in vitro optimization strategies for nutraceutical production is lacking. Therefore, this chapter focuses on in vitro plant cell culture technique and optimization of in vitro culture conditions and their role in plant growth and nutraceutical compounds production.

15.2 In Vitro Plant Cell Culture: A Technique from Field to Flasks

In vitro cell culture technique is economically feasible and important method to produce high yield of desired valuable metabolites from culture cells. It was first introduced at the end of 1960 as a preferable technique for studying and improving the production of plant bioactive compounds. There are two main systems of in vitro plant cell culture: the close system and the open system. The close system is more commonly applied in the research laboratories. In this technique, cell growth is achieved by adding fixed mass of cells and the biomass increases until one of the environmental factors becomes limiting. At this stage, the metabolites can be recovered from the cell mass directly. In the open systems, instead, the nutritional factors are supplied to the cells by continuously adding fresh medium and chemo state is used to monitor nutrient medium component. Therefore, in vitro cell culture system could be applied for the large-scale production of plant cell biomass from which nutraceuticals can be commercially obtained. It has been proven more advantageous over old conventional methods for the industrial production of plants-related bioactive secondary metabolites. Advantages of in vitro plant cell culture technique are as follows: (a) in cell culture methods, the choice of plant compounds can be generated independently of external factors (like soil and climatic conditions); (b) decrease the threatened attacks of pathogens; (c) cells of any plant (rare/endangered) can easily be maintained in order to produce their bioactive compounds; and (d) robotic-driven regulation of nutraceuticals production with low costs (DiCosmo and Misawa 1995; Efferth 2019). The advantage of plant cell culture in nutraceuticals production is that it can ultimately provide a sustainable and reliable source of natural products. Many researchers all over the world are working on the development of in vitro cultivation methods for important nutraceutical plants for obtaining uniform bioactive compounds in a rapid way with higher yields than ones obtained in the natural processes (Table 15.1).

15.3 Optimization of In Vitro Plant Cell Culture: A Powerful Technique for Industrial Scale Production of Nutraceutical Compounds

In plant cell culture technique, we can produce nutraceuticals in controlled environmental conditions independent from plethora of natural climatic factors with the elimination of microbial contamination. However, selection of productive cell and

Table 15.1 In vitro plant cell culture of some important plants having nutraceutical properties

Nutraceutical plant	Common name	Family	Medicinal importance	Nutraceutical compounds	Plant cell culture type	Reference
Aegle marmelos(L.) Corr.	Bael	Rutaceae	Anti-microbial, anti-diarrheal, gastroprotective, anti-ulcerative colitis, and anti-diabetic activity	Alkaloids and flavonoids	Callus culture	Sainulabdeen and Sulekha (2008)
Allium sativum L.	Garlic	Alliaceae	Anti-microbial and anti-cancer activity	Alkaloids, carbohydrates, cardiac glycosides, fats & oils, flavonoids, saponins, and steroidal terpenoids	Callus culture	Mostafa et al. (2020)
Aloe Saponaria Ait.	Zebra aloe	Asphodelaceae	Analgesic and anti-inflammatory activities on burn injuries	Anthraquinones, alkaloids, flavonoids and vitamins A, C, E, and B12, folic acid	Suspension culture	Yagi et al. (1983)
Asparagus racemosus wild.	Satavar	Liliaceae	Antidepressant activity, antidiarrhoeal, and analgesic activity	Saponins (shatavarins)	Suspension culture	Pise et al. (2011)
Bacopa monnieri(L.)	Brahmi	Plantaginaceae	Used in epilepsy, manic psychosis, depression, promoting memory	Saponin, bacoside	Suspension culture	Rahman et al. (2002)
Celastrus paniculatus wild.	Malkangani	Celastraceae	Used in intellect promoting, sharpening/increase memory, diabetes mellitus	Sugars, anthraquinones, phenols, flavonoids, alkaloids, steroids, terpenoids, phlobotannins, glycosides, saponins, and tannins	Callus culture	Anusha et al. (2016)
Centella asiatica (L.)	Gotu kola	Apiaceae	Used in chronic rheumatism, increase memory, epilepsy, hysteria	Rich source of Vit C, Vit A, carotene, potassium, and calcium	Callus and suspension culture	Nath and Buragohain (2005)

Convolvulus pluricaulis Choisy	Shankhpushpi	Convolvulaceae	Used in epilepsy, uterine disorders, anxiety, hypertension, chronic headache	Kaempferol, ursolic acid, betulinic acid, stigmasterol, lupeol, convolvine	Suspension culture	Nazir (2019)
Curcuma longa L.	Haldi	Zingiberaceae	Used in bronchitis, allergies, high blood pressure, high cholesterol, skin diseases	Curcumin	Cell suspension culture	Jie et al. (2019)
Daucus carota L.	Carrot	Apiaceae	Used in kidney stones, bladder problems, water retention, and excess uric acid in the urine	Cyanidin 3-O-lathyroside, 4-coumaric, ferulic, 4-hydroxybenzoic, sinapic acid	Suspension culture	Gläßgen et al. (1992)
Eclipta alba L.	Bhringraj	Asteraceae	Used in gastrointestinal disorders, respiratory tract disorders (including asthma), fever, hair loss and graying of hair, liver disorders	Wedelolactone	Suspension culture	Salma et al. (2018)
Erythrina variegata L.	Indian coral tree	Fabaceae	Used as nervine sedative, anti-asthmatic, antiepileptic, antiseptic, and astringent	Carbohydrates, glycosides, proteins, volatile oils, tannins, flavonoid	Shoot regeneration	Javed et al. (2017)
Glycine max(L.) Merr.	Soybean	Fabaceae	Used in oedema, dysuria, chest fullness, decreased perspiration, the initial stages of flu and arthralgia	Isoflavonoids, coumestrin	Suspension culture	Lee et al. (2020)
Ipomoea batatas(L.) lam.	Sweet potato	Convolvulaceae	Used in diabetes, hypertension, dysentery, constipation, fatigue, arthritis, rheumatoid diseases, hydrocephaly, meningitis, kidney ailments, and inflammations problems	Phenolic acid, flavonoids	Suspension culture	Liu et al. (2001)

(continued)

Table 15.1 (continued)

Nutraceutical plant	Common name	Family	Medicinal importance	Nutraceutical compounds	Plant cell culture type	Reference
Lycopersicon esculentum mill.	Tomato	Solanaceae	Used for burns/scalds/ sunburn, protect from heart attacks and the difficulties in urination	Carotenoids, Brassinosteroids, 24-EPI-24-EPI-Castasterone	Suspension culture	Hai et al. (1996)
Medicago truncatula Gaertn.	Medick	Fabaceae	Used in central nervous and digestive system disorders	Phytase	Suspension culture	Pires et al. (2008)
Nardostachys grandiflora DC.	Jatamansi	Caprifoliaceae	Used in hypertension, epilepsy, hysteria, hepatitis, migraine, depression	Flavonoid, tannin, total phenolics	Suspension culture	Rawat et al. (2019)
Panax ginseng C.A. Mey.	Asiatic ginseng	Araliaceae	Antioxidation, anti-inflammatory, vasorelaxation, antiallergic, anti-diabetic, and anticancer activity	Ginsenoside	Suspension culture	Thanh et al. (2005)
Pueraria tuberosa (Willd.)	Kudzu	Fabaceae	Used as tonic, aphrodisiac, antirheumatic, diuretic, and galactagogue	Isoflavonoids	Suspension culture	Sharma et al. (2009)
Terminalia bellerica (Gaertn.) Roxb.	Baheda	Combretaceae	Used to protect the liver and to treat respiratory conditions, including respiratory tract infections, cough, and sore throat	Gallic acid, ellagic acid	Callus culture	Mehta et al. (2012)
Terminalia chebula Retz.	Harra	Combretaceae	Used in dementia, constipation, and diabetes	Terpenoids	Callus culture	Anjaneyulu et al. (2004)
Withania somnifera (L.)	Aswagandha	Solanaceae	Used in stress reduction, migraine, headache, neuralgia, epilepsy	Withanolides	Suspension culture	Nagella and Murthy (2010)

Actinidia deliciosa (A.Chev.)	Kiwi	Actinidiaceae	Anti-hepatotoxic, anti-pyorrheal, gingival inflammation, mild laxative, natural blood thinner, treatment of stones in the urinary tract, rheumatoid arthralgia, cancers of liver and oesophagus	Flavonoids (rutin and quercetin).	Callus culture	Akbaş et al. (2009)
Momordica dioica Roxb. Ex Willd.	Jangli karela, Kakrol	Cucurbitaceae	Good anti-diabetic activity	Proteins, vitamin (ascorbic acid, thiamine, riboflavin, and niacin), fatty acid (oleic acid and linoleic acid)	Shoot regeneration	Choudhary et al. (2017)

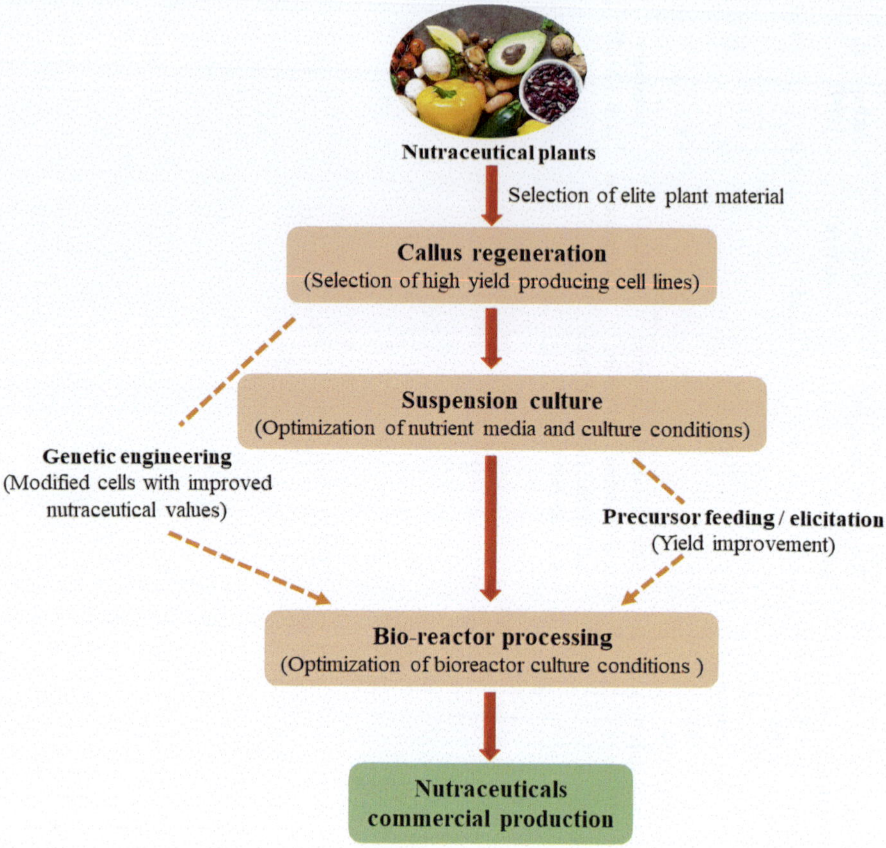

Fig. 15.1 Systematic view of in vitro plant cell culture method and their optimization for nutraceutical compounds production

optimization of in vitro culture conditions are imperative to produce high content of nutraceuticals (Fig. 15.1). A large number of physical and chemical factors are identified that directly affect biomass and nutraceutical compounds production under in vitro conditions such as selection of elite plant, explants, nutrient media composition, mineral content, phytohormone supplementation, pH, temperature, humidity, aeration, agitation, etc. All these parameters are important to define the quality of in vitro plant culture to obtain natural compounds of nutraceutical interest. According to Verpoorte et al. (2002), screening of high yield producing genotypes, optimization of culture medium, and culture environment for plant cell culture may increase production of bioactive nutraceuticals by 20 to 30-fold. Apart from that, precursor feeding and elicitation strategies are also advantageous in plant cell culture. A number of studies have been reported about role of optimization of these parameters for in vitro cell culture of nutraceutical plants (Table 15.2).

Table 15.2 Optimization of in vitro cell culture conditions for biomass and nutraceutical compounds production

Name of plant	Optimized factors	Impact of optimization	References
Actinidia deliciosa (A. Chev.)	Type of explant	Leaf explants produce best callus regeneration	Akbaş et al. (2009)
Allium hirtifolium Boiss.	Explant type and growth regulator	Interaction of explant and growth regulator improves callus regeneration	Farhadi et al. (2017)
Allium sativum (L.)	Genotypes and explant type	Tips as explant produce significant callus regeneration	Mostafa et al. (2020)
Asparagus racemosus Willd.	Plant growth regulators and pH	High 2,4-D promoted biomass and saponin production with optimal pH ~3.3 to 5.5	Pise et al. (2012)
Brassica napus L.	Light intensity	Optimized dark conditions beneficial for callogenesis from cotyledon sections as explants, but for hypocotyl sections light conditions are preferable	Afshari et al. (2011)
Citrus jambhiri lush.	Type of explant	Nodal explants produce maximum callus regeneration response	Savita et al. (2010)
Cucumis melo L.	Genotype and light intensity	Genotype and the intensity of light significantly affected callus regeneration and somatic embryogenesis	Kintzios and Taravira (1997)
Glycine max (L.)Merr.	Plant growth regulators	The ratio of NAA to BAP found to have profound effect on both callus growth and isoflavonoid accumulation	Downey et al. (2013)
Glycine max (L.) Merr.	Genotypes	Callus regeneration frequency alters with genotypes	Sairam et al. (2003)
Panax ginseng C. A. Mey.	Media strength and plant growth regulators supplementation	Maximum biomass with 2,4-D supplementation and maximum saponin production with IBA or NAA	Lian et al. (2002)
Spilanthes acmella Murr.	Agitation	*Spilanthes* cells highly sensitive to increase in agitation speed	Singh and Chaturvedi (2012)
Stevia rebaudiana (Bertoni)	Culture medium, phytohormone supplementation, and inoculums density	Callus cultured on MS basal liquid medium without growth regulators showed no growth initiation response, while with BAP and NAA supplementation maximum stevioside produced. Inoculum size and mineral content in media regulate metabolites production	Mathur and Shekhawat (2013)

(continued)

Table 15.2 (continued)

Name of plant	Optimized factors	Impact of optimization	References
Vitis vinifera L.	Inoculum size	Fluctuation in terms of inoculum sizes had a greater effect on the stability of anthocyanin biosynthesis in suspension cultures of *V. vinifera*. Among all the subculture conditions investigated, 7-d subculture cycle and 1.60-g inoculum size were the best for maintaining relatively stable anthocyanin production	Jun-Ge et al. (2006)
Zingiber officinale roscoe	Explant types and plant growth regulators	Callus regeneration was maximum in shoot tip explants inoculated in 2,4-D fortified medium	El-Nabarawy et al. (2015)

15.3.1 Selection of Parent Plant and Optimization of their Physiological Factors

The selection of healthy mother plant is prerequisite, followed by the optimization of explant age, position, and developmental stage to improve the nutraceuticals production in plant cell culture. In vitro morphogenic response is highly dependent on genotype and physiological state of the explants. Therefore, researchers need to optimize explants-related physiological factors along with the sterilization procedure to improve the regeneration ability. Many studies have explained the role of parent plant selection and explant optimization in nutraceuticals plants biomass production (Farhadi et al. 2017; Savita et al. 2010). For instance, Chevreau and Leblay (1992) have explained that pre-treatment of parent plant and explants position (on plant) alter the regeneration potential of pear leaves. Similarly, it has been found that physiological status of explant tissues and endogenous hormone level can significantly modify callus regeneration in garlic (Mostafa et al. 2020).

15.3.2 Optimization of Culture Medium and Mineral Content

Optimization of culture nutrient medium is as important as selection of mother plant and explants and is necessary to achieve high nutraceutical production. Although standard plant cell culture medium provides a platform for the growth and maintenance of plant cell lines, enhanced production of bioactive compounds can be obtained by further medium modification. Significant benefits can be gained by identifying the optimum culture medium type and nutrients content. The optimum concentration and proportion of mineral salts are a critical determinant in controlling the growth of cells in suspension cultures (Rao and Ravishankar 2002). In 2011, Behbahani et al. reported the impact of nutrient media on lycopene production from cell suspension culture of *Barringtonia racemosa*. Kintzios et al. (2001) reported the impact of vitamins and micronutrients content on callus regeneration of important nutraceutical plant *Capsicum annuum*. Similarly, optimization of medium strength and growth regulators supplementation has been reported in *Panax ginseng* for improved biomass and saponin production (Lian et al. 2002). Also, it has been found that sucrose and nitrogen supplementation improve both biomass and saponin production in *Panax*.

15.3.3 Optimization of Phytohormones

Plant cell cultures induced under sterile condition with phytohormone supplementation have shown high rate of cell proliferation and tremendous potentialities of stable production of valuable bioactive secondary metabolites. Screening of phytohormones supplementation and optimization of their concentrations may increase the production of biomass and nutraceuticals by several folds. In this context, many studies have been reported about optimization of plant growth regulators for high yield of nutraceutical compounds (Mathur and Shekhawat 2013; El-Nabarawy et al. 2015). Component and concentration of phytohormones within the media can be optimized more rapidly with employing a statistical design of experiments approach. This statistical approach has been shown to be advanced for the optimization of nutrient media composition in the production of a human antibody in a BY-2 tobacco cell suspension culture (Ochoa-Villarreal et al. 2016). Similar optimization strategy has also been applied to increase biomass and geraniol production in tobacco cell suspension cultures (Vasilev et al. 2014).

Nutraceuticals production in in vitro cell culture has been well-investigated. For instance, Aguilar-Camacho et al. (2019) reported the potential role of different concentrations of phytohormones for production of bioactive compounds (such as glucosinolates, isothiocyanates, phenolic compounds, and ascorbic acid) in *Brassica oleracea* L. var. *Italica*. Similarly, Downey et al. (2013) showed importance of optimization of phytohormones concentrations and combinations in *Glycine max* (L.) biomass and nutraceuticals production. They suggested that combination of NAA with BAP is highly beneficial for bioflavonoids production. Ginsenoside production has also been found improved in bioreactor root regeneration method with the optimization of phytohormones concentrations (Hahn et al. 2003). Seth et al. (2020) optimized culture conditions and phytohormone supplementation by central composite design for bacoside A production in *Bacopa monnieri* L. and found that 5.4 pH, 18–6 light/dark photoperiod, and combination of BAP and NAA fortification are necessary for the maximum plant biomass and bacoside A content.

15.3.4 Optimization of Light and Temperature

Light and temperature strongly influence many aspects of plant growth and biomass production (Park et al. 2004). The effect of light on nutraceutical bioactive products synthesis is quite varied. Light usually induces chloroplast differentiation, which leads to elevation of secondary metabolism in plants. Light acts as a stimulus that activates a broad range of genes related to photosynthesis and photomorphogenesis. The temperature of cultivation influences both plant cell growth and secondary metabolism to some extent. Also, temperature regulates biosynthesis of natural bioactive products with the regulation of various morphogenesis-related enzyme activities (Georgiev et al. 2009). Therefore, it is necessary to optimize these physical culture conditions to improve nutraceuticals production. Previous study has explained role of optimization of light and temperature in ginseng cell suspension culture (Wu and Zhong 1999). Light was reported as optimal factor to produce high content of lycopene from cell culture of *Barringtonia racemosa* (Behbahani et al. 2011).

15.3.5 Optimization of Inoculum Size

With the use of a suitable initial inoculum size, higher production and accumulation of bioactive compounds could be achieved as it has been found to affect the production of secondary metabolites and cell biomass in many plant cell cultures (Lee et al. 2006; Zhang et al. 2002; Zhao et al. 2001). It has been reported that in plant cell culture low inoculum size induces longer lag phase (Carvalho and Curtis 1999). Wu et al. (2006) reported that high inoculum size is good for biomass accumulation for *Echinacea angustifolia*; however, it was responsible for decreasing the accumulation of secondary metabolites. With this study, they explained the requirement of optimization of inoculum density for maximum biomass production along with high yield of metabolites. Lo et al. (2012) reported the potential role of inoculum concentration in artemisinin production from *Artemisia annua* L and found that increasing initial inoculum size reduces the time at which the *A. annua* cells achieved maximum growth along with the high content of artemisinin.

15.3.6 Optimization of Oxygen Level and Agitation

Agitation and oxygen concentration are also determinant factors in bioreactor-based plant cell culture-based secondary metabolites production. Agitation is an essential process because it promotes homogeneity with respect to the plant cells mass and nutrients and enhances mass and heat transfer in bioreactors. Singh and Chaturvedi (2012) reported that speed of agitation directly affects the growth and viability of cells in culture. Also, it is a very important parameter for establishment of plant cell suspension cultures. Oxygen supply has also shown significant effects on bioactive metabolites production in cell cultures. Therefore, Thanh et al. (2006) optimized the impact of oxygen supply on saponin production in bioreactor cultures of *Panax ginseng*. Similarly, positive effects of oxygen supply on plant cell growth of tobacco suspension cultures have been reported (Gao and Lee 1992). Thanh et al. (2014)

optimized bioreactor parameters (such as oxygen supply, inoculum density, aeration, etc.) for the production of biomass and ginsenosides.

15.4 Other Approaches to Improve Plant Cell Culture and Bioactive Compounds Production: Precursor Feeding, Elicitation, and Genetic Engineering

The primary challenges in regular commercial application of plant cell culture method are low and variable yields of secondary metabolites accumulation. It has been reported that some bioactive compounds do not accumulate in appreciable amount in undifferentiated cells (Kolewe et al. 2008). In these cases, several other strategies have been applied to boost the yield of targeted metabolites with higher biomass production in in vitro plant cell culture, such as precursor feeding, elicitation, gene editing, etc. (Biswas et al. 2020). Exogenous supply of a metabolites biosynthetic precursor to nutrient medium may increase the content of the desired bioactive compounds. This strategy is very effective when the precursors are low in cost. This concept of precursor feeding is based on the idea that any intermediate compound of a secondary metabolite biosynthetic route stands a good chance of improving the yield of the final bioactive compounds. In the past, a lot of efforts have been made to enhance the production of nutraceuticals by supplying precursor (Fett-Neto et al. 1994).

Bioactive compounds are produced in natural plants as a defence mechanism. Elicitors are signals that trigger the hypersensitive activity and thus enhanced synthesis of secondary metabolites in plants. Nowadays, use of elicitors has been one of the most effective strategies for improving the yield of bioactive secondary metabolites (Ramirez-Estrada et al. 2016; Narayani and Srivastava 2017). In elicitation mechanism, elicitor concentration, cell age, and the stage of the given culture are other important factors that are necessary to optimize for improving production of nutraceutical compounds (Ochoa-Villarreal et al. 2016). In many plants, elicitation strategy has been used to increase nutraceuticals production, such as tomato, ginseng, broccoli, Ashwagandha, etc. (Rodríguez et al. 2018; Rahimi et al. 2015; Natella et al. 2016; Sivanandhan et al. 2014).

Genome editing technology has also been used to manipulate plant nutraceuticals production (Kumar et al. 2018). State-of-the-art genomics tools can be combined with metabolic profiling to identify regulatory key genes of biosynthesis that could be engineered to produce improved nutraceuticals. In 2007, Li et al. successfully attempt to engineer flavonoid biosynthetic pathway in *Malus domestica* Borkh. (apple) with the overexpression of leaf colour regulatory gene from maize. However, genome editing technology such as CRISPR-Cas9 can be used to produce the desired nutraceutical plant with health promoting bioactive metabolites in near future.

15.5 Conclusion and Future Prospective

This chapter outlines the role of in vitro plant cell culture technology in nutraceuticals production and optimization of in vitro culture conditions for improving yield of nutraceutical compounds. In vitro plant cell culture technique is an important biotechnological tool to produce the plant-based nutraceutical compounds at industrial level. It has gained attention due to its simple growth medium requirements, genetic stability, and capacity for scale-up of nutraceuticals production. However, using plant cell culture technique, the major limitation for commercial production of nutraceuticals is the development of appropriate bioreactor production platform and in vitro culture conditions. Therefore, for the commercialization of nutraceutical compounds, it is utmost required to select favourable genotype/explants and optimize suitable culture conditions. Optimization of in vitro plant cell culture conditions could allow to enhance biomass and secondary metabolites production from nutraceutical plants by 20 to 30 folds. At present, the main disadvantage of plant cell culture is that this platform is still considered as an alternative to existing conventional technology and not being used as primary production route for nutraceuticals production (excluding some cases like ginseng production). Therefore, in near future, plant cell culture techniques can be applied as a primary selected method (with the optimization of procedure) to produce large-scale plant biomass and dietary nutraceutical compounds for human welfare.

Acknowledgement The authors express their gratitude towards the Director, G. B. Pant National Institute of Himalayan Environment and Sustainable Development, Uttarakhand, for providing all the necessary facilities. Deepika Tripathi is thankful to Department of Biotechnology for financial assistance as awarding DBT-RA fellowship.

References

Afshari RT, Angoshtari R, Kalantari S (2011) Effects of light and different plant growth regulators on induction of callus growth in rapeseed (Brassica napus L.) genotypes. Plant Omics 4(2): 60–67

Aguilar-Camacho M, Welti-Chanes J, Jacobo-Velázquez DA (2019) Combined effect of ultrasound treatment and exogenous phytohormones on the accumulation of bioactive compounds in broccoli florets. Ultrason Sonochem 50:289–301

Akbaş F, Işikalan Ç, Namli S (2009) Callus induction and plant regeneration from different explants of Actinidia deliciosa. Appl Biochem Biotechnol 158(2):470–475

Anjaneyulu C, Shyamkumar B, Giri CC (2004) Somatic embryogenesis from callus cultures of Terminalia chebula Retz.: an important medicinal tree. Trees 18(5):547–552

Anusha TS, Joseph MV, Elyas KK (2016) Callus induction and elicitation of total phenolics in callus cell suspension culture of Celastrus paniculatus–willd, an endangered medicinal plant in India. Pharm J 8(5):471

Behbahani M, Shanehsazzadeh M, Hessami MJ (2011) Optimization of callus and cell suspension cultures of Barringtonia racemosa (Lecythidaceae family) for lycopene production. Sci Agric 68(1):69–76

Biswas T, Mathur A, Gupta V, Luqman S, Mathur AK (2020) Elicitation and phenylalanine precursor feeding-based modulation of in vitro anthocyanin production, enzyme activity and gene expression in an Indian ginseng congener-Panax sikkimensis Ban. Ind Crop Prod 145: 111986

Carvalho EB, Curtis WR (1999) The effect of inoculum size on the growth of cell and root cultures of Hyoscyamus muticus: implications for reactor inoculation. Biotechnol Bioprocess Eng 4(4): 287–293

Chevreau E, Leblay C (1992, June) The effect of mother plant pretreatment and explant choice on regeneration from in vitro pear leaves. In: II international symposium on in vitro culture and horticultural breeding, vol vol 336, pp 263–268

Choudhary SK, Patel AK, Shekhawat S, Shekhawat NS (2017) An improved micropropagation system, ex vitro rooting and validation of genetic homogeneity in wild female Momordica dioica: an underutilized nutraceutical vegetable crop. Physiol Mol Biol Plants 23(3):713–722

DiCosmo F, Misawa M (1995) Plant cell and tissue culture: alternatives for metabolite production. Biotechnol Adv 13(3):425–453

Downey PJ, Levine LH, Musgrave ME, McKeon-Bennett M, Moane S (2013) Effect of hypergravity and phytohormones on isoflavonoid accumulation in soybean (Glycine max. L.) callus. Microgravity Sci Technol 25(1):9–15

Dureja H, Kaushik D, Kumar V (2003) Developments in nutraceuticals. Indian J Pharm 35(6): 363–372

Efferth T (2019) Biotechnology applications of plant callus cultures. Engineering 5(1):50–59

El-Nabarawy MA, El-Kafafi SH, Hamza MA, Omar MA (2015) The effect of some factors on stimulating the growth and production of active substances in Zingiber officinale callus cultures. Ann Agric Sci 60(1):1–9

Farhadi N, Panahandeh J, Azar AM, Salte SA (2017) Effects of explant type, growth regulators and light intensity on callus induction and plant regeneration in four ecotypes of Persian shallot (Allium hirtifolium). Sci Hortic 218:80–86

Fett-Neto AG, Melanson SJ, Nicholson SA, Pennington JJ, DiCosmo F (1994) Improved taxol yield by aromatic carboxylic acid and amino acid feeding to cell cultures of Taxus cuspidata. Biotechnol Bioeng 44(8):967–971

Gao J, Lee JM (1992) Effect of oxygen supply on the suspension culture of genetically modified tobacco cells. Biotechnol Prog 8(4):285–290

Georgiev MI, Weber J, Maciuk A (2009) Bioprocessing of plant cell cultures for mass production of targeted compounds. Appl Microbiol Biotechnol 83(5):809–823

Gläßgen WE, Wray V, Strack D, Metzger JW, Seitz HU (1992) Anthocyanins from cell suspension cultures of Daucus carota. Phytochemistry 31(5):1593–1601

Hahn EJ, Kim YS, Yu KW, Jeong CS, Paek KY (2003) Adventitious root cultures of Panax ginseng CV Meyer and ginsenoside production through large-scale bioreactor system. J Plant Biotechnol 5(1):1–6

Hai T, Schneider B, Porzel A, Adam G (1996) Metabolism of 24-epi-castasterone in cell suspension cultures of Lycopersicon esculentum. Phytochemistry 41(1):197–201

Javed SB, Alatar AA, Basahi R, Anis M, Faisal M, Husain FM (2017) Copper induced suppression of systemic microbial contamination in Erythrina variegata L. during in vitro culture. Plant Cell Tiss Org Cult 128(2):249–258

Jie EY, Ahn MS, Lee J, Cheon YI, Kim CY, Kim SW (2019) Establishment of a high-frequency plant regeneration system from rhizome-derived embryogenic cell-suspension cultures of Curcuma longa L. Plant Biotechnol Rep 13(2):123–129

Jun-Ge QU, Zhang W, Quan-Li HU, Mei-Fang JIN (2006) Impact of subculture cycles and inoculum sizes on suspension cultures of Vitis vinifera. Chin J Biotechnol 22(6):984–989

Kintzios SE, Taravira N (1997) Effect of genotype and light intensity on somatic embryogenesis and plant regeneration in melon (Cucumis melo L.). Plant Breed 116(4):359–362

Kintzios S, Drossopoulos JB, Lymperopoulos C (2001) Effect of vitamins and inorganic micronutrients on callus growth and somatic embryogenesis from leaves of chilli pepper. Plant Cell Tiss Org Cult 67(1):55–62

Kolewe ME, Gaurav V, Roberts SC (2008) Pharmaceutically active natural product synthesis and supply via plant cell culture technology. Mol Pharm 5(2):243–256

Kumar A, Mosa KA, Ji L, Kage U, Dhokane D, Karre S et al (2018) Metabolomics-assisted biotechnological interventions for developing plant-based functional foods and nutraceuticals. Crit Rev Food Sci Nutr 58(11):1791–1807

Lee EJ, Mobin M, Hahn EJ, Paek KY (2006) Effects of sucrose, inoculum density, auxins, and aeration volume on ceil growth of Gymnema sylvestre. J Plant Biol 49(6):427

Lee EJ, Jiménez Z, Seo KH, Nam GB, Kang YG, Lee TR et al (2020) Mass production of coumestrol from soybean (Glycine max) adventitious roots through bioreactor: effect on collagen production. Plant Biotechnol Rep 14(1):99–110

Lian ML, Chakrabarty D, Paek KY (2002) Effect of plant growth regulators and medium composition on cell growth and saponin production during cell-suspension culture of mountain ginseng (Panax ginseng CA Mayer). J Plant Biol 45(4):201–206

Liu QC, Zhai H, Wang Y, Zhang DP (2001) Efficient plant regeneration from embryogenic suspension cultures of sweet potato. In Vitro Cell Dev Biol Plant 37(5):564–567

Lo K, Nadali BJ, Chan LK (2012) Investigation on the effect of subculture frequency and inoculum size on the artemisinin content in a cell suspension culture of 'Artemisia annua' L. Aust J Crop Sci 6(5):801

Mathur S, Shekhawat GS (2013) Establishment and characterization of Stevia rebaudiana (Bertoni) cell suspension culture: an in vitro approach for production of stevioside. Acta Physiol Plant 35(3):931–939

Mehta J, Sain M, Mathuriya BL, Naruka R, Kavia A, Sharma DR (2012) Rapid micropropagation and callus induction of Terminalia bellerica Roxb.-an endangered plant. Asian J Plant Sci 2(3): 364–368

Mostafa HH, Wang H, Song J, Li X (2020) Effects of genotypes and explants on garlic callus production and endogenous hormones. Sci Rep 10(1):1–11

Nagella P, Murthy HN (2010) Establishment of cell suspension cultures of Withania somnifera for the production of withanolide A. Bioresour Technol 101(17):6735–6739

Narayani M, Srivastava S (2017) Elicitation: a stimulation of stress in in vitro plant cell/tissue cultures for enhancement of secondary metabolite production. Phytochem Rev 16(6): 1227–1252

Natella F, Maldini M, Nardini M, Azzini E, Foddai MS, Giusti AM et al (2016) Improvement of the nutraceutical quality of broccoli sprouts by elicitation. Food Chem 201:101–109

Nath S, Buragohain AK (2005) Establishment of callus and cell suspension cultures of Centella asiatica. Biol Plant 49(3):411–413

Nazir R (2019) Biotechnological strategies for production of anti-Alzheimer compounds from Valeriana spp. J Gujarat Res Soc 21(8):653–671

Ochoa-Villarreal M, Howat S, Hong S, Jang MO, Jin YW, Lee EK, Loake GJ (2016) Plant cell culture strategies for the production of natural products. BMB Rep 49(3):149

Park YG, Kim SJ, Kang YM, Jung HY, Prasad DT, Kim SW et al (2004) Production of ginkgolides and bilobalide from optimized the Ginkgo biloba cell culture. Biotechnol Bioprocess Eng 9(1): 41–46

Pires AS, Cabral MG, Fevereiro P, Stoger E, Abranches R (2008) High levels of stable phytase accumulate in the culture medium of transgenic Medicago truncatula cell suspension cultures. Biotechnol J 3(7):916–923

Pise M, Rudra J, Bundale S, Begde D, Upadhyay A (2011) Shatavarin production from in vitro cultures of Asparagus racemosus Wild. J Med Plant Res 5(4):507–513

Pise M, Rudra J, Bundale S, Begde D, Nashikkar N, Upadhyay A (2012) Asparagus racemosus cell cultures: a source for enhanced production of shatavarins and sarsapogenin. In Vitro Cell Dev Biol Plant 48(1):85–91

Rahimi S, Kim YJ, Yang DC (2015) Production of ginseng saponins: elicitation strategy and signal transductions. Appl Microbiol Biotechnol 99(17):6987–6996

Rahman LU, Verma PC, Singh D, Gupta MM, Banerjee S (2002) Bacoside production by suspension cultures of Bacopa monnieri (L.) Pennell. Biotechnol Lett 24(17):1427–1429
Ramirez-Estrada K, Vidal-Limon H, Hidalgo D, Moyano E, Golenioswki M, Cusidó RM, Palazon J (2016) Elicitation, an effective strategy for the biotechnological production of bioactive high-added value compounds in plant cell factories. Molecules 21(2):182
Rao SR, Ravishankar GA (2002) Plant cell cultures: chemical factories of secondary metabolites. Biotechnol Adv 20(2):101–153
Rawat V, Ghildiyal A, Singh L, Jugran AK, Bhatt ID, Nandi SK, Pande V (2019) Methyl jasmonate induced polyphenols and antioxidant production in callus suspension culture of Nardostachys jatamansi. Plant Biosyst 154:851–859
Rodríguez GG, Palomeque BE, Ríos PC, Reséndez AM, Escobedo LL, Galván HS, Mata JS (2018) Influence of rhizobacteria in production and nutraceutical quality of tomato fruits under greenhouse conditions. Rev Mex Cienc Agríc 9:368
Sainulabdeen HM, Sulekha GR (2008) Callus induction and plantlet regeneration in Aegle marmelos (L.) Corr. using cotyledon explants. J Trop Agric 46:79–84
Sairam RV, Franklin G, Hassel R, Smith B, Meeker K, Kashikar N et al (2003) A study on the effect of genotypes, plant growth regulators and sugars in promoting plant regeneration via organogenesis from soybean cotyledonary nodal callus. Plant Cell Tiss Org Cult 75(1):79–85
Salma U, Kundu S, Ali MN, Mandal N (2018) Elicitor mediated enhancement of wedelolactone in cell suspension culture of Eclipta alba (L.) Hassk. Plant Cell Tiss Org Cult 134(3):409–421
Sampaio KB, do Nascimento YM, Tavares JF, Cavalcanti MT, de Brito Alves JL, Garcia EF, de Souza EL (2020) Development and in vitro evaluation of novel nutraceutical formulations composed of Limosilactobacillus fermentum, quercetin and/or resveratrol. Food Chem 2020: 128264
Savita V, Virk GS, Nagpal A (2010) Effect of explant type and different plant growth regulators on callus induction and plantlet regeneration in Citrus jambhiri Lush. Environ Int J Sci Technol 5: 97–106
Seth B, Sahoo KK, Aravind KR, Sahu BB, Singh VR, Patra N (2020) Statistical optimization of bacoside A biosynthesis in plant cell suspension cultures using response surface methodology. 3 Biotech 10(6):264
Sharma V, Goyal S, Ramawat KG (2009) Scale up production of isoflavonoids in cell suspension cultures of Pueraria tuberosa grown in shake flasks and bioreactor. Eng Life Sci 9(3):267–271
Singh M, Chaturvedi R (2012) Evaluation of nutrient uptake and physical parameters on cell biomass growth and production of spilanthol in suspension cultures of Spilanthes acmella Murr. Bioprocess Biosyst Eng 35(6):943–951
Sivanandhan G, Selvaraj N, Ganapathi A, Manickavasagam M (2014) Enhanced biosynthesis of withanolides by elicitation and precursor feeding in cell suspension culture of Withania somnifera (L.) Dunal in shake-flask culture and bioreactor. PLoS One 9(8):e104005
Thanh NT, Murthy HN, Yu KW, Hahn EJ, Paek KY (2005) Methyl jasmonate elicitation enhanced synthesis of ginsenoside by cell suspension cultures of Panax ginseng in 5-l balloon type bubble bioreactors. Appl Microbiol Biotechnol 67(2):197–201
Thanh NT, Murthy HN, Yu KW, Jeong CS, Hahn EJ, Paek KY (2006) Effect of oxygen supply on cell growth and saponin production in bioreactor cultures of Panax ginseng. J Plant Physiol 163(12):1337–1341
Thanh NT, Murthy HN, Paek KY (2014) Optimization of ginseng cell culture in airlift bioreactors and developing the large-scale production system. Ind Crop Prod 60:343–348
Vasilev N, Schmitz C, Grömping U, Fischer R, Schillberg S (2014) Assessment of cultivation factors that affect biomass and geraniol production in transgenic tobacco cell suspension cultures. PLoS One 9(8):e104620
Verpoorte R, Contin A, Memelink J (2002) Biotechnology for the production of plant secondary metabolites. Phytochem Rev 1(1):13–25
Wu J, Zhong JJ (1999) Production of ginseng and its bioactive components in plant cell culture: current technological and applied aspects. J Biotechnol 68(2-3):89–99

Wu CH, Dewir YH, Hahn EJ, Paek KY (2006) Optimization of culturing conditions for the production of biomass and phenolics from adventitious roots of Echinacea angustifolia. J Plant Biol 49(3):193

Yagi A, Shoyama Y, Nishioka I (1983) Formation of tetrahydroanthracene glucosides by callus tissue of Aloe saponaria. Phytochemistry 22(6):1483–1484

Zhang CH, Wu JY, He GY (2002) Effects of inoculum size and age on biomass growth and paclitaxel production of elicitor-treated Taxus yunnanensis cell cultures. Appl Microbiol Biotechnol 60(4):396–402

Zhao D, Xing J, Li M, Lu D, Zhao Q (2001) Optimization of growth and jaceosidin production in callus and cell suspension cultures of Saussurea medusa. Plant Cell Tiss Org Cult 67(3): 227–234

Genetic Engineering of Cell Cultures for Enhanced Production of Nutraceuticals 16

Andrey Marchev, Kristiana Amirova, and Milen Georgiev

Abstract

Driven by various health concerns, there is growing demand for plant-derived ingredients, used as pharmaceutical alternatives that claim certain physiological benefits for the human body. As such, nutraceuticals are defined as bioactive natural substances in the tissue of many plant species and mainly used to prevent human diseases, due to their anti-inflammatory and antioxidant properties. The emerging trend for manufacturing nutraceuticals in the form of dietary supplements or functional foods in a natural and sustainable manner has opened new avenues of plant cell culture technology over the past years. The advantages of this technology opposed to the wild harvesting or field cultivation are the controlled production according to the demand, preservation of the global biodiversity, as well as avoidance of associated ecological restrictions. Moreover, genetic engineering is a tool that provides a great opportunity for metabolic manipulation through overexpression, silencing, or disruption of one or more genes of the biosynthetic pathway of interest. In this chapter, we attempt to provide a framework of the cellular and molecular tools and their application regarding the enhanced production of nutraceuticals from plant cell cultures with special focus on flavonoids as ubiquitous metabolites in plant foods, including vegetables, fruits, nuts, and beverages possessing many beneficial effects over the human body.

A. Marchev · K. Amirova · M. Georgiev (✉)
Laboratory of Metabolomics, The Stephan Angeloff Institute of Microbiology, Bulgarian Academy of Sciences, Plovdiv, Bulgaria

Center of Plant Systems Biology and Biotechnology, Plovdiv, Bulgaria
e-mail: amirova@cpsbb.eu; milengeorgiev@gbg.bg

T. Belwal et al. (eds.), *Nutraceuticals Production from Plant Cell Factory*,
https://doi.org/10.1007/978-981-16-8858-4_16

Keywords

Secondary metabolites · Flavonoids · Nutraceuticals · Plant cell cultures · Bioactivity · Transcription factors

16.1 Introduction

In the last decades, there has been an increased interest from both, scientific community and food industry, towards the identification of active molecules from natural sources that have beneficial health effects beyond their nutritional value (Carrizo et al. 2020). In 1989, Dr. Stephan DeFelice has introduced a joint term "nutraceutical" aiming to fill the gap between nutrition and pharmacological therapy. According to this definition, nutraceuticals are foods or part of foods with positive well-being and health promoting effects, including the prevention and/or treatment of a disease (Kalra 2003). However, compared to therapeutic drugs, nutraceuticals are not subject of patent protection or governmental regulations and could be used as preventive agents or adjuvants in medicinal therapy (Carrizo et al. 2020). Due to the lack of clear definition, nutraceuticals are frequently accepted as functional foods (since they are consumed as a part of the diet and improve the health status), but unlike them nutraceuticals are consumed under dosage forms like conventional drugs in the form of tablets, capsules, or solutions (Coppens et al. 2006). In this regard, nutraceuticals also differ from dietary supplements, whose main role is to supplement the diet by increasing the total dietary uptake of vitamins, amino acids, minerals, or botanicals without aiming the prevention or treatment of diseases (Ross 2000). From a nutritional perspective, the array of nutraceuticals has been extended to a source of both nutrients (carbohydrates, proteins, and fats) and non-nutrients (pro- and prebiotics, as well, as phytochemicals) and comprises food constituents, plant extracts, and food derivatives, including vitamins, amino acids, and minerals (Bergamin et al. 2019). Nutraceuticals may considerably vary in their chemical properties (molar mass, structure, polarity, functional groups, and charge) and biological activities. They could be classified as antioxidants (vitamins, enzymes, and minerals), dairy-based ingredients, plant extracts and phytochemicals, nutritional lipids and oils, fibers and carbohydrates, probiotics, prebiotics, amino acids, and proteins (Jain and Pundir 2013).

Phytochemicals comprise a great variety of chemicals structures and have dual purpose, such as defensive role in plants and specific pharmacological effect in humans. Many of these bioactive phytochemicals, such as alkaloids, triterpenes, and polyphenols (flavonoids, phenolic acids, anthocyanins, etc.), are essential source of nutraceuticals (Ho et al. 2021). However, when these nutraceuticals are derived from wild- or field-grown plants, their yield and consistence in quality are dependent on plant physiology state, harvest and storage processing, as well as, numerous environmental and geographical factors. Along with that, their specific application requires high purity, but when using whole plants there is risk from herbicides or pesticides contamination (Açıkgöz 2020). Therefore, plant-field harvesting might be environmentally destructive for native plants population, especially for rare and endangered plant species, and on the other hand, economically impractical due to the high operational costs, frequently accompanied with very low and inconsistent yields of the final product (Marchev et al. 2020). Alternatively, plant biotechnology based on plant in vitro systems cultivation offers a reliable and sustainable opportunity for the biotechnological production of plant-derived secondary metabolites (SMs) in strictly controlled aseptic conditions, avoiding the influence of climate and geographical conditions and the possibility to obtain products with consistent yield and quality. Major advances, here, are that plant in vitro systems have the potential to biosynthesize the full spectrum of metabolites typical for the mother plant and have the ability to adapt their metabolism according to the environment (Szopa et al. 2019). Therefore, plant in vitro systems are susceptible to genetic manipulations of the biosynthetic pathways or changes in cultural environment (additions of precursors, elicitors, and nutrient medium optimization) and allow commercialization through transfer of the biotechnological process from flask to large-scale production (Cui et al. 2014).

Here, we attempt to summarize recent examples for the biosynthesis of nutraceutical ingredients, with focus on flavonoid molecules derived from plant in vitro cultures. Their significance in cancer prevention and treatment has been described. Recent advances for manipulation of their biosynthetic pathway aiming their enhanced production are highlighted. Opportunities for lab- and large-scale production through bioreactor cultivation are discussed.

16.2 Role of Nutraceuticals in Prevention and Therapy of Human Malignancies

Many plants and herbs consumed by humans are known to contain relevant amounts of polyphenols which have been demonstrated to have many beneficial effects, such as antitumor, anti-inflammatory, antiallergic, antidiabetic, cardioprotective activity, and many others. They are ubiquitous in plant foods, including vegetables, fruits, nuts, and beverages, such as wine, tea, or cocoa. Specifically, the most abundant sources of flavonoids are apples, citrus fruits, berries, etc. In USA and Netherlands, the human flavonoid uptake is approximately 170 and 23 mg/day, expressed as aglycones, based on the content of five flavonoids, such as quercetin, kaempferol, myricetin, luteolin, and apigenin. The dietary consumption of polyphenols consists mainly of 80% flavanols, 8% flavonols, 6% flavanones, 5% anthocyanins, and less than 1% isoflavones and flavones. Therefore, it is not surprising that research has progressively shifted from cure to prevention, from drugs to lifestyle concept, including diet and food (Cas and Ghidoni 2018).

Cancer, the second leading cause of death globally, has accounted for nearly 9.6 million mortalities in 2018. According to World Health Organization (WHO) reports, one in six deaths globally is due to cancer. The major risk factors of cancer are low fruit and vegetable consumption, sedentary lifestyle, obesity, increased exposure to cancer-causing agents, including tobacco and alcohol consumption,

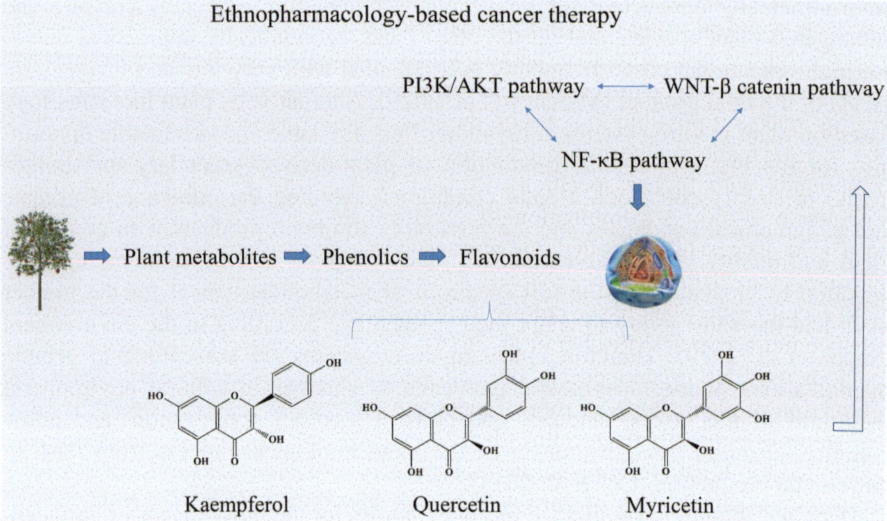

Fig. 16.1 Signalling pathways in cancer. Active phytochemicals in extract from medicinal plants are the main source in prevention and therapy of cancer. Bioactivity of flavonoids influences the cascade mechanism of NF-κB, Wnt/β-catenin, and PI3K/AKT signaling pathways, leads to activation of apoptosis of the malignant cells, or reduces the cell cycle in current phase G2/M

just to name a few (Nalini et al. 2019). Numerous in vitro, in vivo, and epidemiological studies have shown the benefits of flavonoids and flavonoid-rich extracts in preventing or curing cancer, including lung (Amararathna et al. 2020), liver and breast (Ajji et al. 2020), ovarian (Carmi et al. 2020), colorectal (Dong et al. 2020), etc. and the main mechanisms of the anticancer activity were cell cycle arrest and inhibition of several signaling pathways, including extracellular signal-regulated kinase 1/2 (ERK1/2), phosphoinositide-3-kinase/protein kinase B (PI3K/AKT), nuclear factor-kappa B (NF-κB), mitogen-activated protein kinase (MAPK), and Wnt/β-catenin (Amararathna et al. 2020). An illustration of the main signaling pathways affected by flavonoids is presented on Fig. 16.1.

Cell migration of cancer cells increases their motility and invasion and very often metastasis accounts as cause for over 90% of cancer-related mortality. The binding of nutraceuticals, such as flavonoids to actin (the main protein responsible for shape, adhesion, and motility of the cells), is one of the potential mechanisms that can limit actin polymerization and prevent cancer cell movement and invasion (Budryn et al. 2018). Quercetin-induced cytotoxicity in a dose-dependent manner in leukemic (CEM and K562) and breast cancer (T47D and EAC) cells is mediated through DNA fragmentation, cell cycle arrest and activated intrinsic pathway of apoptosis (Srivastava et al. 2016), as well as, suppressed expression of survivin and induced G0/G1 phase cell cycle arrest in breast cancer (MCF-7) cells (Deng et al. 2013). Among all the tested flavonoids (naringenin, quercetin, and naringin), quercetin had the lowest IC_{50} value of 32 μg/mL and reduced HepG2 and MCF-7 cell viability, increased the activation of caspase-3 and -8, and induced apoptosis through upregulation of proapoptotic genes Bax and Bid and downregulation of antiapoptotic genes BCL-2 and BCL-XL (Ajji et al. 2020). Quercetin and kaempferol blocked the growth of human colorectal cancer cell lines HCT116 and SW480 at G0/G1 phase. The key targets were cyclin-D1, MAPK8, and epidermal growth factor receptor (EGFR). These compounds induced cell apoptosis through modulation of PI3K signaling pathway and BCL-2 family proteins (Dong et al. 2020). As a flavonoid able to cross the blood brain barrier, quercetin was used to obtain synthetic quercetin derivatives that were able to induce cell cycle arrest and apoptosis in C6 glioma cells through inhibition of NF-κB and capsase-3 activation. The obtained derivatives appeared not cytotoxic to hippocampal organotypic cultures, used as a model of healthy neural cells (Kiekow et al. 2016). The Wnt/β-catenin signaling pathway has a key role in cell proliferation, cell motility, and stem cell differentiation and its aberrant activation can lead to tumorigenesis, cancer progression, and tumor cell metastasis. A quercetin-rich extract of *Telectadium dongnaiense* Pierre ex Cost (IC_{50} 2 μg/mL) inhibited cell proliferation of HCT116 colon cancer cells, which correlated with its inhibitory effect on the expression of Wnt target genes, such as cyclin-D1, survivin, and c-myc (Kim et al. 2017). Quercetin could be used in combination with other plant-derived molecules, such as paclitaxel or resveratrol for the successful treatment of multidrug-resistant breast (Liu et al. 2020a) or prostate cancer (Singh et al. 2020). The hybrid combination (paclitaxel and quercetin) nanoparticles significantly decreased the IC_{50} values from 108.85 μg/mL (when paclitaxel was used as a single drug) to 11.76 μg/mL, which further resulted in higher apoptosis rate, G2M phase arrest, and stronger microtubule destruction in MCF7/ADR. This evidence revealed that quercetin increased the sensitivity of the cell line to paclitaxel and/or other therapeutic approaches, including chemotherapy (Liu et al. 2020a). The combination quercetin-resveratrol modulates the genes involved in cell cycle, apoptosis, fatty acid metabolism, transcription factors, androgen response, PI3K/AKT, and PTEN signaling. The decreased levels of EGFR, early growth response 3 (EGR3) and interleukin 6 (IL6), and increased levels of insulin growth factor binding protein 7 (IGFBP7) and NK3 homeobox-1 (NKX3.1) overall supported the antiprostate cancer effect of quercetin-resveratrol (Singh et al. 2020). The combinatorial application of quercetin and curcumin inhibited the cell proliferation of human melanoma cells (A375) through downregulation of Wnt/β-catenin signaling pathway proteins, cyclooxygenase 2 (COX-2), cyclin-D1, and Axin2. The IC_{50} value of this combination was significantly lower (1.5 μM) in comparison when both substances were used alone (50.0 and 8.5 μM; Srivastava and Srivastava 2019). Quercetin might be used as an adjuvant drug for treatment of cancer cells resistant to ionization irradiation therapy. The pretreatment of quercetin (20 μM) significantly reduced the ionization radiation to 5 Gy in colon cancer treatment by targeting Notch-1 signaling pathway (Li et al. 2020a, b). Gemcitabine resistance is a major issue for clinicians and patients with advanced cancers, making it crucial to determine ways to bolster its effects. Quercetin had a cytotoxic effect on gemcitabine-resistant pancreatic (PANC-

1) and hepatocellular (HepG2) cancer. Quercetin lead to S phase arrest, tumor protein p53 upregulation, and cyclin-D1 downregulation (Liu et al. 2020b).

As a glucosylquercetin, rutin is more effective in skin carcinogenesis prevention. Rutin is able to prevent the UVB-induced inflammation in mice through inhibition of the induced COX-2 and inducible nitric oxide synthase (iNOS). Rutin inhibited the DNA binding of activator protein-1 (AP-1) and phosphorylation of signal transducer and activator of transcription-3 (STAT3) in mouse skin exposed to UVB (Choi et al. 2019). In addition, rutin significantly increased collagen type I and III and decreased metalloproteinase 1- and − 3 (MMP-1 and 3) in UVB-exposed mice (Her et al. 2020).

Most of the promising nutraceuticals, including flavonoids, have compromised bioavailability, due to their hydrophobic nature and very often remain unabsorbed without providing any therapeutic effect. Therefore, the bioavailability of nutraceuticals needs to be improved by enhancing the physicochemical properties like stability, solubility, and permeability. Hence, nanocarrier-based delivery systems of such supplementary products play an important role in the management and treatment of diseases (Shende and Mallick 2020). Quercetin gold nanoparticles in poly(D,L-lactide-co-glycolide) inhibited the liver cancer cell migration and colony formation. Quercetin also upregulated apoptosis markedly through the cleavage of caspase-9 and caspase-3 and induced the up-releasing cytochrome C (Ren et al. 2017). Quercetin-loaded nanoparticles exhibited improved inhibition on the growth and metastasis of triple negative breast cancer through inhibition of urokinase-type plasminogen activator, which mediates tumor growth and metastasis (Zhou et al. 2020). Nanoencapsulated quercetin was every efficient to combat with the hepatocarcinogenesis induced by diethylnitrosamine in rats (Ghosh et al. 2012).

16.3 Plant Cell Cultures as Potential Sources of Nutraceuticals

Plant SMs are of immense use as potential drugs, nutraceuticals, and food additives. Among the different classes of molecules, plant polyphenols constitute the largest group, of which flavonoids are the most abundant metabolites (Ali and Abbasi 2014) for the human health due to their role in treatment or prevention of cancer, cardiovascular diseases, gastrointestinal disorders, immune/autoimmune diseases, brain dysfunctions, diabetes, and obesity (Taghizadeh et al. 2019). Harvesting wild- or field-grown plant is not able to meet the growing demand for flavonoids for nutraceuticals and commercial use, which raises the awareness for serious ecological concerns (Thiruvengadam et al. 2016). In this regard, the use of plant in vitro system could be a viable alternative for reliable and continuous production of flavonoids, which allows the application of different strategies for manipulation of the growth variables, precursors, and elicitors that might modulate the biosynthetic pathway of the compounds, optimizing their production (Mendoza et al. 2018). Flavonoids, and especially kaempferol, quercetin, and rutin, comprise the main group of biologically active metabolites in many plants and are important biochemical markers authenticating the herbal plant materials and final products (Wang et al. 2015).

They are found in many vegetables, fruits, and medicinal plants and represent a major part of the diet in many societies (Thwe et al. 2013).

Many plant in vitro systems have been established with the aim to produce flavonoids, but they rarely produce sufficient amounts of these SMs. For that reason, different plant in vitro systems, such as cell suspensions (Kubica et al. 2020; Vazquez-Marquez et al. 2016), hairy roots (HRs;), adventitious roots (ARs; Ho et al. 2021; Jiang et al. 2020), and shot type cultures (Szopa et al. 2019; Weremczuk-Jeżyna et al. 2020), have been investigated as potential source of flavonoids. Along with that, different approaches such as nutrient medium optimization (Cui et al. 2011), precursor feeding (Skrzypczak-Pietraszek et al. 2019), and application of biotic/abiotic elicitors have been frequently applied to enhance their biosynthesis (Huang et al. 2016; Li et al. 2020a, b). Elicitors are natural or synthetic molecules that initiate or improve the biosynthesis of SMs when introduced in small amounts into the living cell. Elicitors, such as jasmonic acid (JA), methyl jasmonate (MeJa), or salicylic acid (SA), are signaling molecules that respond to biotic or abiotic stress and induce catalytic reactions leading to enhanced SMs accumulation through activation of specific enzymes involved in their biosynthesis. The elicitor efficiency depends on its type, concentration, and time for treatment (Amani et al. 2021). Due to the smaller cell size (compared to tissue and organ cultures) and close characteristics to bacterial cell, suspension cultures offer a simplified system to study growth and production kinetics and potentially evaluate and implement optimal conditions for the production of SMs (Ali and Abbasi 2014). Flavonoid biosynthesis from cell suspension of *Buddleja cordata* Kunth (Vazquez-Marquez et al. 2016), *Scrophularia striata* Boiss (Ahmadi-Sakha et al. 2018), and *Orostachys cartilaginous* A. Bor has been reported (Wen et al. 2019). The most suitable for their cultivation were the stirred tank bioreactors (STB) operating at low (under 50 rpm) or high agitation speed (400 rpm) depending on the culture type and aeration rate between 0.05 and 0.1 vvm (Kubica et al. 2020; Vazquez-Marquez et al. 2016). However, due to the dedifferentiated nature of cell suspensions, they might lose their stable growth and biosynthetic characteristics (Marchev and Georgiev 2020). Therefore, in such cases plant tissue and organ cultures are the preferred system for production of SMs. Hairy roots obtained via *Agrobacterium rhizogenes*-mediated genetic transformation are genetically and biochemically stable, show rapid growth rate, and have the ability to biosynthesize SMs at levels comparable to that found in the mother plant (Thwe et al. 2013). For that reason, they have been frequently used to investigate the possibilities to manipulate the biosynthetic pathway of flavonoids (Huang et al. 2016). The capacity of the HRs to produce flavonoids might vary due to the random integration of the root-inducing plasmid in the plant genome, as well as, on the strain used for the genetic transformation (Thwe et al. 2016). The biosynthetic potential of HRs might be enhanced through application of elicitors (MeJa, UV-light) or overexpression of transcriptional factors (TFs) and genes encoding enzymes from the biosynthetic pathway discussed in more details further in the chapter (Huang et al. 2016; Li et al. 2020a, b; Yao et al. 2020).

16.4 Genetic Engineering of Plant Cell Cultures for Enhanced Production of Flavonoids

Flavonoids are a family of approximately 6000 nutraceuticals widely distributed in plants with great variety of significant biological functions, biosynthesized through shikimic acid, phenylpropanoid, and flavonoid pathways (Chung et al. 2018). All flavonoids possess a generic C6-C3-C6 structure comprising of 15 C-atoms, thus forming two aromatic A and B rings connected to one heterocyclic benzopyran C ring, which contains one oxygen. Depending on the modifications, especially hydroxylation and methylation reactions in the C ring are leading to the formation of more than 9000 derivatives (Marín et al. 2018). In this chapter, we have focused on a specific subtype of flavonoids classified as flavonols, whose main representatives, such as quercetin, kaempferol, myricetin, and rutin, present the major intake of the dietary flavonoids in most societies.

Among all the biosynthetic genes, the key ones are the phenylalanine ammonia lyase (PAL; EC 4.3.1.24), chalcone synthase (CHS; EC 2.3.1.74), chalcone isomerase (CHI; EC 5.5.1.6), and flavanone 3-hydroxylase (F3H; Bai et al. 2019). The biosynthetic pathway of flavonoids is illustrated in Fig. 16.2.

The entry metabolite and predecessor for the flavonoid biosynthesis in the phenylpropanoid pathway is the amino acid phenylalanine, which is deaminated by PAL to form trans-cinnamic acid that is further hydroxylated into para-coumaric acid by cinnamate 4-hydroxylase (C4H; EC 1.14.14.91). In plants, this enzyme catalyzes the initial steps in the production of diverse polyphenyl compounds and is mainly involved in defense mechanisms (Pervaiz et al. 2017). As an alternative in some plant species, para-coumaric acid can be produced by tyrosine from tyrosine ammonia lyase (TAL; EC 4.3.1.23). Further, the para-coumaric acid, which appeared to be the most limiting substrate in this pathway, is conjugated with coenzyme A by the enzyme 4-coumarate-CoA ligase (4CL; EC 6.2.1.12) to obtain para-coumaroyl-CoA. This is actually a branching point to other biosynthetic pathways, including coumarins, monolignols, or polyketides, such as flavonoids, isoflavonoids, and stilbenes (Chaves-Silva et al. 2018; Zhang et al. 2019). Para-coumaroyl-CoA is the initial substrate of the flavonoid pathway from which is generated naringenin chalcone through condensation of three molecules of malonyl-CoA by the first specific enzyme in the flavonoid biosynthetic pathway, named chalcone synthase (CHS; EC 2.3.1.74). At the same time, catalyzed by acetyl CoA carboxylase, acetyl CoA is converted to malonyl-CoA. Therefore, the A and C rings are formed by the acetate pathway, whereas B ring is derived from the shikimate pathway (Pervaiz et al. 2017). Subsequently, chalcone isomerase (CHI; EC 5.5.1.6) converts stereospecifically the chalcone to its isomer naringenin in a ring-closing step that forms the heterocyclic C ring. Naringenin appears to be a central intermediate, since it gives the start of different side branches for the synthesis of diverse classes of flavonoid molecules derived by the activity of functionalizing enzymes that might hydroxylate, reduce, alkylate, oxidize, and glycosylate the phenylpropanoid core structure (Huang et al. 2016). Further, the B ring of naringenin is hydroxylated by flavonoid 3′-hydroxylase (F3'H; EC

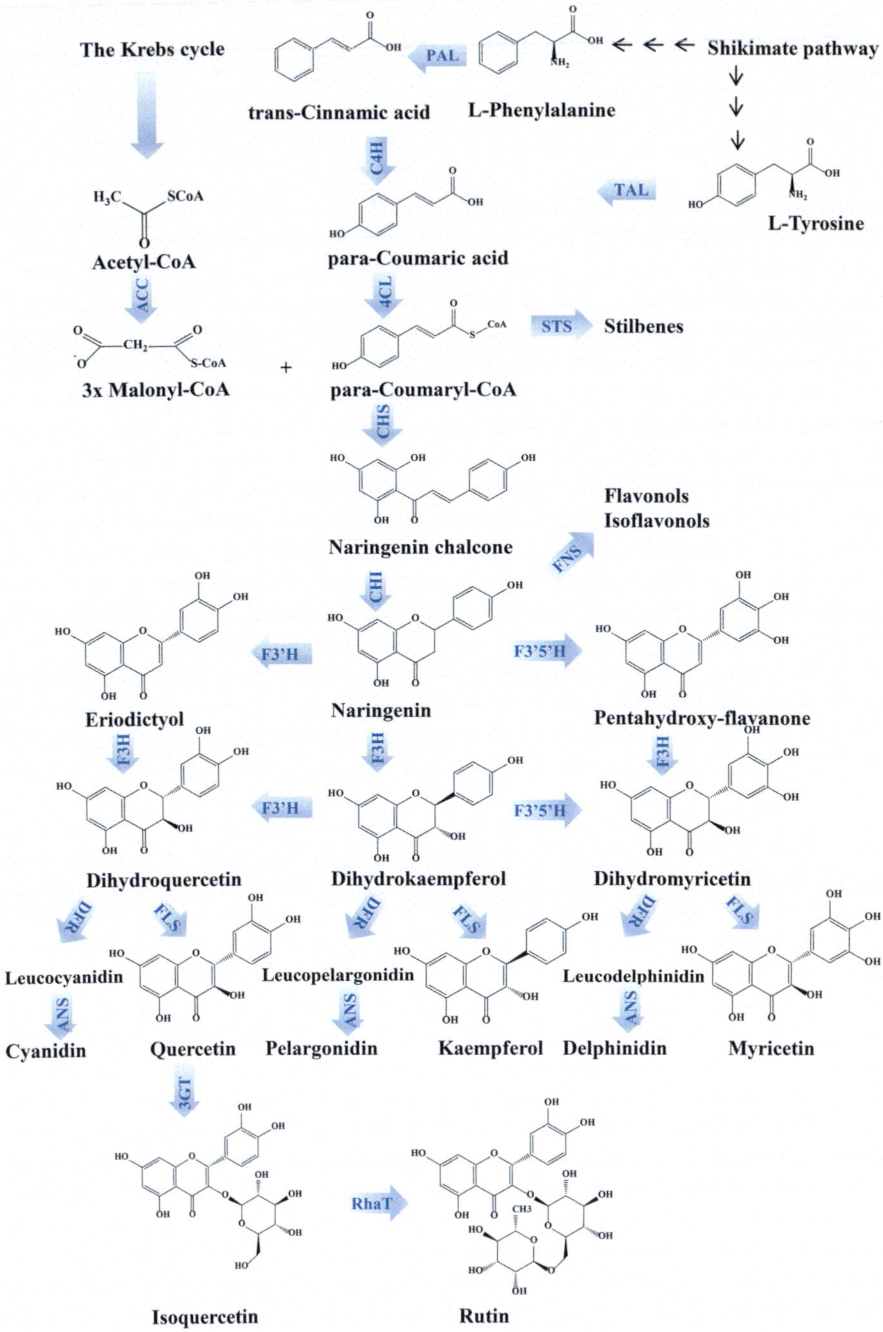

Fig. 16.2 Illustration, presenting the biosynthetic pathway of flavonoids. *PAL* phenylalanine ammonia lyase (EC 4.3.1.24), *TAL* tyrosine ammonia lyase (EC 4.3.1.23), *C4H* cinnamate 4-hydroxylase (EC 1.14.14.91), *ACC* acetyl-CoA carboxylase (EC 6.4.1.2), *4CL* 4-coumarate-CoA ligase (EC 6.2.1.12), *CHS* chalcone synthase (EC 2.3.1.74), *STS* stilbene synthase (EC 2.3.1.95), *CHI* chalcone isomerase (EC 5.5.1.6), *F3′H* flavonoid 3′-hydroxylase

1.14.13.21) or flavonoid 3′5′-hydroxylase (F3′5'H; EC 1.14.13.88) to produce eriodyctiol or pentahydroxy-flavanone. All these (2S)-flavanones are further modified by the catalysis of flavanone 3β-hydroxylase (F3H; EC 1.14.11.9) in dihydroflavonols, such as dihydroquercetin, dihydrokaempferol, and dihydromyricetin. Dihydrokaempferol could be potentially directly oxidized by F3'H and F3′5'H to dihydroquercetin and dihydromyricetin (Zhang et al. 2019). Further, the enzyme flavone synthase (FLS; EC 1.14.11.22) catalyzes the hydroxylation of dihydroflavonols to their respective flavonols quercetin, kaempferol, and myricetin (Irmish et al. 2019). This enzyme is a member of the large enzyme family of 2-oxoglutarate-dependent dioxygenases (2OGDs) and is a key enzyme of the flavonoid biosynthetic pathway, acting at the diverging point into the flavonol subclass branch (Li et al. 2013). The other branching pathway is the biosynthesis of anthocyanins, according to which the obtained dihydroflavonols are reduced to colorless leucoanthocyanidins by dihydroflavonol 4-reductase (DFR; EC 1.1.1.219), which are then oxidized to colored anthocyanidins by the activity of anthocyanidin synthase (ANS; EC 1.14.20.4, synonyms leucocyanidin oxygenase: LDOX), which requires NADP as a cofactor and is oxoglutarate-dependent Fe2+/Fe3+ dioxygenase. These reactions are also high stereospecific with 2S, 3S, 4S configuration (Liu et al. 2014). After the biosynthesis of the flavonols, they might be subjected to glycosylation. For example, under the activity of flavonol 3-O¬-glucosyltransferase (3GT; EC 2.4.1.91) and flavonol-3-O-glucoside L-rhamnosyltransferase (RhaT; EC 2.4.1.159), quercetin is subsequently converted to isoquercetin and rutin (Abdel-Salam et al. 2021).

The main strategies adopted to alter metabolic pathways and improve phytochemical production might be summarized as: (1) gain-of-function of a gene of the pathway; (2) loss-of-function of a gene of the pathway; (3) ectopic expression of transcription factors (TFs) responsible for expressing all or most pathway genes, and (4) overexpression of a single gene encoding a rate-limiting enzyme (Nielsen et al. 2019). The SMs biosynthetic pathway is generally governed by its structural genes, which control the activity of enzymes involved in the pathway, as well as, regulatory genes that encode the TFs. On the other hand, the TFs regulate the structural genes activity through direct activation or repression through binding to the promoter region of the structural genes (Hidalgo et al. 2017). Flavonoid biosynthesis is regulated at transcriptional level by different families of TFs, in particular the multimeric unit formed by the members of the major TFs, such as myeloblastosis (MYB), Myc (encoding basic helix-loop-helix), and WDR (WD-repeat proteins), which are essential regulators in cell development and expression of genes involved in the flavonoid biosynthesis (Amani et al. 2021). Although the biosynthetic activity

Fig. 16.2 (continued) (EC 1.14.13.21), *F3′5′H* flavonoid 3′5′-hydroxylase (EC 1.14.13.88), *FNS* flavones synthase (EC 1.14.11.22), *F3H* flavanone 3-hydroxylase (EC 1.14.11.9), *DFR* dihydroflavonol 4-reductase (EC 1.1.1.219), *ANS* anthocyanidin synthase (EC 1.14.20.4), *FLS* flavone synthase (EC 1.14.11.22), *3GT* flavonol 3-O-glucosyltransferase (EC 2.4.1.91), *RhaT* flavonol-3-O-glucoside L-rhamnosyltransferase (EC 2.4.1.159)

is dependent on the formation of an active WBM complex, the MYB TFs are sufficient to regulate the structural genes of the flavonoid branches, whereas the cofactors encoding the MYC and WDR have been exclusively associated with the anthocyanin accumulation (Czemmel et al. 2017). The activity of the TFs is regulated by a variety of abiotic/biotic elicitors or light exposure, which means that they are regulatory hubs between the elicitors or light-induced pathways and the SMs regulation (Huang et al. 2016; Li et al. 2020a, b).

The most exploited plant in vitro system to study the flavonoid biosynthesis are the HRs and in particular from the plant *Fagopyrum tataricum* Gaerth. and *F. esculentum* Moenh., since these plants have been recognized as a rich source of quercetin, kaempferol, myricetin, and especially rutin (Huang et al. 2016; Park et al. 2012). The UV-B stress treatment (used as abiotic elicitor) of *F. tataricum* HRs resulted in 5.18-fold increased content of rutin (from 0.93 to 4.82 mg/g DW). The observed effect was due to the enhanced expression of the structural genes FtPAL, FtCHI, FtCHS, FtF3H, and especially FtFLS-1, which was upregulated approximately 40-fold (Huang et al. 2016). The contents of rutin and quercetin in HR line Hokkai T8 were 0.25 and 0.03 mg/g DW, while their respective amounts in HR line T10 were 59.02 and 0.43 mg/g DW. This difference was due to the particular higher expression of *FtPAL, FtCHI, FtCHS, FtF3H, FtF3'H,* and *FtFLS* observed in HR line T10 (Thwe et al. 2013). Further, it was established that the *A. rhizogenes* strain is influencing the gene expression of the structural genes. The strain R100 showed higher expression levels of *FtFLS-1, FtFLS-2,* and *F3'H2,* while A4 strain showed significant expression of the key enzymes related to the anthocyanin biosynthetic pathway, such as ANS. This fact resulted in rutin accumulation of 22.31 μg/mg DW in the HR lines obtained with strain R100, while its content in the A4 HR lines was 13.04 μg/mg DW (Thwe et al. 2016). The potential of several flavonoid-specific TFs, such as *FtMYB6* and *AtMYB12,* has been investigated in the management of the flavonoid pathway (Park et al. 2012; Yao et al. 2020). The overexpression of *FtMYB6* in the HRs of tartary buckwheat promoted the activity of *Ft4CL, F4C4H, FtCHI, FtF3H,* and *FtFLS-1* and doubled content of rutin and quercetin, while the amounts of kaempferol and myricetin increased six- and threefold, respectively. At the same time, the expression of *FtDFR* leading to anthocyanin synthesis remained unchanged (Yao et al. 2020). In a similar way, the overexpression of *AtMYB12* elevated the expression levels of the same structural genes in *F. esculentum* leading to enhanced flavonoid production, without affecting the *AtDFR* and *AtANS* (Park et al. 2012). The exposure of *F. tataricum* HRs with blue light overexpressed the *FtMYB116* TF and its targeted genes from the flavonoid pathway, especially *FtCHI, FtF3H, FtFLS, FtF3'H,* and *FtRhaT,* leading to five- and fourfold increased production of quercetin and rutin. The same TF was able to suppress the *FtDFR* and *FtLAR* (leucoanthocyanidin reductase; EC 1.17.1.3) and the production of anthocyanins, respectively (Zhang et al. 2019). The ectopic expression of *VviMYBF1* promoted the flavonols and glycosylated flavonol biosynthesis, such as quercetin and quercetin derivatives. A positive correlation between *VviMYBF1* and the expression of the structural genes *VviFLS, Vvi3GT,* and *VviRhaT* in "Chardonnay" HRs has been observed. Additionally, the expression of *VviMYBF1* also altered the expression of the UV-B light-responsive HY5 homolog (*VviHY5*). Collectively, these results support the theory that the UV-B response machinery exists in grapes to propel flavonol accumulation through the activation of the regulatory network consisting of both MYB and bZIP TFs, such as *VviMYBF1* and *VviHY5* (Czemmel et al. 2017). Accordingly, the catechin and epicatechin biosynthesis has been boosted in cacao cell suspension after the formation of active complex between *MYB12* and *HY5*, which overexpression was induced by blue light (Gallego et al. 2018). The effect of different abiotic and biotic elicitors over the flavonoid biosynthesis has been investigated as well. The total flavonoid content has been increased 1.57 and 1.80-fold when *GlMYB4* and *GlMYB88* have been overexpressed in *Glycyrrhiza glabra* Fisch cell suspension 9 h after the addition of 100 μM MeJa (Li et al. 2020a, b). Elicitation with 100 mg/L Ag nanoparticles in *Brassica rapa* ssp. *rapa* HRs resulted in the overexpression of *BrMYB34* TF, which enhanced the expression levels of *PAL, CHI,* and *FLS* and increased the content of myricetin, quercetin, kaempferol, and rutin with 1.1 to 1.4-fold (Chung et al. 2018). As a low-cost alternative, chitosan has been used as an elicitor to enhance the biosynthesis of several flavonoids in *Isatis tinctoria* L. HRs. After treatment with chitosan at 150 mg/L for 36 h, the total flavonoid content has been increased 7.08-fold. In particular, the content of multiple hydroxyl substitute flavonoids rutin, quercetin, and kaempferol has been increased 8.27, 13.05, and 2.12-fold compared to the control. The most sensitive to chitosan were *CHS* and *F3'H*, the transcriptional abundance of which increased 45.13 and 41.04-fold after 12 h of treatment. The increased expression of CHS is important since this a gatekeeper enzyme in flavonoid biosynthesis, leading to the accumulation of naringenin chalcone as a basic C6-C3-C6 skeleton used in the flavonoid biosynthesis (Jiao et al. 2018). The cell wall elicitor (1% final concentration) from *Piriformospora indica* increased the content of catechin, kaempferol, luteolin, and myricetin up to twofold after 72 or 120 h treatment of *Linum albul* Kotschy ex Boiss. HRs. From the genes participating in the phenylpropanoid pathway, the highest expression level was observed for *PAL* (Tashackori et al. 2018). The same fungal elicitor was able to enhance the accumulation of rutin and apigenin and quercetin 5.7, 27.6, and 8.8-fold when used to treat the HRs of *Ficus carica* L. The highest expression level was observed for the *PAL* and *CHS* and *3GT* genes, which activity was induced by *MYB3* TF (Amani et al. 2021).

Along with the overexpression of single or multiple genes from the flavonoid biosynthetic pathway or the selective stimulation of flavonoid-specific TFs, another strategy could be the inhibition of the competitive pathways leading to the biosynthesis of other SMs. The suppression of *NtDFR1* and *2* by RNA interference (Ri)-mediated posttranscriptional gene silencing in pink-flowered tobacco resulted in 72–97% reduction of the anthocyanin content. On the other hand, suppression of *DFR* increased the total levels of flavonoids due to (dihydro) flavonol biosynthesis, revealing the possibility to fine-tune the precursors flux to flavonoid biosynthesis (Lim et al. 2016).

16.5 Scale-up Strategies for Nutraceuticals Production

The cell transfer from flasks to bioreactor is an important step and main concern toward the large-scale production and hence commercialization of valuable SMs (Ahmadi-Sakha et al. 2018; Cui et al. 2011). Several biological (e.g., cell size, cell heterogeneity, and genetic instability) and technological barriers (sensitivity to shear stress, intensive mixing or aeration, foaming, cell adhesion, and aggregation) might result in relatively low amounts of the target metabolites, which is a bottleneck for their large-scale production (Marchev et al. 2020). Therefore, a suitable bioreactor design must solve these challenges and assure homogenous cell suspending and mixing that guarantee optimal mass and energy transfer, as well as, lower shear stress environment, despite the cell tendency to grow as aggregates and sediment during cultivation (Marchev et al. 2020).

Different approaches have been used to increase the biosynthetic capacity of flavonoids during bioreactor cultivation, such as selection of an appropriate bioreactor design (Cui et al. 2020; Kubica et al. 2020), type of impellers (Vazquez-Marquez et al. 2016), nutrient medium optimization (Jiang et al. 2020; Ho et al. 2021), agitation and aeration rate (Vazquez-Marquez et al. 2016), inoculum size (Cui et al. 2011), and cell culture elicitation (Wen et al. 2019). Examples for bioreactor production of flavonoids are presented in Table 16.1. Most of the bioreactor cultivations aiming the biosynthesis of flavonoids are based on ARs (Ho et al. 2021) or shoot type cultures in pneumatically driven bioreactors (Weremczuk-Jeżyna et al. 2020). The effect of nutritional factors (inorganic, organic compounds, and plant growth regulators) and initial inoculum concentration over the SMs production in ARs of *Eurycoma longifolia* Jack during cultivation in a 3-L balloon-type bubble bioreactor (BTBB) has been investigated. Nutrient medium constituents' optimization is essential for maximizing the biomass growth and SMs biosynthesis. Among all the used media, only Murashige and Skoog (MS) and Gamborg's B5 medium (B5) lead to increased lateral root branching and increased biomass accumulation. After optimization of all the nutrient components (3/4 MS medium, 30 g/L sucrose, and 3.0 mg/L indole-3-butyric acid) and inoculum amount (6 g/L), after 7 weeks of cultivation the final biomass increased double (3.22 g/L) and the total phenolics and total flavonoid amounts reached up to 8.04 and 2.95 mg/g DW, respectively. The BTBB are one of the most suitable constructions for cultivation of tissue and organ cultures, since the lower shear stress due to the lack of impellers (Cui et al. 2020). Optimum auxin concentration is critical for ARs growth and flavonoid accumulation. Indole-3-butyric acid (IBA) was more effective for lateral root branching and growth than 1-naphthaleneacetic acid (NAA) during the cultivation of *Polygonum multiflorum* Thunb. ARs in a 3-L BTBB. After selectin, the optimum nutrient medium components (2 mg/L IBA and 5% sucrose and 1X MS) achieved the highest root biomass (13.46 g/L DW), total phenolic (53.08 mg/g DW), and flavonoid compounds (25.10 mg/g DW). Salt strength higher than 1X may decrease the water potential and inhibit the absorption of water and mineral nutrients leading to inhibited root growth. Higher sucrose concentrations may cause loss of cell viability, reduced growth, and modified root morphology, due

Table 16.1 Bioreactor production of flavonoids, used as nutraceuticals

Plant species	Bioreactor type and volume	Bioreactor operating conditions	Secondary metabolite	Metabolite production (mg/L)	Reference
Adventitious roots					
Oplopanax elatus (Nakai) Nakai	5-L balloon-type airlift	25 °C; 400 mL/min flow rate; dark	Quercetin Kaempferidine	3.77 1.67	Jiang et al. 2020
Hypericum perforatum L.	3-L balloon-type airlift	22 °C; 0.1 vvm flow rate; dark	Quercetin Hyperoside	0.02 0.22	Cui et al. 2011
Hypericum perforatum L.	500-L balloon-type airlift	22 °C; 0.1 vvm flow rate; dark	Hyperin Hypericin Quercetin	0.15 0.60 0.15	Cui et al. 2014
Cell suspension					
Schisandra chinensis (Turcz.)	3-L nutrient sprinkle	26 °C; 16 h/8 h light/dark regime; 30 s pump operating time/10 min breaks	Kaempferol Quercetin Rutoside	4.5 144.5 1.30	Szopa et al. 2019
Scrophularia striata Boiss	10-L stirred tank	26 °C in dark; 110–170 rpm agitation; 0.5–1.5 L/min flow rate	Rutin Kaempferol Luteolin Myricetin Quercetin	330 3.50 10.14 19.38 196.92	Ahmadi-Sakha et al. 2018
Shoot culture					
Dracocephalum forrestii W.W.	10-L nutrient sprinkle	26 °C; 16 h/8 h light/dark regime; 25 s pump operating time/2.5 min breaks	Apigenin p-Coumaroylrhamnoside	11.89	Weremczuk-Jeżyna et al. 2020

to the increased osmolarity of the culture medium, which is promoting the diffusion of metabolites from tissue to media (Ho et al. 2021). For the fed-batch cultivation of *Oplopanax elatus* (Nakai) Nakai ARs in a 5-L balloon-type airlift bioreactor, the optimum medium feeding factors were 60 g/L sucrose, 1X MS medium, 0.94 mM phosphorus, 30 mM nitrogen, and 3.0 mg/L IBA. Although phosphorus and nitrogen did not affect the root growth, they greatly influenced the flavonoid accumulation.

This medium was used as a feeding medium after 20 days of cultivation, after which the process continued for 20 days more. The fed-batch process has some advantages compared to the standard batch cultivation. Several restrictions, such as prolonged lag phase in the early culture stage, nutrient exhaustion, and product inhibition in the late stage, could be overcome. Thus, at the end of the cultivation, the total flavonoid content was 265.71 mg/g DW, of which 0.27 and 0.12 mg/g DW were quercetin and kaempferol (Jiang et al. 2020). After optimizing the inoculum density and aeration volume (3 g/L and 0.1 vvm), the biomass of *Hypericum perforatum* L. ARs contained 0.80 mg/g DW total flavonois, of which 1.33 and 14.01 μg/g DW were quercetin and hyperoside. It was established that higher inoculum densities (6.0–12.0 g/L) were not beneficial and resulted in decreased biomass content at the end of the cultivation. Higher aeration rates created turbulent flow leading to lower biomass accumulation, increased shear stress, and key volatiles, such as ethylene (Cui et al. 2011). The most suitable cultivation system among the temporary immersion systems (TIS), such as RITA® and Plantfrom™, as well as, nutrient sprinkle bioreactor (NSB) has been chosen for the shoots cultivation of *Dracocephalum forrestii* W.W. The highest biomass growth index (55.67 g/DW) was observed for the shoots in the RITA® bioreactor, while the NSB achieved the highest phenol content (24.15 mg/g DW). Both bioreactor systems are beneficial for cultivation of plant in vitro systems because they are not exposed to continuous immersion with medium or share forces, as frequently observed in mechanically agitated bioreactors. In both cases, the cultures are supported on a porous base, and in TIS, the plant cultures are immersed with medium at specific period of time at specific intervals, while in NSB the cultures are periodically sprayed with medium from the top of the vessel. The main advantage of these bioreactors are the increased culture multiplication and high biomass accumulation due to high nutrient availability, as well as, reduced hyperhydricity and physiological stress (Weremczuk-Jeżyna et al. 2020). The highest flavonoid content of 29.02 mg/100 g DW, 21.00 mg/100 g DW of which was rutin, was achieved during the cultivation of *Schisandra chinensis* Turcz shoot cultures in NSB. These amounts were 1.40, 1.32, 1.36, and 1.42-fold higher when compared with the flavonoid content accumulated in two continuous immersion systems (a cone type bioreactor and cylindrical type bioreactor) and two TIS (RITA® and Plantform™; Szopa et al. 2019). The Plantfrom™ bioreactor system working at 6 min immersion/24 h ensured 1.4-fold higher relative growth and enhanced biosynthesis of total phenolic (23.6 mg GAE/g DW) and flavonoid content (10.9 mg RE/g DW) compared to the semisolid cultivation of *Lycium barbarum* L. (Goji) shoot cultures (Ruta et al. 2020).

A frequently applied strategy to boost the flavonoid production during bioreactor cultivation is the elicitation of the plant in vitro systems. The elicitation of *Panax ginseng* C. A. Mey ARs with 200 μM SA had a superior effect in total flavonoid accumulation (88% compared to the non-treated control) than MeJa, during their cultivation in a 5-L STB. The effect of the elicitor was due to the caused abiotic stress through the generation of H_2O_2. The observed changes in the secondary metabolism were as consequences of the changes observed in the primary metabolism. The activities of the enzymes, glucose 6-phosphate dehydrogenase, PAL, substrate specific peroxidases, shikimate dehydrogenase, and β-glucosidase that generate precursors in the phenylpropanoid pathway, were highly expressed (Ali et al. 2007). The highest flavonoid content (up to 22.4-fold increase) was observed when shoot cultures of *Centella asiatica* (L.) Urban were treated with 50 μM MeJa during their cultivation in RITA®. The elicitor increased the content of the main flavonoid rutin 4.3-fold, while the highest kaempferol levels (22.4-fold increase) were observed by combined elicitor treatment of 50 μM MeJa and 50 μM ethephon. The amino acid L-phenyl alanine had a superior effect on the apigenin biosynthesis, the concentration of which was increased ninefold compared to the control (Skrzypczak-Pietraszek et al. 2019).

Several investigations report the biosynthesis of flavonoids from plant cell suspensions as well. After optimizing the inoculum size (0.9 g/L), the growth characteristics and biosynthetic potential of *Verbena officinalis* L. cell suspension in STB and BTBB were compared. At the end of the cultivation, the biomass in the STS and BTBB was 5.09 and 4.55-fold higher than the initial value. Along with that, the total phenolics in STB were 36.78 mg/100 g DW, while in the BTBB were 19.99 mg/100 g DW. This difference is due to the fact that STB assures more homogenous stirring and agitation of the cell suspension, which reflect in facilitated mass transfer, oxygen, and nutrient to the cells. The shear stress sensitivity was decreased by choosing low values of agitation and aeration, such as 33 rpm and 0.5 vvm (Kubica et al. 2020). The flavonoid production from *Buddleja cordata* Kunth was influenced from the agitation speed. The stirrer speed of 400 rpm resulted in production of 5.02 mg QE/ g DW total phenolics, while at 120 rpm the total phenolics were 4.43 mg QE/ g DW. The authors conclude that even at this stirring speed, the cells do not experience shear stress and there is sufficient medium and oxygen transfer that favors the SMs production (Vazquez-Marquez et al. 2016). To overcome problems with cell sedimentation with increasing cell density, the flow rate and agitation speed were increased from 0.5 to 1.5 L/min and from 110 to 170 rpm during the cultivation of *Scrophularia striata* Boiss cell suspension in a 10-L STB. The total flavonoid content was tenfold higher than that in the shake flask experiment, and in addition, rutin was biosynthesized only in the bioreactor (Ahmadi-Sakha et al. 2018). The elicitation with 100 μM SA doubled the content of total flavonoids compared to the control sample during the cultivation of *Orostachys cartilaginous* A. Bor cell suspension in 5-L balloon-type airlift bioreactor (Wen et al. 2019).

At a pilot-scale, flavonoid biosynthesis has been performed by cultivation of *H. perforatum* ARs in a 500-L horizontal drum type and 500-L airlift BTBB. At optimal cultivation parameters (1/2 MS medium with ammonium and nitrate ratio 5: 25, 0.1 mg/L IBA, 0.1 mg/L kinetin, 30% sucrose, and 3 g/L inoculum), the airlift construction offered better growth and biosynthetic characteristics than the drum type bioreactor. The biomass yield was 1.4-fold higher, while the flavonoid content was 1.5-fold higher in the airlift BTBB (Cui et al. 2014).

16.6 Conclusions

Plant in vitro systems represent a superior source of bioactive SMs, especially flavonoids, which have been demonstrated to possess many beneficial effects, such as antineoplastic, anti-inflammatory, antiallergic, antidiabetic, cardioprotective activity, and many others. Although the examples given for the biological activities are based on cell lines or animals, it is of a critical importance to perform clinical investigations to elucidate the daily doses and toxicity effect for humans. Flavonoids are ubiquitous in plant foods, including vegetables, fruits, nuts, and beverages, such as wine, tea, or cocoa, and comprise main part of the human diet. Harvesting wild- or field-grown plant is not able to meet the growing demand for flavonoids for nutraceuticals and commercial use, which raises the awareness for serious ecological concerns. In this regard, the cultivation of plant in vitro systems could be a viable alternative for reliable and continuous production of flavonoids, which allows the application of different strategies for manipulation of the growth variables, precursors, and elicitors that might modulate the biosynthetic pathway of the compounds, optimizing their production. However, supreme results are achieved through selective overexpression or suppression of structural genes or TFs that can trigger to substrate pool to the desired metabolites or suppress competitive biosynthetic branches.

Essential for obtaining high yields of the desired SMs in bioreactors is the application of different optimization on process and cultivation parameters. The selection of an appropriate bioreactor configuration is a complex task in which the cultivation parameters (especially mixing and aeration) must be precisely equilibrated to achieve optimal growth and product synthesis.

Acknowledgments This research received funding from the European Union's Horizon 2020 research and innovation programme project PlantaSYST (SGA no. 739582 FPA no. 664620), and the BG05M2OP001-1.003-001-C01 project, financed by the European Regional Development Fund through the "Science and Education for Smart Growth" Operational Programme.

References

Abdel-Salam E, Faisal M, Alatar A et al (2021) Genome-wide transcriptome variation landscape in *Ruta chalepensis* organs revealed potential genes responsible for rutin biosynthesis. J Biotechnol 325:43–56

Açıkgöz M (2020) Establishment of cell suspension cultures of *Ocimum basilicum* L. and enhanced production of pharmaceutical active ingredients. Ind Crop Prod 148:e112278

Ahmadi-Sakha S, Sharifi M, Niknam V et al (2018) Phenolic compounds profiling in shake flask and bioreactor system cell cultures of *Scrophularia striata* Boiss. In Vitro Cell Dev Biol Plant 54:444–453

Ajji PK, Walder K, Puri M (2020) Combination of balsamin and flavonoids induce apoptotic effects in liver and breast cancer cells. Front Pharmacol 11:574496

Ali M, Abbasi B (2014) Light-induced fluctuations in biomass accumulation, secondary metabolites production and antioxidant activity in cell suspension cultures of *Artemisia absinthium* L. J Photochem Photobiol B Biol 140:223–227

Ali M, Hahn E-J, Paek K-Y (2007) Methyl jasmonate and salicylic scid induced oxidative stress and accumulation of phenolics in *Panax ginseng* bioreactor root suspension cultures. Molecules 12: 607–621

Amani S, Mohebodini M, Khademvatan S, Jafari M, Kumar V (2021) *Piriformospora indica* based elicitation for overproduction of phenolic compounds by hairy root cultures of *Ficus carica*. J Biotechnol 327:43–53

Amararathna M, Hoskin D, Rupasinge V (2020) Cyanidin-3-O-Glucoside-Rich haskap berry administration suppresses carcinogen-induced lung tumorigenesis in A/JCr mice. Molecules 25:3823

Bai S, Tao R, Tang Y et al (2019) BBX16, a B-box protein, positively regulates light-induced anthocyanin accumulation by activating MYB10 in red pear. Plant Biotechnol J 17:1985–1997

Bergamin A, Mantzioris E, Cross G et al (2019) Nutraceuticals: reviewing their role in chronic disease prevention and management. Pharmaceut Med 33:291–309

Budryn G, Grzelczyk J, Pérez-Sánchez H (2018) Binding of red clover isoflavones to actin as a potential mechanism of anti-metastatic activity restricting the migration of cancer cells. Molecules 23:2471

Carmi Y, Mahmoud H, Khamaisi H et al (2020) Flavonoids restore platinum drug sensitivity to ovarian carcinoma cells in a phospho-ERK1/2-dependent fashion. Int J Mol Sci 21:6533

Carrizo A, Izzo C, Forte M et al (2020) A novel promising frontier for human health: the beneficial effects of nutraceuticals in cardiovascular diseases. Int J Mol Sci 21:8706

Cas M, Ghidoni R (2018) Cancer prevention and therapy with polyphenols: sphingolipid-mediated mechanisms. Nutrients 10:940

Chaves-Silva S, dos Santos A, Júnior C et al (2018) Understanding the genetic regulation of anthocyanin biosynthesis in plants-tools for breeding purple varieties of fruits and vegetables. Phytochemistry 153:11–27

Choi K-S, Kundu J, Chun K-S et al (2019) Rutin inhibits UVB radiation-induced expression of COX-2 and iNOS in hairless mouse skin: p38 MAP kinase and JNK as potential targets. Arch Biochem Biophys 559:38–45

Chung IM, Rekha K, Rajakumar G, Thiruvengadam M (2018) Influence of silver nanoparticles on the enhancement and transcriptional changes of glucosinolates and phenolic compounds in genetically transformed root cultures of *Brassica rapa* ssp. *rapa*. Bioprocess Biosyst Eng 41: 1665–1677

Coppens P, Silva M, Pettman S (2006) European regulations on nutraceuticals, dietary supplements and functional foods: a framework based on safety. Toxicology 221:5974

Cui X-H, Murthy H, Jin Y-X et al (2011) Production of adventitious root biomass and secondary metabolites of *Hypericum perforatum* L. in a balloon type airlift reactor. Bioresour Technol 102: 10072–10079

Cui X-H, Murthy H, Paek K-Y (2014) Pilot-scale culture of *Hypericum Perforatum* L. adventitious roots in airlift bioreactors for the production of bioactive compounds. Appl Biochem Biotechnol 174:784–792

Cui XH, Murthy HN, Zhang JD et al (2020) Effect of nutritional factors on the accretion of secondary metabolites in Malaysian ginseng adventitious root cultures. Plant Biotechnol Rep 14:381–386

Czemmel S, Höll J, Loyola R et al (2017) Transcriptome-wide identification of novel UV-B- and light modulated flavonol pathway genes controlled by VviMYBF1. Front Plant Sci 8:1084

Deng X, Song H, Zhou Y et al (2013) Effects of quercetin on the proliferation of breast cancer cells and expression of survivin in vitro. Exp Ther Med 6(5):1155–1158

Dong B, Yang Z, Ju Q et al (2020) Anticancer effects of fufang yiliu yin formula on colorectal cancer through modulation of the PI3K/Akt pathway and BCL-2 family proteins. Front Cell Dev Biol 8:704

Gallego A, Tojas L, Parra O et al (2018) Transcriptomic analyses of cacao cell suspensions in light and dark provide target genes for controlled flavonoid production. Sci Rep 8:13575

Ghosh A, Ghosh D, Sarkar S et al (2012) Anticarcinogenic activity of nanoencapsulated quercetin in combating diethylnitrosamine-induced hepatocarcinoma in rats. Eur J Cancer Prev 21:32–41

Her Y, Lee T-K, Kim J et al (2020) Topical application of *Aronia melanocarpa* extract rich in chlorogenic acid and rutin reduces UVB-induced skin damage via attenuating collagen disruption in mice. Molecules 25:4577

Hidalgo D, Georgiev M, Marchev A et al (2017) Tailoring tobacco hairy root metabolism for the production of stilbenes. Sci Rep 7:17976

Ho TT, Le KC, Kim SW et al (2021) Culture condition optimization and FT-IR analysis of *Polygonum multiflorum* Thunb. Adventitious root cultures grown in an air-lift bioreactor system. Plant Cell Tiss Org Cult 144:371–381

Huang X, Yao J, Zhao Y et al (2016) Efficient rutin and quercetin biosynthesis through flavonoids-related gene expression in *Fagopyrum tataricum* Gaertn. Hairy root cultures with UV-B irradiation. Front Plant Sci 7:63

Irmish S, Ruebsam H, Jancsik A et al (2019) Flavonol biosynthesis genes and their use in engineering the plant antidiabetic metabolite montbretin A. Plant Physiol 180:1277–1290

Jain P, Pundir R (2013) Nutraceuticals: recent developments and future prospectives. In: Shahid M et al (eds) Recent trends in biotechnology and therapeutic applications of medicinal plants. Springer, Dordrecht, pp 213–223

Jiang X, Jin M, Piao X et al (2020) Fed-batch culture of *Oplopanax elatus* adventitious roots: feeding medium selection through comprehensive evaluation using an analytic hierarchy process. Biochem Eng J 167:107927

Jiao J, Gai Q-Y, Wang X et al (2018) Chitosan elicitation of Isatis tinctoria L. hairy root cultures for enhancing flavonoid productivity and gene expression and related antioxidant activity. Ind Crop Prod 124:28–35

Kalra E (2003) Nutraceutical—definition and introduction. AAPS PharmSci 5(3):25

Kiekow C, Figueiró F, Dietrich F et al (2016) Quercetin derivative induces cell death in glioma cells by modulating NF-κB nuclear translocation and caspase-3 activation. Eur J Pharm Sci 84:116–122

Kim W, Bach D-H, Ryu H et al (2017) Cytotoxic activities of *Telectadium dongnaiense* and its constituents by inhibition of the Wnt/β-catenin signaling pathway. Phytomedicine 34:136–142

Kubica P, Szopa A, Kokotkiewicz A et al (2020) Production of verbascoside, isoverbascoside and phenolic acids in callus, suspension, and bioreactor cultures of *Verbena officinalis* and biological properties of biomass extracts. Molecules 25(23):5609

Li X, Kim Y-B, Kim Y et al (2013) Differential stress-response expression of two flavonol synthase genes and accumulation of flavonols in tartary buckwheat. J Plant Physiol 170:1630–1636

Li Y, Chen X, Wang J et al (2020a) Two responses to MeJA induction of R2R3-MYB transcription actors regulate flavonoid accumulation in *Glycyrrhiza uralensis* Fisch. PLoS One 15(7): e0236565

Li Y, Wang Z, Jin J et al (2020b) Quercetin pretreatment enhances the radiosensitivity of colon cancer cells by targeting Notch-1 pathway. Biochem Biophys Res Commun 523:947–953

Lim S-H, You M-K, Kim D-H et al (2016) RNAi-mediated suppression of dihydroflavonol 4-reductase in tobacco allows fine-tuning of flower color and flux through the flavonoid biosynthetic pathway. Plant Physiol Biochem 109:482–490

Liu Z, Shi M, Xie D-Y (2014) Regulation of anthocyanin biosynthesis in *Arabidopsis thaliana* red pap1-D cells metabolically programmed by auxins. Planta 239(4):765–781

Liu M, Fu M, Yang X et al (2020a) Paclitaxel and quercetin co-loaded functional mesoporous silica nanoparticles overcoming multidrug resistance in breast cancer. Colloids Surf B Biointerfaces 196:11284

Liu Z-J, Xu W, Han J et al (2020b) Quercetin induces apoptosis and enhances gemcitabine therapeutic efficacy against gemcitabine-resistant cancer cells. Anticancer Drugs 31(7):684–692

Marchev A, Georgiev M (2020) Plant *in vitro* systems as a sustainable source of active ingredients for cosmeceutical application. Molecules 25(9):2006

Marchev A, Yordanova Z, Georgiev M (2020) Green (cell) factories for advanced production of plant secondary metabolites. Crit Rev Biotechnol 40(4):443–458

Marín L, Gutiérrez-del-Río I, Entrialgo-Cadierno R et al (2018) De novo biosynthesis of myricetin, kaempferol and quercetin in *Streptomyces albus* and *Streptomyces coelicolor*. PLoS One 13(11):e0207278

Mendoza D, Cuaspud O, Arias J et al (2018) Effect of salicylic acid and methyl jasmonate in the production of phenolic compounds in plant cell suspension cultures of *Thevetia peruviana*. Biotechnol Rep 19:e00273

Nalini D, Selvaraj J, Kumar G (2019) Herbal nutraceuticals: safe and potent therapeutics to battle tumor hypoxia. J Cancer Res Clin Oncol 146(1):1–18

Nielsen E, Temporiti M, Cella R (2019) Improvement of phytochemical production by plant cells and organ culture and by genetic engineering. Plant Cell Rep 38(10):1199–1215

Park N, Li X, Thwe A et al (2012) Enhancement of rutin in *Fagopyrum esculentum* hairy root cultures by the *Arabidopsis* transcription factor AtMYB12. Biotechnol Lett 34:577–583

Pervaiz T, Songtao J, Faghihi F et al (2017) Naturally occurring anthocyanin, structure, functions and biosynthetic pathway in fruit plants. J Plant Biochem Physiol 5:2

Ren K-W, Li Y-H, Wu G et al (2017) Quercetin nanoparticles display antitumor activity via proliferation inhibition and apoptosis induction in liver cancer cells. Int J Oncol 50:1299–1111

Ross S (2000) Functional foods: the food and drug administration perspective. Am J Clin Nutr 71: 1735–1738

Ruta C, De Mastro G, Ancona S et al (2020) Large-scale plant production of *Lycium barbarum* L. by liquid culture in temporary immersion system and possible application to the synthesis of bioactive substance. Plan Theory 9(7):844

Shende P, Mallick C (2020) Nanonutraceuticals: a way towards modern therapeutics in healthcare. J Drug Delivery Sci Technol 58:101838

Singh C, Chhabra G, Ndiaye M et al (2020) Quercetin-resveratrol combination for prostate cancer management in TRAMP mice. Cancer 12:2141

Skrzypczak-Pietraszek E, Urbanska A, Zmudzki P (2019) Elicitation with methyl jasmonate combined with cultivation in the Plantform™ temporary immersion bioreactor highly increases the accumulation of selected centellosides and phenolics in *Centella asiatica* (L.) Urban shoot culture. Eng Life Sci 19:931–943

Srivastava S, Srivastava R (2019) Curcumin and quercetin synergistically inhibit cancer cell proliferation in multiple cancer cells and modulate Wnt/β-catenin signaling and apoptotic pathways in A375 cells. Phytomedicine 52:117–128

Srivastava S, Somasagara R, Hegde M et al (2016) Quercetin, a natural flavonoid interacts with DNA, arrests cell cycle and causes tumor regression by activating mitochondrial pathway of apoptosis. Sci Rep 6:240–249

Szopa A, Kokotkiewicz A, Bednarz M et al (2019) Bioreactor type affects the accumulation of phenolic acids and flavonoids in microshoot cultures of *Schisandra chinensis* (Turcz.) Baill. Plant Cell Tiss Org Cult 139:199–206

Taghizadeh M, Nasibi F, Kalantari K et al (2019) Evaluation of secondary metabolites and antioxidant activity in *Dracocephalum polychaetum* Bornm. Cell suspension culture under magnetite nanoparticles and static magnetic field elicitation. Plant Cell Tiss Org Cult 136: 489–498

Tashackori H, Sharifi M, Chashmi N et al (2018) Piriformospora indica cell wall modulates gene expression and metabolite profile in *Linum album* hairy roots. Planta 248:1289–1306

Thiruvengadam M, Pekha K, Rajakumar J et al (2016) Enhanced production of anthraquinones and phenolic compounds and biological activities in the cell suspension cultures of *Polygonum multiflorum*. Int J Mol Sci 17:1912

Thwe AA, Kim JK, Li X et al (2013) Metabolomic analysis and phenylpropanoid biosynthesis in hairy root culture of tartary buckwheat cultivars. PLoS One 8(6):e65349

Thwe A, Valan Arasu M, Li X et al (2016) Effect of different *agrobacterium rhizogenes* strains on hairy root induction and phenylpropanoid biosynthesis in tartary buckwheat (*Fagopyrum tataricum* Gaertn). Front Microbiol 7:318

Vazquez-Marquez AM, Zepeda-Gómez C, Burrola-Aguilar C et al (2016) Effect of stirring speed on the production of phenolic secondary metabolites and growth of *Buddleja cordata* cells cultured in mechanically agitated bioreactor. Plant Cell Tiss Org Cult 139:155–166

Wang J, Quian J, Yao L et al (2015) Enhanced production of flavonoids by methyl jasmonate elicitation in cell suspension culture of *Hypericum perforatum*. Bioresour Bioprocess 2:5

Wen T, Hao Y-J, An X-L et al (2019) Improvement of bioactive compound accumulation in cell cultures of *Orostachys cartilaginous* A. Bor. through elicitation with salicylic acid and effect of cell extract on bioactive activity. Ind Crop Prod 139:111570

Weremczuk-Jeżyna I, Lisiecki P, Gonciarz W et al (2020) Transformed shoots of *Dracocephalum forrestii* W.W. Smith from different bioreactor systems as a rich source of natural phenolic compounds. Molecules 25(19):4533

Yao P, Huang Y, Dong Q et al (2020) FtMYB6, a light-induced SG7 R2R3-MYB transcription factor, promotes flavonol biosynthesis in tartary buckwheat (*Fagopyrum tataricum*). J Agric Food Chem 68(47):13685–13696

Zhang D, Jiang C, Huang C (2019) The light-induced transcription factor FtMYB116 promotes accumulation of rutin in Fagopyrum tataricum. Plant Cell Environ 42(4):1340–1351

Zhou Y, Chen D, Xue G et al (2020) Improved therapeutic efficacy of quercetin-loaded polymeric nanoparticles on triple-negative breast cancer by inhibiting uPA. RSC Adv 10:34517

Transfer of Plant Biosynthetic Pathways to Microbes for the Production of Nutraceuticals

17

Fatima M. Alessa

Abstract

Secondary metabolites of plants are rich source of highly diverse chemical compounds with biological characteristics for food and pharmaceutical use. Among polyphenolic compounds, flavonoids are potential naturally occurring group present in fruit, vegetables, and beverages which are derived from plant sources. Structurally, flavonoids' skeleton consists of C-15 carbon with framework of C_6-C_3-C_6. Favones, favonols, favanones, favanonols, favan-3-ols, chalcones, anthocyanidins, and isofavones are important subgroups of flavonoids. The linkage position between the B ring and the C ring delineates the basic flavonoids. The biosynthesis of such bioactive compounds to meet the increasing market demands is difficult due to the high chemical structural complexity, lack of direct extraction techniques from plants, use of harsh chemicals, and slow plants' growth. Therefore, microbial biosynthesis is superior alternative to sustainable and economic production of flavonoids due to rapid and easy growth of microbes, genetically tradable, friendly approach to the environment, and less use of harsh chemical. *Escherichia coli and Saccharomyces cerevisiae* are well-known and studied organisms have been used in food, beverages, and pharmaceutical industries. This chapter explains the importance of flavonoids compounds and their contribution to health, the advantages of utilizing microbes in biosynthesizing flavonoids over plants, and explaining the microbial biosynthesis pathways of several abundant flavonoids' compounds. Finally, the important advances of microbial biosynthesis of flavonoids in food and beverages industries are discussed.

F. M. Alessa (✉)
Department of Food and Nutrition Science, College of Agriculture and Food Sciences, King Faisal University, Al-Ahsa, Saudi Arabia
e-mail: falissa@kfu.edu.sa

T. Belwal et al. (eds.), *Nutraceuticals Production from Plant Cell Factory*,
https://doi.org/10.1007/978-981-16-8858-4_17

Keywords

Flavonoids · Microbial biosynthesis · Bioactive compounds · Nutraceuticals

17.1 Introduction

Nutraceuticals are the significant class of molecules due to their potential effects in health-promoting, treating, and preventing human's diseases. Nutraceuticals involve bioactive compounds that have distinct functional and structural features and can trigger health benefits and physiological effect in long term (Wang et al. 2016). They can be produced from diverse sources including plants (phytochemicals and vitamins), animals (polysaccharides), and microorganisms (poly amino acids), and they are superior antioxidants, anti-inflammatory, and prevent gastrointestinal, arthritis, osteoporosis, cancer, cardiovascular, and chronic diseases. Nutraceuticals market has been greatly increased due to the growing demand on promoting human health through diet rather than drugs. Global Information Inc. estimated that nutraceuticals market in 2019 will reach $241.1 billion, and by 2022, $1121 (Jain and Ramawat 2013; Yuan and Alper 2019). Flavonoids are the secondary metabolic phenolic substances, which are widely produced in all vascular and medicinal plants (Karak 2019; Jucá et al. 2020). Flavonoids are the major diverse class of phenolic compounds in plants kingdom (Nabavi et al. 2020). They contribute several health benefits to human such as, anti-inflammatory, antioxidant, anticancer, antimicrobial functions, and add color and taste to food and beverages (Karak 2019; Jucá et al. 2020). However, various issues need to be addressed in order to fulfill the expanding demand of flavonoid nutraceuticals market. For example, limited direct extraction approaches, high raw materials cost, short-term sustainability and stability (Wang et al. 2016; Yuan and Alper 2019), and low amount of produced and purified (flavonoids) nutraceuticals. Therefore, microbial production of nutraceuticals (flavonoids) can be promising strategy as an attractive alternative approach of chemical synthesis and extractions (Yuan and Alper 2019). *E-coli* strains and food grade *S. cerevisiae* are great examples of factorial cells that have been widely utilized in developing nutraceuticals (Wang et al. 2016). Therefore, biosynthetic pathways of microbes to produce nutraceuticals (flavonoids) can provide nonfood lignocellulosic feedstocks, low cost of raw materials, and safe use in pharmaceutical and food industries (Yuan and Alper 2019). Taditional host of microorganisms such as *Escherichia coli*, some strains of *Saccharomyces cerevisiae, and Corynebacterium glutamicum* have been investigated. Furthermore, they are generally recognized as safe microbes other than traditional microbes such as, the oleaginous organisms *Yarrowia lipolytica*, which are commercially available for their ability to produce high quantities of omega-3 polyunsaturated fatty acid. Microbial cocultures or synthetic consortia were recognized as the best way to produce a highly complex molecule of nutraceuticals (flavonoids) such as, glycosides (Yuan and Alper 2019). Plants, fungi, and microorganisms are living organisms that can produce crucial compounds and are called primary metabolites. Primary metabolites are important for various vital processes involving photosynthesis, production/expenditure of energy, metabolism process of carbohydrate, fat, and protein. Beside fundamental primary metabolites, there are broad secondary metabolites such as, polyketides, terpenoids, phenylpropanoids, and alkaloids.

Administered enzymatic reactions facilitate secondary metabolites' biosynthesis in cellular systems. It is crucial to understand the metabolic process to control the potential of desired plant phytochemicals' biosynthesis in food and pharmaceuticals industries. Secondary metabolites production can be regulated by investigating the compromised genes in the metabolic pathways and environmental and physiological processes (Nabavi et al. 2020). Thus, this chapter will discuss the importance of flavonoids, the general biosynthetic pathway of flavonoids in plants, the microbial biosynthesis of several flavonoid substances including naringenin, apigenin and genkwanin, flavone, isoflavones, and anthocyanin. The importance of flavonoids microbial biosynthesis in food, beverages, and pharmaceutical industries is discussed, and the recent modifications in the field of flavonoids microbial biosynthesis are highlighted.

17.2 Overview of Flavonoids

Phenolic compounds are the highly diverse phytochemicals existed in plant kingdom. Benzoic and cinnamic acid, coumarins, tannins, lignins, lignans, and flavonoids are simple phenols found in plant food. They offered significant health benefits because they prevent oxidative stress, which eventually reduces the risk of inflammation, cancer, diabetics, heart diseases, and cells mutagenesis. The global market of flavonoids is estimated by 2022 to reach \$1121 due to their physiological effects and diverse structures (Yuan and Alper 2019). Ecological and physiological pressures can partly synthesize phenolic compounds in plants (Khoddami et al. 2013). Flavonoids are the secondary metabolic phenolic substances, which are widely produced in all vascular plants such as, fruit and vegetables, and medicinal plants (glycoside and methylated derivatives) (Karak 2019; Jucá et al. 2020). Flavonoids are the major diverse class of phenolic compounds in plants kingdom (Marranzano et al. 2018; Nabavi et al. 2020). They possess several biological activities due to the fact that they have diverse complex chemical structures. It has been reported that flavonoids offer strong antioxidant activities and protect plants and human cells against free radical activities (ROS). The functional group arrangement of flavan nucleus is responsible for strong antioxidant properties (Heim et al. 2002; Kukić et al. 2006; Zhang et al. 2014; Marranzano et al. 2018; Jucá et al. 2020).

Flavonoids protect cells and tissues from injury, pathogen infections, damage, chemical irradiation, and overall immune system, and they potentially prevent several diseases including leukemia, sepsis, asthma, sclerosis, atherosclerosis, psoriasis, allergic rhinitis, ileitis/colitis, and rheumatoid arthritis (Marranzano et al. 2018). In addition, flavonoids and anthocyanidins have demonstrated anticancer properties, which reduce or prevent the chance of cancer's initiation, promotion,

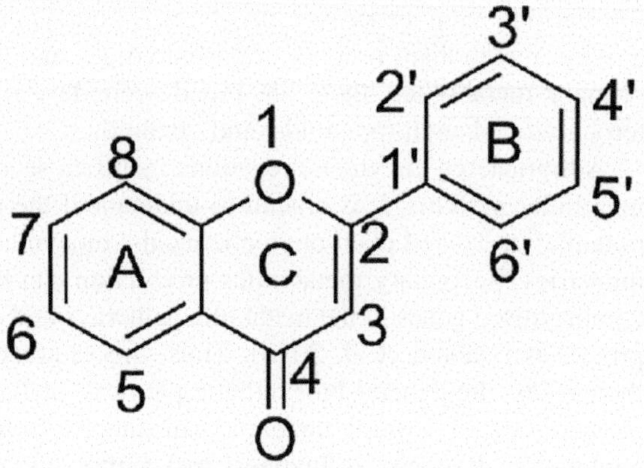

Fig. 17.1 General structure of flavonoids. This figure is reprinted with permission from international journal of current research (open access)

and progression due to their pharmaceutical characteristics (Prasad et al. 2010; Marranzano et al. 2018; Jucá et al. 2020). Several studies have demonstrated antimicrobial effects of several flavonoids including Apigenin, galangin, flavone and flavonol glycosides, isoflavones, flavanones, and chalcones due to the fact they are formed in plants as a result of microbial infections. Flavonoids can inactivate adhesion, protein transportation, enzymes of microorganisms, and interrupt microbial membranes (Mishra et al. 2009; Pandey et al. 2012; Mishra et al. 2013; Marranzano et al. 2018; Jucá et al. 2020). Moreover, they significantly contribute color and taste characteristics to the food and beverages are derived from plants industries, and they have potential effect in improving the nutritional properties of food. Flavonoids are the hydroxylated substances from a large group of phenolic compounds. In response to microbial infections in plants, flavonoids are synthesized Aglycones, glycosides, and methylated derivatives that are the three forms of flavonoids in plans (Karak 2019; Jucá et al. 2020).

From chemical standpoint, flavonoids are structurally built from 15-carbon skeleton and A and B benzenic ring linked by 3-carbon. Favonoids subdivision into diverse groups relies on the carbon that attach (C ring to B ring), degree of C ring' unsaturation, and oxidation. For example, the three-carbon linking chain characterized as C6-C3-C6 compounds and called chalcones (Fig. 17.1). Chalcones are precursors for most classes of flavonoids. Isoflavone can be formed by linking the carbon at position three of (C ring) to (B ring) (Fig. 17.2). However, a heterocyclic pyran or pyrone ring (C ring) can be further formed from the linking chain in many flavonoids (Fig. 17.2) (Nabavi et al. 2020). Flavones are found in steam, leaves, and roots, and they are converted by favone synthase (FNS) from favanones via the present of single double bound between the carbon atoms at second and third position. Anthocyanins are another important class of natural colorant and preservatives flavonoids, which are derived from favonols and possess the favylium ion basic structure at the lack of a ketone oxygen fourth position (Shah et al. 2019).

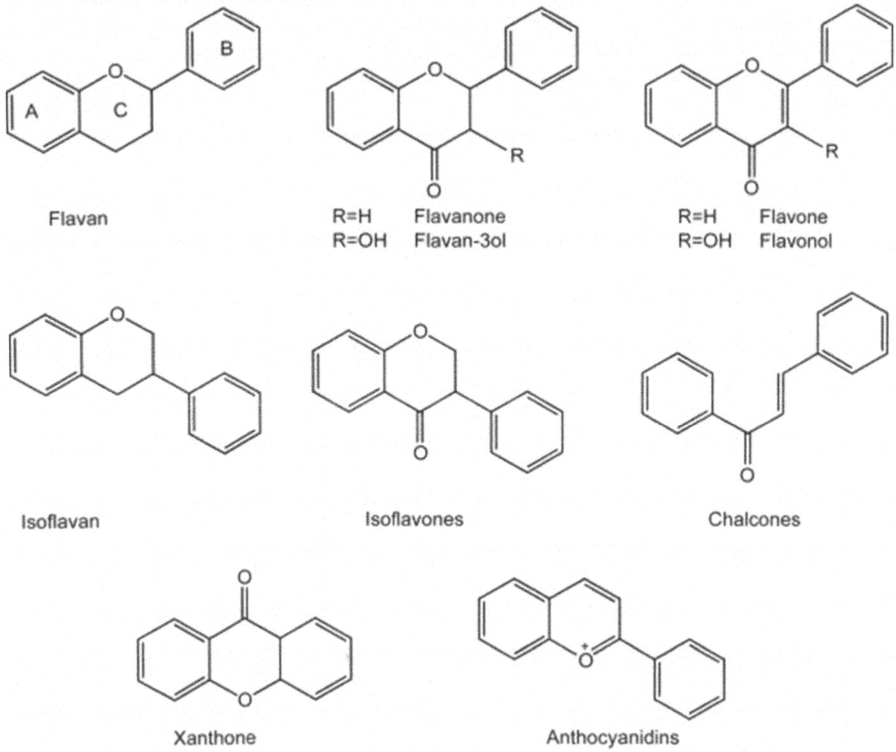

Fig. 17.2 Some major kinds of flavonoids. This figure is reprinted with permission from International journal of current research (open access)

17.3 Biosynthetic Pathways of Flavonoids in Plants

The acetate (A ring) and the shikimate pathways (B ring) with the linking chain (C ring) forming the C6-C3 components are the two biosynthesis pathways of flavonoids (Nabavi et al. 2020). Figure 17.3 shows the full biosynthetic pathway of flavonoids in plants. The flavylium ion, which is the core of all flavonoids that exist in the upstream pathway transformations of glucose, generates three malonyl-CoA molecules that form (A ring) and one of 4-coumaroyl-CoA (Petrussa et al. 2013). Phenylalanine produces 4-coumaroyl-CoA through shikimate pathways to form (B ring) (Nabavi et al. 2020). Two enzymes are involved in the condensation of two steps to produce naringenin chalcone synthase (CHS) and chalcone isomerase (CHI). The dihydrokaempferol (colorless dihydroflavonol) is produced via the oxidation of naringenin, which is a colorless flavanone via flavanone 3-hydroxylase (F3H). Subsequently, flavonoid 3′-hydroxylase (F3′H) or flavonoid 3′,5′-hydroxylase (F3′5′H) produces dihydroquercetin or dihydromyricetin, respectively, by hydroxylation of dihydrokaempferol on the 3′ or 5′ position of (B ring).

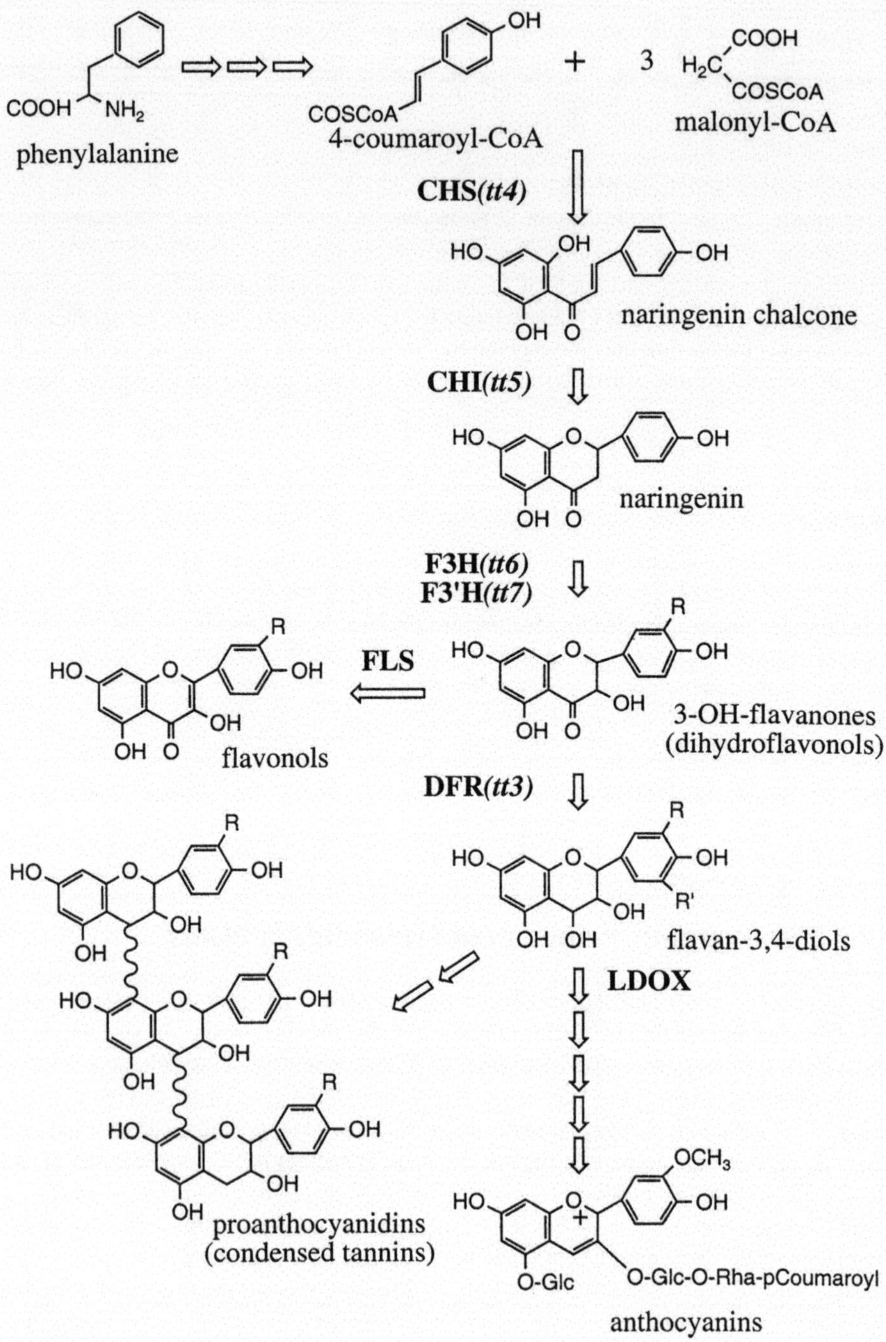

Fig. 17.3 An overview of general biosynthesis pathway of Flavonoids. This figure is reprinted with permission from Winkel

F3′H or F3′5′H can directly hydroxylated naringenin to deliver eriodictyol and pentahydroxy-flavanone, respectively.

Eriodictyol and pentahydroxy-flavanone are further hydroxylated to produce dihydroquercetin and dihydromyricetin. Unsaturation and oxidation of (C ring) forms the further divers' structures of flavonoids such as, (2-phenylchromenyliums (anthocyanidins/anthocyanins); 2-phenylochromones (flavones, flavonols, flavanones, di-OH-flavonols); 2-phenylchromanes (flavans, flavan-3-ols, and flavan-3,4-diols (proanthocyanidins)); chalcones/dihydrochalcones; 2-benzylidene coumaranones (Fig. 17.3) (Nabavi et al. 2020). Anthocyanidins are formed via dihydroflavonol reductase (DFR) and leucoanthocyanidin oxidase (LDOX), which catalyze reactions, "The (DFR) converts dihydroquercetin, dihydrokaempferol, and dihydromyricetin to leucocyanidin, leucopelargonidin, and leucodelphinidin (colorless flavan-3,4-cis-diols), respectively. Subsequently, (LDOX) catalyzes the oxidation of leucocyanidin, leucopelargonidin, and leucodelphinidin to cyanidin (red-magenta anthocyanidin), pelargonidin (orange anthocyanidin), and delphinidin (purple-mauve anthocyanidin), respectively"(Petrussa et al. 2013).

Chalcone is produced by condensation of (A ring) and (B ring). The later compound goes through isomerase-catalyzed cyclization to form flavanone, which is the budding blocs for other flavonoids. Approximately 7000 of flavonoids have the same basic structure since most of them are biosynthesized in this pathway. However, the simple structure of flavonoids that includes (C ring) is flavan. Therefore, flavonoids are characterized into two categories due to their structural diversity including flavanones, flavanols, flavones, and flavonol; on the other hand, isoflavones, biflavonoids, flavonolignans, prenylflavonoids, flavonoid glycosidoesters, aurones, and chalcones. Two different forms of flavonoids are found in plants' free aglycones form and glycoside-bound form. The later form is the major ingested form of flavonols and flavones by human (Nabavi et al. 2020). However, the production of flavonoids from plant sources is very difficult to be accomplished for large-scale production due to the long extraction process, expensive extraction approaches, and harsh chemical and environmental conditions. Thus, microbial biosynthesis of flavonoids can be a superior and sustainable alternative.

17.4 Importance of Microbial Biosynthetic Pathways of Flavonoids

The flavonoids as secondary metabolites are having several distinguishable characteristics among substances of primary metabolites. First, they have an ability to accumulate on organs or tissues such as, the accumulation of flavanols on skin of grabs (Mu et al. 2014). Second, they have limited distribution of specific taxonomic entities. As an example, the biosynthesis of isoflavones in *Fabaceae* plant species (Reynaud et al. 2005). Third, flavonoids production can be regulated when the tissue cultures are unable to form the secondary metabolites since the important genetic information possess in plant cells. Fourth, they demonstrated several biological activities either in the interaction between organism-organism or in diverse organisms (Trantas et al. 2015). Fifth, the culture suspensions have lower complexity than solid matrices of plants as well as low separation cost and carbon footprint. Finally, levels of flavonoids production from plants relay on crop's season. On the other hand, microbial productions have very short life cycles and do not rely on season. Microbial production requires simple feedstock such as, carbon source (glucose and oxygen and/or carbon dioxide of initial input and nutrients and will be able to meet the demand of production) (Ng et al. 2019).

The metabolomics improvements can offer a great chance to discover new chemicals formed through organisms. Moreover, the biosynthetic of flavonoids has been improved. Due to the fact that the advancement of structural and biology including deputation of gene and mutation, which are the main responsible process for the enzymatic modification in microbial flavonoid's biosynthesis. Diverse array of compounds could be produced through this enzymatic action and complex of enzymes on the metabolic pathway basic structure. As it is obvious, the enzymes involving in the flavonoid's biosynthesis pathway are progressing in a way to result in catalysis to glycosylation, acylation, prenylation, sulfation, methylation, isomerization, or condensation to specific region (Trantas et al. 2015).

A core molecule can be synthesized by the flavonoids, which are produced by organisms such as, flavanone naringenin. Several enzymes (such as, hydroxylases, isomerases) further proceeded the core molecule downstream to achieve the end product. Utilizing enzymes that can act on different substrates is a common feature of different organisms. For instance, the oxidation of flavanonol to flavono is catalyzed via flavonol synthase utilizing substrate dihydrokaempferol or dihydroquercetin to produce kaempferol or quercetin, respectively (Trantas et al. 2015). Bacterial, yeast, or cell biotransformation of plants is an innovative emerging strategy for production of rare, natural, inexpensive, and low environmental pollution products in in vivo biosynthesis under fully controlled conditions; moreover, it helps in easier recovery of the formed products than the extractions of natural products or chemically synthesized products due to generation of less waste and side products. A lot of efforts have been made to produce array of diverse natural compounds, analogs, and useful intermediates class including isoprenoids, flavonoids, stilbenes, polysaccharides and glycoproteins, and alcohols, and these compounds possess potential pharmaceutical applications (Abdullah et al. 2008;) (Otero and Nielsen 2010; Van Summeren-Wesenhagen and Marienhagen 2013; Trantas et al. 2015) such as, antimicrobial, anticancer, cardioprotective, antioxidant, anti-inflammatory, and immune system-promoting effects (Cazarolli et al. 2008; Tungmunnithum et al. 2018).

For instance, the number of hydroxyl substituents that constituted the flavonoids structure is highly associated with the increased antioxidants' capacity and scavenging of free radical caused by reactive oxygen species in human body (Jucá et al. 2020). For example, (*Ocimum basilicum L.*) and oregano (*Origanum vulgare L.*) are natural antioxidant and preservatives extracted from basil (Pitaro et al. 2012; Jucá et al. 2020). In addition, cardiovascular disease caused by atherosclerosis' clinical events led to increased morbidity and mortality rate, such as myocardial infarction, embolisms, and cerebrovascular accidents which can be prevented via antioxidant and free radical scavenging properties of flavonoids. The chronic or acute inflammatory process could be prevented by the presence of substances such as, the active compound campferol3-O-methyl of ginger extract (*Zingiber zerumbet*), which decrease itric oxide and prostaglandin E2 production (Soares et al. 2015; Jucá et al. 2020). Flavonoids are natural chemical factors that can promote the immune system and body defense against diseases. It has been scientifically proven that propolis possesses antimicrobial properties against microorganisms due to the presence of flavonoids galangina and pinocembrin. Obviously, the oxidative stress and free radical activities are the major causes of most disease' initiation and progression. Therefore, high antioxidant capacity of flavonoids can promote human health and prevent or treat several diseases such as, aging, cancer, cardiovascular diseases, arteriosclerosis, and neurodegenerative diseases (Tapas et al. 2008; Sandhar et al. 2011; Lima et al. 2015; Jucá et al. 2020).

17.5 Microbial Biosynthetic Pathways of Flavonoids

Traditional laboratories, expensive extraction methods, environmental and seasonal conditions of plants, toxic catalysts utilization, harsh chemical conditions, and presence of many compounds, such as chiral centers and labile connectivities, cannot meet the growing demand on flavonoids production from plants source in large scale (Chouhan et al. 2017; Nabavi et al. 2020). Also, individual compound of polyphenols cannot be achieved in large quantities due to the complex mixture of plants' polyphenols (Milke et al. 2018). Therefore, utilizing microorganisms in biotechnological production of plant secondary metabolites could be an ideal solution (Nabavi et al. 2020), and they can be grown with high rate very easily (Chouhan et al. 2017). High quality, economical, ecofriendly, short time, less harsh chemical conditions, less of intermediate pathways loss, glycerol, and cellulose use as cheap carbon source are converted to high chemicals, and a diverse generation of synthesized natural products has been produced by microorganisms using biology tools and novel enzymes (Chouhan et al. 2017).

Biotechnological production of plants secondary metabolites requires several techniques involving, choosing the most effective host strains, gene manipulation objects determination, and clear understanding of biosynthesis enzymes. The primary metabolism of plants and microorganism is very similar, which means polyphenol precursor molecules including malonylCoA and aromatic amino acids can be produced by microbial metabolism (Milke et al. 2018). *Saccharomyces cerevisiae and Escherichia coli* are the major considered microorganisms as cellular factories of flavonoids such as, flavones, flavonols, flavanones, and isoflavones, and other species such as *Streptomyces venezuelae* could be used in some cases (Trantas et al. 2015; Nabavi et al. 2020).

The platform of flavonoids production mainly requires manipulation and controlling the pathways of two rings:(A) malonyl-CoA pathway and (B) 4-coumaroyl-CoA pathway (Pandey et al. 2016; Trantas et al. 2015; Nabavi et al. 2020). Flavonoids can be originated using microorganism by the universal phenylpropanoid route leading to form (B ring) starting from aromatic amino acid phenylalanine or tyrosine (Yuan and Alper 2019; Nabavi et al. 2020). The phenylalanine ammonia-lyase (PAL), Cinnamic acid 4-hydroxylase (C4H), coumarate CoA ligase (4CL), chalcone synthase (CHS), and chalcone isomerase (CHI) genes are required in the case of using *Saccharomyces cerevisiae* organism to complete the pathway and form naringenin.

17.6 Naringenin

Naringenin, a (2S)-flavanone, is the major citrus bioactive compound found in grabs and oranges and is demonstrated to have several health benefits such as, antioxidant and anti-inflammatory effects. In addition, it can act as a modulator of immune system and promoter of carbohydrates metabolisms (Yıldız et al. 2009; Leonardi et al. 2010; Qin et al. 2011; Thapa et al. 2019). In order to achieve hydroxylation step from cinnamate to an electron-donor (p-coumarate), a cytochrome P450 reductase (CPR) requires the presence of C4H as coexpression. Naringenin can be also biosynthesized in *Escherichia coli* via the aromatic amino acids (a precursor tyrosine) with the presence of the coumarate CoA ligase (4CL), chalcone synthase (CHS), and chalcone isomerase (CHI) genes (Nabavi et al. 2020). The formation of several flavonoid compounds (naringenin, apigenin, kaempferol, and quercetin) utilizing p-coumarate as a precursor has been extensively reported (Leonard et al. 2007, 2008; Nabavi et al. 2020). Coumarate CoA ligase (4CL), chalcone synthase (CHS), and chalcone isomerase (CHI) genes are used to achieve kaempferol and quercetin. Flavanone 3β-hydroxylases (FLS, F3H) are the further genes used in the pathway for kaempferol, while quercetin can further use (FLS, F3H and F3′,5′H lavonoid-3′,5′- hydroxylase. F3′,5′H) bonded to a cytochrome P450 reductase (CPR), which requires to obtain hydroxylation steps on (B ring) (Nabavi et al. 2020). Because of the enzymes and enzyme complex involving in flavonoids production, specific region catalyst condensation, glycosylation, acylation, prenylation, sulfation, methylation, or isomerization can be achieved, and these enzymes enable to form flavanone naringenin. In order to produce the flavonoids' end products, the core molecules flavanone naringenin via multiple enzymes such as, hydroxylases and isomerases further regulated downstream (Trantas et al. 2015).

Another approach of flavonoids (naringenin and pinocembrin) production via yeast or bacteria is D-glucose as a simple source of carbon (Nabavi et al. 2020) as it is shown in (Fig. 17.4) since it decreases the cost of production in large scale (Jones et al. 2016). In the case of using the glucose as a supplementation for the microbial cells, shikimate pathway originates phenylalanine and tyrosine (aromatic amino acid precursors). The shikimate route in microorganisms could be imposed accordingly, and the enzyme versions' resistant will be introduced to feedback inhibition evacuated via phenylalanine or tyrosine when the 3-deoxy-D-arabino-heptulosonate-7-phosphate (DAHP) synthase level is reached. For phenylalanine, the chorismate is mutase/prephenate dehydratase (CM/PDT), while tyrosine is the chorismate mutase/prephenate dehydrogenase (CM/PDH). Modular metabolic

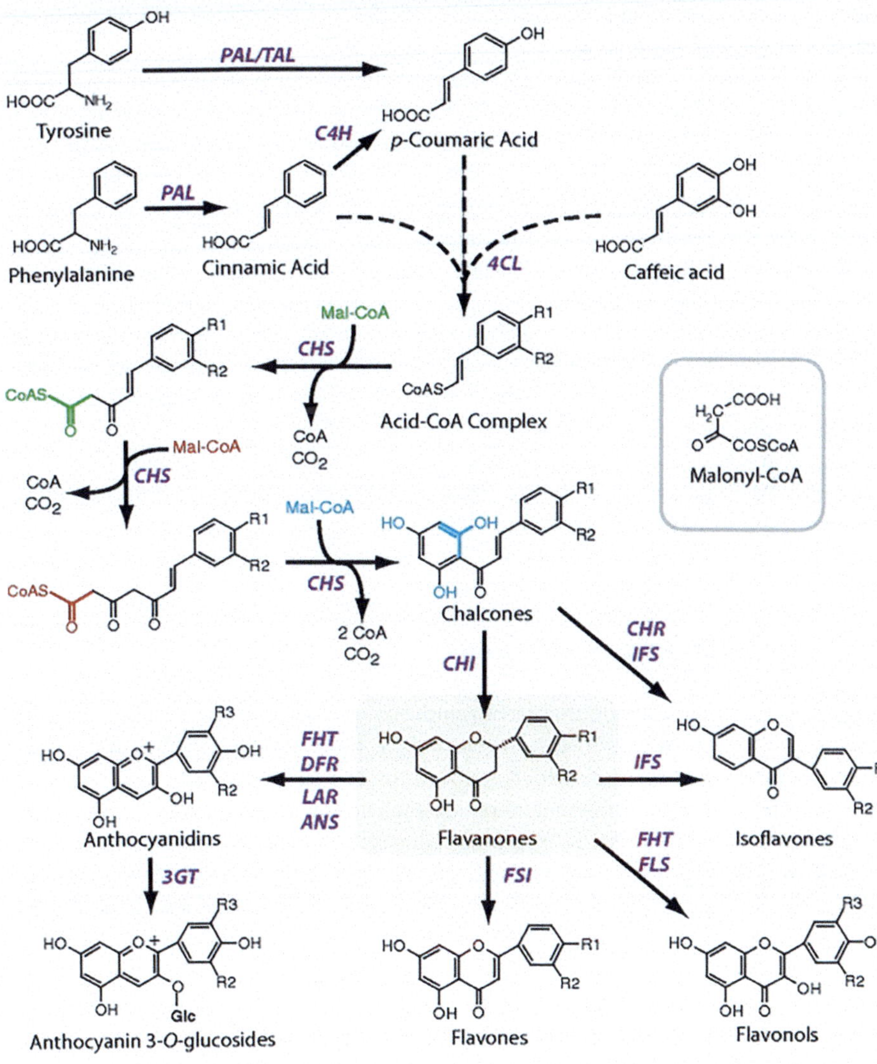

Fig. 17.4 Flavonoids biosynthesis from D-Glucose. This figure is reprinted with permission from Springer

strategy could optimize the production of (2S)-pinocembrin via D-glucose in *E. coli* through four molecules comprising 1. First, "the aroF gene encoding a 3-deoxy-D-arabino-heptulosonate-7-phosphate (DAHP) synthase and the pheA gene encoding a resistant version of chorismate mutase/prephenate dehydratase (CM/PDT). Second, phenylalanine ammonia-lyase (PAL) and 4-coumarate CoA ligase (4CL). Third, chalcone synthase (CHS) and chalcone isomerase (CHI). Forth, matC encoding a malonate carrier protein and matB encoding a malonyl-CoA synthase to increase the intracellular pool of malonyl-CoA" (Nabavi et al. 2020).

In another study, a combination of three molecules, which are module 1: tyrosine ammonia-lyase (TAL) and 4-coumarate CoA ligase (4CL), module 2: chalcone synthase (CHS) and chalcone isomerase (CHI), module 3: matC and matB, and Two genes, aroG encoding (DAHP) synthase and tyrA encoding a resistant version of chorismate mutase/prephenate dehydratase (CM/PDH), were used to optimize *E. coli* de novo synthesis of (2S)-naringenin from D-glucose (Wu et al. 2014). Controlling the malonylCoA intracellular pool is a very important feature in the formation of (A ring) in flavonoids biosynthesis because the malonylCoA is the intersection compound that links pathway of flavonoids and biosynthesis of fatty acids. Upstream and downstream pathways could be controlled via the intracellular pool of malonyl-CoA regulation (Nabavi et al. 2020). Overexpression of the acetate assimilation pathway' key enzyme, which is the acetyl-CoA carboxylase complex (ACC), can increase malonyl-CoA concentration. In flavonoids' production from *E. coli*, the ACC gene and the biotin ligase (BirA) gene were used (Leonard et al. 2007; Fowler and Koffas 2009; Nabavi et al. 2020).

17.7 Apigenin and Genkwanin

Apigenin is a typical flavone that is biosynthesized from naringenin using flavone synthase (FNS), which acts as a substrate for (S)-naringenin. O-methylation is the derivative of Apigenin, from which genkwanin can be generated. Apigenin and genkwanin have several biological activities such as anti-inflammatory, antioxidant, anticancer etc. (Lee et al. 2015). The biosynthesis of genkwanin from apigenin was accomplished by glucose source using *E. coli* with presence of four genes (Os4CL, PeCHS, MtCHI, and FNS) as it is shown. Four enzymes, 4-coumarate CoA ligase. (4CL), chalcone synthase (CHS), chalcone isomerase (CHI), and flavone synthase (FNS), and four genes (Os4CL, PeCHS, MtCHI, and FNS) are introduced to synthesize apigenin via p-coumaric acid pathway. Subsequently, naringenin chalcone converts to naringenin. Three genes (Os4CL, PeCHS, and FNS) are introduced to *E. coli* as well as the CHI. The naringenin chalcone transformed to (S)-naringenin, and the CHI could be utilized as FNS' substrate. 4CL and CHS coordinated expression is an important step for apigenin end product. For genkwain's biosynthesis from *E. coli,* two additional genes tyrosine ammonium lyase.

(TAL) and apigenin 7-Omethyltransferase (POMT7)) are required. Tyrosine ammonia-lyase (TAL) is critical to produce p-coumaric acid from tyrosine, while (POMT7) is required to catalyze the reaction of apigenin conversion to genkwanin (Lee et al. 2015).

17.8 Flavones

Flavones are the first generated compounds of all other flavonoids, which represent pharmaceutical characteristics. Also, it is demonstrated as the largest group of flavonoids compounds and provides several biological, physiological, and nutritional functions (Jiang et al. 2016). The structure of flavones is shown in (Fig. 17.2). The flavone backbone biosynthesis in most species generates from phenylpropanoid pathway followed by the biosynthetic branch of flavonoid (Fig. 17.4) The phenylalanine ammonia-lyase (PAL) deaminates the phenylalanine to cinnamic acid. Cinnamic acid 4-hydroxylase (C4H) can catalyze hydroxyl group introduction to the phenyl ring, which forms p-coumaric acid from cinnamic acid. P-coumaroyl-coenzyme A (CoA) forms through activation of the carboxyl group of p-coumaric acid. A thioester bond with (CoA) to form p-coumaroyl-CoA ligase and (4CL) catalyzes the process (Jiang et al. 2016). The chalcone synthase (CHS) enzyme generates chalcone through condensation reaction of three molecules of Malonyl-CoA, which is the first step for core flavonoids molecule production (Fig. 17.4). The (CHS) is acting as a starter molecule. Acysteine residue attaches to p-coumaroyl-CoA at chalcone synthase (CHS) active site. Subsequently, a series of decarboxylative condensation reactions produces tetraketide intermediate. The intramolecular cyclization of the tetraketide intermediate forms the chalcone (4,2′,4′,6′-tetrahydroxychalcone). (Ververidis et al. 2007; Jiang et al. 2016; Trantas et al. 2015). Then, chalcone isomerase (CHI) is applied for consequent isomeration of chalcone into flavones.

Flavones group is produced via flavone synthase I (FNS I) action. In further steps of flavonoids biosynthetic metabolism of flavanones downstream, a flavanone 3β–hydroxylase (F3H) enzyme creates Dihydroflavonols. This enzyme (F3H) assists in transformation of flavanones to flavonols via a flavonol synthase (FLS) enzymatic action (Ververidis et al. 2007; Jiang et al. 2016; Trantas et al. 2015).

17.9 Isoflavones

Isoflavones are known for their ability to attach oestrogen receptors (ER) based on their structures. They share common structural features with the mammalian oestrogen oestradiol-17β such as, similar distance separation of hydroxyl groups and they exhibit oestrogenicity (Cassidy et al. 2000). The structure of isoflavones is shown in (Fig. 17.2). To date, more than 1600 derivatives of isoflavones have been investigated. For heterologous isoflavonoids production, the most common hosts are *E. coli* and *S. cerevisiae.* The main biosynthetic region of isoflavones in the flavonoids is (B ring). After the formation of chalcone as it is shown in (Fig. 17.4), a combination of an isoflavone synthase (IFS)-encoded gene expressed in soybean with cytochrome P450 oxidoreductase (CPR) enzymatic action isoflavones group is generated (Ververidis et al. 2007; Trantas et al. 2015) via introduction to yest *S. cerevisiae*. Then, 7,40 - dihydroxyflavone (liquiritigenin) and naringenin are converted through a probable 2-hydroxyiso-favanone intermediate to daidzein and genistein, respectively. In *S. cerevisiae,* Glycine max co-expressed isoflavone synthase (IFS), cytochrome P450 oxidoreductase (CPR), chalcone isomerase, and (CHI), which led to transfer chalcone substrates to 2-hydroxyiso-favanones and isoflavones group, respectively, due to CHI and IFS functions. (Song et al. 2014). Also, isoflavones could be biosynthesized in *E. coli* with improved turnover rate via the fusion of isoflavone synthase (IFS) and cytochrome P450 oxidoreductase (CPR) enzymes and *E. coli* (P450) and (IFS) in bacterial P450BM-3 in order to mimic its intrinsic architecture, and daidzein and genistein were produced with high yields (Leonard and Koffas 2007; Song et al. 2014). Cocultures can significantly improve the yield of daidzein and genistein in either yeast or microorganisms. For example, *E. coli* and *S. cerevisiae* co-incubation is a great strategy since *S. cerevisiae* possessed (IFS), while *E. coli* possessed phenylalanine ammonia lyase (PAL), ScCCL, chalcone synthase (CHS), chalcone isomerase (CHI), and acetyl-CoA carboxylase (ACC) (Song et al. 2014; Teplova et al. 2018).

17.10 Anthocyanins

It is one of useful and desirable flavonoids compounds groups in food industry since it provides color pigments of red, orange, purple, and blue, and it is a common natural colorant in food and beverages industries. Moreover, it could contribute several health benefits to human such as, prevention and protective effect against cardiovascular diseases, improvement of vision, scavenging of free radical activities, and protection from cancer, diabetics, and inflammation (Pojer et al. 2013; Cortez et al. 2017; Eichenberger et al. 2018; Nabavi et al. 2020). The structure of anthocyanin is shown in (Fig. 17.2). Depending on the initial precursor, dihydroflavonol 4-reductase (DFR) and anthocyanin synthase (ANS) combined action or ANS sole of action require to build flavylium cation (Fig. 17.4). Leucoanthocyanidin reductase (LAR) and Flavanone 3β-hydroxylase (FHT) are catalyzing enzymes that involve the biosynthesis to produce anthocyanin. Decorating enzymes are important to control the (B ring) of anthocyanin such as, anthocyanin-O-glucosyltransferases (GT) and anthocyanin-O-methyltransferases (AOMT). Respectively, uridine diphosphate.

(UDP)-glucose (UDPG) and S-adenosyl-L-methionine (SAM) are used as donors of glucose or methyl (Nabavi et al. 2020). For instance, in *E. coli* strain, Cn 3-G were produced from naringenin by the action of three genes, which are ANS, DFR, and 3-GT (Yan et al. 2005; Eichenberger et al. 2018; Nabavi et al. 2020); further modifications of anthocyanidins to form anthocyanin through glycosylation, acylation, and methylation. Then, the anthocyanin is transported to vacuoles. The color of anthocyanin is highly dependent on the flavonoid's structures and the flavones and flavanols present. For anthocyanin heterologous production, it is difficult to express the required multiple plant CYP enzymes for full biosynthesis pathway in bacterial hosts such as, *E. coli*; therefore, yeast such as, *Saccharomyces cerevisiae* is considered the best (Eichenberger et al. 2018).

17.11 Microbial Biosynthesis in Food and Beverages Industries

There are growing demand of protective nutraceuticals such as flavonoids in a form of food or drugs due to the fact that flavonoids potentially can reduce the risks of several chronic and aging diseases such as, Alzheimer, cancer, and inflammation. Flavonoids' global market has been estimated to reach US $200 yearly by 2022 (Xu et al. 2020), due to the fact that they are considered potential natural colorant, preservatives, and antimicrobial candidates in food, beverages, and pharmaceutical industries (Marranzano et al. 2018).

Rei Ng et al. (2019) analyzed the effect of secreted phenolic metabolites from a naringenin, which resulted from *Saccharomyces cerevisiae* strain (a GRAS organism), and their potentials as preservatives with high antimicrobial and antioxidant characteristics. Phenolic metabolites biosynthesized using *Saccharomyces cerevisiae* strain were compared to naturally produced flavonoid naringenin and its prenylated derivatives. The results showed that the biosynthesized flavonoid naringenin from yeast exhibited superior antimicrobial against major pathogens, which cause food borne illness such as, (*Campylobacter sp., Salmonella sp., S. aureus, E. coli, Listeria monocytogenes, C. botulinum*). Phenylacetaldehyde, homogentisic acid, and phloretic acid are the key identified metabolites that increase the bioactivities. In addition, it showed superior antioxidant effect via DMPD. $^{+}$ scavenger the free radical against the naturally extracted flavonoid naringenin. Therefore, the biosynthesized phenolic metabolites naringenin from yeast could be innovative natural preservative since it offers economic, maintainable, and safe to environment over large-scale extraction process from plants.

Also, complex of flavonoids, which are produced by complex cluster of genes and precursors, can be biosynthesized via microbial cocultures (Ganesan et al. 2017; Akdemir et al. 2019). Naringenin has been rapidly produced through the engineered coculture of two strains of *Escherichia coli.* Coculture biosynthetic pathway results in decreasing the metabolic stress and improved catalytic reaction, and associated coculture strains biosynthetic steps were simply regulated (Ganesan et al. 2017). *Escherichia coli* cocultures were able to increase flavan-3-ol titer up to 970-fold over monoculture strain when the carbon source, temperature, point of indication, compatibility of strains, and ratio of inculcation were optimized (Jones et al. 2016).

For example, hydroxyphenyl-pyranoanthocyanins are important natural colorants that have been used in low-to-medium acidic beverages and food industries, and they extensively present in fruit and red wine. They are more stable under variation of pH than due to the fact that they consist of (D ring) or pyran their anthocyanin precursors. Extraction of pyranoanthocyanins is a very difficult task because they present in very low concentrations. Cocultures of two strains of *Escherichia coli* were accomplished, and 4-vinylphenol and 4-vinylcatechol producer modules were engineered.

Then, each strain was cocultured with yanidin-3-*O*-glucoside producer recombinant cells to achieve pyranocyanidin-3-*O*-glucoside-phenol (cyanidin-3-*O*-glucoside with vinylphenol adduct) and pyranocyanidin-3-*O*-glucoside-catechol (cyanidin-3-*O*-glucoside with vinylcatechol adduct). The results show that pyranoanthocyanins was produced with higher yield, titer, and stability in compression to the one produced from plants (Akdemir et al. 2019).

3-Deoxyanthocyanidins is another rare type of natural colorant derived from anthocyanin with interesting chemical and biochemical characteristics, including antioxidant, anticancer, and antimicrobial, and stable under environmental stress and conditions. 3-Deoxyanthocyanidins pigmies have potential applications in food and beverages industries due to the fact that C-3 position lacks hydroxyl group (Xiong et al. 2019). Coculture is a potential strategy for increasing yields of complex chemical compounds such as, flavonoids allowing division and optimization to achieve the complete pathways, controlling, and balancing the metabolic efflux ratio between the strains. The functional overexpression of genes can be optimized in all pathways and enhanced the utilization of substrates to achieve maximum yields of flavonoids. Therefore, simple mixing and matching of different microbial strains facilitate various chemical compounds production (Wang et al. 2020).

In Jones et al. (2016) study, phenylpropanoic acid precursors were used in the production of flavan-3-ols via six enzymatic phases in *E-Coli*, which were 4-coumaroylCoA ligase (4CL), chalcone synthase (CHS), chalcone isomerase (CHI), flavanone 3β-hydroxylase (F3H), dihydroflavonol 4-reductase (DFR), and leucoanthocyanidin reductase (LAR) as shown in (Fig. 17.5). The upstream and downstream modules were partitioned to complete the pathway with three genes. The separation of the pathways was according to malonylCoA (upstream) and NADPH (downstream) because they are important cofactors in the pathway. Coculture genetic optimization of each strain enhances the important cofactor efflux, substrates, and reduce the efflux of unwanted side of interest. 4CL, CHS, and CHI were used in upstream strain, and F3H, DFR, and LAR were used in downstream strain genetic optimization. Source of carbon, temperature, compatibility of strain, and point of indication are important factors in genetic optimization for efficient flavonoids production in vivo (Jones et al. 2016).

Lv et al. (2019)' study utilized the chassis (*Y. lipolytica)* to biosynthesized flavonoids by constructing two models: synthesis and hydroxylation modules. Synthesis module consists of malonyl-CoA and chalcone precursors to produce the shikimic acid. The hydroacylation module consists of cytochrome c P450 (CYP) flavonoid 3′-hydroxylase 129 (F3'H) and cytochrome c P450 reductase (CPR) since they are important in phenyl ring oxidation and producing hydroxylated flavonoids. P-coumaric acid-CoA ligase (4CL), chalcone synthase (CHS), and chalcone isomerase (CHI) were used in synthesis pathway. The results showed that for heterologous *Y. lipolytica, it* is an ideal platform to produce flavonoids with high value.

Thus, microbial biosynthesis is a great alternative for producing sustainable, cost-effective, enhanced functionality, and nutritional values of bioactive compounds (flavonoids).

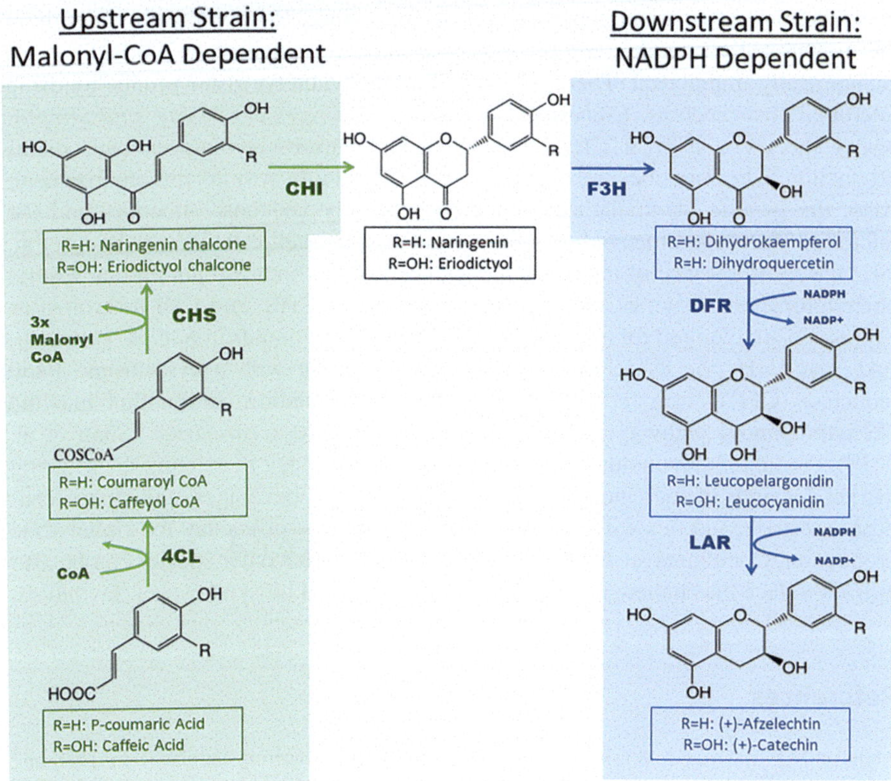

Fig. 17.5 The upstream strain (Malonyl-CoA) and downstream strain (NADPH) genetic optimization of flavan-3-ols. This figure is reprinted with permission from Metabolic engineering

17.12 Conclusion and Recommendation

Plants' Secondary metabolites are rich source of highly diverse chemical compounds with biological characteristics for food, beverages, and pharmaceutical use (Song et al. 2014). Among polyphenolic compounds, flavonoids are potential naturally occurring group present in fruit, vegetables, and beverages that are derived from plant sources. Due to the fact that flavonoids have essential role in heath promoting, treating, and preventing several diseases, they are involved in diverse food, beverages, and pharmaceutical applications in industries as nutraceuticals. In addition, they produce series of substances including flavones, flavonols, flavanones, isoflavones, flavanols, and anthocyanins. These compounds exhibit great biological activities such as, antioxidants and free radical scavenging, anticancer, anti-inflammatory, antidiabetic, antiviral, antimicrobial, and antiallergic activities (Karak 2019). Due to high antioxidant capacity of flavonoids, they have received great attention as promising pharmaceutical compounds. However, the extraction of these chemically complex compounds is difficult for large-scale production and economically impractical. Therefore, microbial biosynthesis of flavonoids by using microbial hosts such as, *Escherichia coli, Saccharomyces cerevisiae, and Streptomyces* species is a great alternative and offers various advantages over plants production. The microbes are rapidly grown in friendly way to the environment. Also, the genetic manipulation of microbes is easy and cost effective, and the microbes' developed metabolic engineering tools are well-established (Song et al. 2014). Recent advance of microbial biosynthesis of flavonoids reported that several phenylpropanoid enzymes such as, PAL, C4H, 4CL, CHS, and CHI heterologous expression are required for microbial biosynthesis of flavonoids in *E. Coli.* However, PAL and C4H (the cytochrome P450 enzyme), along with a cytochrome P450 reductase (CPR), were reported to be sufficient in feeding carbon flux into the phenylpropanoid pathway in the case of *Saccharomyces cerevisiae* (Shah et al. 2019). Producing flavonoids via microbes is a great strategy in meeting the growing market demands of such bioactive compounds for food, beverages, and pharmaceutical industries, and this will offer sustainability and cost-efficiency for global scale production. Cocultures such as, *E. coli-E. coli* and *E. coli*-other species can be also utilized to face the challenges and improve the microbial biosynthesis of flavonoids.

References

Abdullah MA, Rahmah A, Sinskey AJ, Rha CK (2008) Cell engineering and molecular pharming for biopharmaceuticals. Open Med Chem J 2:49–61. https://doi.org/10.2174/1874104500802010049

Akdemir H, Silva A, Zha J, Zagorevski DV, Koffas MA (2019) Production of pyranoanthocyanins using Escherichia coli co-cultures. Metab Eng Commun 55:290–298. https://doi.org/10.1016/j.ymben.2019.05.008

Cassidy A, Hanley B, Lamuela-Raventos RM (2000) Isoflavones, lignans and stilbenes–origins, metabolism and potential importance to human health. J Sci Food Agric 80(7):1044–1062. https://doi.org/10.1002/(SICI)1097-0010(20000515)80:7<1044::AID-JSFA586>3.0.CO;2-N

Cazarolli LH, Zanatta L, Alberton EH, Bonorino Figueiredo MSR, Folador P, Damazio RG, Barreto Silva FRM (2008) Flavonoids: prospective drug candidates. Mini Rev Med Chem 8(13): 1429–1440. https://doi.org/10.2174/138955708786369564

Chouhan S, Sharma K, Zha J, Guleria S, Koffas MA (2017) Recent advances in the recombinant biosynthesis of polyphenols. Front Microbiol 8:2259. https://doi.org/10.3389/fmicb.2017.02259

Cortez R, Luna-Vital DA, Margulis D, Gonzalez de Mejia E (2017) Natural pigments: stabilization methods of anthocyanins for food applications. Compr Rev Food Sci Food Saf 16(1):180–198. https://doi.org/10.1111/1541-4337.12244

Heim KE, Tagliaferro AR, Bobilya DJ (2002) Flavonoid antioxidants: chemistry, metabolism and structure-activity relationships. Nutr Biochem 13(10):572–584. https://doi.org/10.1016/S0955-2863(02)00208-5

Eichenberger M, Hansson A, Fischer D, Dürr L, Naesby M (2018) De novo biosynthesis of anthocyanins in Saccharomyces cerevisiae. FEMS Yeast Res 18(4):foy046. https://doi.org/10.1093/femsyr/foy046

Fowler ZL, Koffas MA (2009) Biosynthesis and biotechnological production of flavanones: current state and perspectives. Appl Microbiol Biotechnol 83(5):799–808

Ganesan V, Li Z, Wang X, Zhang H (2017) Heterologous biosynthesis of natural product naringenin by co-culture engineering. Synth Syst Biol 2(3):236–242. https://doi.org/10.1016/j.synbio.2017.08.003

Jain N, Ramawat KG (2013) Nutraceuticals and antioxidants in prevention of diseases. In: Natural products. Springer, Berlin, pp 2559–2580. https://doi.org/10.1007/978-3-642-22144-6_70

Jiang N, Doseff AI, Grotewold E (2016) Flavones: from biosynthesis to health benefits. Plan Theory 5(2):27. https://doi.org/10.3390/plants5020027

Jones JA, Vernacchio VR, Sinkoe AL, Collins SM, Ibrahim MH, Lachance DM et al (2016) Experimental and computational optimization of an Escherichia coli co-culture for the efficient production of flavonoids. Metab Eng 35:55–63. https://doi.org/10.1016/j.ymben.2016.01.006

Jucá MM, Cysne Filho FMS, de Almeida JC, Mesquita DDS, Barriga JRDM, Dias KCF et al (2020) Flavonoids: biological activities and therapeutic potential. Nat Prod Res 34(5):692–705. https://doi.org/10.1080/14786419.2018.1493588

Karak P (2019) Biological activities of flavonoids: an overview. Int J Pharm Sci 10(4):1567–1574. https://doi.org/10.13040/IJPSR.0975-8232.10(4).1567-74

Khoddami A, Wilkes MA, Roberts TH (2013) Techniques for analysis of plant phenolic compounds. Molecules 18(2):2328–2375. https://doi.org/10.3390/molecules18022328

Kukić J, Petrović S, Niketić M (2006) Antioxidant activity of four endemic Stachys taxa. Biol Pharm Bull 29(4):725–729. https://doi.org/10.1248/bpb.29.725

Lee H, Kim BG, Kim M, Ahn JH (2015) Biosynthesis of two flavones, apigenin and genkwanin in Escherichia coli. J Microbiol Biotechnol 25(9):1442–1448. https://doi.org/10.4014/jmb.1503.03011

Leonard E, Lim KH, Saw PN, Koffas MA (2007) Engineering central metabolic pathways for high-level flavonoid production in Escherichia coli. Appl Environ Microbiol 73(12):3877–3886. https://doi.org/10.1128/AEM.00200-07

Leonard E, Koffas MA (2007) Engineering of artificial plant cytochrome P450 enzymes for synthesis of isoflavones by Escherichia coli. Appl Environ Microbiol 73(22):7246–7251. https://doi.org/10.1128/AEM.01411-07

Leonardi T, Vanamala J, Taddeo SS, Davidson LA, Murphy ME, Patil BS, Turner ND (2010) Apigenin and naringenin suppress colon carcinogenesis through the aberrant crypt stage in azoxymethane treated rats. Exp Biol Med 235:710–717. https://doi.org/10.1258/ebm.2010.009359

Leonard E, Yan Y, Fowler ZL, Li Z, Lim CG, Lim KH, Koffas MA (2008) Strain improvement of recombinant Escherichia coli for efficient production of plant flavonoids. Mol Pharm 5(2):257–265. https://doi.org/10.1021/mp7001472

Lima EBC, Sousa CNS, Meneses LN, Ximenes NC, Júnior S, Vasconcelos GS et al (2015) Cocos nucifera (L.)(Arecaceae): a phytochemical and pharmacological review. Braz J Med Biol Res 48(11):953–964. https://doi.org/10.1590/1414-431x20154773

Lv Y, Marsafari M, Koffas M, Zhou J, Xu P (2019) Optimizing oleaginous yeast cell factories for flavonoids and hydroxylated flavonoids biosynthesis. ACS Synth Biol 8(11):2514–2523. https://doi.org/10.1021/acssynbio.9b00193

Marranzano M, Rosa RL, Malaguarnera M, Palmeri R, Tessitori M, Barbera AC (2018) Polyphenols: plant sources and food industry applications. Curr Pharm Des 24(35):4125–4130. https://doi.org/10.2174/1381612824666181106091303

Milke L, Aschenbrenner J, Marienhagen J, Kallscheuer N (2018) Production of plant-derived polyphenols in microorganisms: current state and perspectives. Appl Microbiol Biotechnol 102(4):1575–1585. https://doi.org/10.1007/s00253-018-8747-5

Mishra AK, Mishra A, Kehri HK, Sharma B, Pandey AK (2009) Inhibitory activity of Indian spice plant Cinnamomum zeylanicum extracts against Alternaria solani and Curvularia lunata, the pathogenic dematiaceous moulds. Ann Clin Microbiol Antimicrob 8(1):1–7. https://doi.org/10.1186/1476-0711-8-9

Mishra A, Kumar S, Pandey AK (2013) Scientific validation of the medicinal efficacy of Tinospora cordifolia. Sci World J. https://doi.org/10.1155/2013/292934

Mu L, He J-J, Pan Q-H, He F, Duan C-Q (2014) Tissue-specific accumulation of flavonoids in grape berries is related to transcriptional expression of VvF3 H and VvF3 5 H. S Afr J Enol Vitic 35(1):68–81

Nabavi SM, Šamec D, Tomczyk M, Milella L, Russo D, Habtemariam S et al (2020) Flavonoid biosynthetic pathways in plants: versatile targets for metabolic engineering. Biotechnol Adv 38: 107316. https://doi.org/10.1016/j.biotechadv.2018.11.005

Ng KR, Lyu X, Mark R, Chen WN (2019) Antimicrobial and antioxidant activities of phenolic metabolites from flavonoid-producing yeast: potential as natural food preservatives. Food Chem 270:123–129. https://doi.org/10.1016/j.foodchem.2018.07.077

Otero JM, Nielsen J (2010) Industrial systems biology. Biotechnol Bioeng 105:439–460. https://doi.org/10.1002/bit.22592

Pandey AK, Mishra AK, Mishra A (2012) Antifungal and antioxidative potential of oil and extracts derived from leaves of Indian spice plant Cinnamomum tamala. Cell Mol Biol 58(1):142–147. https://doi.org/10.1170/T933

Pandey RP, Parajuli P, Koffas MA, Sohng JK (2016) Microbial production of natural and non-natural flavonoids: pathway engineering, directed evolution and systems/synthetic biology. Biotechnol Adv 34(5):634–662. https://doi.org/10.1016/j.biotechadv.2016.02.012

Petrussa E, Braidot E, Zancani M, Peresson C, Bertolini A, Patui S, Vianello A (2013) Plant flavonoids—biosynthesis, transport and involvement in stress responses. Int J Mol Sci 14(7): 14950–14973. https://doi.org/10.3390/ijms140714950

Pitaro SP, Fiorani LV, Jorge N (2012) Potencial antioxidante dos extratos de manjericão (Ocimum basilicum Lamiaceae) e orégano (Origanum vulgare Lamiaceae) em óleo de soja. Rev Bras Plantas Med 14(4):686–691. https://doi.org/10.1590/S1516-05722012000400017

Pojer E, Mattivi F, Johnson D, Stockley CS (2013) The case for anthocyanin consumption to promote human health: a review. Compr Rev Food Sci Food Saf 12(5):483–508. https://doi.org/10.1111/1541-4337.12024

Prasad KN, Xie H, Hao J, Yang B, Qiu S, Wei X et al (2010) Antioxidant and anticancer activities of 8-hydroxypsoralen isolated from wampee [Clausena lansium (Lour.) Skeels] peel. Food Chem 118(1):62–66. https://doi.org/10.1016/j.foodchem.2009.04.073

Qin L, Jin L, Lu L, Lu X, Zhang C, Zhang F, Liang W (2011) Naringenin reduces lung metastasis in a breast cancer resection model. Protein Cell 2:507–516. https://doi.org/10.1007/s13238-011-1056-8

Reynaud J, Guilet D, Terreux R, Lussignol M, Walchshofer N (2005) Isoflavonoids in non-leguminous families: an update. Nat Prod Rep 22(4):504–515. https://doi.org/10.1039/B416248J

Sandhar HK, Kumar B, Prasher S, Tiwari P, Salhan M, Sharma P (2011) A review of phytochemistry and pharmacology of flavonoids. Int Pharm Sci 1(1):25–41

Shah FLA, Ramzi AB, Baharum SN, Noor NM, Goh HH, Leow TC et al (2019) Recent advancement of engineering microbial hosts for the biotechnological production of flavonoids. Mol Biol Rep 46(6):6647–6659. https://doi.org/10.1007/s11033-019-05066-1

Soares DJ, de Moura Neto LG, Damaceno MN, de Souza PA, Braga RC, Brasil IM (2015) Atividade antiinflamatória de frutas. Rev Saude Publica 15(39):33–45. https://doi.org/10.15600/2238-1244/sr.v15n39p33-45

Song MC, Kim EJ, Kim E, Rathwell K, Nam SJ, Yoon YJ (2014) Microbial biosynthesis of medicinally important plant secondary metabolites. Nat Prod Rep 31(11):1497–1509. https://doi.org/10.1039/c4np00057a

Tapas AR, Sakarkar DM, Kakde RB (2008) Flavonoids as nutraceuticals: a review. Trop J Pharm Res 7(3):1089–1099. https://doi.org/10.4314/tjpr.v7i3.14693

Teplova VV, Isakova EP, Klein OI, Dergachova DI, Gessler NN, Deryabina YI (2018) Natural polyphenols: biological activity, pharmacological potential, means of metabolic engineering. Appl Biochem Microbiol 54(3):221–237. https://doi.org/10.1134/S0003683818030146

Thapa SB, Pandey RP, Bashyal P, Yamaguchi T, Sohng JK (2019) Cascade biocatalysis systems for bioactive naringenin glucosides and quercetin rhamnoside production from sucrose. Appl Microbiol Biotechnol 103(19):7953–7969. https://doi.org/10.1007/s00253-019-10060-5
Trantas EA, Koffas MA, Xu P, Ververidis F (2015) When plants produce not enough or at all: metabolic engineering of flavonoids in microbial hosts. Front Plant Sci 6:7. https://doi.org/10.3389/fpls.2015.00007
Tungmunnithum D, Thongboonyou A, Pholboon A, Yangsabai A (2018) Flavonoids and other phenolic compounds from medicinal plants for pharmaceutical and medical aspects: an overview. Medicines 5(3):93. https://doi.org/10.3390/medicines5030093
Van Summeren-Wesenhagen PV, Marienhagen J (2013) Putting bugs to the blush: metabolic engineering for phenylpropanoid-derived products in microorganisms. Bioengineered 4:355–362. https://doi.org/10.4161/bioe.23885
Ververidis F, Trantas E, Douglas C, Vollmer G, Kretzschmar G, Panopoulos N (2007) Biotechnology of flavonoids and other phenylpropanoid-derived natural products. Part I: chemical diversity, impacts on plant biology and human health. Biotechnol J 2(10):1214–1234. https://doi.org/10.1002/biot.200700084
Wang J, Guleria S, Koffas MA, Yan Y (2016) Microbial production of value-added nutraceuticals. Curr Opin Biotechnol 37:97–104. https://doi.org/10.1016/j.copbio.2015.11.003
Wang R, Zhao S, Wang Z, Koffas MA (2020) Recent advances in modular co-culture engineering for synthesis of natural products. Curr Opin Biotechnol 62:65–71. https://doi.org/10.1016/j.copbio.2019.09.004
Wu J, Du G, Zhou J, Chen J (2014) Systems metabolic engineering of microorganisms to achieve large-scale production of flavonoid scaffolds. J Biotechnol 188:72–80. https://doi.org/10.1016/j.jbiotec.2014.08.016
Xiong Y, Zhang P, Warner RD, Fang Z (2019) 3-Deoxyanthocyanidin colorant: nature, health, synthesis, and food applications. Compr Rev Food Sci Food Saf 18(5):1533–1549. https://doi.org/10.1111/1541-4337.12476
Xu P, Marsafari M, Zha J, Koffas M (2020) Microbial coculture for flavonoid synthesis. Trends Biotechnol 38(7):686–688. https://doi.org/10.1016/j.tibtech.2020.01.008
Yan Y, Kohli A, Koffas MA (2005) Biosynthesis of natural flavanones in Saccharomyces cerevisiae. Appl Environ Microbiol 71(9):5610–5613
Yıldız SZ, Küçükislamoğlu M, Tuna M (2009) Synthesis and characterization of novel flavonoid-substituted phthalocyanines using (±) naringenin. J Org Chem 694:4152–4161. https://doi.org/10.1016/j.jorganchem.2009.09.012
Yuan SF, Alper HS (2019) Metabolic engineering of microbial cell factories for production of nutraceuticals. Microb Cell Fact 18(1):46. https://doi.org/10.1186/s12934-019-1096-y
Zhang J, Wu Y, Zhao X, Luo F, Li X, Zhu H et al (2014) Chemopreventive effect of flavonoids from Ougan (Citrus reticulata cv. Suavissima) fruit against cancer cell proliferation and migration. J Funct Foods 10:511–519. https://doi.org/10.1016/j.jff.2014.08.006

In Vitro Production of Nutraceutical: Challenges and Opportunities

18

Muneera Q. Al-Mssallem

Abstract

There is a remarkable acceleration in the production of nutraceuticals which is increasingly associated with escalated commercial demands. Several techniques are being applied for the production of nutraceuticals. The in vitro plant culture system is regarded as a fundamental technique for the production of fast multiplication of extraordinary plant genotypes, disease-free plants, and plant genome transformation. However, each technique possesses its limitations. Fortunately, most of the obstacles of in vitro plant techniques can now be overcome through the promising new technology CRISPR/Cas9, which allows change of specific regions of the genome with an increased precision of the insertion and offering excellent reproducibility. This chapter discusses the production of nutraceuticals and briefly summarizes its challenges and opportunities.

Keywords

Abiotic stress · Biosynthesis · Biotechnological techniques · Cell culture · Callogenesis · Micropropagation · Nutraceutical · Organogenesis

Abbreviation

CRISPR/Cas9 Clustered regularly interspaced short palindromic repeats/cascade

M. Q. Al-Mssallem (✉)
Department of Nutrition and Food Sciences, College of Agriculture and Food Sciences, King Faisal University, Hofuf, Al-Ahsa, Saudi Arabia
e-mail: mmssallem@kfu.edu.sa

T. Belwal et al. (eds.), *Nutraceuticals Production from Plant Cell Factory*,
https://doi.org/10.1007/978-981-16-8858-4_18

18.1 Introduction

There is a growing demand on nutraceuticals production for their medicinal applications such as improving human health and preventing or delaying chronic and degenerative diseases. Nowadays, nutraceuticals are available on the market in the form of processed products, dietary supplements, and nutrients (Guan et al. 2019). Practically, they can be extracted from their natural sources such as plants, microorganism, and animals. In fact, the nutraceuticals are mostly based on plant-derived metabolites. In fact, plants can be used to produce valuable nutraceuticals which are naturally produced in response to environmental stresses including biotic and abiotic stresses (Huang et al. 2013; Ramakrishna and Ravishankar 2011; Wang et al. 2018). It is evident that the bioactive phytochemicals derived from plants are recognized as a main ingredient for the nutraceuticals (Atanasov et al. 2015; Jain and Ramawat 2013).

Traditionally, nutraceutical compounds can be extracted and purified from their natural sources or chemically synthesized. However, this approach has its limit in which the large-scale production of nutraceuticals is restricted due to the difficulty of their extraction from natural sources. In addition, this traditional approach is not environmentally sustainable (Guan et al. 2019). Alternatively, micropropagation techniques are also introduced and effectively applied for in vitro production of nutraceuticals at large-scale. This approach has its merits including reproducibility and elimination of the need to depend on wild plants (Chen et al. 2014; Debnarh et al. 2006; Kaul et al. 2013). However, micropropagation system has its own limitation such as high cost and the uncertainty of the markets requirements (Espinosa-Leal et al. 2018; Pence 2011). Novel approaches have been recently developed for in vitro production of nutraceuticals by applying the most recent biotechnological techniques such as gene editing through CRISPR/Cas9 technique, which allows change of specific regions of the genome with an increased precision of the insertion and offering excellent reproducibility (Altpeter et al. 2016; Voytas and Gao 2014). This chapter will summarize the in vitro production of nutraceuticals and its major challenges and opportunities facing the in vitro production of nutraceuticals (Fig. 18.1).

18.2 In Vitro Production of Nutraceutical

The conventional approach of nutraceuticals production was based on an extraction process, where they can directly be isolated and purified from their natural sources such as plants, animals, and microorganisms. This direct extraction approach has its limitations as its isolation and purification procedures are time-consuming and require intensive efforts (Lam 2007; Gil-Chavez et al. 2013). In addition, the purification difficulties include the fact that the nutraceuticals are often presented as conjugates and are usually synthesized in small quantities (Lam 2007; Gil-Chavez et al. 2013). Moreover, the availability of raw materials is limited (Wang et al. 2016a, b).

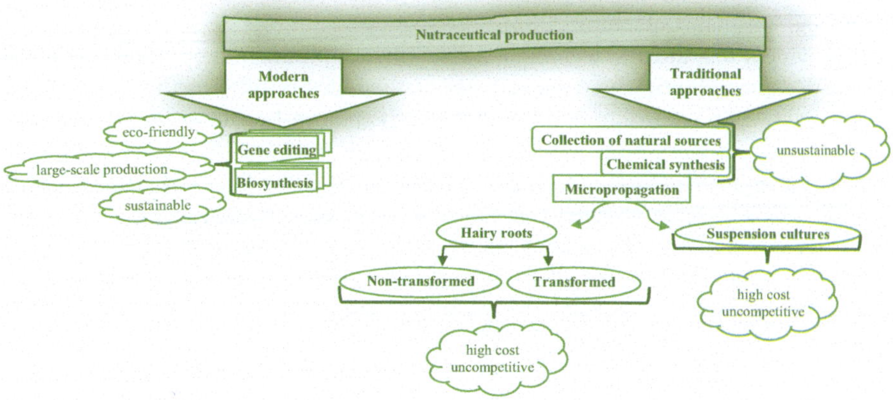

Fig. 18.1 Challenges and opportunities of conventional and modern approaches of nutraceutical production

Nutraceutical compounds can be also produced by using chemical synthesis approach. This traditional approach has its limitations including high cost, insufficient crude materials, low yield, and environmental pollution (Liu et al. 2017; Cruz et al. 2013). In addition, chemical synthesis is restricted to producing simple biochemicals and it is unattainable for complex biochemicals (De Luca et al. 2012). Because of the complexity and diversity of natural products, the chemical synthesis of nutraceutical compound is considered unprofitable (Lam 2007; Gil-Chavez et al. 2013). Therefore, this technique is being replaced by biosynthesis for production of nutraceuticals including the use of microbial strains (Liu et al. 2017). In this case, strains must be safe enough and should be selected carefully for the production of nutraceuticals by following the Food and Drug Authority protocols. Alternatively, production of nutraceuticals with high quality and yield has been achieved using biosynthesis approach (Guan et al. 2019). In fact, biosynthesis technique is an attractive method which is relied on genetic manipulation and optimization of the culture conditions or fermentation processes; metabolic engineering makes it possible to scale up the production of nutraceuticals (Singh et al. 2017). Enzymatic synthesis has been also applied for production of nutraceutical compounds with higher specificity (Sorour et al. 2012; Figueroa-Espinoza and Villeneuve 2005). A combination of chemical and enzymatic technologies has been also successfully applied for nutraceuticals production (Reddy et al. 2010; Vázquez et al. 2017).

In fact, in vitro plant tissue culture techniques are considered as efficient ways to produce nutraceuticals (Fischer et al. 2015). The most common plant tissue culture methods used are organogenesis and callogenesis. Organogenesis attributes to the production of plant shoots or roots. Organogenesis can be achieved straightly from meristems or indirectly from callus (dedifferentiated cells) and the products can later be applied for the micropropagation through hairy root culture system. Callogenesis produces unshaped mass of cells which can then be used for the production of nutraceuticals through cell suspension culture systems (Espinosa-Leal et al. 2018; Sévon and Oksman-Caldentey 2002).

It is evident that the best technique for the in vitro production of nutraceuticals is cell suspension culture, where the calli are first induced in solid media and then the grown cells will be transferred to liquid media (Espinosa-Leal et al. 2018). Suspension cultures and hairy roots are frequently used for the large-scale manufacture of plant components (Mehrotra et al. 2020; Su and Lee 2007; Xu et al. 2012). These technologies are environmentally sustainable processes. However, they are more expensive compared to chemical synthesis technology (Espinosa-Leal et al. 2018). In fact, the scale-up of these culture methods is restricted as these technologies still require the employ of sterilizable bioreactors (Espinosa-Leal et al. 2018; Nogueira et al. 2018; Buyel et al. 2017).

There are a plenty of options of plants for in vitro production of nutraceutical, ranging from model plants to wild-type species plants. In fact, the selection of the appropriate plant material depends mainly on the purpose of the study and the availability of the plant (Espinosa-Leal et al. 2018).

All plant tissue culture approaches proceed with a series of steps, starting from selecting the interested plant based on the purpose of the study and disease-free plants; second, sterilization of explants with chemicals; then placement and incubation of the explants in the appropriate culture media; and finally, hardness is carried out on micropropagated plants allowing them to acclimate with the external conditions (Ahloowalia et al. 2003; Espinosa-Leal et al. 2018).

The low production of nutraceuticals in plant cell culture can be as a result of a lack of cell differentiation. However, organized culture of shoots or roots can be an alternative strategy. Interestingly, hair root culture can be obtained by either infection of roots with *Agrobacterium rhizogenes* and then transfer of the Ri plasmid or by wounding shoot explants using needles dipped in *Agrobacterium rhizogenes* (Espinosa-Leal et al. 2018; Ron et al. 2014; Grzegorczyk-Karolak et al. 2018). Hairy root culture has some advantages such as the biochemical and genetic stability of the cultures and high growth levels without requirement of growth regulators (Espinosa-Leal et al. 2018; Mora-Pale et al. 2014). However, there is a limitation of the hairy root culture which requires that the produced nutraceutical compound has to be normally synthesized within the source plant roots (Espinosa-Leal et al. 2018; Ochoa-Villarreal et al. 2016).

It is a fact that in vitro production of nutraceuticals can be produced at faster growth rate through improving in vitro conditions such as using precursors and elicitors. Additionally, in vitro production of nutraceutical can be increased by manipulation of environmental factors. Although these conditions such as drought, ultraviolet, low or high temperature, salinity, heavy metals, and alkalinity are probably deteriorating the plants, they often increase the capacity of the production of nutraceutical compounds (Espinosa-Leal et al. 2018; Lajayer et al. 2017; Moon et al. 2017; Puente-Garza et al. 2017).

There are some difficulties that can affect the behavior and yield of plant cell cultures such as somaclonal variations and variability in product biosynthesis (Martínez-Estrada et al. 2017; Yun et al. 2012). Somaclonal variations can be avoided by the determination of the optimal number of subcultures, while variability in product biosynthesis can be also avoided through using undifferentiated cambial meristematic cells (Espinosa-Leal et al. 2018; Martínez-Estrada et al. 2017; Ochoa-Villarreal et al. 2016). The latter technique offers a great stability in product accumulation and it was established for large-scale production of natural products including medicinal and nutritional products (Espinosa-Leal et al. 2018; Ochoa-Villarreal et al. 2016).

Because of the inherent limitations of natural products, the new technologies have been developed and introduced recently to compensate for these limitations. Applying these advanced techniques provides a remarkable opportunity to reinstitute natural products as a major source for nutraceuticals (Lam 2007). The application of the novel molecular tools includes the use of targeted genome engineering such as genome editing-mediated where the genetic material can be stored and then be used for the large-scale production of the desired nutraceuticals (Espinosa-Leal et al. 2018; Yuan and Alper 2019; Buyel et al. 2017). It is evident that the application of novel technologies such as gene editing and environmental factor manipulation will reach its maximum potential in the future (Espinosa-Leal et al. 2018).

18.3 Challenges

In vitro production of nutraceuticals is carried out through several stages and encounters some difficulties that may limit the speed or efficiency of production and may expose it to failure. Isolation and purification of nutraceuticals from natural resources are environmentally and economically unsustainable (Wu et al. 2019). Additionally, some approaches of in vitro production are not cost-effective and uncompetitive. In this section, some of these difficulties are summarized underneath.

18.3.1 Nutrients Availability

In vitro production of nutraceutical through plant tissue culture relies on several factors mainly including nutrients provided for explant growth. In fact, the growth of the explants and the production of the nutraceutical required optimum concentration of nutrient which is considered a crucial determinant in the explant thriving (Espinosa-Leal et al. 2018; Fargoso Monfort et al. 2018; Nagella and Murthy 2010; Murthy et al. 2014).

18.3.2 High Costs

The expenses associated with in vitro techniques are considered an important obstacle for in vitro production of nutraceuticals. These obstacles include the expenses combined with media components and electricity. The cost of media mainly involves the expenses of carbon sources, growth regulators, and gelling agent. Reducing the cost of a culture medium can be accomplished using substituted materials for sources of carbon and agar. For the electricity, a modification in the laboratories can be done for adapting the use of solatube which alters sunlight from rooftops through reversing tubing (Espinosa-Leal et al. 2018).

18.3.3 Genetic Instability

Under in vitro culture stress conditions, some alterations in sensitive area of the plan genome can be induced by generating genetic instability in cultured plant. This occurrence is known as a somaclonal variation (Espinosa-Leal et al. 2018). When somaclonal variations occur during in vitro production of nutraceuticals, this can lead to enormous commercial consequences and bring a serious difficulty in utilizing plant tissue approaches for production of such active nutraceutical compounds (Bhattacharyya et al. 2017; Devi et al. 2014).

18.3.4 Environmental Factors

Some abiotic environmental factors can heavily affect biosynthesis regulation of nutraceuticals such as high or low temperature, water availability, toxic gases, heavy metals, pesticides, and light intensity (Huang et al. 2013). The latter factor is an essential component for plant photosynthesis, but high levels of ultraviolet radiation can be detrimental (Espinosa-Leal et al. 2018; Verma and Shukla 2015). Other key factors affecting the in vitro production are the type of culture media and the concentration and type of the growth regulators and salt used. It is essential to select a suitable medium with an appropriate type and concentration of the salt and growth regulators (Fargoso Monfort et al. 2018; Espinosa-Leal et al. 2018).

18.3.5 Growth Rates

In vitro plant culture techniques possess a number of limitations associated with the production of nutraceuticals such as variability of the yields and slow rate of growth. However, these restrictions can be overcome and the production of nutraceuticals can be increased with some modifications to the in vitro culture media including addition of precursors or elicitors (Espinosa-Leal et al. 2018).

18.3.6 Identification and Isolation Techniques

The isolation and identification of the nutraceuticals generate requirement for a method for their continuous production. There are some factors that influence the extraction of nutraceuticals from their source species such as the diversity and complexity of their chemical structures, slow growth rates, the presence of biotic or abiotic stress to induce biosynthesis, and low concentrations of the bioactive compounds of interest (Atanasov et al. 2015; Espinosa-Leal et al. 2018; Ochoa-Villarreal et al. 2016).

18.3.7 Variability in the Quantification and Purification

Despite the advantages of plant biotechnological approach in nutraceutical production, this technique cannot precisely supervise the expression level of transgenes in plants. Therefore, the quantity of nutraceutical produced may differ in each plant. The variability also has been shown in the protocols for the purification of the nutraceuticals from plants. Each plant requires its own purification protocol that is equipped to the in vitro plant production system. There are some factors that can cause difficulties in the purification such as pesticides, plant pathogens, and fertilizers (Yao et al. 2015).

18.4 Opportunities

The development of new in vitro techniques has provided exciting opportunities for in vitro production of nutraceuticals. This section will discuss some of these promising techniques.

18.4.1 Novel Molecular Techniques

The advent of modern molecular tools now presents new possibilities for the production of important nutraceuticals applying in vitro plant systems such as using CRISPR/Cas9 in genome editing. This technique is considered as an auspicious solution to most of the complications associated with the currently available methods (Baltes and Voytas 2015; Nogueira et al. 2018). In fact, genome editing creates potentiality of producing varieties of new plants by allowing a modification of specific regions of the genome with an elevated precision of the insertion (Voytas and Gao 2014). The application of genome editing can be carried out through inserting new genes in predetermined areas of the genome, replacing an allele with an existing one, or alteration of a limited number of nucleotides (Abdallah et al. 2015). Genome editing can potentially be applied for promoter replacement or the introduction of new pathways resulting in the creation of novel expression of favorable bioactive compounds (Nogueira et al. 2018).

18.4.2 Micropropagation

In vitro plant propagation has been developed for producing nutraceuticals (Atanasov et al. 2015; Espinosa-Leal et al. 2018; Ochoa-Villarreal et al. 2016). In fact, micropropagation has grown a commercially profitable achievement compared to traditional plant propagation through promoting the production of enormous quantities of homogenous and disease-free plants (Lucchesini and Mensuali-Sodi 2010; Espinosa-Leal et al. 2018). A substantial number of micropropagation protocols are currently available for nutraceuticals production (Atanasov et al. 2015; Bhattecheryya et al. 2014; Chen et al. 2014; Debnarh et al. 2006; Kaul et al. 2013). Despite these advantages of using micropropagation, it has its limitation at commercial level in terms of its high cost and the uncertainty of the market requirements (Espinosa-Leal et al. 2018; Pence 2011; Sahu and Sahu 2013).

18.4.3 Transgenic Approach

Transgenic plant is an important alternative means for nutraceuticals production. A transgenic way possesses a number of advantages, including reduced risk of contamination with pathogens and decreased costs (Espinosa-Leal et al. 2018; Thomas et al. 2011). However, some difficulties are associated with this method including purification obstacles and low yields (Espinosa-Leal et al. 2018; Yao et al. 2015). However, some of these hurdles such as intensive purification can be reduced by appropriate selection of the plant material. In addition, preservation of useful traits, such as high expression of the desired nutraceutical, can be accomplished by using appropriate breeding technique and judicious selection of the desired plant material (Espinosa-Leal et al. 2018; Pniewski et al. 2017).

Growth regulators have been also bioengineered. Growth regulators are extremely important for the plant cellular processes' regulation and play a crucial role in directing the responses of the plant to environmental abiotic stresses (Espinosa-Leal et al. 2018; Wani et al. 2016).

18.4.4 Environmental Factor Manipulation

Manipulation of environmental factors is an alternative technique for increasing the production of plant nutraceutical compounds. It is well-known that in vitro plan cell cultures are heavily influenced by abiotic stresses which in turn regulate the biosynthesis of nutraceuticals through response mechanism. Such environmental factors include modifications in abiotic stresses such as salinity, drought, UV-radiation, high or low temperatures, and light intensity (Huang et al. 2013). Several phenolic compounds have been produced through this technique such as phenolic acids, flavonoids, rosmarinic acids, polyphenols, saponins, and tocopherols (Espinosa-Leal et al. 2018; Raduisene et al. 2012).

18.5 Conclusions and Recommendations

There is a rapid increase in the market demand of the nutraceuticals for their physiological properties and beneficial human health to meet the improvement of human quality lives. Natural sources of nutraceuticals include plant, animals, and microorganisms. Precious nutraceuticals can be produced from plants and microbial synthesis through several approaches. Conventional extraction of natural sources and chemical synthesis of nutraceuticals were applied for collection of wild plants and it is regarded as an inexpensive approach but it is environmentally unsustainable process. Additionally, biosynthesis for production of nutraceuticals with high quality and yield has been applied and it is considered an efficient and environmentally friendly production technique. Moreover, the in vitro plant culture systems for the production of nutraceutical offer important advantages, including rapid production, low costs, and low burden of human pathogens. However, these techniques are not sufficient to sustain an appropriate large-scale production of nutraceutical compounds. In fact, in vitro production of nutraceuticals encounters some challenges that have been overcome. However, there are numerous opportunities which are already available and must be exploited and employed to serve the production of these vital compounds. Recently, genome editing technique is considered as a promising technology for creation of novel expression of favorable nutraceutical compounds. This novel molecular tool offers an excellent technique which allows change of specific regions of the genome with an increased precision of the insertion and offering excellent reproducibility.

References

Abdallah NA, Prakash CS, McHughen AG (2015) Genome editing for crop improvement: challenges and opportunities. GM Crops Food 6:183–205. https://doi.org/10.1080/21645698.2015.1129937

Ahloowalia BS, Prakash J, Savangikar VA, Savangikar C (2003) Plant tissue culture. Low cost options for tissue culture technology in developing countries. In: Proceedings of a technical meeting organized by the joint FAO/IAEA division of nuclear techniques in food and agriculture and held in Vienna, 26–30 August 2002. International Atomic Energy Agency, Vienna, Austria, pp 3–10. ISBN 92–0–115903–X

Altpeter F, Springer NM, Bartley LE, Blechl AE, Brutnell TP, Citovsky V et al (2016) Advancing crop transformation in the era of genome editing. Plant Cell 28:1510–1520. https://doi.org/10.1105/tpc.16.00196

Atanasov AG, Waltenberger B, Pferschy-Wenzig E-M, Linder T, Wawrosch C, Uhrin P et al (2015) Discovery and resupply of pharmacologically active plant-derived natural products: a review. Biotechnol Adv 33(8):1582–1614. doi.org/10.1016/j.biotechadv.2015.08.001

Baltes NJ, Voytas DF (2015) Enabling plants synthetic biology through genome engineering. Trends Biotechnol 33:120–131. https://doi.org/10.1016/j.tibtech.2014.11.008

Bhattacharyya P, Kumar V, Van Staden J (2017) Assessment of genetic stability amongst micropropagated *Ansellia africana*, a vulnerable medicinal orchid species of Africa using SCoT markers. S Afr J Bot 108:294–302. https://doi.org/10.1016/j.sajb.2016.11.007

Bhattecheryya P, Kumaria S, Diengdoh R, Tandon P (2014) Genetic stability and phytochemical analysis of the in vitro regenerated plants of *Dendrobium nobile* Lindl., an endangered medicinal orchid. Meta Gene 2:489–504. https://doi.org/10.1016/j.mgene.2014.06.003

Buyel JF, Twyman RM, Fischer R (2017) Very-large-scale production of antibodies in plants: the biologization of manufacturing. Biotechnol Adv 35:458–465. https://doi.org/10.1016/j.biotechadv.2017.03.011

Chen C-C, Chang C-L, Agrawal DC, Wu C-R, Tsay H-S (2014) In vitro propagation and analysis of secondary metabolites in *Glossogyne tenuifolia* (Hsiang-Ju)—a medicinal plant native to Taiwan. Bot Stud 55:45. https://doi.org/10.1186/s40529-014-0045-7

Cruz L, Mateus N, de Freitas V (2013) First chemical synthesis report of an anthocyanin metabolite with in vivo occurrence: cyanidin-40-O-methyl-3-glucoside. Tetrahedron Lett 54:2865–2869. https://doi.org/10.1016/j.tetlet.2013.03.100

Debnarh M, Malik CP, Baisen PS (2006) Micropropagation: a tool for the production of high quality plant based medicines. Curr Pharm Biotechnol 7:33–49

De Luca V, Salim V, Atsumi SM, Yu F (2012) Mining the biodiversity of plants: a revolution in the making. Science 336:1658–1661. http://science.sciencemag.org/content/336/6089/1658

Devi SP, Kumaria S, Rao SR, Tandon P (2014) Single primer amplification reaction (SPAR) methods reveal subsequent increase in genetic variations in micropropagated plants of *Nepenthes kha- siana Hook. f.* maintained for three consecutive regenerations. Gene 538:23–29. https://doi.org/10.1016/j.gene.2014.01.028

Espinosa-Leal CA, Puente-Garza CA, García-Lara S (2018) In vitro plant tissue culture: means for production of biological active compounds. Planta 248(1):1–18. https://doi.org/10.1007/s00425-018-2910-1

Fargoso Monfort LE, Vilela Bertolucci SK, Fabri Lima A, Alves de Carvalho A, Mohammed A, Fitzgerard Blank A, JEB PP (2018) Effects of plant growth regulators, different culture media and strength MS on production of volatile fraction composition in shoot cultures of *Ocimum basilicum*. Ind Crop Prod 116:231–239. https://doi.org/10.1016/j.indcrop.2018.02.075

Figueroa-Espinoza MC, Villeneuve P (2005) Phenolic acids enzymatic lipophilization. J Agric Food Chem 53(8):2779–2787. https://doi.org/10.1021/jf0484273

Fischer R, Vasilev N, Twyman RM, Schillberg S (2015) High-value products from plants: the challenges of process optimization. Curr Opin Biotechnol 32:156–162. https://doi.org/10.1016/j.copbio.2014.12.018

Gil-Chavez GJ, Villa JA, Ayala-Zavala FJ, Heredia JB, Sepulveda D, Yahia EM et al (2013) Technologies for extraction and production of bioactive compounds to be used as nutraceuticals and food ingredients: an overview. Compr Rev Food Sci Food Saf 12:5–23. https://doi.org/10.1111/1541-4337.12005

Grzegorczyk-Karolak I, Kuźma Ł, Skala E, Kiss A (2018) Hairy root cultures of Salvia viridis L. for production of polyphenolic compounds. Ind Crops Prod 117:235–244. https://doi.org/10.1016/j.indcrop.2018.03.014

Guan N, Li J, Du G, Chen J, Liu L (2019) Microbial production of nutraceuticals: challenges and prospects. In: Liu L, Chen J (eds) Systems and synthetic biotechnology for production of nutraceuticals. Springer, Singapore. https://doi.org/10.1007/978-981-15-0446-4_9

Huang J, Levine A, Wang Z (2013) Plant abiotic stress. Sci World J 2013:432836. https://doi.org/10.1155/2013/432836

Jain N, Ramawat KG (2013) Nutraceuticals and antioxidants in prevention of diseases. In: Ramawat K, Mérillon JM (eds) Natural products. Springer, Berlin, Heidelberg, pp 2559–2580. https://doi.org/10.1007/978-3-642-22144-6_70

Kaul S, Das S, Srivastava PS (2013) Micropropagation of *Ajuga bracteosa*, a medicinal herb. Physiol Mol Biol Plants 19:289–296. https://doi.org/10.1007/s12298-012-0161-3

Lajayer BA, Ghorbanpour M, Nikabadi S (2017) Heavy metals in contaminated environment: destiny of secondary metabolite biosynthesis, oxidative status and phytoextraction in medicinal plants. Ecotoxicol Environ Saf 145:377–390. https://doi.org/10.1016/j.ecoenv.2017.07.035

Lam K (2007) New aspects of natural products in drug discovery. Trends Microbiol 15(6):279–289

Liu L, Guan N, Li J, Shin HD, Du G, Chen J (2017) Development of GRAS strains for nutraceutical production using systems and synthetic biology approaches: advances and prospects. Crit Rev Biotechnol 37(2):139–150. https://doi.org/10.3109/07388551.2015.1121461

Lucchesini M, Mensuali-Sodi A (2010) Plant tissue culture-an opportunity for the production of nutraceuticals. Adv Exp Med Biol 698:185–202. https://doi.org/10.1007/978-1-4419-7347-4_14

Martínez-Estrada E, Caamal-Velázquez JH, Salina Ruíz J, Bello-Bello JJ (2017) Assessment of somaclonal variation during sugarcane micropropagation in temporary immersion bioreactors by inter- simple sequence repeat (ISSR) markers. In Vitro Cell Dev Biol Plant 53:553–560

Mehrotra S, Mishra S, Srivastava V (2020) Hairy roots biotechnology unzipped: a journey of reality and promises. In: Srivastava V, Mehrotra S, Mishra S (eds) Hairy root cultures based applications. Rhizosphere biology. Springer, Singapore. https://doi.org/10.1007/978-981-15-4055-4_1

Moon SH, Mistry B, Kim HD, Pandurangan M (2017) Antioxidant and anticancer potential of bioactive compounds following UV-C light-induced plant cambium meristematic cell cultures. Ind Crop Prod 109:762–772. https://doi.org/10.1016/j.indcrop.2017.09.024

Mora-Pale M, Sanchez-Rodriguez SP, Linhardt RJ, Dordick JS, Koffas MA (2014) Biochemical strategies for enhancing the in vitro production of natural products with pharmaceutical potential. Curr Opin Biotechnol 25:86–94. https://doi.org/10.1016/j.copbio.2013.09.009

Murthy HN, Lee EJ, Paek KY (2014) Production of secondary metabolites from cell and organ cultures: strategies and approaches for biomass improvement and metabolite accumulation. Plant Cell Tiss Org Cult 118:1–16. https://doi.org/10.1007/s11240-014-0467-7

Nagella P, Murthy HN (2010) Establishment of cell suspension cultures of *Withania somnifera* for the production of withanolide A. Bioresour Technol 101:6735–6739. https://doi.org/10.1016/j.biortech.2010.03.078

Nogueira M, Enfissi EM, Almeida J, Fraser PD (2018) Creating plant molecular factories for industrial and nutritional isoprenoid production. Curr Opin Biotechnol 49:80–87. https://doi.org/10.1016/j.copbio.2017.08.002

Ochoa-Villarreal M, Howat S, Hong S, Jang MO, Jin YW, Lee EK, Loake GJ (2016) Plant cell culture strategies for the production of natural products. BMB Rep 49:149–158. https://doi.org/10.5483/BMBRep.2016.49.3.264

Pence VC (2011) Evaluating costs for in vitro propagation and preservation of endangered plants. In Vitro Cell Dev Biol Plant 47:176–187. https://doi.org/10.1007/s11627-010-9323-6

Pniewski T, Czyz M, Wyrwa K, Bociag P, Krajewski P, Kapusta J (2017) Micropropagation of transgenic lettuce containing HBsAg as a method of mass-scale production of standardised plant material for biofarming purposes. Plant Cell Rep 36:49–60. https://doi.org/10.1007/s00299-016-2056-1

Puente-Garza CA, Meza-Miranda C, Ochoa-Martínez D, García-Lara S (2017) Effect of in vitro drought stress on phenolic acids, flavonols, saponins and antioxidant activity in *Agave salmiana*. Plant Physiol Biochem 115:400–407. https://doi.org/10.1016/j.plaphy.2017.04.012

Raduisene J, Karpaviciene B, Stanius Z (2012) Effect of external and internal factors on secondary metabolites accumulation in St. John's wort. Bot Lith 18:101–108. https://doi.org/10.2478/v10279-012-0012-8

Ramakrishna A, Ravishankar GA (2011) Influence of abiotic stress signals on secondary metabolites in plants. Plant Signal Behav 6:1720–1731. https://doi.org/10.4161/psb.6.11.17613

Reddy KK, Shanker KS, Ravinder T, Prasad RBN, Kanjilal S (2010) Chemo-enzymatic synthesis and evaluation of novel structured phenolic lipids as potential lipophilic antioxidants. Eur J Lipid Sci Technol 112(5):600–608. https://doi.org/10.1002/ejlt.200900200

Ron M, Kajala K, Pauluzzi G, Wang D, Reynoso MA, Zumstein K et al (2014) Hary root transformation using *Agrobacterium rhizogenes* as a tool for exploring cell type-specific gene expression and function using tomato as a model. Plant Physiol 166:455–469. https://doi.org/10.1104/pp.114.239392

Sahu J, Sahu RK (2013) A review on low cost methods for in vitro micropropagation of plant through tissue culture technique. Pharm Biosci J 1:38–41. https://doi.org/10.20510/ukjpb/1/i1/91115

Sévon N, Oksman-Caldentey KM (2002) Agrobacterium rhizogenes-mediated transformation: root cultures as a source of alkaloids. Planta Med 68(10):859–868. https://doi.org/10.1055/s-2002-34924

Singh R, Kumar M, Mittal A, Mehta PK (2017) Microbial metabolites in nutrition, healthcare and agriculture. 3 Biotech 7:15. https://doi.org/10.1007/s13205-016-0586-4

Sorour N, Karboune S, Saint-Louis R, Kermasha S (2012) Enzymatic synthesis of phenolic lipids in solvent-free medium using flaxseed oil and 3,4-dihydroxyphenyl acetic acid. Process Biochem 47(12):1813–1819. https://doi.org/10.1016/j.procbio.2012.06.020

Su WW, Lee K-T (2007) Plant cell and hairy root cultures process characteristics, products, and applications. In: Yang S-T (ed) Bioprocessing for value-added products from renewable resources. Elsevier, Boston, pp 263–292

Thomas DR, Penney CA, Majumder A, Walmsley AM (2011) Evolution of plant-made pharmaceuticals. Int J Mol Sci 12:3220–3236. https://doi.org/10.3390/ijms12053220

Vázquez JA, Ramos PR, Mirón J, Valcarcel J, Sotelo CG, Pérez-Martín RI (2017) Production of chitin from Penaeus vannamei by-products to pilot plant scale using a combination of enzymatic chemical processes and subsequent optimization of the chemical production of chitosan by response surface methodology. Mar Drugs 15(6):180. https://doi.org/10.3390/md15060180

Verma N, Shukla S (2015) Impact of various factors responsible for fluctuation in plant secondary metabolites. J Appl Res Med Aromat Plants 2:105–113. https://doi.org/10.1016/j.jarmap.2015.09.002

Voytas DF, Gao C (2014) Precision genome engineering and agriculture: opportunities and regulatory challenges. PLoS Biol 12:1–6. doi.org/10.1371/journal.pbio.1001877

Wang J, Guleria S, Koffas MAG, Yan Y (2016a) Microbial production of value-added nutraceuticals. Curr Opin Biotechnol 37:97–104. https://doi.org/10.1016/j.copbio.2015.11.003

Wang K, Zhang ZJ, Stewart CN (2016b) Advancing crop transformation in the era of genome editing. Plant Cell 28:1510–1520. https://doi.org/10.1105/tpc.16.00196

Wang W, Li Y, Dang P, Zhao S, Lai D, Zhou L (2018) Rice secondary metabolites: structures, roles, biosynthesis, and metabolic regulation. Molecules 23(12):3098

Wani SH, Kumar V, Shriram V, Sah SK (2016) Phytohormones and their metabolic engineering for abiotic stress tolerance in crop plants. Crop J 4:162–176. https://doi.org/10.1016/j.cj.2016.01.010

Wu Y, Gu Y, Tian R, Du G, Chen J, Liu L (2019) Construction of microbial cell factories by systems and synthetic biotechnology. In: Liu L, Chen J (eds) Systems and synthetic biotechnology for production of nutraceuticals. Springer, Singapore. https://doi.org/10.1007/978-981-15-0446-4_2

Xu JF, Dolan MC, Medrano G, Cramer CL, Weathers PJ (2012) Green factory: plants as bioproduction platforms for recombinant proteins. Biotechnol Adv 30:1171–1184. https://doi.org/10.1016/j.biotechadv.2011.08.020

Yao J, Weng Y, Dickey A, Wang KY (2015) Plants as factories for human pharmaceuticals applications and challenges. Int J Mol Sci 16:28549–28565. https://doi.org/10.3390/ijms161226122

Yuan S, Alper HS (2019) Metabolic engineering of microbial cell factories for production of nutraceuticals. Microb Cell Fact 18:46. https://doi.org/10.1186/s12934-019-1096-y

Yun UW, Yan Z, Amir R, Hong S, Jin YW, Lee EK, Loake GJ (2012) Plant natural products: history, limitations and the potential of cambial meristematic cells. Biotechnol Genet Eng Rev 28:47–59. https://doi.org/10.5661/bger-28-47

The manufacturer's authorised representative in the EU is Springer Nature Customer Service Centre GmbH, Europaplatz 3, 69115 Heidelberg, Germany. If you have any concerns regarding our products, please contact ProductSafety@springernature.com

Printed and bound by CPI Group (UK) Ltd, Croydon, CR0 4YY
30/09/2026
02225241-0012